Indigenous Fruit Trees in the Tropics

Domestication, Utilization and Commercialization

Front cover

Pouteria caimito (Ruiz et Pavon) Radlk (Sapotaceae), known as abiu, caimito or egg fruit, is native to Amazonia and was domesticated by Native Amazonians long before European conquest. The wild populations have small (20–40 g) fruit with abundant latex in the rind; domesticated populations have medium to large (up to 500 g) fruit with less latex. In western Amazonia there are two landraces, differentiated by fruit shape: the more common has the typical egg shape and can weigh as much as 400 g; the less common has a spherical shape and weighs more than 500 g. Early chronicles highlight the popularity of abiu throughout Amazonia, where it was reported to be commonly present in the homegardens of both Native Amazonians and colonists. Today it is still a popular homegarden fruit, although not very important in the market. (Photo credit: F.K. Akinnifesi.)

Indigenous Fruit Trees in the Tropics

Domestication, Utilization and Commercialization

Edited by

Festus K. Akinnifesi

World Agroforestry Centre (ICRAF), Lilongwe, Malawi

Roger R.B. Leakey

James Cook University, Australia

Oluyede C. Ajayi and **Gudeta Sileshi**

World Agroforestry Centre (ICRAF), Lilongwe, Malawi

Zac Tchoundjeu

World Agroforestry Centre (ICRAF), Cameroon

Patrick Matakala

World Agroforestry Centre (ICRAF), Maputo, Mozambique

and

Freddie R. Kwesiga

Forum for Agricultural Research in Africa (FARA), Ghana

in association with the
World Agroforestry Centre (ICRAF)

CABI is a trading name of CAB International

CABI Head Office
Nosworthy Way
Wallingford
Oxfordshire OX10 8DE
UK

Tel: +44 (0)1491 832111
Fax: +44 (0)1491 833508
E-mail: cabi@cabi.org
Website: www.cabi.org

CABI North American Office
875 Massachusetts Avenue
7th Floor
Cambridge, MA 02139
USA

Tel: +1 617 395 4056
Fax: +1 617 354 6875
E-mail: cabi-nao@cabi.org

A catalogue record for this book is available from the British Library, London, UK.

Library of Congress Cataloging-in-Publication Data

Indigenous fruit trees in the tropics : domestication, utilization and commercialization / edited by Festus K. Akinnifesi ... [*et al.*].

p. cm.

Includes bibliographical references and index.

ISBN 978-1-84593-110-0

1. Tree crops--Tropics. 2. Fruit trees--Tropics. 3. Endemic plants--Tropics. I. Akinnifesi, Festus K. II. Title.

SB171.T73I53 2007
634--dc22

2007020221

ISBN: 978 1 84593 110 0

Produced and typeset by Columns Design Ltd, Reading, UK
Printed and bound in the UK by Biddles Ltd, King's Lynn

Contents

Contributors ix

Preface xiii

Foreword by Dennis Garrity xvii

Acknowledgements xix

PART I: SETTING THE SCENE

1 Setting Priorities among Indigenous Fruit Tree Species in Africa: Examples from Southern, Eastern and Western Africa Regions 1
S. Franzel, F.K. Akinnifesi and C. Ham

2 Towards a Domestication Strategy for Indigenous Fruit Trees in the Tropics 28
R.R.B. Leakey and F.K. Akinnifesi

3 Challenges to Stimulating the Adoption and Impact of Indigenous Fruit Trees in Tropical Agriculture 50
N. Haq, C. Bowe and Z. Dunsiger

PART II: INDIGENOUS FRUIT TREE DOMESTICATION IN ASIA, LATIN AMERICA AND OCEANIA

4 Domestication of Trees or Forests: Development Pathways for Fruit Tree Production in South-east Asia 70
K.F. Wiersum

5 **Homegarden-based Indigenous Fruit Tree Production in Peninsular India** 84
B.M. Kumar

6 **Native Fruit Tree Improvement in Amazonia: An Overview** 100
C.R. Clement, J.P. Cornelius, M.H. Pinedo-Panduro and K. Yuyama

7 **The Domestication of Fruit and Nut Tree Species in Vanuatu, Oceania** 120
V. Lebot, A. Walter and C. Sam

PART III: REGIONAL DOMESTICATION RESEARCH AND DEVELOPMENT IN AFRICA

8 **Creating Opportunities for Domesticating and Commercializing Miombo Indigenous Fruit Trees in Southern Africa** 137
F.K. Akinnifesi, O.C. Ajayi, G. Sileshi, P. Matakala, F.R. Kwesiga, C. Ham, I. Kadzere, J. Mhango, S.A. Mng'omba, T. Chilanga and A. Mkonda

9 **Domestication, Utilization and Marketing of Indigenous Fruit Trees in West and Central Africa** 171
Z. Tchoundjeu, A. Atangana, E. Asaah, A. Tsobeng, C. Facheux, D. Foundjem, C. Mbosso, A. Degrande, T. Sado, J. Kanmegne, P. Mbile, H. Tabuna, P. Anegbeh and M. Useni

10 **Improving Rural Livelihoods through Domestication of Indigenous Fruit Trees in the Parklands of the Sahel** 186
A. Kalinganire, J.C. Weber, A. Uwamariya and B. Kone

11 **The Role of Indigenous Fruit Trees in Sustainable Dryland Agriculture in Eastern Africa** 204
Z. Teklehaimanot

PART IV: THE BIOPHYSICAL AND SOCIO-ECONOMIC CONTEXT OF MIOMBO FRUIT TREES

12 **Marketing of Indigenous Fruits in Southern Africa** 224
T. Ramadhani and E. Schmidt

13 **Economics of On-farm Production of Indigenous Fruits** 237
D. Mithöfer and H. Waibel

14 Opportunities for Commercialization and Enterprise Development of Indigenous Fruits in Southern Africa 254
C. Ham, F.K. Akinnifesi, S. Franzel, D. du P.S. Jordaan, C. Hansmann, O.C. Ajayi and C. de Kock

15 The Feasibility of Small-scale Indigenous Fruit Processing Enterprises in Southern Africa 273
D. du P.S. Jordaan, F.K. Akinnifesi, C. Ham and O.C. Ajayi

16 Product Development: Nutritional Value, Processing and Utilization of Indigenous Fruits from the Miombo Ecosystem 288
J.D.K. Saka, I. Kadzere, B.K. Ndabikunze, F.K. Akinnifesi and B.P.M. Tiisekwa

17 The Role of Institutional Arrangements and Policy on the Conservation, Utilization and Commercialization of Indigenous Fruits in Southern Africa 310
P.A. Oduol, O.C. Ajayi, P. Matakala and F.K. Akinnifesi

18 Ecology and Biology of *Uapaca kirkiana, Strychnos cocculoides* and *Sclerocarya birrea* in Southern Africa 322
P.W. Chirwa and F.K. Akinnifesi

19 Germplasm Supply, Propagation and Nursery Management of Miombo Fruit Trees 341
F.K. Akinnifesi, G. Sileshi, A. Mkonda, O.C. Ajayi, J. Mhango and T. Chilanga

20 Pest Management in Miombo Fruit Trees 369
G. Sileshi, P. Barklund, G. Meke, R.R. Bandeira, C. Chilima, A.J. Masuka, R.K. Day and F.K. Akinnifesi

PART V: LESSONS FOR COMMODITIZING INDIGENOUS FRUIT AND NUT TREES IN THE TROPICS

21 Accelerated Domestication and Commercialization of Indigenous Fruit and Nut Trees to Enhance Better Livelihoods in the Tropics: Lessons and Way Forward 392
F.K. Akinnifesi, G. Sileshi, O.C. Ajayi and Z. Tchoundjeu

Index 429

Contributors

Oluyede C. Ajayi, *PhD, is an Agricultural Economist with the World Agroforestry Centre (ICRAF), PO Box 30798, Lilongwe, Malawi.*

Festus K. Akinnifesi, *PhD, is a Senior Tree Scientist with the World Agroforestry Centre (ICRAF), PO Box 30798, Lilongwe, Malawi.*

Paul Anegbeh, *PhD, is with the World Agroforestry Centre (ICRAF) West and Central Africa Region, Humid Tropic Node, IITA Station, Onne-Wharf Road, PMB 008, Nchia, Eleme, Port Harcourt, Rivers State, Nigeria.*

Ebenezar Asaah *is with the World Agroforestry Centre (ICRAF) West and Central Africa Region, Humid Tropic Node, BP 16317, Yaounde, Cameroon.*

A. Atangana *is with the World Agroforestry Centre (ICRAF) West and Central Africa Region, Humid Tropic Node, BP 16317, Yaounde, Cameroon.*

R.R. Bandeira *is with the Eduardo Mondelane University, Maputo, Mozambique.*

Pia Barklund, *PhD, is an Associate Professor at the Swedish University of Agricultural Sciences (SLU), Department of Forest Mycology and Pathology, Box 7026, SE 75007 Uppsala, Sweden.*

Colm Bowe, *PhD, is with the Centre for Underutilised Crops, Environmental Sciences Division, School of Civil Engineering and the Environment, University of Southampton, Southampton SO17 1BJ, UK.*

Thomson Chilanga *is a Horticulturalist with Bvumbwe Agricultural Research Station, Box 5748, Limbe, Malawi.*

Clement Chilima, *PhD, is a Pest Scientist with the Forestry Research Institute of Malawi, Zomba, Malawi.*

Paxie W. Chirwa, *PhD, is a Senior Lecturer at Stellenbosh University, Department of Forest and Wood Science, Stellenbosh 7602, South Africa.*

Charles R. Clement, *PhD, is a Senior Scientist with the Instituto Nacional de Pesquisas da Amazônia, Av. André Araújo, 2936 Aleixo, 69060-001 Manaus, Amazonas, Brazil.*

Jonathan P. Cornelius, *PhD, is a Senior Scientist with the World Agroforestry Centre (ICRAF), CIP, Apartado 1558, Lima 12, Perú.*

Roger K. Day, *PhD, is a Pest Scientist with the CAB International Africa Regional Centre, PO Box 633, United Nations Avenue, Nairobi, Kenya.*

Caroline de Kock *is the CEO of Speciality Foods of Africa Pvt Ltd, Harare, Zimbabwe.*

Ann Degrande, *PhD, is a Social Scientist with the World Agroforestry Centre (ICRAF) West and Central Africa Region, Humid Tropic Node, BP 16317, Yaounde, Cameroon.*

Zoë Dunsiger, *PhD, is with the Centre for Underutilised Crops, Environmental Sciences Division, School of Civil Engineering and the Environment, University of Southampton, Southampton SO17 1BJ, UK.*

Charly Facheux *is with the World Agroforestry Centre (ICRAF) West and Central Africa Region, Humid Tropic Node, BP 16317, Yaounde, Cameroon.*

Divine Foundjem *is with the World Agroforestry Centre (ICRAF) West and Central Africa Region, Humid Tropic Node, BP 16317, Yaounde, Cameroon.*

Steven Franzel, *PhD, is a Principal Agricultural Economist at the World Agroforestry Centre (ICRAF), PO Box 30677, Nairobi, Kenya.*

Cori Ham *is a Lecturer and Co-ordinator of the Commercial Product of the Wild (CP Wild), Department of Forest and Wood Science, University of Stellenbosch, Private Bag X1, Matieland 7602, South Africa.*

Chris Hansmann, *PhD, is a Senior Researcher with the Agricultural Research Centre, Infruitec-Nietvoorbij, Private Bag X5013, Stellenbosch 7599, South Africa.*

Nazmul Haq, *PhD, is the Director of the Centre for Underutilised Crops, Environmental Sciences Division, School of Civil Engineering and the Environment, University of Southampton, Southampton SO17 1BJ, UK.*

Danie du P.S. Joordan is a Lecturer at the Department of Agricultural Economics, Extension and Rural Development, University of Pretoria, Pretoria 0002, South Africa.

Irene Kadzere, *PhD, is a Post Harvest Scientist with the World Agroforestry Centre (ICRAF), c/o Division of Agricultural Research and Extension, 5th Street Extension, PO Box CY594, Causeway, Harare, Zimbabwe.*

Antoine Kalinganire, *PhD, is a Senior Scientist with the World Agroforestry Centre (ICRAF) West and Central Africa Region, BP 320, Bamako, Mali.*

Jacques Kanmegne *is with the World Agroforestry Centre (ICRAF) West and Central Africa Region, Humid Tropic Node, BP 16317, Yaounde, Cameroon.*

Brehima Kone *is with the World Agroforestry Centre (ICRAF) West and Central Africa Region, BP 320, Bamako, Mali.*

B.M. Kumar, *PhD, is a Professor at the Department of Silviculture and Agroforestry, College of Forestry, Kerala Agricultural University, Thrissur 680656, Kerala, India.*

Freddie R. Kwesiga, *PhD, is the Coordinator of the Sub-Saharan African Challenge Programme, PMB CT 173 Cantonments, 2 Gowa Close, Roman Ridge, Accra, Ghana.*

Roger R.B. Leakey, *PhD, is a Professor of Agroecology and Sustainable Development, Agroforestry and Novel Crops Unit, School of Tropical Biology, James Cook University, PO Box 6811, Cairns, Qld 4870, Australia.*

Vincent Lebot, *PhD, is a Senior Researcher with CIRAD, PO Box 946, Port-Vila, Vanuatu.*

Anxious J. Masuka *is with the Kutsaga Research Company (Pvt) Ltd, Airport Ring Road, PO Box 1909, Harare, Zimbabwe.*

Patrick Matakala, *PhD, is the Principal Scientist with the World Agroforestry Centre (ICRAF) Mozambique, Zaixa Postal 3658, Av. Das FPLM, 2698 Mavalane, Maputo 8, Mozambique.*

Peter Mbile *is with the World Agroforestry Centre (ICRAF) West and Central Africa Region, Humid Tropic Node, BP 16317, Yaounde, Cameroon.*

Charlie Mbosso *is with the World Agroforestry Centre (ICRAF) West and Central Africa Region, Humid Tropic Node, BP 16317, Yaounde, Cameroon.*

Gerald Meke *is a Pest Scientist with the Forestry Research Institute of Malawi, Zomba, Malawi.*

Jarret Mhango *is a Lecturer with the Mzuzu University, Forestry Department, Private Bag 201, Luwinga, Mzuzu, Malawi.*

Dagmar Mithöfer, *PhD, is an Agricultural Economist with the International Centre of Insect Physiology and Ecology, PO Box 30772-00100, Nairobi, Kenya.*

Alfred Mkonda *is a Horticulturalist, with the World Agroforestry Centre (ICRAF) Zambia, Msekera Research Station, PO Box 510046, Chitapa, Zambia.*

Simon A. Mng'omba *is a PhD Research Fellow with the Department of Plant Production and Soil Science, Faculty of Natural and Agricultural Sciences, University of Pretoria, 0002 Pretoria, South Africa.*

Bernadette K. Ndabikunze, *PhD, is a Senior Lecturer at the Department of Food Science and Technology, Sokoine University of Agriculture, PO Box 3006, Morogoro, Tanzania.*

Peter A. Oduol, *PhD, is a Scientist with the World Agroforestry Centre (ICRAF) Mozambique, Av. Das FPLM 2698, Caixa Postal 1884, Maputo, Mozambique.*

Mario H. Pinedo-Panduro *is a Researcher with the Instituto de Investigaciones de la Amazonia Peruana, Avda. Abelardo Quiñones, km 2.5, Iquitos, Perú.*

Tunu Ramadhani, *PhD, is Marketing Researcher at the University of Hanover, Herrenhauserstrasse 2, 30419 Hanover, Germany.*

Thadée Sado *is with the World Agroforestry Centre (ICRAF) West and Central Africa Region, Humid Tropic Node, BP 16317, Yaounde, Cameroon.*

John D.K. Saka, *PhD, is a Professor at the Chemistry Department, Chancellor College, University of Malawi, PO Box 280, Zomba, Malawi.*

C. Sam *is with the MQAFF, National Herbarium, PMB Port-Vila, Vanuatu.*

Erich Schmidt, *PhD, is a Professor at the University of Hanover, Herrenhauserstrasse 2, 30419 Hanover, Germany.*

Gudeta Sileshi, *PhD, is a Pest Management Scientist with the World Agroforestry Centre (ICRAF), PO Box 30798, Lilongwe Malawi.*

Honoré Tabuna *is a Marketing Researcher with the World Agroforestry Centre (ICRAF) West and Central Africa Region, Humid Tropic Node, BP 16317, Yaounde, Cameroon.*

Zac Tchoundjeu, *PhD, is a Principal Tree Scientist with the World Agroforestry Centre (ICRAF) West and Central Africa Region, Humid Tropic Node, BP 16317, Yaounde, Cameroon.*

Zewge Teklehaimanot, *PhD, is a Senior Lecturer at the School of the Environment and Natural Resources, University of Wales Bangor, Gwynedd LL57 2UW, UK.*

Bendantunguka P.M. Tiisekwa, *PhD, is a Professor at the Department of Food Science and Technology, Sokoine University of Agriculture, PO Box 3006, Morogoro, Tanzania.*

Alain Tsobeng *is with the World Agroforestry Centre (ICRAF) West and Central Africa Region, Humid Tropic Node, BP 16317, Yaounde, Cameroon*

Marcel Useni *is with the World Agroforestry Centre (ICRAF) West and Central Africa Region, Humid Tropic Node, ICRAF Office, Kinshasa, Democratic Republic of Congo.*

Annonciata Uwamariya *is a Consultant in Agroforestry and Environmental Management, BP E570, Bamako, Mali.*

Hermann Waibel, *PhD, is a Professor at the Faculty of Economics and Management, Leibniz University of Hanover, Königsworther Platz 1, 30167 Hanover, Germany.*

Annie Walter *is with the IRD (Institut de Recherche pour le Développement), Agropolis, Montpellier, France.*

John C. Weber, *PhD, is a Senior Consultant with the World Agroforestry Centre (ICRAF) West and Central Africa Region, BP 320, Bamako, Mali.*

K. Freerk Wiersum, *PhD, is a Senior Lecturer with the Forest and Nature Conservation Policy Group, Department of Environmental Sciences, Wageningen University, PO Box 47, 6700AA Wageningen, The Netherlands.*

Kaoru Yuyama *is a Researcher with the Instituto Nacional de Pesquisas da Amazônia, Av. André Araújo, 2936 Aleixo, 69060-001 Manaus, Amazonas, Brazil.*

Preface

Agriculture is at the core of rural livelihoods. The viability of rural livelihoods in many developing nations is threatened by several interconnected factors which reduce options for smallholder farmers and rural community dwellers. These include: inadequate food and nutritional imbalance in diets; narrow opportunities for off-farm and off-season income, especially for women and children; and depleting production systems. These negative impacts of modern agriculture are further exacerbated by the loss of biodiversity of wild forest resources due to deforestation, overexploitation and the 'tragedy of the commons'.

Historically, fruit trees were the earliest source of food known to mankind and wild-harvesting of indigenous fruit trees predated hunting and settled agriculture. There are also strong links with culture and religion, starting with the biblical 'fruit tree of life' in the Garden of Eden. Jewish history documents how the spies sent to the land of Canaan in 1490 BC had seen and collected fruits from cultivated or domesticated grapes, pomegranates and figs from the Valley of Eshcol (Numbers 13:23). An Indian king was reported to have encouraged the cultivation of mangoes, jackfruits and grapes in 273–232 BC, while in AD 300–400, Vatsyana commented on the importance of fruit trees in his book on Hindu aesthetics (Chapter 4, this volume). Coconut cultivation was documented by a Persian traveller visiting the Malabar Coast of Kerala in 300–100 BC (Chapter 4, this volume). More recently, mangosteens from Asia were the preferred official fruit at banquets in England during the reign of Queen Victoria (1837–1901) (Silva and Tassara, 2005). Today, these fruits are cultivated, commercialized and found in supermarkets world-wide.

The indigenous fruit and nut trees (IFTs) of the tropics, have been described as 'Cinderella species' because they have been overlooked by science and development (Leakey and Newton, 1994). They now represent a unique asset that could be developed, domesticated and owned by farmers. In this sense, IFTs differ from the conventional tropical and subtropical horticultural tree crops such as mango (*Mangifera indica* L.), orange (*Citrus sinensis* L.), coconut

(*Cocos nucifera* L.), breadfruit (*Artocarpus* spp.), mangosteen (*Garcinia mangostana* L.), rambutan (*Nephelium lappaceum* L.), banana (*Musa* spp.), papaya (*Carica papaya* L.), cashew (*Anarcardium occidentale* L.), cacao (*Theobroma cacao* L.), avocado (*Persea americana* Mill.), guava (*Psidium guajava* L.), oil palm (*Elaeis guineense* Jacq.) and coffee (*Coffea arabica* L.), which are not covered by this book. These species were typically developed as crops by research institutes as part of the gains of the colonial era, a process which has continued since independence, although often with substantially lower levels of public investment. These fruits have become 'cash crop' commodities in the global market, and have been improved more with the needs of overseas consumers in mind than those of the local people who produce them. It is often assumed that there is no need for new tree crops, but in recent years the domestication of IFTs has been seen as an opportunity to meet the needs of poor smallholder farmers and thereby to enhance the livelihoods of more than 50% of the world population living on less than US$2 per day. To realize this potential role of IFTs in rural development and to raise their international profile as new crops, Simons and Leakey (2004) proposed that the products of these new tree crops should be called Agroforestry Tree Products (AFTPs) to distinguish them from Non-Timber Forest Products (NTFPs), which are extractive resources from natural forests.

Up until the late 1980s there was almost no scientific work on indigenous fruit trees and little was known about their biology, ecology or social impact on rural populations, and it was generally thought that they are not amenable to cultivation. However, subsistence farmers have been the custodians of these IFTs, and in many societies these species have deep cultural significance, often associated with taboos and community regulations around their conservation and use. Since the early 1990s, the domestication of IFTs has become a new and active field of research and development led by the World Agroforestry Centre (ICRAF) and partners around the world. New concepts and approaches have been developed, case studies have been produced, and evidence-based research is being undertaken on the potential and feasibility of domesticating IFTs and commercializing their products, for the benefit of rural communities.

In this book we focus on the 'underutilized' indigenous fruit trees in order to further promote recognition of the role they can play in meeting the rural development goals of the new millennium. This role is currently being highlighted by the International Assessment of Agricultural Science and Technology for Development (IAASTD), which recognizes that sustainable agriculture is dependent on the multifunctionality of farming systems, supporting environmental and social sustainability, providing food, enhancing health and nutrition, while at the same time promoting economic growth. Although the concepts and principles presented in this book are not unique to the tropics, we have chosen to concentrate on the IFT species of the tropics and subtropics, as they have the greatest underutilized potential. This volume has gathered together contributions providing state-of-the-art information on IFT research and development to complement existing knowledge – principally from proceedings of conferences and technical meetings (Leakey and Newton, 1994; Maghembe, 1995; Leakey and Izac, 1996; Shumba *et al.*, 2000), special issues of scientific journals (Leakey and Page, 2006), and a range of research articles, species monographs and agroforestry textbooks. The studies reported in this book span a wide range of approaches and practices. The authors have experience in all facets of the fruit-tree supply chain – from wild collection to the nursery (propagation, cultivation), utilization and marketing; from academic research to the practical needs of farmers, marketeers, industry, policy makers and investors.

The 21 chapters cover a wide spectrum of topics. The book begins with general principles, methods and practices that are cross-cutting (Chapters 1–3). and a series of case studies across subcontinents (Asia – Chapters 4 and 5; Latin America – Chapter 6; Oceania – Chapter 7; and Africa by region: Southern Africa – Chapter 8; West and Central Africa – Chapter 9; the Sahel zone – Chapter 10; and Eastern Africa – Chapter 11). These are followed by a set of studies (Chapters 12–21) in southern Africa aiming to understand the AFTP supply chain from production to the market (markets, economics, nutritional value, enterprise development and feasibility assessments of IFTs – Chapters 12–16), institutional policy and indigenous knowledge in utilization of IFTs (Chapter 17), ecology and biology, germplasm production and pest management (Chapters 19–20). The final chapter (Chapter 21) provides a synthesis of the concepts, principles and methods, practices and results, and how IFT research can be of direct benefit to farmers, scientists, development communities and investors when developing, managing and commercializing these natural assets. Readers need to appreciate that IFT domestication must be viewed within the context of wider natural resource management at the farm and landscape level.

This book links the exploration, husbandry and domestication of IFTs with the markets, and broader transdisciplinary concern on the promotion of growth and poverty reduction for rural farmers. We hope it will improve understanding of smallholder needs, constraints and priorities, as well as the practices for the development of technological and market-oriented solutions aimed at improving the livelihoods of smallholders. Four questions may assist the reader: (i) How can farmers and entrepreneurs benefit from IFT domestication and commercialization? (ii) How can the experience and knowledge gained so far be integrated into smallholder farming systems? (iii) What research topics are emerging as priorities for the future? (iv) What are the drivers of change in the domestication and commercialization of IFTs in the tropics?

As the development sector refocuses its attention on poverty, food security and malnutrition, it is important to think again about the potential contribution of the 'Cinderella fruits' of the tropics. In what we hope is a turning point in tropical IFT research and development, this book brings together authors from Africa, Asia, Oceania, Latin America and Europe, with rich experience on domestication and commercialization of IFTs, in order to broaden and deepen our understanding of the challenges and opportunities facing these 'hidden commodities', and the constraints that prevent poor farmers from getting out of poverty. The book is an appropriate example of how cooperation between research, investors and communities can signal a beacon of hope to farmers in the development of new crops.

We hope that this book will not only provide information on the extent to which indigenous fruit trees have been researched and understood, but also show the extent of domestication and technological solutions, the relevance of creating and expanding markets, and how institutional property rights and policy interventions can facilitate the process, and encourage further research and investment in this important area.

Festus K. Akinnifesi and Roger R.B. Leakey

References

Leakey, R.R.B. and Izac, A.-M.N. (1996) Linkages between domestication and commercialization of non-timber forest products: implications for agroforestry. In: Leakey, R.R.B., Temu, A.B., Melnyk, M. and Vantomme, P. (eds) *Domestication and Commercialisation of Non-timber Forest Products in Agroforestry Systems.* Non-forest Products No. 9, Food and Agriculture Organization of the United Nations, Rome, pp. 1–6.

Leakey, R.R.B. and Newton, A.C. (1994) *Tropical Trees: Potential for Domestication and the Rebuilding of Forest Resources.* HMSO, London, 284 pp.

Leakey, R.R.B. and Page, T. (2006) The 'ideotype concept' and its application to the selection of cultivars of trees providing agroforestry tree products. *Forests, Trees and Livelihoods* 16, 5–16.

Maghembe, J.A. (1995) Achievement in the establishment of indigenous fruit trees of the miombo woodlands of southern Africa. In: Maghembe, J.A., Ntupanyama, Y. and Chirwa, P.W. (eds) *Improvement of Indigenous Fruit Trees of the Miombo Woodlands of Southern Africa.* ICRAF, Nairobi, pp. 39–49.

Shumba, E.M., Lusepani, E. and Hangula, A. (2000) *The Domestication and Commercialization of Indigenous Fruit Trees in the SADC Region.* SADC Tree Seed Network, 85 pp.

Silva, S. and Tassara, H. (2005) *Frutas Brasil Frutas.* Empresa das Artes, São Paulo, Brazil.

Simons, A.J. and Leakey, R.R.B. (2004) Tree domestication in tropical agroforestry. *Agroforestry Systems* 61, 167–181.

Foreword

This book on indigenous fruit trees in the tropics documents some of the ways in which the use of indigenous fruit trees in traditional and modern agroforestry systems can substantially contribute to achieving the Millennium Development Goals in low-income countries, by creating new cultivars of high-value trees, new enterprises and market opportunities for tree products, and crop diversification and maintenance of biodiversity on-farm.

Domesticating indigenous fruit trees is an alternative avenue to unlocking the potentials of genetic materials and indigenous knowledge in rural communities. A research and development strategy is advocated that reduces their dependency on the few primary agricultural commodities, based on creating new and superior fruit tree crops, with potential for establishing value-added products and tapping into emerging market opportunities.

Indigenous tree crops are a major opportunity for asset building for smallholder farmers. Most market studies reported in the book have pointed to women and children as the major beneficiaries of indigenous fruits and fruit product enterprises. Domestication and commercialization interventions will help increase returns and market shares to this segment of the rural population and help address gender inequality in income and farm opportunities.

The domestication of kiwi fruit (*Actinidia chinensis*) was a classic case of a new horticultural fruit of international significance. It was first grown commercially in New Zealand in the 1930s, despite its more than 1000 years' history in China. That success was achieved by farmer-led domestication and commercialization efforts. The selection of the macadamia nut (*Macadamia integrifolia*) from Australia also began in 1934, motivated by promising market interests. The domestication of many other trees of the tropics was triggered by globalization, especially during the colonial conquests, followed by growing market demand that promoted research and cultivation.

For tropical indigenous fruit trees, farmer-driven and market-led participatory domestication can cut short the long cycles of improvement and slow market development to create novel cash crops and opportunities for smallholders. This was the major thinking behind the Agroforestry Tree Domestication Programme pioneered by the World Agroforestry Centre (ICRAF) since the early 1990s. Tree domestication is now one of the key pillars of the global programmes of the centre, building on the efforts of smallholder farmers and our partners. The centre implements its domestication research and development work in eastern, southern, western and central Africa, south and South-east Asia, and Latin America.

The World Agroforestry Centre defines tree domestication as encompassing the socio-economic and biophysical processes involved in the identification, characterization, selection, multiplication and cultivation of high-value tree species in managed ecosystems. The term has now been expanded to include not just species but also landscape domestication. This concept encompasses the whole set of activities required to cultivate, conserve and manage indigenous fruit trees in an ecosystem, including agroforest and forest gardens of the humid regions of the tropics.

Readers of this book will recognize the importance of indigenous fruit trees in providing economic benefits to smallholder farmers and small-scale entrepreneurs in low-income countries in the tropics. The market advances of new crops such as peach palm (*Bactris gasipaes*), guaraná (*Paullinia cupana*), camu camu (*Myrciaria dubia*), açai (*Euterpe oleracea*) and Brazil nut (*Bertholletia excelsa*) in Latin America, durian (*Durio zibethinus*) and tamarind (*Tamarindus indica*) in Asia, and shea (*Vitellaria paradoxa*), marula (*Sclerocarya birrea*), safou (*Dacryodes edules*) and cola nut (*Cola esculentum*) in Africa, are a few of the success stories of indigenous fruit tree domestication. The range of work on domestication and commercialization of these 'hidden treasures' of the wild, and how they have been brought into cultivation and marketed in local, national, regional and international markets, has rarely been documented in one single book.

This volume systematically reviews and documents tree domestication experiences in the tropics. The authors are renowned scientists on indigenous fruit tree domestication and agroforestry. The book provides a solid foundation on which new science, partnerships and market opportunities can be further developed. It is geared to researchers, academics and students in agroforestry, horticulture, and forestry interested in creating new opportunities for smallholder farming communities. As the 21 chapters of the book show, there is much to be learnt and discovered. The hard work and dedication of the authors have resulted in a work of such comprehensiveness that is assured to stand for many years as an important landmark in the literature of indigenous fruit domestication, utilization and commercialization.

Dennis Garrity
Director General
World Agroforestry Centre
23 April 2007

Acknowledgements

We acknowledge funding from two projects (1997–2005) by the Federal Ministry of Economic Cooperation (BMZ/GTZ), which supported the two successive phases of indigenous fruit tree (IFT) research implemented by ICRAF Southern Africa. The publication of the book was also made possible by funding from BMZ/GTZ Project Phase 2 (Project No. 2001.7860.8-001.00 and Contract No. 81051902). The Canadian International Development Agency (CIDA) is thanked for funding the Zambezi Basin Agroforestry Projects in the last 10 years (Projects 2001.7860.8-001 and 050/21576) which provided the enabling environment (especially in staffing) for implementation of other regional IFT projects and the production of the book.

We wish to thank and acknowledge the support from the senior leadership of ICRAF, and especially the Director General, Dr Dennis Garrity, and the DDG designate Dr Tony Simons for their encouragement and allowing us to research and document our experiences and put this book together. We sincerely thank all colleagues who gave their time to review various chapters of this book in their respective expertise. The text would not have been readable and substantial without their help and amendments.

In this book, these results have been set in a wider context through contributions from authors working in other continents. We thank all the authors from Asia, Latin America, Oceania and Africa and their co-authors. Special thanks are due to agribusiness experts from South Africa, including Mr Cori Ham, Professor Van Vyk, Dr Mohamed Karaan of the Stellenbosch University, and Dr Chris Hansmann of the ARC, Danie Joordan of the University of Pretoria, South Africa, and to postharvest experts from the USA, especially Professors Chris Watkins and Ian Merwin of Cornell University. In addition, we sincerely appreciate the contributions and interactions from colleagues and collaborators in southern and eastern Africa, including: Professor Jumane Maghembe, Dr Diane Russell, Professor John D.K. Saka, Dr Paxie Chirwa, Dr Bernadette Ndabikunze, Professor Tiisekwa, Professor

Moses Kwapata, Professor Elsa du Toit, the late Charles Mwamba, the late Dr M. Ngulube, Dr Dennis Kayambazinthu, Professor E. Sambo, Dr Steve Franzel, Dr Ramni Jamnadas, Dr Jean-Marc Boffa, Professor August Temu, Dr Irene Kadzere, Mr Alfred Mkonda, Ms Jarret Mhango, Mr Thomson Chilanga, Ms Caroline Cadu, Mr Remen Swai, Ms Evelina Sambane and Ms Patient Dhliwayo. We especially recognize the contributions of several PhD students including: Dr Tunu Ramadhani, Dr Dagmar Mithöfer, Dr Susan Kaaria, Dr Khosi Ramachela, Simon Mng'omba, Weston Mwase; and graduate students: Thorben Kruse, Lars Fielder, George German, Jacob Mwita and Lovemore Tembo. Without the contributions and efficient interactions between many researchers on indigenous fruit trees this book would not have become a reality. All contributors to this book are acknowledged.

Lorraine Itaye and Fannie Gondwe are sincerely thanked for their administrative support during the preparation of the draft book. Our sincere thanks also go to Claire Parfitt, CABI development editor, for initial guidance and Sarah Hulbert, who took over to see us through the various stages of completing this book when Claire was on maternity leave, and Meredith Carroll, CABI editorial assistant. We also thank our technical editor Nancy Boston, production editor Tracy Ehrlich, typesetter Alan Cowen and indexer Krys Williams for their valuable contributions.

Finally, we thank our wives and children for their usual patience, support, inspiration and understanding as we spent long evenings and weekends over the draft chapters instead of with them.

1 Setting Priorities among Indigenous Fruit Tree Species in Africa: Examples from Southern, Eastern and Western Africa Regions

S. Franzel,[1] F.K. Akinnifesi[2] and C. Ham[3]

[1]*World Agroforestry Centre (ICRAF), Nairobi, Kenya;* [2]*World Agroforestry Centre, ICRAF, Lilongwe, Malawi;* [3]*Department of Forest and Wood Science, University of Stellenbosch, Matieland, South Africa*

1.1 Introduction

Priority setting in agricultural research has received considerable attention over the years (Contant and Bottomley, 1988; Alston *et al.*, 1995; Kelly *et al.*, 1995; Braunschweig *et al.*, 2000). The objective of prioritizing agroforestry tree species is to determine the species for which domestication research (that is, research on the selection, management and propagation of a plant (see Simons and Leakey, 2004)) would be likely to have the highest impact. Impact, in turn, needs to be defined in terms of specific objectives, such as increasing the incomes of resource-poor farmers or conserving biodiversity. Formal priority-setting procedures based on the calculation of producer and consumer surpluses have been applied to a range of agricultural products (Alston *et al.*, 1995). However, such procedures require time-series data on the quantities and values produced and consumed. Such data are not readily available for most agroforestry products in the developing world.

In classical plantation forestry, the selection of species for improvement is straightforward – a single end-product is involved, much information is available about economically important species, and the clients, who are mostly companies and governments, have close control over the genetic improvement process. In agroforestry, in contrast, the choice of species is much more complex and the clientele is very heterogeneous, consisting of many small-scale farmers with differing needs. Farmers use many different tree species, and little scientific or economic information is available for most of them. Moreover, farmers may use a single species in several different ways and the products (e.g. fruits or

 Indigenous Fruit Trees in the Tropics: Domestication, Utilization and Commercialization (eds F.K. Akinnifesi *et al.*)

timber) and services (e.g. windbreak or shade) are often difficult to value (Franzel *et al.*, 1996).

In the past, researchers' own interests and opinions on the importance of particular species were probably the most important criteria in setting research priorities when choosing between tropical agroforestry tree species. This in turn led to a focus on a few exotic species, e.g. *Eucalyptus* spp. and *Leucaena leucocephala* (Lam.) de Wit, in many areas at the expense of valuable, but little known, indigenous species. The priority-setting exercises conducted in Africa and reported in this chapter, in contrast, provide a more objective and systematic approach to dealing with setting priorities and arriving at a best possible set of research activities. The priority-setting exercises encouraged the participation and integrated the views and expertise of various stakeholders: rural households, research scientists, development practitioners and policy makers.

This chapter presents examples of the setting of priorities among indigenous fruit species for domestication research in three regions of Africa: the humid lowlands of West Africa (Franzel *et al.*, 1996), the semi-arid Sahelian zone of West Africa (ICRAF, 1996), and the miombo woodlands of southern Africa (Maghembe *et al.*, 1998). An example is also given from southern Africa of setting priorities among indigenous fruit products. First, the methods are discussed. Next, we present the results from each of the three areas. Finally, the priority-setting process is assessed and we present the lessons learned.

1.2 Approaches for Priority Setting

Priority setting is not just an analytical process – it also seeks to bring about agreement and consensus among the various stakeholders involved in research on indigenous fruits (Franzel *et al.*, 1996). Simplicity, transparency and collaborative appraisal are three features of an effective priority-setting approach. These features encourage the participation and support of the various partners in the priority-setting process, including scientists of national and international research institutions, policy makers, donor agencies, and farmers. The challenge of priority setting is to design a procedure that combines simplicity, transparency, participation and analytical rigour in order to ensure that the right decisions are made and that suitable conditions are created for successfully implementing them.

The methods used in the priority-setting procedures presented in this chapter are summarized in Franzel *et al.* (1996). The priority-setting exercises took place in the humid lowlands of West Africa in 1994/1995, in the Sahel in 1995, and in the miombo woodlands of southern Africa in 1997/1998. Over time, alterations to the priority species have been made, as conditions and markets have changed. The priority-setting procedure involves seven key steps, although there has been some variation in the methods used in each zone (Fig. 1.1).

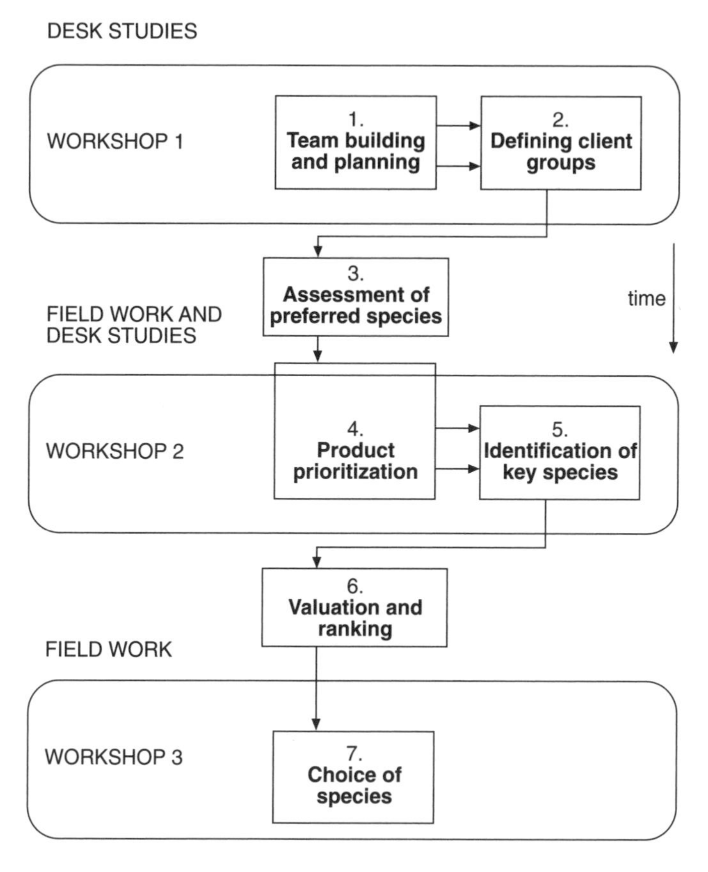

Fig. 1.1. The priority-setting process for indigenous fruit species. Note that steps 2 and 4 can be initiated before the workshop, but the consensus on the results of these steps should be achieved in the workshop. Feedback is important at each step.

1.2.1 Team building and planning

In each region, workshops were held to build an effective team among the participants and different institutions. Most participants were from national research institutes, national extension services, universities, and ICRAF. Participants developed a consensus on the application of the priority-setting approach and the modifications that would be required. National teams designed and implemented the surveys in their respective countries. The teams met periodically in regional meetings to compare findings. The team in southern Africa decided to restrict their initial surveys to preferences among indigenous fruit trees, whereas participants in the Sahel and the humid lowlands of West Africa examined preferences among all indigenous trees. Recent surveys in southern Africa have assessed preferences among both indigenous and exotic fruit trees.

1.2.2 Defining client groups

A review of secondary information was used to define user groups, and identify their main problems and the agroforestry products that may best meet their needs. User groups were defined by agroecological zone, e.g. the humid lowlands of West Africa extend from Senegal to Congo, including parts of 11 countries with altitudes below 1000 mm, rainfall above 1500 mm, and growing periods over 220 days. User groups were also defined by socio-economic variables. For example, in the Sahel, three distinct user groups within the Sahelian agroecological zone were identified: women, young men, and older men. Survey data were disaggregated by these three categories.

1.2.3 Assessment of species preferred by clients

Since no data on user preferences were available, semi-structured field surveys were conducted in which randomly selected farmers listed the trees they valued most and explained why. Stratified random sampling methods were used, with criteria for stratification varying by region. In the humid lowlands of West Africa, eight land-use systems were defined using criteria such as population density, crops grown, and access to forests and markets. Sample villages were then selected using area sampling methods. In Malawi, in southern Africa, researchers selected one district from each of the country's eight silvicultural zones. The districts were selected based on the knowledge that indigenous fruits were abundant there. Extension agents then helped to select a village in each district, using the same criteria. Local leaders or extension staff prepared lists of all households in the villages and 15 names were randomly selected for interview (Malembo *et al.*, 1998). Sample size varied considerably between regions, from 94 to 470, depending primarily on resource availability (Table 1.1).

In all three regions, representatives from each country developed standardized survey protocols for interviewing farmers about their preferred indigenous fruit species and the traits they wished researchers to improve. Farmers were first asked to list the indigenous species in their area. Next they

Table 1.1. Sample sizes in surveys to assess farmer preferences among tree species.

Region	Number of farmers interviewed	Proportion of female farmers interviewed (%)	Countries where farmers were interviewed
Humid lowlands of West Africa	94	25	Nigeria, Cameroon, Ghana
Sahelian zone of West Africa	470	39	Senegal, Mali, Niger, Burkina Faso
Miombo woodlands of southern Africa	451	24	Zambia, Zimbabwe, Malawi and Tanzania[a]

[a] Mozambique was included in a separate survey.

were asked which species were most important to them and why. They also ranked them in order of importance and provided information on managing, utilizing and processing indigenous fruit. Finally, they specified the fruit and tree traits they wished researchers to improve.

Focus group discussions were also conducted with traditional leaders, children and other key informants. Participatory tools were used in the group discussions, such as village maps to show the locations of indigenous fruit trees, and seasonal calendars to show the times that fruit trees flowered and were harvested. The data were analysed by country so that the priority species of each could be identified and compared. Two measures were used to indicate the degree of species importance (importance and preference are used interchangeably in this chapter): the percentage of farmers mentioning the species as important, and the average preference score (10 = most preferred, 9 = second most preferred, etc.). Average preference scores were computed only for those mentioning the species as important. The scores thus measure the degree of preference among those who find the species important.

1.2.4 Product prioritization

Here, tree products and services are ranked in order of their current and potential importance and value to clients. Participants used information about trends and market opportunities, including some that most farmers were probably not aware of. In the humid lowlands of West Africa, this exercise was conducted in a stakeholder workshop (Franzel *et al.*, 1996). In the other two regions, product prioritization was not conducted because the priorities were only being set for indigenous fruit trees.

1.2.5 Identification of key species

In addition to information on clients' preferences, participants in the priority-setting process needed additional information for identifying priority species. In the humid lowlands of West Africa, participants refined their list further by ranking species according to three other criteria:

1. Their 'researchability', that is, the potential of research to achieve impact in domesticating the species.
2. Expected rates of adoption, based on ease of regeneration, adaptability across the region, and commercial potential.
3. The degree to which women in the community were likely to benefit from domestication of the species.

1.2.6 Valuation and ranking of priority species

In the humid lowlands of West Africa, researchers conducted household surveys to collect detailed data from farmers and markets to estimate the value of products of the remaining species and to update the information on farmer preferences, researchability and adoption. The survey involved 152 farmer interviews in Nigeria and Cameroon. These surveys were not undertaken in the Sahel or in southern Africa because they were considered too costly to undertake.

1.2.7 Choice of species to focus domestication efforts on

Here, the results of each of the regional priority-setting exercises were synthesized in a workshop and the choice of priority species was reviewed and approved.

1.2.8 Choice of commercial products on which to focus research and development activities

This step was not included in the original priority-setting procedures but was added to the process in southern Africa in 2003. In that year, researchers organized a workshop in Harare, Zimbabwe, of stakeholders from the private sector, NGOs, and academia to determine the fruit species and products they preferred to process. Next, workshops were held in Magomero, Malawi, for local fruit-processing groups from Malawi and Zambia, and in Tabora, Tanzania, for groups there to assess their preferences among fruits and products to process. Participants set priorities among all fruits in the Malawi and Tanzania workshops but only among indigenous ones at the Harare workshop (Ham, 2004). A total of 97 people, mostly women, attended the three workshops.

The product prioritization process for the Harare workshop was based on a process developed by FAO (Lecup and Nicholson, 2000) in which products are evaluated according to indicators in four categories:

1. *Ecological*, including availability of trees, ease of regeneration, pest resistance, variation in fruit taste between trees, and willingness to plant trees.
2. *Socio-economic*, including suitability for community processing, experience with processing, opportunities for adding value, employment creation potential, and gender impact.
3. *Market*, including extent and quality of existing information, market demand, market readiness of the product, financial viability, and competition.
4. *Technical*, including ease of processing, availability of equipment and infrastructure, product shelf life, availability of skills, personnel, and packaging material.

1.3 Regional and Country Priority-setting Results

1.3.1 Humid lowlands of West Africa

Farmer preference survey

There was considerable variability among farmers' priority species within and between the countries surveyed: Cameroon, Nigeria and Ghana (Table 1.2).

Table 1.2. Farmers' most important tree species in the humid lowlands of Nigeria, Cameroon and Ghana.

	Nigeria			Cameroon			Ghana		
Rank	Species	%	APS	Species	%	APS	Species	%	APS
1	*Irvingia gabonensis*	86	7.8	*Irvingia gabonensis*	86	6.0	*Artocarpus communis*	67	4.5
2	*Chrysophyllum albidum*	56	6.8	*Baillonella toxisperma*	77	6.0	*Annona squamosa*	58	3.5
3	*Dacryodes edulis*	64	6.2	*Dacryodes edulis*	67	4.6	*Chrysophyllum albidum*	46	3.0
4	*Milicia excelsa*	64	5.8	*Ricinodendron heudelotii*	67	4.1	*Tetrapleura tetraptera*	54	2.9
5	*Vernonia amygdalina*	62	5.6	*Alstonia boonei*	63	3.6	*Garcinia afzelji*	29	2.0
6	*Garcinia kola*	39	4.6	*Guibourtia demeusi*	37	3.4	*Ricinodendron heudelotii*	42	1.9
7	*Treculia africana*	28	3.9	*Endrandsphaga cylindricum*	60	2.6	*Dacryodes klaineana*	25	1.9
8	*Dennetia tripelata*	30	3.7	*Milicia excelsa*	37	1.8	*Synsepalum dulcificum*	42	1.8
9	*Neubouldia laevis*	30	3.0	'Esok'	33	1.7	*Spondias mombin*	42	1.8
10	*Occimum gratissimum*	11	2.5	*Raphia* spp.	23	*1.5*	*Solanum indicum*	*33*	1.8

% = percentage of households mentioning the species; APS = average preference scores for those households mentioning the species, '10' indicates a ranking of 1st, '9' a ranking of 2nd, etc.

The number of species mentioned in each of the three countries as among the ten priority species by farmers ranged from 60 in Cameroon to 172 in Nigeria. No single species ranked among the top ten in all three countries. Three species, discussed below, ranked among the top four in at least two of the countries. All three were important as both food and cash earners (Adeola *et al.*, 1998).

- *Irvingia gabonensis* ranked first in both Nigeria and Cameroon. Also called 'bush mango', it is prized for its cotyledons, which are used as a constituent of sauces. In addition, the fruits of some varieties are eaten. *I. gabonensis* was not ranked among the top ten species in Ghana, although its fruits are being widely harvested (and grown in a few areas) for export to Nigeria.
- *Dacryodes edulis*, also called 'bush butter' or 'prune', ranked third in both Nigeria and Cameroon. The fruit is boiled or roasted and is an important food during its brief fruiting season. The species is not known in Ghana.
- *Chrysophyllum albidum*, also called 'star apple', ranked second in Nigeria and third in Ghana, but is unknown in Cameroon. It is consumed as a fresh fruit.

Two other species, *Ricinodendron heudelotii* and *Milicia excelsa*, featured among the top ten species in at least two of the countries. *R. heudelotii* kernels are ground and used as a constituent of sauces; *Milicia* is a timber species. *Garcinia kola* was the only species that ranked among the top 15 in all three countries. *G. kola* nuts are prized as a stimulant.

Ranking of products

The farmer preference survey confirmed that food was by far the most important use of priority species; out of 30 species comprising the top ten species in the three countries, food was named as the main use for 21 (69%) species. Timber and medicine were each the main use of four (13%) species.

In the stakeholder workshop, food received the highest rating of any product. The potential for making impact through research on food was considered to be high, as principles developed for industrial tree crops could be applied. For example, extending the harvesting period or improving fruit quality could be achieved in a reasonable period of time. Moreover, the mandates of the institutions participating in the priority-setting exercise gave high priority to food production. Other products and services considered in this ranking included soil fertility, fodder, medicine and timber. Soil fertility and fodder received low ratings, mainly because of their low present value to farmers. Medicine scored low because it had little relevance to the institutional mandates of the government services participating in the workshop. Timber scored low because of the long period required to produce it.

Team members advanced ten species to the next stage. Of the six species mentioned above as being rated highest in the farmer preference survey, only one, *Milicia*, was excluded – because the long time required to reach maturity, up to 50 years, would reduce farmers' interest in planting it. Five other food-producing species were included – *Artocarpus communis* and *Annona squamosa*, because they were the highest-ranking species in Ghana, and *Coula edulis*, *Tetrapleura tetraptera* and *Spondias mombin*, because they were believed to be important across the region, even though they did not rank particularly high in the farmer preference survey.

Screening on basis of researchability

D. edulis and *I. gabonensis* rated highest on researchability in the survey of researchers (Table 1.3), as they did in the farmer preference survey. Both species are known to have a high level of genetic variability, and the knowledge base concerning them is high relative to the other species. Both have short reproduction cycles; they can each bear fruit within 5 years (*I. gabonensis* only through vegetative propagation). *C. albidum* ranked fifth, receiving low to medium ratings on germplasm availability, knowledge base, and the level of

Table 1.3. Weights of criteria on researchability and mean scores of tree species on criteria.

Criteria	Level of genetic variability[a]	Knowledge base[b]	Germplasm availability[c]	Speed of reproduction[d]	Uniqueness of research efforts[e]	Total score[f]	Ordinal score
Weight[g]	3	2	2	2	1		
Mean scores[h]							
Dacryodes edulis	3.0	2.0	2.2	3.0	2.3	25.7	High
Irvingia gabonensis	2.8	1.7	2.0	3.0	2.7	24.5	High
Spondias mombin	2.3	1.25	1.7	3.0	3.0	21.8	Medium
Tetrapleura tetraptera	2.4	1.25	1.7	2.0	3.0	20.1	Medium
Chrysophyllum albidum	2.0	1.25	1.0	3.0	3.0	19.5	Medium
Artocarpus communis	1.5	1.3	1.0	3.0	3.0	18.1	Medium
Coula edulis	2.5	1.25	1.5	1.0	3.0	18.0	Medium
Annona squamosa	1.0	1.0	1.0	3.0	3.0	16.0	Low
Garcinia kola	1.5	1.3	1.6	1.0	3.0	15.3	Low
Ricinodendron heudelotii	1.0	1.4	1.25	2.0	2.8	15.1	Low

[a] Includes number of provenances, phenotypes and differences in fruit size, quality and growth rates.
[b] Includes all information available on the species, including knowledge about sexuality, flowering/fruiting habits, pests and diseases. Indicators include number of publications and number of researchers working on it.
[c] Indicates whether collections have taken place, numbers of accessions in gene banks, ease of collection, fruiting frequency, and storeability of seed.
[d] 0–5 years = 3, 6–10 years = 2, 11+ years = 1. Through vegetative or other methods of propagation.
[e] Considers whether other research groups are already working on the domestication of the species and whether there is a risk of duplication of effort.
[f] Each criterion score is multiplied by the weight of the criterion to arrive at a weighted criterion score.
[g] As determined by the priority-setting team.
[h] 3 = high, 2 = medium, 1 = low. Twelve researchers rated the species but not all rated each species on each criterion.

genetic variability. *G. kola* and *R. heudelotii* ranked ninth and tenth, respectively. They received low to medium ratings on the above three criteria, and have medium to long reproduction cycles. The range in ratings was considered to be narrow enough so that priority was still given to the five farmer-preferred species.

Valuation and ranking of five priority species

In the humid lowlands of Nigeria, *I. gabonensis* was grown or collected by the greatest percentage of households (75%) and had the highest value of production per grower/collector per year, US$124 (US$ = Naira 40, June 1994) (Table 1.4). However, values were extremely variable and coefficients of variation for the value of production were more than 70%. Nationwide, *D. edulis* ranked second and *C. albidum* third in percentage of households and value per grower/collector (Table 1.4). In Cameroon, *D. edulis* ranked slightly higher than *I. gabonensis* in terms of percentage of farmers growing/collecting and value per grower/collector. *R. heudelotii* ranked second in percentages of households growing or collecting, but values per grower/collector were only one-fourth those of *I. gabonensis* and *D. edulis*. *G. kola* was grown or collected by only 37% of the households but had a high value per grower/collector, over double that of *R. heudelotii*. Two methods were used to estimate quantities harvested: farmers' recall, usually expressed in terms of local measures, and multiplying the numbers of trees by tree yields obtained from secondary data. The two methods yielded fairly similar estimates; for example, in Cameroon, mean values for *D. edulis* quantities differed by 8% between the two methods, for *I. gabonensis* by 25%, and for *G. kola* by 36%.

By extrapolating sample values to the estimated number of agricultural households of the humid lowlands of Nigeria, *I. gabonensis*'s farm-gate value for the humid lowlands is estimated at US$145 million, *D. edulis* at US$78 million and *C. albidum* at US$16 million (Table 1.4). Because of the high variability in the data, there was no significant difference between the values for *I. gabonensis* and *D. edulis*, though both were higher than

Table 1.4. Region-wide values of production for main tree products, humid lowlands of Nigeria and Cameroon.

	Humid lowlands of Nigeria			Central, south and east provinces of Cameroon		
		90% confidence interval			90% confidence interval	
	Mean (US$ million)	Upper limit	Lower limit	Mean (US$ million)	Upper limit	Lower limit
Irvingia gabonensis	145	230	59	17	24	11
Dacryodes edulis	78	101	55	22	28	15
Chrysophyllum albidum	16	21	11	0	0	0
Ricinodendron heudelotii	0	0	0	5	11	1

C. albidum ($P < 0.05$). For the central, south and east provinces of Cameroon, the areas covered in the survey, values were US$21.6 million for *D. edulis*, US$17.5 million for *I. gabonensis*, and US$4.9 million for *R. heudelotii* (Table 1.4). Again there were no significant differences between *D. edulis* and *I. gabonensis*, but both were higher than the third-ranking species, *R. heudelotii* ($P < 0.05$). All five of the priority species are found in other countries of western and central Africa (Aiyelaagbe *et al.*, 1997; Ndoye *et al.*, 1998), but data on their value are scant. In an extensive nationwide survey of Gabon, farmers ranked *I. gabonensis* first and *D. edulis* second in importance among agroforestry and forestry species (D. Boland, 1995, personal communication). None of the other species mentioned above ranked among the top seven species. In Ivory Coast, *I. gabonensis* is grown in the west, south and south-west regions of the country and is widely traded in urban areas (Bonnehin, 1998). *I. gabonensis* is also an important component of regional trade; exports to Nigeria are reported from Cameroon, Central African Republic, Benin, Ghana and Ivory Coast. Ivory Coast also exports to Liberia and Sierra Leone. Gabon imports from Cameroon and Equatorial Guinea (ICRAF, 1995). *D. edulis* fruits and *G. kola* bark, an additive to palm wine, are exported from Cameroon to Gabon (Ndoye *et al.*, 1998). All species had additional uses, e.g. for firewood, fodder, timber and medicine, but the value of these appeared to be relatively low.

Concerning expected rates of adoption, *I. gabonensis* and *D. edulis* ranked about equally. *I. gabonensis* was the most widely grown species in Nigeria, whereas *D. edulis* was the most widely grown in Cameroon. In terms of spread, *I. gabonensis* plants were grown or collected by over 50% of the farmers in all six surveyed land-use systems, *D. edulis* in five. *I. gabonensis* was the preferred species for future planting in Nigeria (49% of the farmers versus 18% favouring *D. edulis*) while *D. edulis* was preferred in Cameroon (51% versus 27% for *I. gabonensis*). *G. kola* ranked third in both countries. Gender roles varied somewhat across the different species and tasks. Men, women and children were generally all involved in harvesting, and females dominated processing. In Nigeria, all 12 key informants reported that females received cash from *I. gabonensis* sales, whereas only five reported that men also received cash. Results were similar in Cameroon, where all five key informants who responded to the question reported that females received more cash from fruit sales than did men. Concerning seed sales, five of eight informants reported that women received more cash from seed sales than men did. In Cameroon, all nine respondents claimed that women or children were the main beneficiaries of *R. heudelotii* sales. Men tended to receive most of the cash earned from *G. kola* and *C. albidum*, while receipts from *D. edulis* appeared to be shared about equally. The results were consistent across countries, although there was some variation within each country.

Farmers' improvement objectives

Farmers had very clear ideas on how researchers could improve their preferred trees. Concerning *Irvingia*, farmers in Nigeria were mostly interested in reducing the height of the tree; in Cameroon interest was mainly in bigger fruit and earlier maturing trees. Nigerian farmers' interest in shorter trees probably reflects the scarcity of land and thus greater competition of these trees with crops in the crop fields and homegardens. The long period before maturity, 10–15 years, explains farmers' interest in reducing this period. Concerning *D. edulis*, reduced height was an important consideration in Nigeria and Ghana because the trees are frequently grown in homegardens. Bigger fruits were desired in all three countries. Fruit size is especially important in southern Cameroon, where farmers receive a high price premium for fruits being exported to Gabon. Other important criteria included taste (some trees give sweeter tasting fruit than others do) and early maturity (trees take about 5–7 years to bear fruit).

Choice of priority species

I. gabonensis and *D. edulis* had the highest overall ratings on value of expected benefits from domestication (Table 1.5). *I. gabonensis* had the highest financial value in Nigeria, ranked second in Cameroon, and was extensively traded among countries in the region. In addition, it had high researchability, a high expected rate of adoption, and it directly benefited women. *Dacryodes* ranked first in Cameroon on financial value and second in Nigeria and had similar ratings to *I. gabonensis* on other criteria, except that it benefited males and females about equally. *C. albidum*, *R. heudelotii* and *G. kola* each received an overall rating of medium. *G. kola* had a broader geographical spread and range of products but relatively few farmers used it in any particular region compared with *C. albidum* and *R. heudelotii*. *C. albidum* rated higher on researchability and *R. heudelotii* had greater benefits for women.

In 2006, 10 years after the initial priority-setting exercise was completed, the list of priority species included the same five indigenous fruit species mentioned above, plus several others (Table 1.6). One indigenous fruit species had been added, three medicinal trees, one vegetable, two spices, and one species used for edible oil. New species were added on the basis of farmers' interest and the

Table 1.5. Value of expected benefits from domestication, humid lowlands of West Africa.

	Current annual value of main product	Current annual value of other products	% increase in value expected from domestication	Expected rate of adoption	Gender receiving most benefits	Overall value of expected benefits
Irvingia gabonensis	High	Medium	High	High	Female	High
Dacryodes edulis	High	Low	High	High	About equal	High
Chrysophyllum albidum	Medium	Low	Medium	Medium	Male	Medium
Ricinodendron heudelotii	Medium	Low	Low	Medium	Female	Medium
Garcinia kola	Medium	Medium	Low	Medium	Male	Medium

Adapted from Franzel *et al.* (1996).

Table 1.6. Priority indigenous species for the African humid tropics, 2006.

Scientific name	Common name(s)
Fruit trees	
Irvingia gabonensis	Andok, bush mango
Dacryodes edulis	African plum, atanga, safou
Ricinodendron heudelotii	Essang, njanssang
Garcinia kola	Bitter cola, onie
Chrysophyllum albidum	White star apple
Cola spp.	Cola
Medicinal plants	
Prunus africana	Pygeum
Pausinystalia johimbe	Yohimbe
Vegetables	
Gnetum africanum	Afang, eru, okasi, okok
Spices and edible oils	
Afrostyrax lepidophyllus	Country onion
Monodora myristica	Faux muscadier
Allanblackia spp.	Nsangomo
Other	
Annickia chlorantha	Moambe jaune
Macronetum combretum	Liane de vers

Source: ICRAF (2006).

potential of the tree product to earn cash income for poor farmers, especially in response to favourable market opportunities (A. Degrande, 2006, personal communication).

1.3.2 Semi-arid Sahelian zone of West Africa

The area surveyed extends across Senegal, Mali, Burkina Faso and Niger at the southern margins of the Sahara desert. Rainfall ranges from 350 to 850 mm/year and altitude from 0 to 350 m above sea level. In listing the top 14 or 15 species in the four countries, the maximum number of species could have been 59 had there been no species in common among countries, whereas with complete agreement between countries, only 15 would have been listed. The actual number of 28 (Table 1.7) shows that there is general interest throughout the region in a number of species and this augurs well for regional tree domestication initiatives. Most striking is that baobab (*Adansonia digitata*) is a clear favourite across the region, closely followed by karité (*Vitellaria paradoxa*) and tamarind (*Tamarindus indica*). Seven species were mentioned in all four countries, one was mentioned in three countries, eight were recorded in two countries, and 12 were recorded in only one country. Interestingly, only two of the 28 species are exotic to West Africa and even those have been widely naturalized there (*Azadirachta indica* and tamarind).

As in the humid lowlands of West Africa, trees providing food dominated farmers' choices among species. In Mali, for example, eight of the top ten

Table 1.7. The 15 top species in each of the four Sahelian countries, ranked according to the proportion of respondents mentioning the species as important.

	Rank in country			
Species	Burkina Faso	Mali	Niger	Senegal
Acacia macrostachya	14			
Acacia nilotica				14
Adansonia digitata	5	2	1	1
Azadirachta indica				10
Balanites aegyptiaca	8	11	2	6
Bombax costatum	6			
Borassus aethiopum		10	6	
Combretum nigricans			13	
Cordyla pinnata		9		2
Detarium microcarpum	11	14	15	9
Diospyros mespiliformis	10		3	
Faidherbia albida	7	6	10	3
Ficus gnaphalocarpa	15			13
Ficus iteophylla				8
Hyphaene thebaica			9	
Khaya senegalensis		8	14	
Landolphia senegalensis		15		
Lannea microcarpa	3	5		
Parinari macrophylla			8	12
Parkia biglobosa	2	3	12	7
Pterocarpus erinaceus		13		
Saba senegalensis	13			
Sclerocarya birrea	12	7		
Sterculia setigera				11
Tamarindus indica	4	4	11	4
Vitellaria paradoxa	1	1	4	
Vitex doniana			5	
Ziziphus mauritiana	9	12	7	5

Source: ICRAF (1996).

priority species were important because of their edible fruits. Two also provided leaves as food (Sidibé *et al.*, 1996). On average, 39% of the interviewees were women and they expressed different preferences from those of the men. Men ranked *Balanites aegyptiaca* and *Faidherbia albida* very highly, whereas women consistently ranked baobab as the most valuable species. Men however, were not a homogeneous group, and older and younger men showed differences in preferences for species such as tamarind and *Ziziphus mauritiana*.

Some of the most threatened species in the region, including *Bauhinia rufescens* and *Prosopiis africana* were not included in the top 15 species in any of the countries, even though they are important fodder species. Work on species that really need to be conserved will have to be carefully balanced against work on those with more immediate prospects for improvement because of farmer-led demand for food species.

1.3.3 Miombo woodlands of southern Africa

Farmer preference surveys

The area surveyed in southern Africa was the miombo woodlands, which is characterized by a single rainy season (800–1200 mm/year) and altitude 600–1200 m above sea level. Four countries were included in the survey: Tanzania, Malawi, Zambia and Zimbabwe. There was considerable agreement in the perceived importance of species across countries (Table 1.8). Three species were widely regarded as important. In all four countries, *Uapaca kirkiana* was mentioned as important by over 50% of farmers, *Parinari curatellifolia* was important for at least 40% and *Strychnos cocculoides* for at least 30%. All three species were among the top five in all four countries, as measured by the frequency with they were mentioned as a priority species. *U. kirkiana* also had the highest average preference score in the three countries where such scores were computed. This shows that it was highly preferred among those mentioning it. *P. curatellifolia* had the third highest preference score in each of the three countries and *S. cocculoides* was rated ninth in Malawi, fifth in

Table 1.8. Overall ranking and importance value of the top 20 miombo fruits selected for domestication by farmers in Malawi, Tanzania, Zambia and Zimbabwe.

	Malawi		Tanzania		Zambia		Zimbabwe	
Species name	%	APS	%	APS	%	APS	%	APS[a]
Uapaca kirkiana	77	9.3	53	9.3	85	9.5	70	
Parinari curatellifolia	55	8.4	43	8.9	80	9.0	57	
Strychnos cocculoides	32	7.0	64	8.2	52	7.7	78	
Anisophyllea boehmii	0	0	0	0	75	7.9	0	
Azanza garckeana	20	6.6	0	0	11	5.4	52	
Flacourtia indica	29	7.4	3	7.0	19	6.5	26	
Syzygium guineense	11	7.4	1	8.0	39	5.4	0	
Strychnos pungens	0	0	0	0	16	5.0	31	
Physalis peruviana	2	6.5	0	0	35	4.7	0	
Uapaca nitida	1	1.0	0	0	35	6.0	0	
Ximenia americana	7	7.6	1	6.0	27	4.0	0	
Diospyros mespiliformis	0	0	0	0	8	8.6	28	
Tamarindus indica	10	5.9	33	8.9	7	8.3	0	
Vangueria infausta	19	7.3	16	6.5	3	9.0	4	
Annona senegalensis	19	8.1	1	7.0	5	8.4	7	
Vitex payos	1	7.0	0	0	0	0	34	
Adansonia digitata	12	8.5	1	8.0	10	9.5	3	
Syzygium spp.	0	0	0	0	0	0	28	
Vitex mombassae	0	0	46	9.0	1	5.0	0	
Vitex doniana	8	6.4	26	8.1	1	5.5	0	

% = percentage of households mentioning the species.
APS = average preference scores for those households mentioning the species: '10' indicates a rank of 1st, '9' a rank of 2nd, etc.
[a] Data on average preference scores were not available for Zimbabwe.
Source: Kadzere *et al.* (1998).

Tanzania, and ninth in Zambia. In fact, many of the species with the highest preference scores were mentioned as important by only a few farmers. For example, *Adansonia digitata* was mentioned as important by only 10% of the sample in Tanzania, reflecting its limited geographical spread. Nevertheless, it tied with *U. kirkiana* for the highest preference score, reflecting its importance among the farmers who used it (Kadzere *et al.*, 1998). *Azanza garckeana* and *Flacourtia indica* were important for at least 10% of farmers in each of three countries in the region: Malawi, Zambia and Zimbabwe. However, they had relatively low preference scores: *Azanza* ranked 11th in Malawi and 13th in Zambia and *Flacourtia* ranked sixth and ninth. A few species were very important in only one or two countries. *Anisophyllea boehmii* was important for 75% of the respondents in Zambia, with the eighth highest preference score, but was not mentioned in any other country. *Vitex mombassae* and *Tamarindus indica* were preferred by 46% and 33% in Tanzania, respectively, and received the second and third highest preference scores in that country (Kadzere *et al.*, 1998).

The survey also revealed important findings about the processing of the different indigenous fruit species. Among the five most important species across the region (*U. kirkiana*, *P. curatellifolia*, *S. cocculoides*, *A. garckeana* and *F. indica*), four are made into jams, and three each are used to prepare alcoholic drinks or juices, or are ground into powder for mixing with other foods (Kadzere *et al.*, 1998). Two are dried, two are used to prepare porridge, and from one (*P. curatellifolia*) oil is extracted from its nut. Of major importance is the finding that all five of the most important species are harvested just before or during the region's 'hunger period' (Akinnifesi *et al.*, 2004). Farmers have confirmed that indigenous fruits are indeed important components of the 'coping strategies' that farmers use during times of famine (Akinnifesi *et al.*, 2004). Indigenous fruit pulp and nuts of *P. curatellifolia* and sometimes *U. kirkiana* are pounded and mixed with small amounts of cereal flour to make the regional food staple ('sadza' or 'nshima') (Kadzere *et al.*, 1998).

Like most agroforestry trees, all five of the most important species had multiple uses; all were used for medicine, crafts, fuelwood and construction. Two each were used for shade, as ornamentals, or for browsing by livestock.

Farmers' improvement objectives

Farmers mentioned fruit and tree traits that they wanted researchers to improve. Concerning the fruits, farmers' priorities focused on improving taste and increasing size. Other characteristics that farmers wanted improved were more species-specific and included *U. kirkiana*'s vulnerability to infestation by fruit-fly maggots, *P. curatellifolia*'s foul smell, and *S. cocculoides*' high ratio of pericarp to fruit pulp. Concerning the trees, farmers' highest priority was to reduce fruit precocity – the period between planting and fruiting – because many of the trees require 10 or more years before fruiting. Other desired improvements included reducing tree size and increasing fruit yield. Farmers want smaller trees because they are easier to harvest and they compete less with adjacent crops. Interestingly, extending the seasonal availability of fruits was not mentioned as important, perhaps because different fruits are available at different seasons or because fruiting seasons of the same species vary by latitude (Kadzere *et al.*, 1998). Other household surveys identified other critical problems for research to address: reduced fibre content in *Azanza garckeana*, pest and disease resistance for *P. curatellifolia,* and longer shelf life for *Flacourtia indica*, *P. curatellifolia*, *U. kirkiana* and *Ficus sycomorus* (Akinnifesi *et al.*, 2006).

Dynamics in users' preferences among priority fruit species: 1995–2004

In order to understand whether farmers' preferences of indigenous fruit tree species are the same throughout the region, we collated priority-setting information collected by different researchers at different times and locations in the region. These included both household surveys and market surveys done between 1995 and 2004 (Table 1.9). A new prioritization exercise was also carried out for Mozambique in Tete and Manica provinces. The results were triangulated to obtain regional priority species. The results showed that priority species vary with location in the region, but a common thread can be seen in the first three priority species that is similar to the earlier results (Table 1.7). The five most preferred indigenous fruit trees across the countries were analysed and ranked as follows: *Uapaca kirkiana*, *Strychnos cocculoides*, *Parinari curatellifolia*, *Ziziphus mauritiana* and *Adansonia digitata.* Of these, *U. kirkiana*, *S. cocculoides* and *P. curatellifolia* remain the three most preferred species in the region, confirming the validity of the previous regional survey in four countries by Maghembe *et al.* (1998). In addition, *Z. mauritiana* and *A. digitata* broke into the top-ranking species in the new analysis. Fruits from these two species are unique in that they would store better than most miombo indigenous fruits. *Z. mauritiana* is well traded in Malawi (Kaaria, 1998; Schomburg *et al.*, 2002) and Zimbabwe (Ramadhani, 2002), and is probably one of the most widely traded local fresh fruits in the region after *U. kirkiana*.

In Malawi, *Adansonia digitata* has been commercialized as 'mlambe' juice which is sold in supermarkets. *Strychnos cocculoides* seems to be more strongly preferred than *Uapaca kirkiana* in Zimbabwe. In Tanzania, *Vitex mombassae* and *S. cocculoides* are important, followed by *Parinari curatellifolia* and *U. kirkiana.* Because of the commercial potential of *Sclerocarya birrea*, it was also included in the domestication programme in southern Africa (Akinnifesi *et al.*, 2006); however, it was not captured as a high-priority species by communities in the five countries. This is the shortcoming of priority setting based on surveys, as communities are seldom knowledgeable about regional or global markets outside their own country.

Many of the species-preference surveys undertaken in the region deliberately omitted exotic fruits from the ranking lists (Kwesiga and Mwanza, 1995; Kadzere *et al.*, 1998; Maghembe *et al.*, 1998; Ramadhani *et al.*, 1998; Rukuni *et al.*, 1998). This was a major shortcoming because it was difficult to identify the relative preferences for different indigenous fruit trees in situations where investors or development agencies were interested in identifying the fruit species most relevant to communities' livelihoods. We carried out two new surveys during 2002–2004 to assess the relative preference for all fruits by smallholder farmers in southern Malawi (Table 1.10). This was also compared

Table 1.9. Priority indigenous fruit trees in southern Africa.

Country	Type of survey	Rank 1	Rank 2	Rank 3	Rank 4	Rank 5	Reference
Malawi	1. Household (n = 128)[a]	*Uapaca kirkiana*	*Parinari curatellifolia*	*Strychnos cocculoides*	*Flacourtia indica*	*Azanza garckeana*	Malembo *et al.* (1998)
	2. Market survey[a]	*Uapaca kirkiana*	*Casimiroa edulis*	*Parinari curatellifolia*	*Azanza garckeana*	*Flacourtia indica*	Minae *et al.* (1995)
	3. Farmer survey (n = 155)[a]	*Uapaca kirkiana*	*Casimiroa edulis*	*Annona senegalensis*	*Ziziphus mauritiana*	*Strychnos cocculoides*	F.K. Akinnifesi (unpublished data)
	4. Farmer survey (n = 223)[b]	*Uapaca kirkiana*	*Ziziphus mauritiana*	*Adansonia digitata*	*Strychnos cocculoides*	*Parinari curatellifolia*	J. Mhango and F.K. Akinnifesi (unpublished data)
Mozambique	Household (n = 156)[a]	*Adansonia digitata*	*Ziziphus mauritiana*	*Tamarindus indica*	*Uapaca kirkiana*	*Strychnos cocculoides*	E.C. Sambane (unpublished data, 2004)
Tanzania	1. Household (n = 70)[a]	*Strychnos cocculoides*	*Uapaca kirkiana*	*Parinari curatellifolia*	*Vitex mombassae*	*Vitex doniana*	Ramadhani *et al.* (1998)
	2. Market survey (n = 118)[c]	*Vitex mombassae*	*Vitex doniana*	*Strychnos cocculoides*	*Parinari curatellifolia*	*Syzygium guineense*	Mumba *et al.* (2004)
Zambia	1. Household (n = 153)[c]	*Uapaca kirkiana*	*Parinari curatellifolia*	*Anisophyllea boehmii*	*Strychnos cocculoides*	*Syzygium guineense*	Simwaza and Lungu (1998)
	2. Market survey (n = 132)[c]	*Ziziphus mauritiana*	*Adansonia digitata*	*Uapaca kirkiana*	*Tamarindus indica*	*Diospyros mespiliformis*	Kwesiga and Mwanza (1995)
Zimbabwe	1. Household survey (n = 118)[a]	*Strychnos cocculoides*	*Strychnos pungens*	*Azanza garckeana*	*Parinari curatellifolia*	*Uapaca kirkiana*	Rukuni *et al.* (1998)
	2. Household survey (n = 113)[a]	*Strychnos cocculoides*	*Uapaca kirkiana*	*Parinari curatellifolia*	*Azanza garckeana*	*Vitex payos*	Kadzere *et al.* (1998)
Regional	Various surveys as above	*Uapaca kirkiana* (39)[d]	*Strychnos cocculoides* (27)	*Parinari curatellifolia* (18)	*Ziziphus mauritiana* (15)	*Adansonia digitata* (11)	F.K. Akinnifesi (unpublished data)

[a] Based on weighted mean; [b] based on mean rank order values as per Franzel *et al.* (1996); [c] based on percentage mentioned as top priority; [d] figure in parenthesis is a weighted average estimated by multiplying the number of times ranked among priority species and the rank nominal value.

Table 1.10. Relative preference of indigenous and exotic fruit species in the southern and central regions of Malawi.

	Southern region[a]			Southern region[b]		Central region[c]	
Rank	Species	Total weighted score	Mean rank order value	Species	Weighted average	Species	Weighted average
1.	*Mangifera indica*	678	3.02	*Mangifera indica*	121	*Citrus sinensis*	259
2.	*Persea americana*	472	2.12	*Persea americana*	98	*Mangifera indica*	177
3.	*Citrus sinensis*	291	1.30	*Citrus sinensis*	93	*Psidium guajava*	136
4.	*Uapaca kirkiana*	269	1.24	*Uapaca kirkiana*	63	*Musa paradisiaca*	116
5.	*Carica papaya*	234	1.09	*Psidium guajava*	55	*Carica papaya*	115
6.	*Musa paradisiaca*	199	0.89	*Musa paradisiaca*	51	*Citrus reticulata*	105
7.	*Psidium guajava*	198	0.91	*Citrus reticulata*	46	*Citrus limon*	73
8.	*Casimiroa edulis*	198	0.89	*Carica papaya*	46	*Uapaca kirkiana*	51
9.	*Citrus reticulata*	164	0.74	*Prunus persea*	35	*Persea americana*	47
10.	*Ziziphus mauritiana*	158	0.71	*Annona comosus*	25	*Strychnos cocculoides*	42
11.	*Annona senegalensis*	89	0.40	*Casimiroa edulis*	20	*Prunus persea*	33
12.	*Prunus persea*	51	0.23	*Passiflora ligularis*	17	*Casimiroa edulis*	18
13.	*Annona comosus*	47	0.21	*Annona senegalensis*	12	*Annona senegalensis*	13
14.	*Flacourtia indica*	42	0.19	*Citrus limon*	10	*Malus domestica*	12
15.	*Malus domestica*	39	0.17	*Ziziphus mauritiana*	9	*Flacourtia indica*	9
16.	*Vangueria infausta*	29	0.13	*Malus domestica*	8	*Syzygium cordatum*	8
17.	*Adansonia digitata*	28	0.13	*Strychnos cocculoides*	4	*Azanza garckeana*	4
18.	*Citrus limon*	24	0.11	*Flacourtia indica*	3	*Ficus natalensis*	3
19.	*Strychnos* spp.	22	0.10	*Tamarindus indica*	3		
20.	*Vitex* spp.	15	0.07				
21.	*Parinari curatellifolia*	13	0.06				
22.	*Azanza garckeana*	13	0.06				
23.	*Tamarindus indica*	10	0.04				
24.	*Cocos nucifera*	9	0.04				

[a] J. Mhango and F.K. Akinnifesi (unpublished) ($n = 223$); [b] F.K. Akinnifesi (unpublished) ($n = 155$), [c] Minae *et al.* (1995).

with the survey conducted by Minae *et al.* (1995) in the central region of Malawi. The two studies in the southern region agreed that the top four fruit species were *Mangifera indica* (mango), *Persea americana* (avocado), *Citrus sinensis* (sweet orange) and *Uapaca kirkiana* (wild loquat), in decreasing order of magnitude. The central region study identified *C. sinenses* as first, *M. indica* as second and *Psidium guajava* (guava) as third preference. In general for all the three studies in Malawi, mango (*M. indica*) is the most preferred, followed by sweet orange (*C. sinenses*), avocado (*P. americana*), guava (*P. guajava*), banana (*Musa paradisiaca*), pawpaw (*Carica papaya*), tangerine (*Citrus reticulata*) and wild loquat (*U. kirkiana*). However, in the two studies conducted in the southern region, *U. kirkiana* was ranked in fourth position of all species and eighth position in the central region, but first among the indigenous fruits. The non-exotic fruit-tree species that were mentioned in the top ten preference list were: *U. kirkiana*, *Z. mauritiana*, *Casimiroa edulis* (Mexican apple) and *Strychnos cocculoides*. It is interesting that *S. cocculoides* was only ranked in 17th–19th position in the southern region, but in tenth position in the central region (Table 1.10). This may be due to the fact that *S. cocculoides* is now rare in the southern region.

Compared to the previous surveys, the results showed that several factors determine users' preferences for different fruit-tree species, including the geographical location and availability of fruits.

Assessing priority products to commercialize

As commercialization of fruit production is becoming a high priority for the development community and farmers, it became necessary to decide which fruits to help communities to commercialize and which products to help them market. Workshop participants in Malawi, Tanzania and Zimbabwe set priorities among products they already knew. At the Zimbabwe workshop each participant was asked to list at least three products and the products mentioned most often were selected as top priority. In the Tanzania and Malawi workshops, participants were first asked to list as many species and product combinations as possible and from this list they ranked those that they most preferred (Table 1.11).

There was considerable variability among countries. At the Malawi and Tanzania workshops, participants assessed both indigenous and exotic species, but while the top four products at the Malawi workshop were exotics (mango and tomato products), all of the top four in Tanzania were indigenous species. The large number of exotic fruits included in the priority lists raises the question about the need to include exotic fruit trees in current domestication programmes. Exotic fruits such as mango could be used as the driver for a domestication and commercialization programme, and indigenous fruits could be introduced as complementary products.

It was noted that marula (*Sclerocarya birrea*) products were strongly promoted by the more commercially orientated participants of the Harare workshop. In contrast, marula products received a much lower rating by the community processing groups who attended the Tanzania and Malawi workshops.

Table 1.11. Participants' priority products in workshops held in Magomero (Malawi), Tabora (Tanzania) and Harare (Zimbabwe).

Magomero	Tabora	Harare
Mango juice	Baobab juice	*Parinari* oil
Mango dried	Groundnut butter	*Strychnos* jelly
Mango jam	*Strychnos* juice	Marula oil
Tomato jam	*Parinari* wine	Marula jelly
Baobab wine	*Vitex* jam	*Ziziphus* fruit leather
Baobab juice	*Syzygium* juice	*Uapaca* jam
Uapaca wine	Marula wine	
Uapaca juice	Flacourtia jam	
Marula wine	Mango juice	
Marula juice	Guava jam	

Source: Ham (2004).

The participants from the Zimbabwe workshop also focused strongly on oil products, while the Tanzania and Malawi participants did not list oils as preferred products. This could possibly be attributed to the higher level of technology required for oil processing, which is not available at community level. The markets for oils seem to be mostly export-orientated and community groups are not aware of these markets or do not have access to them.

Tables 1.12–1.14 show participants' rankings of priority products on four sets of criteria: ecological, socio-economic, market potential and technical requirements. The ecological evaluation was conducted in order to determine the impact of harvesting on the fruit supply, but also to look at aspects such as fruit availability and ease of harvesting. In Malawi, exotic fruits such as mango were rated the highest, while the Tanzania workshop participants felt that indigenous fruits would have a more positive ecological impact. Both workshops gave the lowest ecological ratings to marula products. The differing ecological rankings probably reflect variations in the distribution and availability of the fruit-tree species.

The socio-economic evaluation determines how beneficial these products would be to development and the degree to which they can serve as vehicles

Table 1.12. Overall ranking of products identified during the Magomero, Malawi, workshop.

	Mango jam	Tomato jam	Mango dried	Mango juice	*Uapaca* juice	Baobab juice	*Uapaca* wine	Baobab wine	Marula juice	Marula wine
Ecological	1	4	1	1	5	7	5	7	9	9
Socio-economic	2	1	4	3	6	8	5	7	9	10
Market	2	1	5	7	4	3	10	6	9	8
Technical	1	4	2	6	5	3	7	8	9	10
Sum	6	10	12	17	20	21	27	28	36	37
Overall ranking	1	2	3	4	5	6	7	8	9	10

Source: Ham (2004).

Table 1.13. Overall ranking of products identified during the Tabora, Tanzania, workshop.

	Syzygium juice	Guava jam	*Strychnos* juice	*Vitex* jam	Groundnut butter	Mango juice	Marula wine	*Parinari* wine	Baobab juice	*Flacourtia* jam
Ecological	1	9	3	4	10	6	1	5	6	8
Socio-economic	3	1	3	8	2	5	10	8	6	7
Market	2	4	4	6	1	3	10	9	8	6
Technical	3	1	6	3	10	9	3	2	6	6
Sum	9	15	16	21	23	23	24	24	26	27
Overall ranking	1	2	3	4	5	5	7	7	9	10

Source: Ham (2004).

Table 1.14. Overall ranking of products identified during the Harare, Zimbabwe, workshop.

	Ziziphus fruit leather	Marula jelly	Marula oil	*Parinari* oil	*Uapaca* jam	*Strychnos* jelly
Ecological	2	5	5	3	1	3
Socio-economic	3	4	1	2	6	5
Market	3	2	1	4	4	6
Technical	2	1	5	5	4	2
Sum	10	12	12	14	15	16
Overall ranking	1	2	2	4	5	6

Source: Ham (2004).

for development. The commercial processors attending the Harare workshop rated the high-value oil products as most beneficial to socio-economic development, with jam and jelly products as the least beneficial. The community processors at the Malawi and Tanzania workshops rated the lower-value products, such as jams, as having a higher socio-economic potential than the higher-value wine products. This probably reflects their lack of experience in processing products into wine.

Mango, guava, groundnut and tomato products were rated the highest by community processors. The reasons for this could be that the fruits are easily accessible and abundant and that they were more familiar with the processing of these fruits than with indigenous fruits. This again highlights the need to focus on exotic fruits in addition to indigenous fruits in domestication and commercialization programmes. The Zimbabwe workshop participants had much greater exposure to international markets than the Malawian and Tanzanian participants. Therefore they rated marula products the highest. Marula is one of the best-known natural products in southern African and has been made popular by the Distell Corporation's production of Amarula Cream liqueur in South Africa. The community groups in the Malawi and Tanzania workshops rated marula products last. This indicates a need for a differentiation in marketing strategies when dealing with the marketing of fruit products at local and regional/international levels. The technology evaluation shed some light on the reasons that community processors at the Malawi and Tanzania workshops rated high-value oil and wine products lower than low-value juice and jam products. Community processors are aware of the value of oil and wine products but they do not have access to the skills and technology required to process them. It also seems that people are more familiar with the technologies required to process exotic fruit products than indigenous fruits.

The overall product rating for Malawi indicated that mango products were the most preferred. Only one indigenous product, *Uapaca* juice, was rated amongst the top five products from Malawi. In the Tabora workshop, *Syzygium* juice was rated as the overall most preferred product along with two other indigenous products, *Strychnos* juice and *Vitex* jam, amongst the top five products. The Zimbabwe group rated the high-value oil products highest, followed by *Ziziphus* fruit leather (purèed then dried) and marula jelly.

There were some differences between the 'spontaneous' preferences of participants (Table 1.11) and the overall rankings from the assessment of products on ecological, socio-economic, market and technical aspects (Tables 1.12–1.14). In the Tanzania workshop, baobab juice dropped from first place in the spontaneous ranking to ninth in the more detailed assessment, performing poorly on all four criteria. In contrast, guava jam rose from tenth to second, performing particularly well on socio-economic and technical criteria. In the Zimbabwe workshop, *Ziziphus* fruit leather rose from fifth to first and *Strychnos* jelly fell from second to sixth. In Malawi, there was little difference between the rankings in the two exercises.

1.4 Conclusions

Several important lessons were learned from the priority-setting exercises. These can be divided into two areas: the role of priority setting in domestication research and methods for setting priorities.

1.4.1 The role of priority setting

First, the hypothesis that there would be some degree of consensus among farmers in the different countries of an ecoregion as to which species they preferred proved to be valid. In the Sahel and southern Africa there was considerable consensus. In the Sahel, one species appeared in the top five of all four countries and two were in the top five of three countries. In southern Africa, three species appeared in the top five of all five countries. There was less convergence of views across countries in the humid lowlands of West Africa. No species appeared in the top five of all three countries but four appeared in the top five of two countries. A regional approach to conducting dissemination research is valuable because it allows tasks to be divided among countries and experiences to be shared, making the best use of scarce resources for conducting research.

The prioritization procedure proved to be an effective tool for developing a shortlist of target species for domestication research and for setting priorities among them. It also proved useful for improving linkages between institutions involved in the process and for building a spirit of partnership. These contributed to rapid progress in germplasm collection and propagation studies, involving many of the same people who participated in the prioritization exercises. Through the process of setting priorities, the teams conducting domestication research gained sound evidence for defending their choice of species on which to conduct research. This contributed to greater motivation among team members, stronger linkages with policy makers, and greater confidence among donor agencies that domestication research would yield fruitful results.

1.4.2 Methods for setting priorities

The main method used in the priority-setting exercise – the survey of farmer preferences among species – proved to be popular among participants in the exercise as a means of justifying the choice of species to conduct domestication research on. In contrast, the valuation survey was only conducted in one region, the humid lowlands of West Africa, where it yielded the same rankings as those from the survey of farmer preferences. Researchers in the Sahel and southern Africa thus decided it was not worth the cost of conducting such a survey in their own regions. In fact, while the surveys do give useful information on values of production, it is probable that in most instances these values will reflect the rankings obtained in preference surveys.

One weakness of the priority-setting exercise was that it did not explicitly assess the market potential of different species and products. Survey respondents could not be expected to be aware of marketing opportunities, especially international opportunities, and nor could researchers. In fact, many of the species that have been added to lists of priority species following the priority-setting exercise have been added because of market opportunities. In the humid lowlands of West Africa, for example, *Prunus africana* was added because an extract of its bark could be sold for export to Europe, where it is used in medicines for treating benign prostatic hypertrophy, a common disease in men (Cunningham *et al.*, 2002). The assessment of priority products to commercialize in southern Africa, as discussed above, was important and added new species, albeit exotics that had been excluded from the original surveys. It also highlighted the differences that exist between different locations in a region and between rural processors and urban ones, especially when priority setting is based on product preferences (Ham, 2004). Such an exercise should be part of all priority-setting procedures. In addition, formal assessments of market opportunities could be added to the exercise, and the results of these could be shared with stakeholders, who should ultimately decide whether new species and products should be targeted.

References

Adeola, A.O., Aiyelaagbe, I.O.O., Appiagyei-Nkyi, K., Bennuah, S.Y., Franzel, S., Jampoh, E.L., Janssen, W., Kengue, J., Ladipo, D., Mollet, M., Owusu, J., Popoola, L., Quashie-Sam, S.J., Tiki Manga, T. and Tchoundjeu, Z. (1998) Farmers' preferences among tree species in the humid lowlands of West Africa. In: Ladipo, D.O. and Boland, D.J. (eds) *Bush Mango (Irvingia gabonensis) and Close Relatives: Proceedings of a West African Germplasm Collection Workshop, 10–11 May 1994, Ibadan.* ICRAF, Nairobi, pp. 87–95.

Aiyelaagbe, I.O.O., Adeola, A.O., Popoola, L., Obisesan, K. and Ladipo, D.O. (1997) *Chrysophyllum albidum* in the farming systems of Nigeria: its prevalence, farmer preference, and agroforestry potential. In: *Proceedings of the Workshop on Chrysophyllum albidum, 27 February, 1997, Ibadan.* CENRAD, Ibadan, Nigeria, pp. 119–129.

Akinnifesi, F.K., Kwesiga, F.R., Mhango, J., Mkonda, A., Chilanga, T. and Swai, R. (2004) Domestication priority for miombo indigenous fruit trees as a promising livelihood option for small-holder farmers in southern Africa. *Acta Horticulture* 632, 5–30.

Akinnifesi, F.K., Kwesiga, F.R., Mhango, J., Chilanga, T., Mkonda, A., Kadu, C.A.C., Kadzere, I., Mithofer, D., Saka, J.D.K., Sileshi, G., Ramadhani, T. and Dhliwayo, P. (2006) Towards the development of miombo fruit trees as commercial tree crops in Southern Africa. *Forests, Trees and Livelihoods* 16, 103–121.

Alston, J., Norton, G. and Pardey, P. (1995) *Science Under Scarcity: Principles and Practice for Agricultural Research Evaluation and Priority Setting.* Cornell University Press, Ithaca, New York, 585 pp.

Bonnehin, L. (1998) *Irvingia gabonensis*: economic aspects and on-farm domestication in Cote d'Ivoire. In: Ladipo, D.O. and Boland, D.J. (eds) *Bush Mango (Irvingia gabonensis) and Close Relatives. Proceedings of a West African Germplasm Collection Workshop, 10–11 May 1994, Ibadan.* ICRAF, Nairobi, pp. 188–197.

Braunschweig, T., Janssen, W. and Rieder, P. (2000) Identifying criteria for public agricultural research decisions. *Research Policy* 30, 725–734.

Contant, R. and Bottomley, A. (1988) *Priority Setting in Agricultural Research.* Working Paper No. 10. ISNAR, The Hague, The Netherlands, 86 pp.

Cunningham, A.B., Ayuk, E., Franzel, S., Duguma, B. and Asanga, C. (2002) *An Economic Evaluation of Medicinal Tree Cultivation.* People and Plants Working Paper No. 10, UNESCO, Paris.

Franzel, S., Jaenicke, H. and Janssen, W. (1996) *Choosing the Right Trees: Setting Priorities for Multipurpose Tree Improvement.* ISNAR Research Report No. 8. International Service for National Agricultural Research, The Hague, The Netherlands.

Ham, C. (2004) *Priority Fruit Species and Products for Tree Domestication and Commercialization in Zimbabwe, Zambia, Malawi, and Tanzania.* University of Stellenbosch, Stellenbosch, South Africa.

ICRAF (1995) *Annual Report: 1994.* International Centre for Research in Agroforestry, Nairobi.

ICRAF (1996) *Annual Report, 1995.* International Centre for Research in Agroforestry, Nairobi.

ICRAF (2006) *Priority Indigenous Species.* African Humid Tropics Program, World Agroforestry Centre (ICRAF), Nairobi [see http://www.worldagroforestry.org/aht/ accessed 11 December 2006].

Kaaria, S.W. (1998) The economic potential of wild fruit trees in Malawi. Thesis, Faculty of the Graduate School of the University of Minnesota, Minneapolis, Minnesota, 173 pp.

Kadzere, I., Chilanga, T.G., Ramadhani, T., Lungu, S., Malembo, L., Rukuni, D., Simwaza, P.P., Rarieya, M. and Maghembe, J.A. (1998) Choice of priority indigenous fruits for domestication in southern Africa:

summary of case studies in Malawi, Tanzania, Zambia and Zimbabwe. In: Maghembe, J.A., Simons, A.J., Kwesiga, F. and Rarieya, M. (eds) *Selecting Indigenous Fruit Trees for Domestication in Southern Africa: Priority Setting with Farmers in Malawi, Tanzania, Zambia, and Zimbabwe.* ICRAF, Nairobi, pp. 1–15.

Kelly, T.G., Ryan, J.G. and Patel, B.K. (1995) Applied participatory priority setting in international agricultural research: making trade-offs transparent and explicit. *Agricultural Systems* 49, 177–216.

Kwesiga, F. and Mwanza, S. (1995) Under-exploited wild genetic resources: the case of indigenous fruit trees in eastern Zambia. In: Maghembe, J.A., Ntupanyama, Y. and Chirwa, P.W. (eds) *Improvement of Indigenous Fruit Trees of the Miombo Woodlands of Southern Africa.* ICRAF, Nairobi, pp. 100–112.

Lecup, I. and Nicholson, K. (2000) *Community-based Tree and Forest Product Enterprises: Market Analysis and Development.* Forest, Trees and People Program, FAO, Rome.

Maghembe, J.A., Simons, A.J., Kwesiga, F. and Rarieya, M. (eds) (1998) *Selecting Indigenous Fruit Trees for Domestication in Southern Africa: Priority Setting with Farmers in Malawi, Tanzania, Zambia, and Zimbabwe.* ICRAF, Nairobi.

Malembo, L.N., Chilanga, T.G. and Maliwichi, C.P. (1998) Indigenous fruit selection for domestication by farmers in Malawi. In: Maghembe, J.A., Simons, A.J., Kwesiga, F. and Rarieya, M. (eds) *Selecting Indigenous Fruit Trees for Domestication in Southern Africa: Priority Setting with Farmers in Malawi, Tanzania, Zambia, and Zimbabwe.* ICRAF, Nairobi, pp. 16–39.

Minae, S., Sambo, E.Y., Munthali, S.S. and Ng'ong'ola, S.H. (1995) Selecting priority fruit-tree species for central Malawi using farmers' evaluation criteria. In: Maghembe, J.A., Ntupanyama, Y. and Chirwa, P.W. (eds) *Improvement of Indigenous Fruit Trees of the Miombo Woodlands of Southern Africa.* ICRAF, Nairobi, pp. 84–99.

Mumba, M.S., Simon, S.M., Swai, R. and Ramadhani, T. (2004) Utilization of indigenous fruits of miombo woodlands: a case of Tabora District, Tanzania. In: Rao, M.R. and Kwesiga, F.R. (eds) *Proceedings of the Regional Agroforestry Conference on Agroforestry Impacts on Livelihoods in Southern Africa: Putting Research into Practice.* World Agroforestry Centre, Nairobi, pp. 35–38.

Ndoye, O., Perez, M.R. and Eyebe, A. (1998) *The Markets of Non-timber Forest Products in the Humid Forest Zone of Cameroon.* Rural Development Forestry Network Paper 22c. Overseas Development Institute, London.

Ramadhani, T. (2002) *Marketing of Indigenous Fruits in Zimbabwe.* Socioeconomic Studies on Rural Development 129. Wissenschaftsverlag Vauk, Kiel, Germany.

Ramadhani, T., Chile, B. and Swai, R. (1998) Indigenous miombo fruits selected for domestication by farmers in Tanzania. In: Maghembe, J.A., Simons, A.J., Kwesiga, F. and Rarieya, M. (eds) *Selecting Indigenous Fruit Trees for Domestication in Southern Africa: Priority Setting with Farmers in Malawi, Tanzania, Zambia, and Zimbabwe.* ICRAF, Nairobi, pp. 40–56.

Rukuni, D., Kadzere, I., Marunda, C., Nyoka, I., Moyo, S., Mabhiza, R., Kwarambi, J. and Kuwanza, C. (1998) Identification of priority indigenous fruits for domestication by farmers in Zimbabwe. In: Maghembe, J.A., Simons, A.J., Kwesiga, F. and Rarieya, M. (eds) *Selecting Indigenous Fruit Trees for Domestication in Southern Africa: Priority Setting with Farmers in Malawi, Tanzania, Zambia, and Zimbabwe.* ICRAF, Nairobi, pp. 72–93.

Schomburg, A., Mhango, J. and Akinnifesi, F.K. (2002) Marketing of *masuku* (*U. kirkiana*) and *masawo* (*Ziziphus mauritiana*) fruits and their potential for processing by rural communities in southern Malawi. In: Kwesiga, F., Ayuk, E. and Agumya, A. (eds) *Proceedings of the 14th Southern African Regional Review and Planning Workshop, 3–7 September 2001, Harare.* ICRAF, Nairobi.

Sidibé, M., Kergna, A.O., Camara, A. and Dagnoko, H. (1996) Prioritisation des ligneux a usages multiples dans les savanes parcs de la zone semi aride du

Mali. Institut d'Economie Rurale, Bamako, Mali.
Simons, A.J. and Leakey, R.R.B. (2004) Tree domestication in tropical agroforestry. *Agroforestry Systems* 61, 167–181.
Simwaza, C.P. and Lungu, S. (1998) Indigenous fruits chosen by farmers for domestication in Zambia: identification of priority indigenous fruits for domestication by farmers in Zimbabwe. In: Maghembe, J.A., Simons, A.J., Kwesiga, F. and Rarieya, M. (eds) *Selecting Indigenous Fruit Trees for Domestication in Southern Africa: Priority Setting with Farmers in Malawi, Tanzania, Zambia, and Zimbabwe.* ICRAF, Nairobi, pp. 57–71.

2 Towards a Domestication Strategy for Indigenous Fruit Trees in the Tropics

R.R.B. LEAKEY[1] AND F.K. AKINNIFESI[2]

[1]*Agroforestry and Novel Crops Unit, School of Tropical Biology, James Cook University, Cairns, Queensland, Australia*;
[2]*World Agroforestry Centre, ICRAF, Lilongwe, Malawi*

2.1 Introduction

Increasingly, agroforestry trees are being improved in quality and productivity through the processes of market-driven domestication (Simons, 1996; Simons and Leakey, 2004; Leakey *et al.*, 2005d), based on strategies that consider: (i) the needs of the farmers, their priorities for domestication (Maghembe *et al.*, 1998; Franzel *et al.*, Chapter 1, this volume) and an inventory of the natural resource (Shackleton *et al.*, 2003a); (ii) the sustainable production of agroforestry tree products, including fruits, nuts, medicinals and nutriceuticals, timber, etc.; (iii) the restoration of degraded land and reduction of deforestation; and (iv) the wise use and conservation of genetic resources. These approaches to tree domestication are being implemented in southern and western Africa (Akinnifesi *et al.*, 2006; Tchondjeu *et al.*, 2006).

There are two main pathways within a domestication strategy (Fig. 2.1). Domestication can be implemented on-farm by the farmers (Phase 1), who bring the trees into cultivation themselves (Leakey *et al.*, 2004), or through programmes of genetic improvement on research stations (Leakey and Simons, 1998). In recent years, however, scientific approaches are also being introduced into on-farm domestication through the application of participatory approaches to tree improvement (Phase 2). In this approach, researchers typically act as mentors, helping and advising the farmers, and sometimes jointly implementing on-farm research. Participatory approaches have numerous advantages (Leakey *et al.*, 2003), building on tradition and culture and promoting rapid adoption by growers to enhance livelihood and environmental benefits (Simons and Leakey, 2004). Both these pathways to domestication should be targeted at meeting market opportunities, which should examine traditional as well as emerging markets (Shackleton *et al.*, 2003b).

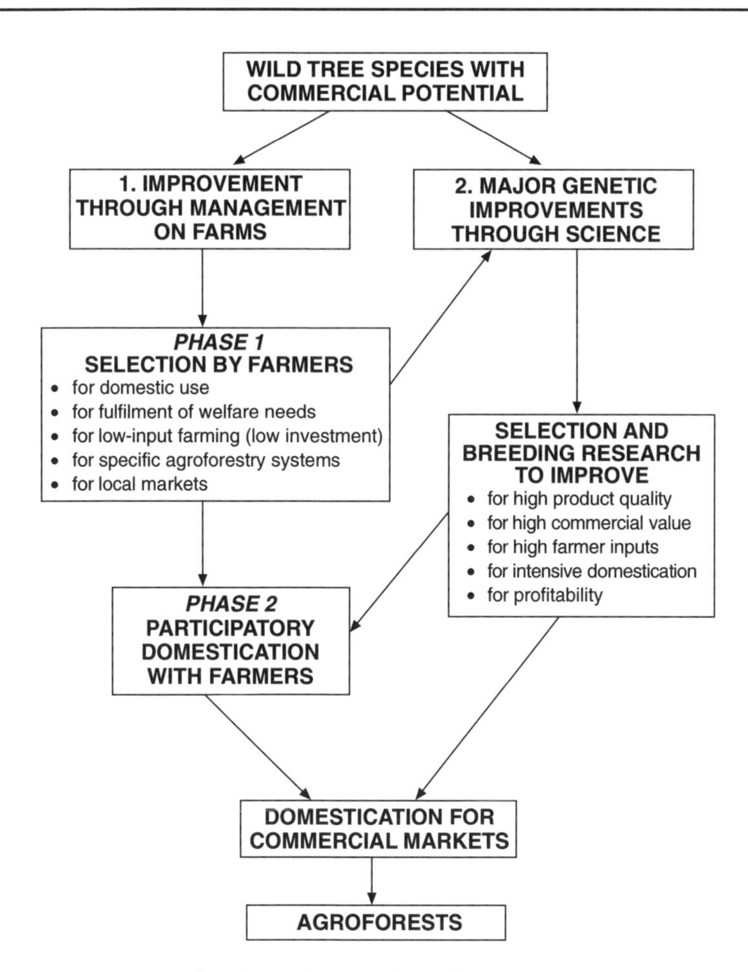

Fig. 2.1. Two pathways for the domestication of agroforestry tree products.

In practice, agroforestry plantings are often constrained by the lack of genetically superior seed sources, the traditional source of planting stock (Simons, 1996). Consequently, one of the first decisions in developing a domestication strategy for a particular species has to be whether to use seed and reproductive processes or vegetative propagation to achieve genetic improvements. Foresters have generally adopted seed-based tree breeding approaches, while horticulturalists have adopted clonal vegetative propagation and the development of cultivars. The following economic and biological situations have been identified as favouring a clonal approach (Leakey and Simons, 2000; Akinnifesi *et al.*, 2006):

- The occurrence of individual trees in a wild population, which have a rare combination of traits such as large fruit size, sweetness, precocity, early fruiting, delayed or extended fruiting season and desirable kernel characteristics.

- The need to combine many desirable traits for simultaneous selection and improvement.
- A requirement for high product uniformity to ensure profitability and to meet market specifications. This contrasts with the genetic heterogeneity that is a characteristic of seedling progenies of outbreeding trees.
- The products are highly valuable and can thus justify the extra expense and care required to ensure quality and productivity, especially when the risks of market saturation are minimal.
- The species to be propagated is a shy seeder (i.e. does not flower and fruit every year or produces only a very small seed crop).
- The propagation material is limited, as for example in: (i) the progeny of a specific controlled pollination made in a breeding programme; (ii) the products of a biotechnological manipulation such as the transmission of genetic material; or (iii) the result of hybridization where segregation would occur in the F_2 generation and beyond (hybrid progenies are also often sterile, so further propagation has to be done vegetatively).
- The timescale in which results are required is insufficient to allow progress through the slower process of breeding. This is particularly relevant in fruit trees with a long juvenile period prior to the attainment of identifiable superiority and sexual maturity, as vegetative propagation can be applied to shoots from trees that have already shown their superiority, and that have already acquired sexual maturity. Such shoots will retain both this superiority and their maturity when propagated vegetatively.
- The seeds of the chosen species have a short period of viability (i.e. they are recalcitrant) or very low viability, and hence cannot be stored for later use.
- Knowledge of proven traits is acquired through either the indigenous knowledge of farmers or a long-term experiment. This situation is plagued by the problems of propagating mature tissues vegetatively. Currently, the usual procedure is to use grafting/budding if the mature traits are required (i.e. fruiting ability) and to coppice the tree if rejuvenation is required.
- A participatory tree domestication programme is planned. This is because farmers do not generally have the time or genetic knowledge to implement a breeding programme.

The antithesis of these situations is that sexual propagation is preferable when the requirements are for large quantities of genetically diverse, low-value plants with unlimited seed supplies.

2.2 Developing a Strategy for Creating New 'Cultivars' Vegetatively

The increased interest in vegetative propagation has arisen from the desire to rapidly acquire higher yields, early fruiting and better quality fruit products in agroforestry trees. The development of cultivars through cloning also results in the uniformity of the products, as all the trees of a given clone are genetically identical. This is beneficial in meeting the market demand for uniform products. A range of vegetative propagation techniques can be utilized to achieve this

(Leakey, 1985; Hartmann *et al.*, 2002), including grafting, stem cuttings, hardwood cuttings, marcotting (air-layering), suckering and *in vitro* techniques, such as meristem proliferation, organogenesis and somatic embryogenesis (Mng'omba *et al.*, 2007a, b).

The decision to pursue the clonal opportunities offered by vegetative propagation necessitates the formulation of a strategy, as there are a number of factors associated with the process of cloning that need to be considered. These factors relate to: (i) the methods of propagation; (ii) the level of technology that is appropriate; and (iii) the effects of using juvenile or mature tissues (Leakey, 1991; Leakey and Simons, 2000).

2.2.1 Methods of propagation and the cloning process

The principal reason for cloning is to take advantage of its ability to capture and fix desirable traits, or combinations of traits, found in individual trees. By taking a cutting or grafting a scion onto a rootstock, the new plant that is formed has an exact copy of the genetic code of the plant from which the tissue was taken. In contrast, following sexual reproduction seedlings are genetically heterogeneous, each seed having inherited different parts of the genetic codes of its parent trees, with segregation of genes among the progeny. Vegetative propagation is thus both a means of capturing and utilizing genetic variation and of producing cultivars to increase productivity and quality (Mudge and Brennan, 1999; Leakey, 2004). Vegetative propagation results in the formation of clones (or cultivars), each of which retains the genetic traits of the original tree from which cuttings or scions were collected. Both on-farm and on-station approaches to tree domestication can involve vegetative propagation and clonal selection, but with a growing interest in the participatory domestication of agroforestry trees (Leakey *et al.*, 2003; Tchoundjeu *et al.*, 2006). There is now great interest in vegetative propagation using stem cuttings (Leakey *et al.*, 1990; Shiembo *et al.*, 1996a, 1996b, 1997; Mialoundama *et al.*, 2002; Tchoundjeu *et al.*, 2002) and grafting for *Uapaca kirkiana* and *Strychnos cocculoides* (Akinnifesi *et al.*, 2006) and *Sclerocarya birrea* (Holtzhausen *et al.*, 1990; Taylor *et al.*, 1996). Vegetative propagation gives the tree improver the ability to multiply, test, select from, and utilize the large genetic diversity present in most tree species. It should be noted that, contrary to some misguided opinion, vegetative propagation does not in itself generate genetically improved material. Only when some form of genetic selection is employed in tandem with propagation will improvement result.

To capture the first asexual propagule from proven and mature field trees that have already expressed their genetic traits, it is necessary to use either grafting/budding or air-layering techniques. Alternatively, coppicing can be used to produce juvenile material. The latter is preferable for clonal timber production, whereas the former is more suitable for fruit trees (Leakey, 1991). Grafting produces many more individuals with less effort than air-layering, although many more individuals can be produced from stem cuttings if the tree resprouts copiously after coppicing. Once clonal stocks have been obtained, the resulting shoots can either be used to provide rooted cuttings for transplanting to the field, additional material for clonal seed orchards, or scions for grafting onto suitable rootstocks. Evidence from simple tests suggests that the majority (probably more than 90%) of tropical trees are amenable to propagation by juvenile stem cuttings (Leakey *et al.*, 1990). In southern Africa, some high-priority indigenous fruit trees are propagated by grafting (e.g. *U. kirkiana*, *Strychnos cocculoides*, *Sclerocarya birrea*, *Adansonia digitata* and *Vitex mombassae*), while others (e.g. *S. birrea*) are easily propagated by large, leafless, hardwood cuttings or stakes/truncheons. *Parinari curatellifolia* is easily propagated by root cuttings.

2.2.2 Appropriate technology

Low-cost non-mist propagators have been developed for the rooting of leafy stem cuttings. These do not require electricity or running water and are extremely effective, meeting the needs of most tree improvement projects in developing countries in both the moist and dry tropics (Leakey *et al.*, 1990). In technologically advanced countries, mist or fogging systems are available for the rooting of cuttings. Alternatively, where laboratory facilities are available *in vitro* culture techniques can be used, but these require highly trained staff and regular power supplies, and are capital-intensive. In many cases, *in vitro* techniques have resulted from empirical testing of different media and plant growth regulators. As a result, inadequate knowledge about the long-term effects of treatments on the field performance has caused some major failures of field performance. In the same way, the vegetative propagation of mature tissues by marcotting, or air-layering, requires a lower level of skill than grafting and budding. The low-technology options are especially appropriate if the participatory approach to domestication is the preferred strategy. In this situation, farmers vegetatively propagate their best trees to create selected cultivars.

2.2.3 Choice between mature and juvenile tissues

An advantage often ascribed to propagation from mature tissues is that by the time the tree is mature it has demonstrated whether or nor it has superior qualities. However, it is not always easy to take advantage of this proven superiority, as propagation by mature stem cuttings is notoriously difficult. This is in contrast with the rooting of juvenile tissues, which is typically easy. Another important advantage of propagating mature tissues is that they are already capable of reproductive processes, and so will flower and fruit within a few years, reducing the time before economic returns start to flow. Plants propagated from mature tissues will also have a lower stature. On the other hand, timber production requires the vigour and form associated with juvenile trees, making propagation by cuttings attractive and appropriate. For timber trees, propagation from mature trees is generally limited to the establishment of clonal seed orchards within breeding programmes.

The use of juvenile tissues

Seedlings, coppice shoots and root suckers are the sources of juvenile tissues. For tree domestication purposes, coppice shoots from the stumps of felled trees have the advantage that it is possible to propagate trees that have already proven to be superior, as it is possible to determine the phenotypic quality of the tree prior to felling. This is highly beneficial in the domestication of trees producing all kinds of agroforestry tree products, but has the added advantage in dioecious fruit tree species that it allows cultivars to be restricted to trees. Nevertheless, there are three reasons why the use of seedlings may still be preferred over coppicing from trees of known phenotype (Leakey and Simons, 2000):

- The population of mature timber trees may be dysgenic because the elite specimens may have been removed by loggers, which means that seedling populations have a better array of genetic variation.
- The felling of large numbers of mature trees for the purpose of generating cultivars may not be acceptable to the owners. In addition, felling the mature trees may be environmentally damaging.
- The use of seedlings allows the screening of far larger populations, with much more diverse origins, maintaining genetic diversity among the cultivars.

Whether using seedling or coppice stumps as stockplants, it is important to ensure that they are managed for sustained, cost-effective and easy rooting. The way in which stockplants are managed is probably one of the most important determinants of the long-term success of a cloning programme. Good rooting ability is maintained by encouraging vigorous orthotropic growth of shoots from regularly pruned stockplants. This requires a much greater level of knowledge than is available for most, if not all, tree species. Good progress has been made in starting to unravel the sources of variation in rooting ability (Leakey, 2004). For example, it has become clear that cuttings taken from different parts of the same shoot differ in their capacity to form roots (Leakey, 1983; Leakey and Mohammed, 1985; Leakey and Coutts, 1989) and that this is influenced by cutting length (or volume). In addition, there are influences on rooting ability that originate from factors between different shoots on the same plant (Leakey, 1983). These factors are also affected by shading, which determines both the amount and the quality of light received by lower shoots. Both the quality and the quantity of light independently affect the physiology, morphology and rooting ability of cuttings from differently illuminated shoots (Leakey and Storeton-West, 1992; Hoad and Leakey, 1996). To further complicate this situation, the nutrient status of the stockplant interacts with the effects of light and shading (Leakey, 1983; Leakey and Storeton-West, 1992), with shading and a high level of nutrients combining to enhance rooting. Light, especially light quality, also affects the relative size and dominance of different shoots on managed stockplants (Hoad and Leakey, 1994). There are also interactions between stockplant factors and the propagator environment.

The complexity of the stockplant factors affecting rooting ability means that for the production of large numbers of cuttings from stockplants it is important to develop a good understanding of these factors. This level of knowledge is not common for tropical and subtropical tree species. However, from work on a few tropical tree species it does appear that some generalizations are possible (Leakey *et al.*, 1994; Leakey, 2004) and that through a modelling approach (Dick and Dewar, 1992) it is possible to predict the best management options for new species. It is, however, usually necessary to start out by propagating a larger number of clones than is needed, as some will be lost while going through the various rooting and multiplication cycles.

From the above, it is clear that, in addition to economic considerations, in formulating a strategy for clonal forestry it is advisable to consider which forms of propagation have the lowest risk of failure. It seems that the low-tech system of rooting stem cuttings is the most robust.

The use of mature tissues

As trees grow they develop a gradient towards maturity (ontogenetic ageing) and after a time reach a threshold above which the newly developing shoots have the capacity to fruit and flower, while those below the threshold are still juvenile. The transition from the juvenile to mature state is called a 'phase change', and the coppicing of mature trees is generally regarded as the best way to return to the juvenile state. Because of the difficulty in rooting cuttings from mature tissues, the most commonly used vegetative propagation methods for mature trees of horticultural and cash crops are grafting and budding techniques (Hartmann *et al.*, 2002). Using these techniques requires skill, as the close juxtaposition of the cambium in the scion and rootstock is necessary if callus growth is to heal the wound and reconnect the vascular tissues. Failures also result from dehydration of the tissues.

These techniques can, however, result in severe and often delayed problems because of incompatibility between the tissues of the rootstock and scion, in which graft unions are rejected and broken, sometimes after 5–10 years of growth. This is a form of tissue rejection, and it is less common between closely related tissues (Jeffree and Yeoman, 1983). Mng'omba *et al.* (2007a) have recently shown that graft incompatibility is caused by the presence of *p*-coumaric acids, and that greater incompatibility can be expected for heterospecific than for homospecific scion/stock combinations. Another common problem with grafting and budding is that shoots can develop and grow from the rootstock. If not carefully managed, growth from the rootstock dominates that of the scion, which dies, resulting in the replacement of the selected mature cultivar with an unselected juvenile plant.

Once mature tissues are successfully established as rooted propagules, be they marcotts, cuttings or grafts, they can be used as stockplants for subsequently harvested cuttings. With good stockplant management, good rooting treatments and an appropriate rooting environment, mature cuttings from these stockplants can usually be rooted easily. Nevertheless, it is clear that the need to propagate from mature tissues does pose a severe constraint and challenge to domestication strategies, especially of fruit trees.

In vitro culture techniques hold some promise of circumventing the problems of maturation, as some rejuvenation of *in vitro* cultures has been reported (e.g. Amin and Jaiswal, 1993), but the mechanism remains unclear. *In vitro* micrografting has also been used to rejuvenate shoots (Ewald and Kretzschmar, 1996).

2.3 Developing a Strategy for Clonal Selection

Cultivar development relies on three processes: selection, testing and breeding. Selection identifies certain genotypes for cultivar development; testing exposes the new cultivars to appropriate environments; and breeding creates new genetic variability. Within the overall domestication strategy, three interlinked populations are conceptualized: the gene resource population, the selection population, and the production population (Fig. 2.2). The gene resource population is often the

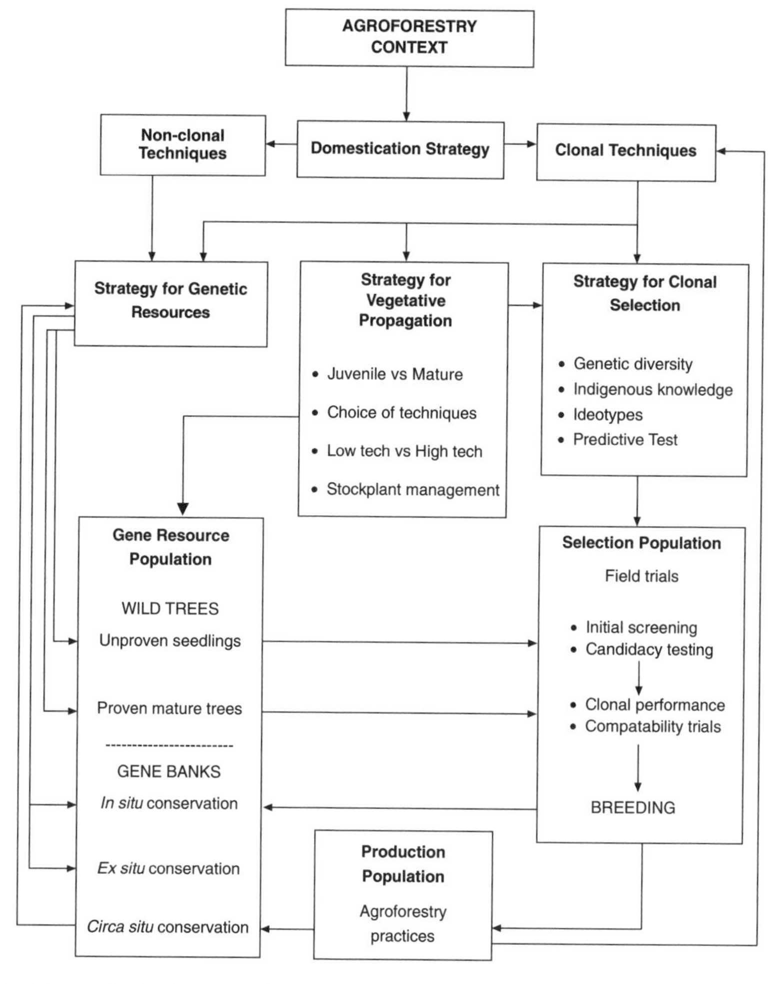

Fig. 2.2. Relationships between a domestication strategy, a genetic resources strategy and strategies for vegetative propagation and clonal selection.

wild or unimproved population from which new selections can be derived. The selection population is the somewhat improved population of genotypes which are being tested and which are used in subsequent breeding programmes to create the next generation of potential cultivars. A wide range of genotypes may be kept in this population as long as each has at least one characteristic of possible future interest. The production population consists of the highly selected genotypes, which are used for planting.

As mentioned earlier, there are two basic approaches to the genetic improvement of trees: the seed-based breeding approach typical of forestry, and the clonal approach typical of horticulture. The seed approach typically involves the selection of populations (provenance testing) and/or families (progeny testing) (Zobel and Talbert, 1984; Leakey, 1991). While this approach could be taken for indigenous fruit trees, it is very likely that an examination of the ten situations outlined in the Introduction would indicate that a clonal approach is more appropriate. There are basically two ways to select the best individual trees for cloning from broad and diverse wild populations: (i) selection from a pool of seedlings of virtually unknown quality in a nursery or field trial (although it may be known that the pool originates from a good provenance or progeny); and (ii) selection of proven mature trees in wild or planted populations (Fig. 2.2). In scenario (i), genetic improvement in yield per hectare will undoubtedly require a series of tests, each spanning many years. Typically, there are four levels of testing (Foster and Bertolucci, 1994):

1. Initial screening with large numbers, preferably tens of thousands, of seedlings or, if seedlings have already been cloned, a few ramets per clone.
2. Candidacy testing with large numbers of cloned genotypes fewer than with initial screening (preferably hundreds or thousands) and two to six ramets per clone.
3. Clonal performance trials with moderate numbers of clones (e.g. fewer than 200) and large numbers (e.g. 0.1 ha plots) of ramets per clone.
4. Compatibility trials with small numbers of clones (e.g. 20–50) with very large plot sizes.

It is important to recognize that there is a trade-off between the accuracy of genetic value estimation and the intensity of selection (i.e. greater accuracy is at the expense of numbers of families, individuals per family, or clones). For cost-effective clonal tree improvement programmes with limited or fixed resources, it has been found that the best strategy is to plant as many clones as possible with relatively few ramets per clone.

To short-circuit the lengthy process of field trials, Ladipo *et al.* (1991a, b) developed a predictive test for timber tree seedlings in which the initial screening is done on young seedlings in the nursery; it is then possible to jump straight into clonal or compatibility trials with some confidence. To date there is no similar opportunity for fruit trees.

Like the predictive test, scenario (i) is an alternative and much quicker option. In this case, mature trees, which have already expressed their genetic potential at a particular site over many years of growth, are selected and propagated vegetatively and the propagules are planted either in clonal performance trials or directly into compatibility trials. This raises the question of how the superior mature trees should be identified, especially if it is desirable to select for multiple traits. This can create a problem, as many traits may be weakly or negatively correlated (e.g. fruit size and kernel size in *S. birrea*; Leakey, 2005). Consequently, as the number of desirable traits increases, the number of genotypes superior for all traits diminishes rapidly. Thus, the selection intensity (and also the number of trees screened) must be substantially increased, or the expected genetic gain will rapidly decline. For this reason, only the few most economically important traits (e.g. fruit flesh or nut mass, taste) should be concentrated on in the early phases of selection.

Two techniques can be used to assist in the identification of superior mature trees (sometimes called 'elite' or 'plus' trees) producing indigenous fruits and nuts. The first is to involve indigenous people in the domestication process and to seek their local knowledge about which trees produce the best products. Local people usually have good knowledge about the whereabouts of elite trees, and this knowledge often extends to superiority in a number of different traits, such as size, flavour and seasonality of production. However, access to this knowledge has to be earned by the development of trust between the holder of the knowledge and the potential recipient. Ideally, the recipient should enter into an agreement that the intellectual property rights of the holder will be formally (and legally) recognized if a cultivar is developed from the selected tree. Unfortunately, at present the process of legally recognizing such cultivars is not well developed and requires considerable improvement.

The second technique for identifying mature elite fruit and nut trees has recently been extended to a study on marula (Leakey, 2005) following its development in Cameroon and Nigeria (Atangana *et al.*, 2002; Leakey *et al.*, 2002, 2005c). This technique involves the quantitative characterization of many traits of fruits and kernels, which are associated with size, flavour, nutritional value, etc. This characterization also determines the extent of the tree-to-tree variation, which is typically three- to sevenfold, as found in marula (Leakey *et al.*, 2005a, b), as well as the frequency distribution, which is typically normal in wild populations but tends to become skewed in populations subjected to some selection. The characterization data can then be used to identify the best combination of traits (the 'ideotype') to meet a particular market opportunity (for an example see Fig. 2.3). The development of single-purpose ideotypes (Leakey and Page, 2006) provides a tool for the development of cultivars with different levels of market focus and sophistication (Fig. 2.4). In addition to advancing the selection process for multiple traits, the ideotypes also provide information about opportunities to select for better partitioning of dry matter between desirable and undesirable traits. For example, in marula 90% of the dry matter in nuts is typically found in the shell and only 10% in the valuable kernel, although the range of variation is 3–16%. The shell : kernel ratio is therefore a trait which could usefully be included in the ideotype. However, the inclusion of each additional trait in a multi-trait selection process greatly increases the number of trees that need to be screened, especially if the traits are independent or only weakly related.

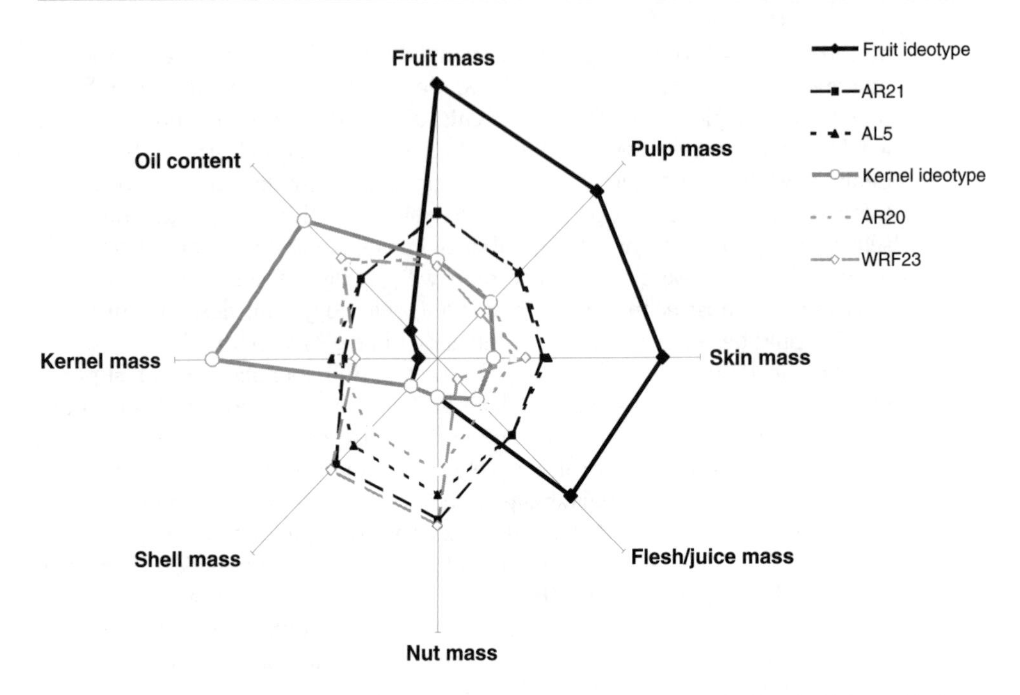

Fig. 2.3. Fruit and kernel ideotypes for marula (*Sclerocarya birrea*) in South Africa, with the best-fit trees (after Leakey, 2005).

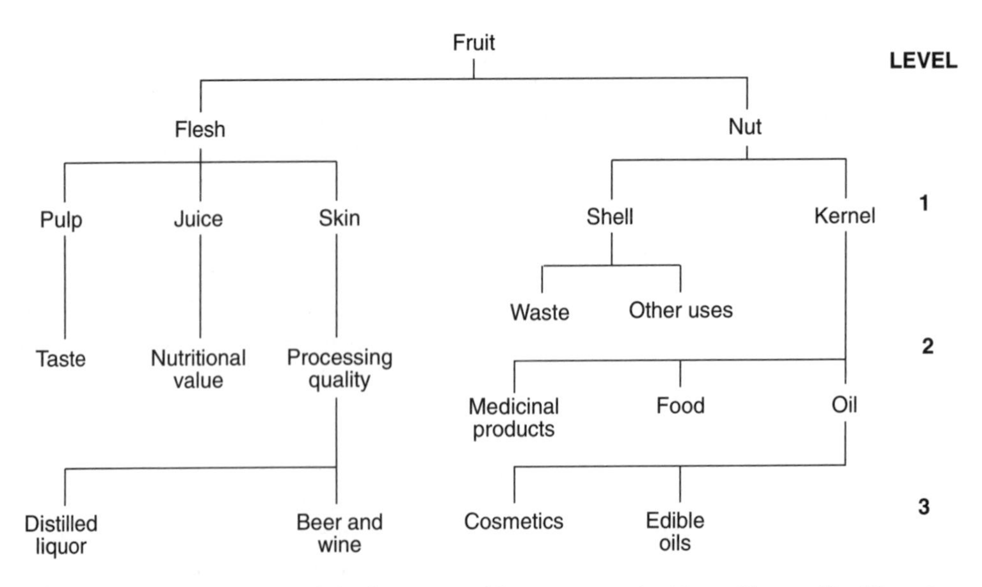

Fig. 2.4. The development of single-purpose ideotypes as a tool for cultivars with different levels of market focus and sophistication.

It is clear from the above discussion that in the early phases of domestication there is sufficient genetic variation in most tropical tree populations to allow considerable progress in the development of cultivars, but a strategy for clonal agroforestry should not forgo any opportunity for creating new variation.

The selection of clones is not a once-and-for-all event. Domestication is a continuous process, which in wheat, rice, maize, oranges and apples, for example, started thousands of years ago and continues today. Thus, a series of clonal selection trials should be established to seek the best individuals from new accessions of genetically diverse populations or progenies (Fig. 2.2). It is also important to discard old clones as they are superseded, although some of these should be retained in the gene resource population. In the first instance, clones may be selected for yield and quality. With time, this can be extended to include nutritional quality, disease/pest resistance, component products (oils, flavourings, thickening agents, etc.). This continued turnover of the selected clonal population will further ensure the diversity of the commercially planted clones and prevent excessive narrowing of the genetic base. Indeed, it can be argued that in this way clonal plantations of 30–50 superior but unrelated clones can be more diverse than seedlings. This is because seed-lots typically originate from a number of related mother trees and share some genetic material.

Because domestication is a continuous process, commercial plantings have to be made with whatever material is best at a given time, knowing that they will be superseded later. Having a succession of increasingly good planting stock is one of the ways in which the diversity of the genetic base can be maintained, although of course this has to be rigorously enforced as one of the objectives of a breeding and selection programme. For species with existing provenance selection and breeding programmes, clones should be derived from seed collections sampling a wide range of the known variation, as it is not uncommon for a few elite trees to be found with poor provenances.

As the selection process intensifies with time, new traits will be introduced into the programme (e.g. seasonality of production, early fruiting, disease and/or pest resistance and drought tolerance). Capturing variation in the seasonality of production and expanding the harvesting season are likely to be among the best ways of supporting market growth. As the price of end-of-season fruits is likely to be better than the mid-season price, it is also a good way to enhance the benefits of producing households that need sources of income throughout the year. The further the domestication process proceeds, the more important it becomes that the combination of traits being selected is targeted at a particular market (Fig. 2.4). This again is where the ideotype concept can be useful, and ideally advice should be sought from industrial partners who are aware of which characteristics are important in the marketplace or in product processing (Leakey, 1999).

Using the example of *Sclerocarya birrea*, fruit-producing cultivars could be developed that have large fruit flesh/juice mass (Leakey *et al.*, 2005a) and are nutritious (Thiong'o *et al.*, 2002) as raw fruits, or are good for traditional beer- or wine-making, or meet the needs of the distilling industry. Likewise, other cultivars could be developed for the size and quality of the kernel, with a low shell : kernel ratio and with either nutritious or medicinal qualities for eating, or with oil yield and quality traits of importance in the cosmetics industry (Leakey, 2005; Leakey *et al.*, 2005b). Similarly, superior phenotypes of *U. kirkiana* with heavy fruit loads, large fruits and high pulp content have been identified by communities during participatory selection in Malawi, Zambia and Zimbabwe (Akinnifesi *et al.*, 2006). Thus, it is clear that in the domestication of multipurpose tree species, highly productive, single-purpose clones or cultivars are probably the best option. To maximize the market recognition of, and farmer interest in, cultivars, it is a good idea to name them. In Namibia every marula tree already has a name, so it is easy to give a name to a cultivar derived from any particular tree. The name can also be used to recognize the person or community holding the rights to the cultivar.

2.4 Opportunities for Introducing New Variation

A clonal selection programme seeks to utilize as much of the existing variation as possible within wild populations, or within progenies from breeding programmes. However, after the initial phases of selection, the opportunity will arise for controlled pollinations between proven elite clones. In this way it may be possible to take advantage of specific combining ability in unrelated superior clones and to produce progeny with heterosis in desirable traits that exceed what is found in the wild populations. The vegetative propagation of new genotypes can become a second clonal generation. The philosophy of adding new variation in clonal populations in the future can also be extended to the possibility of using genetically manipulated materials arising from biotechnology programmes.

Ideally, to make rapid progress in a second-generation breeding programme, means of inducing early flowering in superior clones are needed in order to shorten the generation time (e.g. Longman *et al.*, 1990).

2.5 The Wise Use of Genetic Variability

The genetic resource of a species is the foundation of its future as a wild plant and as a source of products for human use. It is therefore crucial to protect and use this resource wisely. Within the domestication process, whether clonal or not, one of the first requirements of an appropriate strategy is to conserve a substantial proportion of the genetic variability for future use in selection programmes and, subsequently, through breeding to broaden the genetic base of the cultivars in the production population. This also serves as a risk-aversion strategy should it be necessary to breed for resistance to pests and diseases in the future. There are three actions which each contribute to genetic conservation: establishing a gene bank; the wise utilization of the genetic resource in cultivation; and protecting some wild populations. These will be discussed in the following sections.

2.5.1 Establishing a gene bank (*ex situ* conservation)

The strategy for building up the base population will, to some extent, be influenced by whether or not there has been a tree improvement programme for the species in question, and how far it has gone towards identifying genetically superior stock. If, from previous tree improvement work, the population has been subdivided into provenances and progenies, it is easy to ensure that a wide range of unrelated seed sources and clones are set aside for the establishment of a number of living gene banks established at different sites to minimize the risk of loss if a site fails or is destroyed by fire or some other event. If possible, the material conserved in the gene bank should encompass the full geographical range of the species and be obtained from sites with differing soils, rainfall, altitude, etc. The latter is desirable for the subsequent selection of clones that are appropriate for different sites.

In the same way, for species without an existing tree improvement programme, clones for field testing should originate from seed collections spanning the natural range of the species, particularly including populations on the edge of the range and any isolated subpopulations that may be important in a later breeding programme between selected clones. Ideally, each collection area should be represented by identifiable half-sib progenies of individual mother trees. The first round of germplasm collections of this sort were made for *Sclerocarya birrea* and *U. kirkiana* in eight SADC (Southern Africa Development Community) countries in 1996 (Kwesiga *et al.*, 2000) and established in multilocational provenance trials (Akinnifesi *et al.*, 2004, 2006). In both of the above instances, the individual trees to be included in the gene bank should be selected at random from unrelated origins and vegetatively propagated to form as large a clonal population as possible. The unused seedlings should be established at different sites as living gene banks for future use, whether directly by subsequent coppicing and cloning or indirectly through their progeny. The identity of plants within gene banks of this sort should be maintained and different seed-lots should be planted such that cross-pollination between plants of different origins is likely. In this way the gene pool will be maintained with maximum diversity.

Ex situ conservation can also be achieved in species producing seeds, which retain their viability in storage, by creating seed banks. Species differ in their amenability to seed storage at different temperatures and moisture contents, including cryopreservation. The Genetic Resources Unit of the World Agroforestry Centre (ICRAF) contains material of several fruit-tree species. In southern Africa, clonal orchards have been established for the production of superior *U. kirkiana* clones. Numerous studies have also been made on the germination of indigenous fruit tree species (Maghembe *et al.*, 1994; Phofeutsile *et al.*, 2002; Mkonda *et al.*, 2003; Akinnifesi *et al.*, 2006).

It needs to be appreciated that new material should be brought into the gene resource (gene bank), selection and production populations whenever possible (see Fig. 2.2), so that genetic diversity is being continually enhanced.

2.5.2 The wise utilization of genetic resources in cultivation (*circa situ* conservation)

A wise genetic resource strategy will ensure that the production population is based on a diverse set of cultivars, and that this set of cultivars is continually being supplemented with new and better cultivars from the ongoing screening of wild populations, provenances, progenies, a programme of intensive tree breeding, and the regular replacement of cultivars that are no longer the best. In this way selected, highly productive but unrelated clones can be used commercially for agroforestry. Indeed, ten well-selected and unrelated clones may contain as much, or more, genetic variation as a narrowly based sexually reproducing population.

This strategy will also ensure that the cultivars being planted are well adapted to their environment – and thus fully expressing their genetic potential – while at the same time minimizing the risks associated with intensive cultivation. These risks are perhaps greater when indigenous trees, rather than exotics, are planted in the tropics in areas where the complexity of forest ecosystems has been disturbed; for example, by shifting agriculture. The minimization of risk therefore requires that the trees are planted in situations that ensure a minimum of damage to the nutrient and hydrological cycles, food webs and life cycles of the intact ecosystem.

In clonal tree domestication programmes, final yield is strongly influenced by the adaptation of the trees, individually and collectively, to the site. This adaptation has two components: clonal development, and clonal deployment or the stand establishment process (Foster and Bertolucci, 1994). Clonal development approaches, including breeding, testing and selection, largely affect the genetic quality of the resulting clonal population available for planting. Clones must be developed which are highly selected for growth and productivity traits, but which display substantial homeostasis and so can adapt to their changing environment. Clonal deployment, on the other hand, must strike a balance between the need for efficiency of management for the economic production of products, and the need to deploy populations which are genetically buffered against environmental changes, including pests.

Clonal development relies on three processes: breeding, testing and selection. However, despite its overwhelming importance, little research has been conducted to investigate the effects of different deployment strategies on the health, growth and yield of forest stands. This type of research requires a large amount of resources and a long period of assessment. Foster and Bertolucci (1994) identified the following major questions:

- How many clones should be in the production population, and how many should be deployed to a single site?
- Should the clones be deployed at a single site as a mixture or as mosaics of monoclonal stands? This is important both for operational reasons such as planting, and to maximize yield through the optimization of inter- or intraclonal competition.
- What are the key attributes of the clones themselves that cause them to be used either as mixtures or in monoclonal plots?

The question of the number of clones to be deployed has two aspects: the number of clones within the production population, and the number of clones planted per site. Based on probability theory, 7–30 clones provide a reasonable probability of achieving an acceptable final harvest. There is continued debate on the wisdom of releasing a few rather than many clones to optimize gains and minimize risk. The arguments about susceptibility to environmental disasters when only a few clones are deployed may apply more rigorously to agroforestry trees, given the risk-averseness of small-scale farmers and their possible desire to maximize stability of production rather than production per se. Consensus among tree breeders and forest geneticists indicates that production population sizes of 100 genotypes or fewer are acceptable. Actually, the absolute number of clones is less important than the range of genetic diversity among the clones. Ten clones which share the same alleles for a particular trait would express no genetic diversity, in contrast with five clones each of which has different alleles for the same trait. Hence, by sourcing material from diverse origins, the tree breeder must emphasize genetic diversity for traits associated with survival and adaptation while exercising strong selection pressure on production traits. Following rigorous testing, new clones should be added to the production population each year. In the case of dioecious species such as *Sclerocarya birrea*, it is important to include male trees in the production population in order to ensure adequate pollination. This is especially important for kernel production as there is some evidence that the number of kernels per nut may be constrained by inadequate pollination (Leakey *et al.*, 2005b).

There is perhaps one situation in which an exception to the above strategy to broaden the genetic base of the production population may be acceptable. This situation arises when there are good market reasons to preserve regional variations in the quality of the product. For example, in the wine industry, regional attributes of the wine ('appellation') are a result of gene combinations specific to different regions. Nevertheless, within each region processes to maintain a broad genetic base are still important.

2.5.3 Protecting some wild populations (*in situ* conservation)

A conservation programme for any species would be incomplete without a strategy for the protection of wild populations, which represent 'hotspots' of genetic diversity. One advantage of this approach is that a species is conserved together with its symbionts (e.g. mycorrhizal fungi), pollinators and other associated species; something that is not so easy in *ex situ* conservation and which is important for *circa situ* conservation. Molecular techniques now provide a powerful tool to identify these hotspots rapidly (e.g. Lowe *et al.*, 2000), although of course the data can only be as good as the sampling strategy will allow. In southern Africa, molecular studies have been completed for *U. kirkiana* and *S. birrea* (Kadu *et al.*, 2006; Mwase *et al.*, 2006).

2.6 Socio-economic and Environmental Context for this Strategy

In the introduction to this chapter it was stated that a domestication strategy for clonal forestry/agroforestry in the tropics should take into account not only the commercial production of agroforestry tree products but also the need to provide the domestic needs of rural people, encourage sustainable systems of production, and encourage the restoration of degraded land. There is debate in southern Africa about the impacts of poverty on the natural resources of very many African countries, especially those in southern Africa. The domestication of indigenous plants as new, improved crop species for complex agroforestry land uses offers the opportunity to return to more sustainable polycultural systems, to build on traditional and cultural uses of local plants, and to enhance the income of subsistence farmers through the sale of indigenous fruits, medicines, oils, gums, fibres, etc. (Leakey, 2003). Indeed, it has been argued that the biological and economic constraints on the wider use of indigenous trees in agroforestry can be overcome by cloning techniques and that the economic incentives should promote cultivation with the ecologically more important indigenous species. This approach to agroforestry has been recommended as a sound policy for land use in Africa (Leakey, 2001a, b). Potentially, through enhanced food and nutritional security, the domestication of indigenous fruits may even have positive impacts on HIV/AIDS (Swallow *et al.*, 2007), by boosting the immune systems of sufferers (Barany *et al.*, 2003). Consequently, the application of the strategies developed in this chapter is important if the people of Africa are to benefit from the domestication of indigenous trees.

2.7 Conclusions

This chapter presents three interacting, multifaceted strategies for the development of clonal fruit trees in southern Africa. These strategies are the foundation of a sustainable domestication strategy for indigenous fruit trees based on the establishment of three interlinked populations: a gene resource population for genetic conservation; a selection population for the achievement of genetic improvement; and a production population of trees for farmers to grow. The practice of domesticating a species using these strategies is cyclical and therefore continuous.

Vegetative propagation is a powerful means of capturing existing genetic traits and fixing them so that they can be used as the basis of a clonal cultivar, or in a different role as a research variable. The desirability of using clonal cultivars in preference to genetically diverse seedling populations varies depending on the situation and the type of trees to be propagated. However, the advantages of clonal propagules outweigh those of seedlings when the products are valuable, when the tree has a long generation time, and when the seeds are scarce or difficult to keep in storage.

There are many opportunities to enhance agroforestry practices through the wise application of vegetative propagation and clonal selection. These techniques in turn offer many ways of creating new and greatly improved crop plants. The potential for increased profits from clonal techniques arises from their capacity to capture and utilize genetic variation. The consequent uniformity in the crop is advantageous in terms of maximizing quality, meeting market specifications and increasing productivity, but it may also increase the risks of pest and disease problems; consequently, risk avoidance through the diversification of the clonal production population is a crucial component of the strategies presented. The application of these strategies should lead to benefits that encompass many of the rural development goals of development agencies, as specified in the Millennium Development Goals of the United Nations. Achieving these benefits will, however, require the large-scale adoption of the techniques and strategies presented here in ways that will meet the needs both of farmers and also those of new and emerging markets. This makes it important to ensure that policy makers get the message about good domestication strategies (e.g. Wynberg *et al.*, 2003).

References

Akinnifesi, F.K., Chilanga, T.G., Mkonda, A., Kwesiga, F.K. and Maghembe, J.A. (2004) Domestication of *Uapaca kirkiana* in southern Africa: preliminary results of screening provenances in Malawi and Zambia. In: Rao, M.R. and Kwesiga, F.R. (eds) *Agroforestry Impacts on Livelihoods in Southern Africa: Putting Research into Practice.* World Agroforestry Centre (ICRAF), Nairobi, pp. 85–92.

Akinnifesi, F.K., Kwesiga, F., Mhango, J., Chilanga, T., Mkonda, A., Kadu, C.A.C., Kadzere, I., Mithofer, D., Saka, J.D.K., Sileshi, G., Ramadhani, T. and Dhliwayo, P. (2006) Towards the development of miombo fruit trees as commercial tree crops in southern Africa. *Forests, Trees and Livelihoods* 16, 103–121.

Amin, M.N. and Jaiswal, V.S. (1993) In vitro response of apical bud explants from mature trees of jackfruit (*Artocarpus heterophyllus*). *Plant Cell, Tissue and Organ Culture* 33, 59–65.

Atangana, A.R., Ukafor, V., Anegbeh, P.O., Asaah, E., Tchoundjeu, Z., Usoro, C., Fondoun, J.-M., Ndoumbe, M. and Leakey, R.R.B. (2002) Domestication of *Irvingia gabonensis*. 2. The selection of multiple traits for potential cultivars from Cameroon and Nigeria. *Agroforestry Systems* 55, 221–229.

Barany, M., Hammett, A.L., Leakey, R.R.B. and Moore, K.M. (2003) Income generating opportunities for smallholders affected by HIV/AIDS: linking agro-ecological change and non-timber forest product markets. *Journal of Management Studies* 39, 26–39.

Dick, J.McP. and Dewar, R. (1992) A mechanistic model of carbohydrate dynamics during adventitious root development in leafy cuttings. *Annals of Botany* 70, 371–377.

Ewald, D. and Kretzschmar, U. (1996) The influence of micrografting *in vitro* on tissue culture behavior and vegetative propagation of old European larch trees. *Plant Cell, Tissue and Organ Culture* 44, 249–252.

Foster, G.S. and Bertolucci, F.L.G. (1994) Clonal development and deployment: strategies to enhance gain while minimizing risk. In: Leakey, R.R.B. and Newton, A.C. (eds) *Tropical Trees: Potential for Domestication and the Rebuilding of Forest Resources.* HMSO, London, pp. 103–111.

Hartmann, H.T., Kester, D.E., Davis, F.T. and Geneve, R.L. (2002) *Plant Propagation: Principles and Practices*, 7th edn. Prentice-Hall, Upper Saddle River, New Jersey, 880 pp.

Hoad, S.P. and Leakey, R.R.B. (1994) Effects of light quality on gas exchange and dry matter partitioning in *Eucalyptus grandis*

W. Hill ex Maiden. *Forest Ecology and Management* 70, 265–273.

Hoad, S.P. and Leakey, R.R.B. (1996) Effects of pre-severance light quality on the vegetative propagation of *Eucalyptus grandis* W. Hill ex Maiden: cutting morphology, gas exchange and carbohydrate status during rooting. *Trees* 10, 317–324.

Holtzhausen, L.C., Swart, E. and van Rensburg, R. (1990) Propagation of the marula (*Sclerocarya birrea* subsp. *caffra*). *Acta Horticulturae* 275, 323–334.

Jeffree, C.E. and Yeoman, M.M. (1983) Development of intercellular connections between opposing cells in a graft union. *New Phytologist* 93, 481–509.

Kadu, C.A.C., Imbuya, M., Jamnadass, R. and Dawson, I.K. (2006) Genetic management of indigenous fruit trees in Southern Africa: A case study of *Sclerocarya birrea* based on nuclear and chloroplast variation. *Southern Africa Journal of Botany* 72, 421–427.

Kwesiga, F., Akinnifesi, F.K., Ramadhani, T., Kadzere, I. and Saka, J. (2000) Domestication of indigenous fruit trees of the miombo in southern Africa. In: Shumba, E.M., Luseani, E. and Hangula, R. (eds) *Proceedings of a SADC Tree Seed Centre Network Technical meeting, Windhoek, Namibia, 13–14 March 2000.* Co-sponsored by CIDA and FAO, pp. 8–24.

Ladipo, D.O., Leakey, R.R.B. and Grace, J. (1991a) Clonal variation in a four-year-old plantation of *Triplochiton scleroxylon* K. Schum. and its relation to the predictive test for branching habit. *Silvae Genetica* 40, 130–135.

Ladipo, D.O., Leakey, R.R.B. and Grace, J. (1991b) Clonal variation in apical dominance of *Triplochiton scleroxylon* K. Schum. in response to decapitation. *Silvae Genetica* 40, 136–140.

Leakey, R.R.B. (1983) Stockplant factors affecting root initiation in cuttings of *Triplochiton scleroxylon* K. Schum., an indigenous hardwood of West Africa. *Journal of Horticultural Science* 58, 277–290.

Leakey, R.R.B. (1985) The capacity for vegetative propagation in trees. In: Cannell, M.G.R. and Jackson, J.E. (eds) *Attributes of Trees as Crop Plants.* Institute of Terrestrial Ecology, Huntingdon, UK, pp. 110–133.

Leakey, R.R.B. (1991) Towards a strategy for clonal forestry: some guidelines based on experience with tropical trees. In: Jackson M.G.R. (ed.) *Tree Improvement and Breeding.* Royal Forestry Society of England, Wales and Northern Ireland, Tring, UK, pp. 27–42.

Leakey, R.R.B. (1999) Potential for novel food products from agroforestry trees. *Food Chemistry* 64, 1–14.

Leakey, R.R.B. (2001a) Win:win landuse strategies for Africa. 1. Building on experience with agroforests in Asia and Latin America. *International Forestry Review* 3, 1–10.

Leakey, R.R.B. (2001b) Win:win landuse strategies for Africa. 2. Capturing economic and environmental benefits with multistrata agroforests. *International Forestry Review* 3, 11–18.

Leakey, R.R.B. (2003) IPUF: to domesticate or not to domesticate? *South African Ethnobotany: Newsletter of the Indigenous Plant Use Forum* 1(1), 40–42.

Leakey, R.R.B. (2004) Physiology of vegetative reproduction. In: Burley, J., Evans, J. and Youngquist, J.A. (eds) *Encyclopaedia of Forest Sciences.* Academic Press, London, pp. 1655–1668.

Leakey, R.R.B. (2005) Domestication potential of marula (*Sclerocarya birrea* subsp *caffra*) in South Africa and Namibia. 3. Multi-trait selection. *Agroforestry Systems* 64, 51–59.

Leakey, R.R.B. and Coutts, M.P. (1989) The dynamics of rooting in *Triplochiton scleroxylon* cuttings: their relation to leaf area, node position, dry weight accumulation, leaf water potential and carbohydrate composition. *Tree Physiology* 5, 135–146.

Leakey, R.R.B. and Mohammed, H.R.S. (1985) The effects of stem length on root initiation in sequential single-node cuttings of *Triplochiton scleroxylon* K. Schum. *Journal of Horticultural Science* 60, 431–437.

Leakey, R.R.B. and Page, T. (2006) The 'ideotype concept' and its application to the selection of 'AFTP' cultivars. *Forests, Trees and Livelihoods* 16, 5–16.

Leakey, R.R.B. and Simons, A.J. (1998) The domestication and commercialization of indigenous trees in agroforestry for the alleviation of poverty, *Agroforestry Systems* 38, 165–176.

Leakey, R.R.B. and Simons, A.J. (2000) When does vegetative propagation provide a viable alternative to propagation by seed in forestry and agroforestry in the tropics and sub-tropics? In: Wolf, H. and Arbrecht, J. (eds) *Problem of Forestry in Tropical and Sub-tropical Countries: The Procurement of Forestry Seed – The Example of Kenya.* Ulmer Verlag, Germany, pp. 67–81.

Leakey, R.R.B. and Storeton-West, R. (1992) The rooting ability of *Triplochiton scleroxylon* cuttings: the interaction between stock-plant irradiance, light quality, and nutrients. *Forest Ecology and Management* 49, 133–150.

Leakey, R.R.B., Mesén, J.F., Tchoundjeu, Z., Longman, K.A., Dick, J.McP., Newton, A.C., Matin, A., Grace, J., Munro, R.C. and Muthoka, P.N. (1990) Low-technology techniques for the vegetative propagation of tropical trees. *Commonwealth Forestry Review* 69, 247–257.

Leakey, R.R.B., Newton, A.C. and Dick, J.McP. (1994) Capture of genetic variation by vegetative propagation: processes determining success. In: Leakey, R.R.B. and Newton, A.C. (eds) *Tropical Trees: Potential for Domestication and the Rebuilding of Forest Resources.* HMSO, London, pp. 72–83.

Leakey, R.R.B., Atangana, A.R., Kengni, E., Waruhiu, A.N., Usuro, C., Anegbeh, P.O. and Tchoundjeu, Z. (2002) Domestication of *Dacryodes edulis* in West and Central Africa: characterisation of genetic variation. *Forests, Trees and Livelihoods* 12, 57–72.

Leakey, R.R.B., Schreckenberg, K. and Tchoundjeu, Z. (2003) The participatory domestication of West African indigenous fruits. *International Forestry Review* 5, 338–347.

Leakey, R.R.B., Tchoundjeu, Z., Smith, R.I., Munro, R.C., Fondoun, J.-M., Kengue, J., Anegbeh, P.O., Atangana, A.R., Waruhiu, A.N., Asaah, E., Usoro, C. and Ukafor, V. (2004) Evidence that subsistence farmers have domesticated indigenous fruits (*Dacryodes edulis* and *Irvingia gabonensis*) in Cameroon and Nigeria. *Agroforestry Systems* 60, 101–111.

Leakey, R.R.B., Shackleton, S. and du Plessis, P. (2005a) Domestication potential of marula (*Sclerocarya birrea* subsp *caffra*) in South Africa and Namibia. 1. Phenotypic variation in fruit traits. *Agroforestry Systems* 64, 25–35.

Leakey, R.R.B., Pate, K. and Lombard, C. (2005b) Domestication potential of marula (*Sclerocarya birrea* subsp *caffra*) in South Africa and Namibia. 2. Phenotypic variation in nut and kernel traits. *Agroforestry Systems* 64, 37–49.

Leakey, R.R.B., Greenwell, P., Hall, M.N., Atangana, A.R., Usoro, C., Anegbeh, P.O., Fondoun, J.-M. and Tchoundjeu, Z. (2005c) Domestication of *Irvingia gabonensis*. 4. Tree-to-tree variation in food-thickening properties and in fat and protein contents of Dika Nut. *Food Chemistry* 90, 365–378.

Leakey, R.R.B., Tchoundjeu, Z., Schreckenberg, K., Shackleton, S.E. and Shackleton, C.M. (2005d) Agroforestry tree products (AFTPs): targeting poverty reduction and enhanced livelihoods. *International Journal for Agricultural Sustainability* 3, 1–23.

Longman, K., Manurung, R.M. and Leakey, R.R.B. (1990) Use of small, clonal plants for experiments on factors affecting flowering in tropical trees. In: Bawa, K.S. and Hadley, M. (eds) *Reproductive Ecology of Tropical Flowering Plants.* Man and the Biosphere Series No. 7. UNESCO and Parthenon Publishing Group, pp. 389–399.

Lowe, A.J., Gillies, A.C.M., Wilson, J. and Dawson, I.K. (2000) Conservation genetics of bush mango from central/west Africa: implications from random amplified polymorphic DNA analysis. *Molecular Ecology* 9, 831–841.

Maghembe, J.A., Kwesiga, F., Ngulube, M., Prins, H. and Malaya, F.M. (1994) The domestication potential of indigenous fruit trees of the miombo woodlands of southern Africa. In: Leakey, R.R.B. and Newton, A.C.

(eds) *Tropical Trees: Potential for Domestication and the Rebuilding of Forest Resources*. HMSO, London, pp. 220–229.

Maghembe, J.A., Simons, A.J., Kwesiga, F. and Rarieya, M. (1998) *Selecting Indigenous Trees for Domestication in Southern Africa: Priority Setting with Farmers in Malawi, Tanzania, Zambia and Zimbabwe*. ICRAF, Nairobi, 94 pp.

Mialoundama, F., Avana, M.-L., Youmbi, E., Mampouya, P.C., Tchoundjeu, Z., Mbeuyo, M., Galamo, G.R., Bell, J.M., Kopguep, F., Tsobeng, A.C. and Abega, J. (2002) Vegetative propagation of *Dacryodes edulis* (G. Don) H.J. Lam by marcots, cuttings and micropropagation. *Forests, Trees and Livelihoods* 12, 85–96.

Mkonda, A., Lungu, S., Maghembe, J.A. and Mafongoya, P.L. (2003) Fruit- and seed-germination characteristics of *Strychnos cocculoides*, an indigenous fruit tree from natural populations in Zambia. *Agroforestry Systems* 58, 25–31.

Mudge, K.W. and Brennan, E.B. (1999) Clonal propagation of multipurpose and fruit trees used in agroforestry. In: Buck, L.E., Lassoie, J.P. and Fernandes, E.C.M. (eds) *Agroforestry in Sustainable Ecosystems*. CRC Press/Lewis, New York, pp. 157–190.

Mng'omba, S.A., Du Toit, E.S., Akinnifesi, F.K. and Venter, H.M. (2007a) Repeated exposure of jacket plum (*Pappea capensis*) micro-cuttings to indole-3-butyric acid (IBA) improved *in vitro* rooting capacity. *South African Journal of Botany* 73, 230–235.

Mng'omba, S.A., Du Toit, E.S., Akinnifesi, F.K. and Venter, H.M. (2007b) Early recognition of graft compatibility in *Uapaca kirkiana* fruit tree clones, provenances and species using *in vitro* callus technique. *Horticultural Science* 43, 732–736.

Mwase, W.F., Bjornstad, A., Stedje, S., Bokosi, J.M. and Kwapata, M.B. (2006) Genetic diversity of *Uapaca kirkiana* Muel. Arg. populations as revealed by amplified fragment length polymorphisms (AFLP). *African Journal of Biotechnology* 5, 1205–1213.

Phofeutsile, K., Jaenicke, H. and Muok, B. (2002) Propagation and management. In: Hall, J.B., O'Brien, E.M. and Sinclair, F.L. (eds) *Sclerocarya birrea: A Monograph.* School of Agricultural and Forest Sciences Publication No. 19, University of Wales, Bangor, UK, pp. 93–106.

Shackleton, C.M., Botha, J. and Emanuel, P.L. (2003a) Productivity and abundance of *Sclerocarya birrea* subsp. *caffra* in and around rural settlements and protected areas of the Bushbuckridge lowveld, South Africa. *Forests, Trees and Livelihoods* 13, 217–232.

Shackleton, S., Wynberg, R., Sullivan, C., Shackleton, C., Leakey, R., Mander, M., McHardy, T., den Adel, S., Botelle, A., du Plessis, P., Lombard, C., Combrinck, A., Cunningham, A., O'Regan, D. and Laird, S. (2003b) Marula commercialisation for sustainable and equitable livelihoods: synthesis of a southern African case study. *Winners and Losers: Final Technical Report to DFID (FRP Project R7795)*, Volume 4, Appendix 3.5, 57 pp. [see www.nerc-wallingford.uk/research/winners/literature.html].

Shiembo, P.N., Newton, A.C. and Leakey, R.R.B. (1996a) Vegetative propagation of *Irvingia gabonensis* Baill., a West African fruit tree. *Forest Ecology and Management* 87, 185–192.

Shiembo, P.N., Newton, A.C. and Leakey, R.R.B. (1996b) Vegetative propagation of *Gnetum africanum* Welw., a leafy vegetable from West Africa. *Journal of Horticultural Science* 71, 149–155.

Shiembo, P.N., Newton, A.C. and Leakey, R.R.B. (1997) Vegetative propagation of *Ricinodendron heulelotii* (Baill) Pierre ex Pax, a West African fruit tree. *Journal of Tropical Forest Science* 9, 514–525.

Simons, A.J. (1996) ICRAF's strategy for domestication of indigenous tree species. In: Leakey, R.R.B., Temu, A.B., Melnyk, M. and Vantomme, P. (eds) *Domestication and Commercialisation of Non-Timber Forest Products in Agroforestry Systems*. FAO Non-Wood Forest Products No. 9. FAO, Rome, pp. 8–22.

Simons, A.J. and Leakey, R.R.B. (2004) Tree domestication in tropical agroforestry. *Agroforestry Systems* 61, 167–181.

Swallow, B.M., Villarreal, M., Kwesiga, F., Holding, A.C., Agumya, A. and Thangata,

P. (2007) Agroforestry and its role in mitigating the impact of HIV/AIDS. In: *World Agroforestry and the Future: Proceedings of 25th Anniversary Conference, 1–5 November 2003, Nairobi*. ICRAF, Nairobi.

Taylor, F.W., Mateke, S.M. and Butterworth, K.J. (1996) A holistic approach to the domestication and commercialisation of non-timber forest products. In: Leakey, R.R.B., Temu, A.B., Melnyk, M. and Vantomme, P. (eds) *Domestication and Commercialization of Non-timber Forest Products in Agroforestry Systems*. FAO Non-Wood Forest Products No. 9. FAO, Rome, pp. 8–22.

Tchoundjeu, Z., Avana, M.L., Leakey, R.R.B., Simons, A.J., Asaah, E., Duguma, B. and Bell, J.M. (2002) Vegetative propagation of *Prunus africana*: effects of rooting medium, auxin concentrations and leaf area. *Agroforestry Systems* 54, 183–192.

Tchoundjeu, Z., Asaah, E., Anegbeh, P., Degrande, A., Mbile, P., Facheux, C., Tsobeng, A., Atangana, A., Ngo-Mpeck, M.L. and Simons, A.J. (2006) Putting participatory domestication into practice in west and central Africa. *Forests, Trees and Livelihoods* 16, 53–69.

Thiong'o, M.K., Kingori, S. and Jaenicke, H. (2002) The taste of the wild: variation in the nutritional quality of marula fruits and opportunities for domestication. *Acta Horticulturae* 575, 237–244.

Wynberg, R.P., Laird, S.A., Shackleton, S., Mander, M., Shackleton, C., du Plessis, P., den Adel, S., Leakey, R.R.B., Botelle, A., Lombard, C., Sullivan, C., Cunningham, A.B. and O'Regan, D. (2003) Marula policy brief: marula commercialisation for sustainable and equitable livelihoods. *Forests, Trees and Livelihoods* 13, 203–215.

Zobel, B.J. and Talbert, J.T. (1984) *Applied Forest Tree Improvement*. John Wiley & Sons, New York, 505 pp.

3 Challenges to Stimulating the Adoption and Impact of Indigenous Fruit Trees in Tropical Agriculture

N. HAQ, C. BOWE AND Z. DUNSIGER

Centre for Underutilised Crops (CUC), University of Southampton, Southampton, UK

3.1 Introduction

Indigenous tropical fruit trees are distributed over a substantial part of the Earth's surface (the tropics extend from 23°N to 23°S), grow in the most biologically diverse environments in the world, ranging from tropical rainforests to savannahs and drylands, and are well adapted to their environmental conditions. These species are especially important in the drylands, where other fruit tree species cannot be cultivated easily. Fruit trees have been recognized as important components of forestry, horticulture and agriculture for the last 100 years, but unfortunately these trees are under-researched and therefore remain underutilized (Gebauer *et al.*, 2002). However, there have been some recent international initiatives for the domestication of trees producing agroforestry tree products (AFTPs) (Leakey *et al.*, 2005a).

3.1.1 Why stimulate the adoption of indigenous fruit trees?

Many researchers have identified the value of these indigenous fruits as a source of income and household nutrition (Maghembe *et al.*, 1994; Hegde and Daniel, 1995; Doran and Turnbull, 1997; Leakey and Simons, 1998; Huxley, 2001; Stevens *et al.*, 2001; Schreckenberg *et al.*, 2002, 2006; Hughes and Haq, 2003; Lengkeek *et al.*, 2004, Leakey *et al.*, 2005b; Akinnifesi *et al.*, 2006). Fruit trees provide food, improved nutrition, oil and fuel for cooking, and many other products, including fodder, medicine, fibre, mulch, timber, saps and resin. They also maintain the environment by sequestering carbon, trapping pollutant particles from the air, and promoting biodiversity and agroecosystem functions. Fruit trees provide shade, reduce surface runoff of rainwater, and in some species nourish the soil through nitrogen fixation. Fruit trees often have mystical and religious significance for human beings and are used in spiritual ceremonies. Local people still have knowledge of these indigenous tree species, but scientific information on them and the use of their products, is inadequate. Despite recent successful initiatives to promote these species (Lovett and Haq, 2000; Akinnifesi *et al.*, 2006; Tchoundjeu *et al.*, 2006) many of these species remain neglected and underutilized.

People have adopted tropical fruit trees, such as avocado from tropical America, coffee from Africa, and mango and citrus from Asia, to maintain a reliable supply of a variety of foods (Smith *et al.*, 1992). In recent years some effort has been made to encourage a wider use of fruit trees in agricultural systems, as research has shown that mixed cropping systems may be economically less risky than monocropping systems, and thus the land could become more productive (Moss, 1994). The utilization of fruits for product diversification has enhanced the importance of indigenous fruit trees in both developing and developed country marketplaces (Akinnifesi *et al.*, 2005, 2006).

A number of processes are involved in the adoption of indigenous fruit, such as the selection and multiplication of quality planting material, development of economically viable, culturally acceptable production systems, postharvest handling and processing, and correct storage and marketing of good-quality products. In recent years progress has been made towards implementing these processes with a small number of species at a few locations.

There are many constraints which affect all processes within the commodity chain described above and these limit their uptake by the relevant stakeholders. Ways must be found to loosen or remove these constraints in order to promote the adoption of tropical indigenous fruit tree species. This chapter reviews these constraints and the progress made towards solutions to overcome the problems in the context of the role that indigenous fruit trees play in sustainable agricultural systems.

3.2 Constraints to the Adoption of Indigenous Fruit Trees

Hughes and Haq (2003), the International Centre for Underutilised Crops (ICUC, 2003) and Haq (2006) have highlighted some important constraints to adopting indigenous fruit trees. By contrast, Tchoundjeu *et al.* (2006) have reported some success in implementing participatory domestication initiatives across several West African countries. They have addressed and overcome various constraints, as a result of intensive research efforts undertaken in parallel with participatory village-level activities. However, the following constraints still need to be addressed.

3.2.1 Unavailability of high-quality planting material

Access to good-quality planting materials is typically a major constraint for rural and urban farmers. In addition, suitable multiplication methods for many species have not yet been fully developed. The quality of products depends on good-quality planting material, and there is a strong need to carry on production and processing activities in parallel so that good-quality raw materials are available for processing and marketing so as to generate income for household farms.

Recent work on intraspecific variation in indigenous fruits and other AFTPs has shown that there is three- to sixfold variation in almost every trait investigated, whether these traits are morphological, physiological and/or chemical/physio-chemical. A participatory selection of superior germplasm has been undertaken with diverse users across countries in southern Africa (Akinnifesi *et al.*, 2006), and improved cultivars have been established in fruit orchards developed by clonal methods. A similar selection for ideotype candidate trees was also reported in West Africa (Lovett and Haq, 2000). This offers an opportunity for multi-trait selection and the development of market-oriented ideotypes through domestication procedures based on the vegetative propagation of superior trees, and the consequent development of cultivars.

3.2.2 Limitations in production systems

Agroforestry systems developed by researchers, such as alley cropping, have been found to be limited in terms of adoption by farmers (Sumberg and Okali, 1988). Mixed systems with economic crops have been reported as a viable alternative to slash-and-burn agriculture (Palm *et al.*, 2005). Many of these are agroforests that include a range of indigenous fruit trees (Kindt *et al.*, 2004; Schroth *et al.*, 2004; Degrande *et al.*, 2006). There are relatively few fruit-tree-based farming systems that have been well researched, although the use of indigenous fruit trees as an income-generating, shade-producing canopy for cocoa is common in western and central Africa (Leakey and Tchoundjeu, 2001). However, Hocking *et al.* (1997) reported economic yield depression from their investigation in Bangladesh when they included trees in rice and wheat production systems. They reported that there was a variable depression in rice (*Oryza sativa*) and wheat (*Triticum aestivum*) yields, ranging from 16 to 40%, when grown under the tree species *Acacia catechu*, *Artocarpus heterophyllus* and *Mangifera indica*. Similar studies are needed in agrosystems where comparative costs and benefits can be assessed between the loss of crop yields and the gain in tree products of high value, mainly fruits and fodder. Izac and Sanchez (2001) compared the relative benefits of high yield in cash crops and high profitability with high biodiversity in mixed systems. They also reported that, in order to derive the greatest benefit from mixed systems, greater flexibility is needed in the selection of crop species than in traditional farming. Gockowski *et al.* (1997), Gockowski and Dury (1999) and Palm *et al.* (2005) have reported cost benefits from fruit trees and agroforests. However, in all cases there is a need to consider environmental advantages in a cost–benefit analysis.

Agriculture in the tropical world, with a few exceptions, has suffered from a lack of mechanisms to introduce, test and understand less important species. However, when a number of such species were tested, the germplasm base was often narrow (Williams and Haq, 2002). Mannan (2000) reported a large diversity in homegardens in Bangladesh, where fruit trees are selected mostly by farmers. Similar observations have been made by Fondoun *et al.* (2002) and Leakey *et al.* (2004, 2005b).

3.2.3 Limitations in harvesting methods

At present, farmers use traditional methods of fruit harvesting, such as shaking the branches or using sticks to knock the fruits to the ground, which often result in heavy losses. Wilson (2002) reported that indigenous fruits are often picked either when immature or at an advanced stage of maturity, when inherent physiological developments render them more susceptible to injury if subjected to rough handling during harvest and transportation. Kadzere *et al.* (2006) examined the impacts of the time of harvesting on fruit quality, and postharvest handling in *Uapaca kirkiana*.

3.2.4 Limitations in postharvest methods

Heavy postharvest losses of indigenous fruits occur due to poor packaging and inappropriate transport conditions. Although the technology is available for commercially important exotic species, it remains unused for indigenous fruit trees. Commodities in their fresh state are very vulnerable to postharvest losses, both in quantity and quality, during transportation and storage. Postharvest losses occur at all stages of the marketing chain and observations indicate that they could be as high as 40–60% of the harvested crop (Wilson, 2002). It is important, therefore, to develop, adapt and transfer technology that will minimize these losses.

3.2.5 Limitations in processing

In general, indigenous fruits are processed and used locally by families for home consumption and for sale in the locality (Ramadhani, 2002). Nevertheless, many of these fruits have greater potential for processing and marketing of their products. The methods used for home-level processing are likely to be those that have been in use for generations and are based on trial and error rather than on scientific knowledge (Akinnifesi *et al.*, 2006; ICUC, 2006). However, it is likely that the technologies applied to the more commonly marketed fruits can also be adapted to indigenous fruit species. Processing of fresh fruits immediately after harvesting may also reduce losses and transportation costs, as this will require less space.

3.2.6 Limitations in marketing

The short production period of some indigenous fruit trees results in the simultaneous ripening of all fruits, thereby causing a glut in the market and low prices, followed by relative scarcity and high prices. A major problem of homestead farmers is the capacity to market fresh produce and the loss in quality during storage and transportation to the final market. Farmers often have to wait for traders before harvesting, which presents a particular problem during peak production periods and results in losses throughout the market chain. Another problem facing the marketing of indigenous fruits is a consequence of their being wild or semi-domesticated; this leads to very substantial tree-to-tree variation in fruit characteristics and thus to a lack of uniformity in quality. As a result, the wholesalers will not pay a good price because a basket of fruits is sourced from many different trees (Leakey *et al.*, 2002). However, many indigenous fruit trees produce fruits out of season, and this creates the opportunity for the development of cultivars which will extend the season. The 'Noel' cultivar of *Dacryodes edulis* in Cameroon (Leakey *et al.*, 2002) is a good example.

The promotion of commercially viable processing industries for indigenous fruits requires the availability of raw materials (i.e. a regular and reliable supply), product uniformity and consistently good quality, marketability of products, availability of technology to meet processing and market requirements, necessary machinery and equipment, and adequate and easy access to the necessary information and support services, including credit (Wilson, 1998).

Most fruits are sold by farmers immediately after harvest because of lack of storage facilities and the growers' need for cash. The majority of farmers sell their products in weekly markets for food, medicinal, religious and other cultural purposes, although some sell them in daily markets. However, the price received by farmers will depend on whether they take the fruits to market themselves, or sell to an intermediary. The main advantage of the latter practice for farmers is that they receive payment in bulk and also cut down on labour and transport costs. However, the practice does not allow for price setting according to supply and demand. There is usually a large price gap between the grower and the end consumer. In general, households tend to practice this form of selling in response to difficulties with cash flow and are therefore usually in a weak position to negotiate prices. The intermediary may pay a fixed (usually low) price for fruit trees, sometimes approaching the farmer at preharvest, negotiating a price for the tree and paying 50% in advance to the farmer. The intermediary then takes responsibility for the crop and will dictate the harvesting time. This usually coincides with a time when the market price for the fruit is high, rather than according to the maturity index, leading to the sale of poor-quality fruits. However, the role of the intermediary is often very significant in assisting the grower to market the produce. In some countries, such as Nepal, most indigenous fruits are collected from the wild throughout the season and used locally or taken in small quantities to the local market, with only a very few reaching the wholesale markets. This is due primarily to problems of transportation where rural communities are a long way from urban centres.

Product quality and consistency of quality are major factors in the successful marketing of products. Different fruit crops have different characteristics and different uses, even in the same country. Therefore the utilization (including the addition of value), processing and marketing strategies and consumer requirements will be different, and for many small-scale producers, these remain unknown. Farmers will often sell the excess produce from indigenous fruits for extra income, but they have to depend on the market chains because of the present limited commercial uses for these fruits. By combining well-established principles and appropriate equipment with good standards of quality and hygiene, small-scale food processing enterprises can produce high-quality and marketable products (ICUC, 2006).

3.2.7 Financial limitations

Lack of access to credit is also commonly cited as a constraint to small-scale production. The majority of small-scale farmers and processors, especially women, face a variety of problems when seeking credit, including lack of information, high interest rates, lack of collateral, bureaucratic difficulties and misunderstandings, prejudice against women and small-scale farmers and processors, and lack of government support in accessing credit (Azami, 2002). However, progress in domestication and appropriate village-level product development may overcome these problems of financial limitation (Tchoundjeu *et al.*, 2006). Better utilization of indigenous fruit trees will provide opportunities to improve family diets and to raise household income by trading and processing.

3.2.8 Poor agricultural policy

There is a need to develop a strategy for the wider-scale cultivation of indigenous fruit trees. Attention has been given to conservation and better use of the botanical diversity in traditional agroecosystems and natural forest systems, but support (both technical and financial) for sustainable production, product development and marketing at the regional, national and international level has been very limited. Policy guidelines (Wynberg *et al.*, 2003) and the development of AFTP domestication programmes by the World Agroforestry Centre (ICRAF) and other organizations (Leakey *et al.*, 2005a), will encourage policy makers to take account of recent progress. Williams and Haq (2002) have suggested that for any strategic development to succeed, the socio-economic well-being of the farmers and communities needs to be taken into account, and agricultural policy linked to forestry and export policy, which currently provides huge incentives for local people to cut down indigenous tree species for veneer and timber production stimulated by the demands of the furniture industry. Recent studies have examined the impacts of the commercialization of indigenous forest products (including indigenous fruits) on the natural, social and cultural aspects of rural livelihoods and have identified a number of requirements that are needed in order to ensure that communities are 'winners' and not 'losers' from the more intensive use and marketing of products such as indigenous fruits (Shackleton *et al.*, 2003).

3.2.9 Inadequate information, dissemination, knowledge sharing and training

Access to information is lacking throughout the production pathway, from germplasm resources to consumption. Many planners and policy makers are possibly aware of the contribution of indigenous fruit trees to rural development and livelihood, but there is a lack of information on the community-driven benefits. Access to information on technologies is a major constraint to small-scale production. Farmers and potential users of indigenous fruits are unaware of the benefits of the crop and the technologies that may be appropriate to their needs. The particular challenge here is to develop information that can be effective in countries with high levels of illiteracy.

3.3 Methods and Challenges of Stimulation

In order to promote the adoption of indigenous fruits, there is a need to focus on a number of key areas. The following sections focus on these areas.

3.3.1 Research and technology development

Research carried out on indigenous species is variable, and information documented on their germplasm resources, reproductive biology, growing habits, management practices, processing and utilization is scant and scattered. Gundel (2002) has suggested several reasons why various sectors have neglected these underutilized crops, some global and others regional in nature. However, it is the national agricultural research systems (NARS) which influence the crop production in each country (ICUC, 2003). If the NARS includes a species in its national programme, then the funds will be allocated for research and development. Although most have invested resources to improve their staple foods, research has lagged behind on indigenous fruit crops which could improve household farming and generate extra cash. The scientific and development community started to promote the need to investigate underutilized crops from the mid-1970s. However, the number of national programmes involved with indigenous fruit crops is small, albeit growing. Research is often carried out in isolation, with the findings restricted to internal reports, grey literature or academic journals, and the dissemination of information to a wider audience is poor. Recently, ICRAF and its partners have started a systematic research programme to develop technology which will encourage the NARS to include research areas to develop and commercialize indigenous fruit species (ICRAF, 2005).

3.3.2 Selection, domestication and multiplication of high-quality planting materials

Tree habitats, such as bush fallow, homegardens, compound farms, farm wood-lots, etc., have valuable resources, which frequently meet part of the subsistence and income needs of local communities. Hence, the indigenous fruit trees from the forest as well as from other ecosystems can provide household income for both rural people and peri-urban communities. The domestication of fruit trees would provide more opportunities for integrating various fruit tree species into agricultural systems. The species that produce locally valued commodities are new sources for domestication.

However, the domestication process of tree species through selection and breeding takes a long time and is expensive. Commonly, genetic selections are made on the basis of provenance, progeny variation and breeding, whether they are forestry, agroforestry or household farm species (ICUC, 2003). Because of the large investment of time and capital required, the cultivation of most fruit tree species is neglected and they remain underutilized. Recent interest in the ethnobotany and use of forest and agroforestry species has resulted in the need for domestication of some multipurpose indigenous fruit trees. As a result, scientists have been looking into the options for rapid domestication. Leakey and Newton (1994, 1995) described an alternative process for domestication which involves the capture of the existing intraspecific variation within any population, and its maintenance through vegetative propagation of individual favoured types. Thus an individual with superior characteristics can be mass-produced. Through a series of ongoing and increasingly intensive selection processes, it is also possible to achieve rapid and substantial genetic improvements. In recent years, progress has been made in the assessment of the genetic resources of underutilized indigenous trees to capture the genetic diversity for use in the domestication process (ICUC, 2003). It is essential that a systematic programme be established for the domestication of indigenous fruit trees and that studies on the diversity of species and selection of superior clones be conducted. The models developed by Tchoundjeu *et al.* (2006) and Akinnifesi *et al.* (2006) can be tried in other geographical areas in order to scale-up their methods.

Improvement for higher yield and other favourable characteristics is more likely to be achieved if there is high genetic variation. Leakey *et al.* (2004) suggest that intraspecific variation is high at village level and domestication of this variation will allow the selection of superior cultivars. It is important that the domestication of selected trees involves the conservation of genetically variable strains in germplasm collections (*ex situ*) or in the field (*in situ*) coupled with effective management of selected types to protect the environment, avoiding dependence on chemical pesticides and fertilizers and making effective use of water resources. A detailed knowledge of the genetic variation within the species is essential in order to design a strategy to promote the use and conservation of indigenous fruit trees through on-farm cultivation.

For instance, despite germplasm collection and research activities on a number of indigenous fruit trees across southern Africa, little is known concerning genetic structure at a regional level, with no wide-scale molecular work yet undertaken that could help develop regional management programmes (Kadu *et al.*, 2006). Without an awareness of useful traits in indigenous fruits and then the dissemination of information both on desirable types or strains and on their multiplication for farm planting, this variation cannot be used.

Farmer participatory surveys have been carried out in many countries to identify and select planting material stock for multiplication (Azad *et al.*, 2007). It has been found that women play a major role in selection of the desired type(s) for multiplication and production (ICUC, 2003). The women's choices depend on the utilization of the fruits, but they also consider characteristics such as drought tolerance, disease resistance, ability to withstand wide temperature ranges, and growth on marginal lands. Above all they give priority to those fruit species which can be grown in mixed cropping systems. Organizations such as the World Agroforestry Centre (ICRAF), the International Plant Genetic Resources Institute, the International Centre for Underutilised Crops (ICUC), and the Royal Botanic Gardens (Kew) have been involved in community participatory selection schemes for planting materials (ICUC, 2003).

In many cases tree species are multiplied by seed. However, desirable characteristics cannot be adequately controlled or transferred to the offspring through seed propagation. The benefit of known strains or cultivars with consistent growth and yield has still to be emphasized in developing farming systems and must be supported by training in methods of propagation and supply of planting materials (using, for example, the series of manuals published by ICUC and the Commonwealth Secretariat). The growing of planting materials itself can also become a profitable nursery industry (Haq, 2002; Tchoundjeu *et al.*, 2006).

Hence the key aspects of improving the selection and domestication of indigenous fruits are:

- Farmer participatory surveys to identify species of interest for a particular group of producers and consumers. Farmer participatory research to identify good mother stock and for sustainable production and processing technologies so as to promote consumption at the household level. This will empower the farmers to implement the domestication programmes for themselves, to work on species of their own choice, and to develop their own cultivars of those species based on their own selection criteria (with guidance for ideotypes) derived from research. These farmers will be able to protect their rights over the germplasm.
- Rapid methods for propagation of quality planting materials (whether it is seed, seedlings or saplings, cuttings, marcotts or grafts). There is current demand for planting materials and special efforts are required by resource-poor rural farmers to multiply and supply them in order to maintain the growth of indigenous fruit trees.

3.3.3 Production

Indigenous fruit trees need to be assessed for use in crop diversification research programmes, as these satisfy not only the need of small farmers but also of commercial enterprises for diversified products. The inclusion of indigenous fruit trees in production systems reduces the risks inherent to monocultures of staple food crops, such as susceptibility to pests and diseases, soil nutrient depletion, price fluctuations, and reliance on a single crop for income. Hughes and Haq (2003) pointed out that income from underutilized fruit trees in China, Nepal and the Philippines is much higher than from traditional agricultural crops. Furthermore, high cropping intensity risks soil degradation, with an associated reduced production. The development of land-use systems where the soils are managed effectively is an important goal of sustainable agricultural systems. Management schemes need to take account of changes in crop spacing, use of water shading and variation in yield when combinations of crops are grown in mixed systems. This can only be developed with further trials on systems which combine some or all of the following elements: domestic crops, cash crops, and presently underutilized indigenous fruits.

It has to be recognized that the promotion of particular high-yielding strains is likely to bring about changes. For instance, market forces will provide rapid incentives for higher production and the need for new cultivars and, as a consequence, the maintenance of traditional agriculture incorporating underutilized crops may suffer. However, a small swing towards traditional mixed cropping involving indigenous fruits has been noted, and this may increase in the future when more domestication of indigenous fruits occurs.

3.3.4 Marketing research

Key points in the development of indigenous fruit marketing research include:

- Analysis of the marketing environment.
- Size of the potential market.
- Competitive analysis.
- Consumer attitudes towards processed products.
- Attractiveness of processed products to consumers.
- Distribution of processed products.
- Potential viability of agroprocessing ventures.
- Marketing mix development.

Leakey and Izac (1996) suggested that there is a need for simultaneous work on domestication and commercialization for the development of indigenous fruits.

3.3.5 Tools for market development

With effective product development and marketing, indigenous fruits may increase income, as they can be sold easily in local markets. If these perishable underutilized fruits are processed into shelf-stable products at the commercial level, growers can obtain greater financial returns, with additional benefits for those engaged in adding value.

The potential market for indigenous fruit products in India has been summarized by Lobbezoo and Betser (2006). Many of these fruits have high medicinal value and have considerable marketing potential. The high nutritional value and medicinal properties of some fruits can be exploited by producing products for niche markets, such as health foods or natural products. Several products have been developed in some countries from a range of fruits with high medicinal values, such as baobab, jujube and emblica (Parmar and Kaushal, 1982; Pareek, 2001; Sidibe and Williams, 2002).

Suppliers, which includes growers and collectors, have low bargaining power because they enter a highly competitive market with many substitute products, i.e. the same kind of product but from other fruit, where larger brands already have their place. Branding and targeting specific markets both seem to be very important and both require capital input. The potential for providing semi-processed products, such as pulp, to the competitors is probably greater than full processing by the suppliers themselves. In countries where competing substitute products are already available, manufacturers need to make potential buyers aware of the extra value of their products in enhancing health or beauty or saving time (Lobbezoo and Betser, 2006).

3.3.6 Developing information systems

Access to information for the utilization and improvement of indigenous tree species is lacking throughout the production pathway. Information can be presented in different media according to the requirements of the target audience and the technology available. Each has its merits. As mentioned earlier, due to the scattered nature of research findings (reports, grey literature, academic literature), it is often difficult to gather comprehensive information on a potential crops. However, attempts have been made (by the ICUC and others) to develop a series of monographs. In addition, the information on research results should be made available not only to researchers but also to producers and small-scale processors. It must also be made available in a suitable format for policy makers, providing correct and targeted information for those involved in decision making for research, farming and rural development.

In order to put into practice selection, production, processing and marketing, it is necessary to provide practical information or advice on the techniques involved in bringing a product to market through the production chain. This can be achieved through the distillation of research studies in the form of a technology manual. It is necessary to summarize relevant research information and write in an accessible style without inappropriate technical jargon. Leaflets, with processes and techniques illustrated and annotated, are useful and can be used easily for training. If supported by more detailed background information, they can also be a valuable tool for trainers.

It is also important to support those communities where literacy and numeracy is poor or non-existent; training tools can be created using pictorial representations of techniques. Experience of how activities can be illustrated most effectively is useful, and it must be remembered that different communities will interpret pictures differently. This emphasizes the importance of trainers with sensitivity to local attitudes. Information provided in the native language of the audience is particularly important.

A valuable development in the dissemination of information is the increasingly widespread use of computers and the Internet. Information initially made available on CD-ROM, video or DVD can subsequently be more easily distributed than paper products where postal or transport services are poor or costly. The facility to easily browse material in electronic format can be a valuable benefit in searching for information and also for selecting specific areas for training. Where Internet access is available, the difficulties of distribution of research and education materials can be avoided altogether. It is, however, necessary when designing websites, online databases or other digital products, to consider that the older hardware or software likely to be encountered in the target areas can make websites inaccessible, and cross-compatibility between products is vital. Key areas where information should be supplied are:

- Policy briefs to increase awareness among policy makers. The policy should include adequate producer prices and credit facilities aimed at increasing production, storage and marketing of underutilized crops (Peiler, 2004; see also www.agroforestry.net/pubs/PIIFS_Policy_Briefing.pdf).
- Training of community workers (including women's groups) in culturally appropriate methods for food and product processing and marketing. Education programmes for local food preparation and processing, including publication of recipes and ideas for increased consumption.
- Workshops and seminars to increase national and regional awareness.

3.3.7 Policy change

Favourable policy design is central to sustainable agricultural production, land tenure patterns, management of natural resources and poverty alleviation (Wynberg *et al.*, 2003). Short-term project support to indigenous fruit crops research is unlikely to lead to clearer policy definition, especially when experience shows that a relatively longer time frame is needed to develop and commercialize indigenous fruit crops. Also, research requires many inter-sectoral linkages, since programmes may relate to maintenance of agrobiodiversity at the local level, to appropriate production in marginal areas, or extend to supplying market needs for specific products.

Research on indigenous fruit species will only be successful when national priorities are sensitized to the needs of local communities, industries and consumers. Additionally, countries are party to a number of international conventions and agreements, which are often dealt with outside the agricultural sector. It is widely recognized that facilitating mechanisms for underutilized crops has a major role to play in developing national strategies.

3.3.8 Developing cooperatives and finance initiatives

Indigenous fruits are difficult to obtain in adequate quantities and it can be a daunting problem to supply them through collection from the wild because of the quality requirement. One way to overcome this problem is to bulk up quality planting materials and then supply them to the farming community. However, commercial organizations are unlikely to be involved with such materials. It also becomes a problem for the public sector to be involved in the multiplication of planting materials, due to limited budgets which may already be allocated for specific research. None the less, past experience has shown the value of village nurseries and local seed banks, as they are sources for the maintenance of local agrobiodiversity. These are mostly not related to secure national supply systems, but such pragmatic solutions to the problems of sustainability are far more viable when due consideration is given to the pricing of products and provision of credit.

Community enterprise development by non-governmental and community-based organizations and other community groups, such as women's cooperatives in India, are making impacts through local product development and local and regional marketing (ICUC, 2006). The marketing of new products can be difficult because of the lack of availability of storage, processing and distribution facilities. Where producers can be supported at various stages in the production chain, such as by involvement with cooperatives, there is a greater likelihood of a new product being successfully introduced in the market. There is also greater bargaining power for obtaining credit in order to develop enterprises, for negotiating with dealers or processors, or in arranging transportation.

3.4 Impacts

As indicated earlier, the importance of tropical fruit tree species in sustainable livelihoods has been highlighted by many researchers (Hegde and Daniel, 1995; Doran and Turnbull, 1997; Leakey and Simons, 1998; Huxley, 2001; Stevens *et al.*, 2001; Hughes and Haq, 2003; Leakey *et al.*, 2005a, b). With particular reference to food security and nutrition, fruits and nuts provide a variety of foods and food products that are rich in vitamins, minerals, proteins and antioxidants. They improve palatability and contribute to the diets of the poor in improved nutrition and variety. They are particularly valuable sources of food and nutrition during emergency periods of food shortage, such as flood, famine, drought and war. Depending on the seasonal harvest from fruit trees, they may provide an additional dietary supplement during peak times of agricultural labour. They provide oil and fuel for cooking, and a wide range of local medicines that contribute to health. The supply of fuelwood and oils can influence nutrition through their impact on the availability of cooked food (Falconer, 1989). Impacts of the promotion of indigenous fruit trees are discussed in the following sections.

3.4.1 Influence of research outputs

There are a number of examples from major species which can be followed by those promoting indigenous fruits and particularly by those designing and implementing appropriate research, whether on crop diversification, maintenance of agrobiodiversity, or meeting the needs of industries and/or communities.

The research and development of any crop requires a multidisciplinary approach. It is pertinent to investigate a range of species and also to focus on any special needs in the areas where indigenous fruit crops are to be grown. A wide range of species show different degrees of domestication and/or genetic changes from the original wild species or progenitors as domestication pathways vary tremendously and proceed along different lines. There is much to be gained in developing research plans by those working on indigenous fruit crops if they are aware of current advances in research in associated plant sciences and even in human cultural research. To illustrate this, it is helpful to consider species in varying stages of domestication (Rindos, 1984), for instance the studies of baobab (Sidibe and Williams, 2002), shea-butter tree (Lovett and Haq, 2000) and *Moringa* spp. (Harlan *et al.*, 1976) in Africa.

There are many local varieties available, which need systematic characterization and evaluation. Diverse examples are found in the genera *Tamarindus*, *Ziziphus*, *Dacryodes* and *Irvingia* (David, 1976; Gunasena and Hughes, 2000; Pareek, 2001; Kengue, 2002; Anegbeh *et al.*, 2003). Research on these species requires an understanding of the local agroecology and patterns of genetic changes which have occurred and can be used further.

Multidisciplinary research may require a very strong ecological input because data on limitations on the site adaptability are essential when recognizing that some indigenous fruit species will grow well in entirely new environments, while others often need to be adapted by specific management or genetic selection.

The impact of research can influence the policy makers' decisions. For example, influenced by the Global Programme on Fruits for the Future created by ICUC, jackfruit was included in the national research programme of the Nepal Agricultural Research Council (NARC). Similarly the Indian Council for Agricultural Research (ICAR) created a new cell, the All India Co-ordinated Network on Underutilized Fruits, which includes eight underutilized fruit species for research and development.

3.4.2 Increased production and trading

Commercial interest in several underutilized fruits has resulted in increased cultivated areas in many parts of the world, particularly in developing countries (Pareek *et al.*, 1998). The export of fruits from Asia alone has been increased by 12% annually (Singh, 1993) although the potential for export earnings is much higher. The increase in production through systematic programmes, such as the Underutilized Tropical Fruits in Asia Network (UTFANET, 2003), has facilitated the development of high-quality planting materials for jackfruit, mangosteen and pummelo through farmer participatory surveys and the evaluation, collection and characterization of these species. Farmers' choices included characteristics for sweetness, juicy fruits, high yield, small stature and quality timber. Additional characteristics, including tolerance to drought and disease, and the ability to withstand extreme temperatures and grow on marginal lands, were maintained. Approximately 1100 accessions of jackfruit, 230 of mangosteen and 300 of pummelo were collected, with further plants identified in the field (ICUC, 2003). From those, approximately 130 jackfruit, 10 mangosteen and 40 pummelo have been selected for propagation trials (ICUC, 2003).

3.4.3 Improved processing and marketing

Various indigenous fruits are processed at different levels of sophistication. Fruits such as tamarind, jackfruit and pummelo are processed in various forms for domestic consumption. Traditional methods of preservation have developed into cottage industries in some areas, with produce sold in urban domestic markets. Some cottage industry operations also engage in supplying larger processing plants with semi-processed products (ICUC, 2006).

Processing of added-value products can also be lucrative, as demonstrated by a women's group in Thailand. By making use of 'dropped' pummelo fruits, which can attract fruit flies to the orchard, in processing into candies, the group has managed to establish a stable business, expanding its membership from seven to 45 women. The group earns a decent salary and has also started to train other local groups (Muang-Thong, 2002). Women play a major role in the harvesting, processing and marketing of underutilized indigenous tree products (Ruiz *et al.*, 1997). Such extra income, controlled by the women, is likely to be spent on education, nutrition and health, and this brings greater respect from their families and communities.

Some farmers or community groups, however, make the decision not to sell their produce to an intermediary, but to market their own products. The Ikalahan people are a forest-dwelling community living in the northern mountains of the Philippines. After producing quality jams and jellies from locally gathered wild forest fruits, they decided to sell their products in local towns. Through a process of trial and error and with help from the local university, the community has been successful in its endeavour and is producing and successfully marketing locally produced products (Rice, 2002). The Ikalahan people were fortunate to enlist the help of others in their quest to sell their products, and this has demonstrated that training and education in business and development skills for rural communities would improve the earning potential of many indigenous fruit products. By combining well-established principles and appropriate equipment with good standards of quality and hygiene, it has been demonstrated that small-scale food processing enterprises can produce quality, marketable products.

3.5 Conclusions

Most indigenous fruit trees have multiple uses and play a vital role in crop diversification programmes and household systems. Their inclusion in production systems reduces the risks inherent to monocultures of staple food crops, such as pest and disease outbreaks, soil nutrient depletion, price fluctuations, and reliance on a single crop for income. The systematic domestication of indigenous fruit trees can improve their production. Several authors have already reported the progress made in domestication of various African species (Atangana *et al.*, 2001; Leakey *et al.*, 2005a, b, c, d; Tchoundjeu *et al.*, 2006).

A start has been made in understanding the producer-to-consumer chain system so as to stimulate the adoption of indigenous tropical fruit trees. The particular components of this system are: (i) the identification of quality planting materials through farmer participatory surveys, and collection and evaluation of germplasm, readily available to farmers; (ii) knowledge of propagation and production systems to integrate into diverse farming systems; (iii) technology for postharvest processing, product development, storage, packaging, quality assurance and marketing; (iv) appropriate national policy; (v) access to credit; (vi) training in technology relevant to production, food processing and marketing, in particular for small entrepreneurs' development; and, above all, (vii) information dissemination for all these areas. Once we progress on these issues, the incorporation of indigenous fruit trees into agricultural systems to improve rural livelihoods will become easier.

References

Akinnifesi, F.K., Ham, C., Jordaan, D., Mander, M., Mithöfer, D., Ramadhani, T., Kwesiga, F., Saka, J. and Phosiso, S. (2005) Building opportunities for small holder farmers to commoditize indigenous fruit trees and products in southern Africa: processing, rural pilot enterprises and marketing. In: *The Global Food and Product Chain: Dynamics, Innovations, Conflicts, Strategies (International Conference on Research for Development in Agriculture, Forestry and Resource Management). Deutsche Tropentag 2005, 11–13 October, Hohenheim.* University of Hohenheim, Stuttgart, Germany.

Akinnifesi, F.K., Kwesiga, F., Mhango, J., Chilanga, T., Mkonda, A., Kadu, C.A.C., Kadzere, I., Mithofer, D., Saka, J.D.K., Sileshi, G., Ramadhani, T. and Dhliwayo, P. (2006) Towards the development of miombo fruit trees as commercial tree crops in southern Africa. *Forests, Trees and Livelihoods* 16, 103–121.

Anegbeh, P.O., Usoro, C., Ukafor, V., Tchoundjeu, Z., Leakey, R.R.B. and Schreckenberg, K. (2003) Domestication of *Irvingia gabonensis*. 3. Phenotypic variation of fruits and kernels in a Nigerian village. *Agroforestry Systems* 58, 213–218.

Atangana, A.R., Tchoundjeu, Z., Fondoun, J.-M., Asaah, E., Ndoumbe, M. and Leakey, R.R.B. (2001) Domestication of *Irvingia gabonensis*. 1. Phenotypic variation in fruit and kernels in two populations from Cameroon. *Agroforestry Systems* 53, 55–64.

Azad, A.K., Jones, J.G. and Haq, N. (2007) Assessing morphological and isozyme variation of jackfruit (*Artocarpus heterophyllus* Lam.) in Bangladesh. *Agroforestry Systems* doi:10.1007/s10457-007-9039-8.

Azami, S. (2002) Use of information for promotion of underutilised fruit trees. In: Haq, N. and Hughes, A. (eds) *Fruits for the Future in Asia: Proceedings of the Consultation Meeting on the Processing and Marketing of Underutilised Tropical Fruits in Asia*. International Centre for Underutilised Crops (ICUC), Southampton, UK.

David, N. (1976) History of crops and people in the Cameroon to AD 1000. In: Harlan, J.R., de Wet, J.M.J. and Stample, A.B.L. (eds) *Origin of African Plant Domestications*. Moutan, The Hague, The Netherlands, pp. 223–268.

Degrande, A., Schreckenberg, K., Mbosso, C., Anegbeh, P.O., Okafor, J. and Kanmegne, J. (2006) Farmers' fruit tree growing strategies in the humid forest zone of Cameroon and Nigeria. *Agroforestry Systems* 67, 159–175.

Doran, J.C. and Turnbull, J.W. (1997) *Australian Trees and Shrubs: Species for Land Rehabilitation and Farm Planting in the Tropics*. ACIAR, Canberra.

Falconer, J. (1989) *Forestry and Nutrition: A Reference Manual*. FAO, Rome.

Fondoun, J.-M., Ndoumbe, M. and Leakey, R.R.B. (2002) Domestication of *Irvingia gabonensis*. 2. The selection of multiple traits for potential cultivars from Cameroon and Nigeria. *Agroforestry Systems* 55, 221–229.

Gebauer, J., El-Siddig, K. and Ebert, G. (2002) Baobab (*Adansonia digitata* L.): a review on a multipurpose tree with promising future in Sudan. *Gartenbauwissenschaft* 67, 155–160.

Gockowski, J.J. and Dury, S. (1999) The economics of cocoa-fruit agroforests in southern Cameroon. In: Jiménez, F. and Beer, J. (eds) *Multi-strata Agroforestry Systems with Perennial Crops*. CATIE, Turrialba, Costa Rica, pp. 239–241.

Gockowski, J., Tonye, J. and Baker, D. (1997) *Characterisation and Diagnosis of Agricultural Systems in the Alternatives to Slash and Burn Forest Margins Benchmark of Southern Cameroon*. Report to ASB Programme. IITA, Ibadan, Nigeria, 68 pp.

Gunasena, H.P.M. and Hughes, A. (2000) *Tamarind, Tamarindus indica*. International Centre for Underutilised Crops (ICUC), Southampton, UK.

Gundel, S. (2002) *Taking Livelihood Perspective on Neglected Crops*. DFID Discussion Paper. Department for International Development, London.

Harlan, J.R., de Wet, J.M.J. and Stemler, A.B.L. (1976) *Origins of African Plant Domestication*. Mouton, The Hague, The Netherlands, pp. 465–78.

Haq, N. (2002) *Report on SHABJE Project in Bangladesh*. DFID, Dhaka, Bangladesh.

Haq, N. (2006) Underutilised fruits: a resource for sustainable livelihoods. In: Ochatt, S. and Jain, S.M. (eds) *Breeding of Neglected and Under-utilized Crops: Spices and Herbs*. FAO–IAEA Publications, Oxford, and IBH Publishing, New Delhi, India, pp. 1–25.

Hegde, N.G. and Daniel, J.N. (eds) (1995) *Multipurpose Tree Species for Agroforestry in India: Proceedings of the National Workshop held 6–9 April 1994 in Pune, India*. BAIF Development Research Foundation, Pune, India, 137 pp.

Hocking, D., Sarwar, G. and Yousuf, S.A. (1997) Trees on farms in Bangladesh. 4. Crop yields underneath traditionally managed mature trees. *Agroforestry Systems* 35, 1–13.

Hughes, A. and Haq, N. (2003) Promotion of indigenous fruit trees through improved processing and marketing in Asia. *International Forestry Review* 5, 176–181.

Huxley, P.A. (2001) Multipurpose trees: biological and ecological aspects relevant to their selection and use. In: Last, F.T. (ed.) *Tree Crop Ecosystems*. Ecosystems of the World Vol. 19. Elsevier Science, Oxford, UK, pp. 19–74.

ICRAF (2005) *Annual Report 2005: Agroforestry Science to Support the Millennium Development Goals*. World Agroforestry Centre, Nairobi, Kenya.

ICUC (2003) *Improvement of Tropical Underutilised Fruits in Asia*. Final Report to Community Fund, UK. International Centre for Underutilised Crops (ICUC), Southampton, UK.

ICUC (2006) *Technical Manual for Small-scale Processors*. International Centre for Underutilised Crops (ICUC), Southampton, UK, 94 pp.

Izac, I.M.N. and Sanchez, P.A. (2001) Towards a natural resource management paradigm for international agriculture: the example of agroforestry research. *Agricultural Systems* 69, 5–25.

Kadu, C.A.C., Imbuga, M., Jamnadass, R. and Dawson, I.K. (2006) Genetic management of indigenous fruit trees in southern Africa: a case study of *Sclerocarya birrea* based on nuclear and chloroplast variation. *South African Journal of Botany* 72, 421–427.

Kadzere, I., Watkins, C.B., Merwin, I.A., Akinnifesi, F.K., Saka, J.D.K. and Mhango, J. (2006) Fruit variability and relationships between color at harvest and quality during storage of *Uapaca kirkiana* (Muell. Arg.) fruit from natural woodlands. *HortScience* 41, 352–356.

Kengue, J. (2002) *Safou: Dacryodes edulis*. International Centre for Underutilised Crops (ICUC), Southampton, UK, 187 pp.

Kindt, R., Simons, A.J. and van Damme, P. (2004) Do farm characteristics explain differences in tree species diversity among Western Kenyan farms? *Agroforestry Systems* 63, 63–74.

Leakey, R.R.B. and Izac, A.-M.N. (1996) Linkages between domestication and commercialisation of non-timber forest products: implications for agroforestry. In: *Domestication and Commercialisation of Non-timber Forest Products in Agroforestry Systems: Proceedings of an International Conference held in Nairobi, Kenya, 19–23 February*. Non-wood Forest Products No. 9. FAO, Rome, pp. 1–7.

Leakey, R.R.B. and Newton, A.C. (1994) *Tropical Trees: Potential for Domestication and the Rebuilding of Forest Resources*. HMSO, London, 284 pp.

Leakey, R.R.B. and Newton, A.C. (1995) *The Domestication of Timber and Non-timber Forest Products*. MAB Digest 17, UNESCO, Paris, 94 pp.

Leakey, R.R.B. and Simons, A.J. (1998) The domestication and commercialization of indigenous trees in agroforestry for the alleviation of poverty. *Agroforestry Systems* 38, 165–176.

Leakey, R.R.B. and Tchoundjeu, Z. (2001) Diversification of tree crops: domestication of companion crops for poverty reduction and environmental services. *Experimental Agriculture* 37, 279–296.

Leakey, R.R.B., Atangana, A.R., Kengni, E., Waruhiu, A.N., Usuro, C., Anegbeh, P.O. and Tchoundjeu, Z. (2002) Domestication of *Dacryodes edulis* in West and Central Africa: characterisation of genetic variation. *Forests, Trees and Livelihoods* 12, 57–72.

Leakey, R.R.B., Tchoundjeu, Z., Smith, R.I., Munro, R.C., Fondoun, J.-M., Kengue, J., Anegbeh, P.O., Atangana, A.R., Waruhiu, A.N., Asaah, E., Usoro, C. and Ukafor, V. (2004) Evidence that subsistence farmers have domesticated indigenous fruits (*Dacryodes edulis* and *Irvingia gabonensis*) in Cameroon and Nigeria. *Agroforestry Systems* 60, 101–111.

Leakey, R.R.B., Tchoundjeu, Z., Schreckenberg, K., Shackleton, S.E. and Shackleton, C.M. (2005a) Agroforestry tree products (AFTPs): targeting poverty reduction and enhanced livelihoods. *International Journal of Agricultural Sustainability* 3, 1–23.

Leakey, R.B.B., Greenwell, P., Hall, M.N., Atangana, A.R., Usoro, C., Anegbeh, P.O., Fondue, J.-M. and Tchoundjeu, Z. (2005b) Domestication of *Irvingia gabonensis*. 4. Tree-to-tree variation in food-thickening properties and in fat and protein contents of dika nut. *Food Chemistry* 90, 365–378.

Leakey, R.R.B., Shackleton, S. and du Plessis, P. (2005c) Domestication potential of marula (*Sclerocarya birrea* subsp. *caffra*) in South Africa and Namibia. 1. Phenotypic variation in fruit traits. *Agroforestry Systems* 64, 25–35.

Leakey, R.R.B., Pate, K. and Lombard, C. (2005d) Domestication potential of marula (*Sclerocarya birrea* subsp. *caffra*) in South Africa and Namibia. 2. Phenotypic variation in nut and kernel traits. *Agroforestry Systems* 64, 37–49.

Lengkeek, A., Wright, J., Michael, Y.G. and Nielsen, F. (2004) *How to Use Agrobiodiversity and Local Knowledge in the Mitigation of HIV/AIDS*. FAO, Rome.

Lobbezoo, M. and Betser, L. (2006) *Competitive Analysis of the Fruit Processing Industry in Nepal, Vietnam, India, Sri Lanka and Bangladesh Using Porter's Five Forces Model*. Report on FRP Project R8399, Improved Livelihoods Through the Development of Small-scale Fruit Processing Enterprises in Asia, 21 pp.

Lovett, P.N. and Haq, N. (2000) Diversity of sheanut tree (*Vitellaria paradoxa* C.F. Gaertn.) in Ghana. *Genetic Resources and Crop Evolution* 47, 293–304.

Maghembe, J.A., Kwesiga, F., Ngulube, M., Prins, H. and Malaya, F. (1994) Domestication potential of indigenous fruit trees of the miombo woodlands of southern Africa. In: Leakey, R.R.B. and Newton, A.C. (eds) *Tropical Trees: The Potential for Domestication and the Rebuilding of Forest Resources*. HMSO, London, pp. 220–229.

Mannan, A. (2000) Plant biodiversity in the homesteads of Bangladesh and its utilization in crop improvement. PhD thesis, BSMR Agricutlural University, Bangladesh.

Moss, R. (1994) Underexploited tree crops: components of productive and more sustainable farming systems. In: *Recherches-systeme en agriculture et developpement rural: Symposium International, Montpellier, France, 21–25 1994*. Communications, CIRAD-SAR, Montpellier, France, pp. 446–451.

Muang-Thong, N. (2002) Experience in processing and marketing of value-added products in Thailand. In: Haq, N. and Hughes, A. (eds) *Fruits for the Future in Asia: Proceedings for the Consultation Meeting on the Processing and Marketing of Underutilised Tropical Fruits in Asia*. International Centre for Underutilised Crops (ICUC), Southampton, UK, pp. 100–101.

Palm, C.A., Vosti, S.A., Sanchez, P.A. and Erickson, P.F. (eds) (2005) *Slash and Burn: The Search for Alternatives*. Columbia University Press, New York, 463 pp.

Pareek, O. (2001) *Fruits for the Future: Ber*. International Centre for Underutilised Crops (ICUC), Southampton, UK, 294 pp.

Pareek, O., Sharma, S. and Arora, R.K. (1998) *Underutilized Edible Fruits and Nuts: An Inventory of Genetic Resources in their Regions of Diversity*. International Plant Genetic Resources Institute, South Asia Office, 234 pp.

Parmar, C. and Kaushal, M.K. (1982) *Emblica officinalis*. In: *Wild Fruits*. Kalyani Publishers, New Delhi, India, pp. 26–30.

Peiler, E. (2004) *A Briefing Paper on Baobab, Africa's Multipurpose Upside Down Tree, to Improve Rural Livelihoods*. International Centre for Underutilised Crops (ICUC), Southampton, UK.

Ramadhani, T. (2002) Marketing of indigenous fruits in Zimbabwe. Socioeconomic Studies on Rural Development No. 129. PhD thesis, University of Hannover, Germany, 212 pp.

Rice, D. (2002) Community involvement in food processing: the Ikalahan experience. In: Haq, N. and Hughes, A. (eds) *Fruits for the Future in Asia: Proceedings of the Consultation Meeting on the Processing and Marketing of Underutilised Tropical Fruits in Asia*. International Centre for Underutilised Crops (ICUC), Southampton, UK, pp. 93–99.

Rindos, D. (1984) *The Origin of Agriculture: An Evolutionary Perspective*. Academic Press, Orlando, Florida.

Ruiz, P.M., Broekhoven, A.J., Aluma, J.R.W., Iddi, S., Lowroe, J.D., Mutemwa, S.M. and Odera, J.A. (1997) *Research on Non-timber Forest Products in Selected Countries in Southern and Eastern Africa*. CIFOR Working Paper No. 15. CIFOR, Bogor, Indonesia, 21 pp.

Schreckenberg, K., Leakey, R.R.B. and Kengue, J. (eds) (2002) A fruit tree with a future: *Dacryodes edulis* (Safou, the African plum). *Forests, Trees and Livelihoods* 12, 1–152.

Schreckenberg, K., Awono, A., Degrande, A., Mbosso, C., Ndoye, O. and Tchoundjeu, Z. (2006) Domesticating indigenous fruit trees as a contribution to poverty reduction. *Forests, Trees and Livelihoods* 16, 35–51.

Schroth, G., Da Fonseca, G.A.B., Harvey, C.A., Gascon, C., Vasconcelos, H.L. and Izac, A.-M.N. (2004) *Agroforestry and Biodiversity Conservation in Tropical Landscapes.* Island Press, Washington, DC, 576 pp.

Shackleton, S., Wynberg, R., Sullivan, C., Shackleton, C., Leakey, R., Mander, M., McHardy, T., den Adel, S., Botelle, A., du Plessis, P., Lombard, C., Combrink, A., Cunningham, A., Regan, D. and Laird, S. (2003) *Marula Commercialisation for Sustainable and Equitable Livelihoods: Synthesis of a Southern Africa Case Study: Winners and Losers.* Final Technical Report to DFID (FRP Project R7795), Volume 4, Appendix 3.5, 57 pp.

Sidibe, M. and Williams, J.T. (2002) *Baobab: Adansonia digitata.* International Centre for Underutilised Crops (ICUC), Southampton, UK, 99 pp.

Singh, R.B. (1993) *Research and Development of Fruits in the Asia-Pacific.* RAPA Publication 93/9. FA-RAPA, Bangkok.

Smith, N.J.H., Williams, J.T., Plucknett, D.L. and Talbot, J.P. (1992) *Tropical Forests and their Crops.* Cornell University, Ithaca, New York, pp. 4–5.

Stevens, M.L., Bourke, R.M. and Evans, B.R. (2001) *South Pacific Indigenous Nuts.* ACIAR, Canberra.

Sumberg, J. and Okali, C. (1988) Farmers, on-farm research and the development of new technology. *Experimental Agriculture* 24, 333–342.

Tchoundjeu, Z., Asaah, E., Anegbeh, P., Degrande, A., Mbile, P., Facheux, C., Tsobeng, A., Atangana, A. and Ngo-Mpeck, M.L. (2006) AFTPs: putting participatory domestication into practice in west and central Africa. *Forests, Trees and Livelihoods* 16, 53–69.

UTFANET (2003) *Underutilised Tropical Fruits in Asia Network Report.* Scientist Meeting, April 2003, Hanoi, Vietnam. UTFANET.

Williams, J.T. and Haq, N. (2002) *Global Research on Underutilised Crops.* International Centre for Underutilised Crops (ICUC), Southampton, UK, 46 pp.

Wilson, W.R.S. (1998) Identification of problems in processing of underutilized fruits of the tropics and their solutions. *Acta Horticulturae* 518, 237–240.

Wilson, W.R.S. (2002) Status report on the postharvest handling, processing and marketing of underutilized fruit crops in Sri Lanka. In: Haq, N. and Hughes, A. (eds) *Fruits for the Future in Asia: Proceedings of the Consultation Meeting on the Processing and Marketing of Underutilised Tropical Fruits in Asia.* International Centre for Underutilised Crops (ICUC), Southampton, UK, pp. 199–204.

Wynberg, R.P., Laird, S.A., Shackleton, S., Mander, M., Shackleton, C., du Plessis, P., den Adel, S., Leakey, R.R.B., Botelle, A., Lombard, C., Sullivan, C., Cunningham, A.B. and O'Regan, D. (2003) Marula commercialisation for sustainable and equitable livelihoods. *Forests, Trees and Livelihoods* 13, 203–215.

4 Domestication of Trees or Forests: Development Pathways for Fruit Tree Production in South-east Asia

K.F. Wiersum

Wageningen University, Wageningen, The Netherlands

4.1 Introduction

When considering the process of domestication of wild species, wild and domesticated plants are normally juxtaposed, and the changes in biological properties are considered as the main feature of the domestication process. A tree is normally considered 'wild' when it grows spontaneously in self-maintaining populations in natural or semi-natural ecosystems and can exist independently of direct human action (FAO, 1999). It is considered to be domesticated when the tree has been selected purposively for specific genetic characteristics and when it is propagated and cultivated in managed agroecosystems (Leakey and Newton, 1994). Such a dichotomy, however, should be considered as a first approximation only, as in reality several intermediate stages between truly wild and fully domesticated trees exist. Moreover, as illustrated by the above definitions, the process of domestication does not only involve changes in tree properties but also changes in exploitation systems. In this coevolutionary process, several intermediate phases may be distinguished. For instance, the exploitation practices for fruit production may evolve from collecting wild fruits in the forest, to fruit cultivation in enriched fallows and homegardens, to fruit production in orchards and corporate plantations (Verheij, 1991). Consequently, the major questions to consider in fruit tree domestication research concern not only what biological properties need to be adjusted, but also in what type of production system the trees should be cultivated.

In the past, it was often considered that the process of fruit tree domestication involves a linear process from collection of fruits in natural forests to cultivation of improved tree species in specialized tree production systems such as monocrop plantations. Such specialization makes it possible to limit competition by other crops and optimize commercial production, and thus to make the most efficient use of the improved tree species. As argued by Verheij (1991): 'for mixed production systems such as homegardens the strength lies in stability rather than peak performance. The key to a breakthrough in productivity of trees, that have been confined in homegardens because of low productivity in tree crops and hence have been less successful in migrating to more commercial cropping systems, has always been vegetative propagation of superior trees.' Gradually, however, this perspective has started to change as a result of the recognition that the process of domestication does not only concern adjustment of species to improved quality and yield, but also to social and environmental concerns (Leakey *et al.*, 2005). With respect to such concerns, it was gradually recognized that although monocultural tree plantations may provide optimal yields in the case of intensive and professional management, under other conditions they may have several limitations. In many tropical regions specialized forms of monocrop cultivation do not fit into the existing farming systems of predominantly smallholder farming. Especially in the case of less optimal production conditions, farming systems are often characterized by multifunctional production, which focuses on a range of subsistence and commercial products. Such technically non-optimal management may be related to the fact that the tree cultivator lacks the means to obtain all the necessary external inputs needed for intensive tree cultivation. Under such conditions, mixed production systems are often more relevant than highly specialized monocropping systems. Moreover, in case of non-optimal management, monocultural plantations may be subject to ecological hazards such as fluctuating weather conditions or attacks by pests and diseases. Multi-species systems make it possible to make use of synergistic ecological processes, such as: (i) circulation of nutrients as the result of the presence of filters against loss of nutrients; (ii) plant protection as a result of the presence of buffers against damaging agents such as pests and diseases; and (iii) protection by vegetative barriers against potentially degrading forces such as torrential rainfall, surface runoff or strong winds (Schroth *et al.*, 2004a).

As a consequence of the multiple dimensions of the domestication process and the range of social concerns related to production systems, several important questions have to be addressed in research on fruit tree domestication. First, attention should be given to the type of production system used to grow the tree (Leakey *et al.*, 1996) and what, sometimes multipurpose, role the tree is expected to fulfil in this system (Chuntanaparb and MacDicken, 1991; Roshetko and Evans, 1999). Secondly, attention needs to be given to how fruit tree production can best fit into the overall farming conditions of specific groups of cultivators. In view of the fact that fruit production systems are often specifically promoted as a contribution to poverty alleviation under marginal production conditions (Leakey *et al.*, 2005), the specific production requirements of smallholder producers must be considered (Raintree, 1991; Warner, 1995). And thirdly, attention needs to be given to the question of how the fruit tree cultivation system can best meet present environmental and ecological concerns (Schroth *et al.*, 2004a). In order to assess such questions under location-specific conditions, it is essential to have a good understanding of the process of change from the wild to domesticated state. Against this background, the aim of this chapter is to provide an overview of the process of domestication of fruit trees as a coevolutionary process between tree characteristics and production systems. The following questions will be discussed:

- What are the general characteristics of the process of plant domestication?
- What specific forms of coevolution between trees and production systems can be distinguished for fruit trees?
- What is the significance of 'intermediate' phases in this coevolutionary process?
- What conclusions can be drawn regarding new aspects needing attention in the process of domestication of fruit trees?

4.2 The Nature of Plant Domestication

The dichotomy between wild and domesticated species has a long history. In the past, this dichotomy has often been used by archaeologists, anthropologists and historians in a static sense to denote states of being. However, since the 19th century, biologists have started to use the term domestication as a dynamic term referring to a process rather than a state of existence. At present this dynamic interpretation of domestication is scientifically generally accepted (Harris, 1989; McKey *et al.*, 1993).

None the less, different interpretations of the concept of plant domestication still prevail (Wiersum, 1997a). This is not surprising in view of the fact that scientists from diverse disciplinary backgrounds, ranging from botany to anthropology, geography and agricultural sciences, have been involved in describing the process of plant domestication. Some scientists have defined domestication in a relatively restricted sense as a biological process, while others interpret it in a comprehensive sense as an acculturation process characterized by increasing human–plant interactions. These different interpretations can be related to two hierarchical levels. In the biological sense, domestication refers to the processes operating at species level: the cultivation and gradual adaptation of a species' morphological and genetic characteristics for specific uses as well as specific environments. Sometimes the concept of domestication is even restricted to the process of adaptation of the genetic make-up of a crop species. Cultivation in the sense of altering the location or growth habit of a crop may precede such domestication.

In a more comprehensive sense, the concept of domestication refers to processes operating at both species and agroecosystem level. In this interpretation the concept refers to the changes in the plant's morphological and genetic properties brought about by changes in exploitation and management practices. Concomitant with changes in the biological properties, changes in a plant's growing environment occur as well as a gradual intensification in cultivation practices. Thus, in its comprehensive sense, domestication is considered as a multidimensional process in which a progressively closer interaction between people and plant resources takes place (Table 4.1). The process of domestication can be considered as an evolutionary process from gathering to breeding, during which changes at the level of both the production system and of the plant species occur. Concomitantly, a progressively closer interaction takes place between the tree resources and people (Chase, 1989).

To date, most studies on domestication of fruit trees have concentrated on the scope for adjusting tree characteristics, with attention being given to development of improved propagation practices or selection and breeding of particular species (Leakey and Newton, 1994; Leakey *et al.*, 1996; Roshetko and Evans, 1999). This shows that many researchers interpret the process of domestication as involving a change in biological properties of trees rather than as a process of change in human–plant interactions. Other studies, however, have approached the process of domestication as essentially involving a process of change in exploitation systems (Michon and De Foresta, 1996; Wiersum, 1996; Michon, 2005). An important argument for considering the domestication of tree species as a coevolutionary process between species and production systems is the fact that tree species follow a different route into the human domain than arable crops such as cereals. Whereas the agricultural domestication process essentially requires 'open field' environmental conditions, the tree domestication process does not necessarily require such conditions. Due to the ecological nature of fruit trees as essentially forest species rather than 'open field' species, the route of tree domestication involves a much more gradual modification of production conditions (from forest to cultivated fields) than for agricultural crops (Michon and De Foresta, 1997; Wiersum, 1997a). Consequently, in considering the domestication process of fruit trees, it is important to conceptualize this process from an inclusive perspective, including

Table 4.1. Dimensions of crop domestication in the sense of a process of increasing people–plant interactions (Wiersum, 1997a).

1. Modification of a plant's biological characteristics

Human-induced change in a plant's morphological characteristics and genetic make-up

2. Modification (or artificialization *sensu* Michon and De Foresta, 1997) of the biophysical environment

Human-induced manipulation of biophysical environment in which a plant is growing in order to stimulate its production through:

- Homogenization of species composition by selective removal of non-valued species and stimulation of valued crops
- Control of pests and diseases
- Homogenization and enhancement of the physical growth conditions

3. Acculturation of a crop to a social management environment

Increased adaptation of crop species to specific uses

Incorporation of a species in a human-controlled production environment through the formulation of access rules for crop utilization and the formation of management organizations not only changes in tree characteristics, but also changes in forest systems. This coevolutionary process will be further elaborated in the next section.

As discussed above, the inclusive interpretation of domestication also involves the notion of a progressively closer interaction between tree resources and people. In this context, the process of domestication should not only be considered as involving changes in biological characteristics of species and their production systems, but also as involving changes in the fulfilment of human needs with respect to products and environmental services. I return to this dimension of the process of domestication in a later section.

4.3 Coevolution between Trees and Production Systems in the Process of Domestication

The concept of domestication as a coevolutionary process between trees and production systems is particularly apt when considering fruit tree production. Fruit tree exploitation normally starts in the forests, and increasing exploitation involves a gradual modification of forests by changing their composition and structure (Verheij, 1991; Wiersum, 1997b). This process starts with enriching forests with useful crops, often in the form of fruit species. The first phase of domestication of trees involves a process of concentration of naturally occurring useful tree resources in natural forests. In subsequent phases, new species may also be introduced in these forests. At first this will take the form of transplanting of wildlings or seeding of wild species. But as management intensity increases, selected varieties may be introduced as well (Table 4.2). As a result of these dynamics in exploitation systems, fruit trees may be grown in a variety of production systems (Wiersum, 1997a):

- *Natural forests in which wild fruit trees are protected*: specific areas or specific fruit tree species in natural forests that are favoured and protected because of their value for providing useful materials.
- *Resource-enriched natural forests* (Anderson, 1990): natural forests, either old-growth or fallow vegetation, whose composition has been altered by selective protection and incidental or purposeful propagule dispersion of fruit trees. Schroth *et al.* (2004b) characterized these production systems as 'permanent agroforests', which are continuously renovated in a small-scale pattern of replanting and spontaneous regeneration.
- *Reconstructed natural forests* (Anderson, 1990): (semi-)cultivated forest stands with several planted fruit tree species, tolerated or encouraged wild species of lesser value, and non-tree plants (herbs, lianas) composed of mainly wild species. Schroth *et al.* (2004b) characterized these production systems as 'cyclic agroforests', undergoing cycles of distinct management phases, including periodic replanting with agricultural intercropping. The term forest garden is used to refer to both resource-enriched and reconstructed natural forests (Wiersum, 2004).
- *Mixed arboriculture*: cultivated mixed stands, almost exclusively of planted, and often biologically modified, tree species. Typical examples are

Table 4.2. Phases in the process of exploitation of tree crops (Wiersum, 1997a).

Phase 1	
Procurement of wild tree products by gathering/collection	Gradual change from uncontrolled, open-access gathering of forest products to controlled gathering of wild tree products
Phase 2	
Initial production of wild tree products	Gradual change from systematic collection of wild tree products with protective tending of valued tree species to selective cultivation of valued trees by artificial *in situ* regeneration of native trees
Phase 3	
Cultivation of wild trees	Cultivation of selected native tree species in artificially established plantations
Phase 4	
Cultivation of fully domesticated trees	Cultivation of genetically selected or improved tree crops in intensively managed plantations

homegardens with a mixture of fruit trees, other tree crops and vegetables, or mixed smallholder plantations.

- *Mixed or monocultural fruit tree plantations*: cultivated mixed stands, almost exclusively of planted, and often biologically adjusted, fruit tree species. Often this concerns commercial plantations.

The different categories are not discrete: gradual transformation from one category to another may occur. Many of these systems have been developed through the creativity of local farmers (Wiersum, 1997b). The presence of the various indigenously developed fruit production systems has been extensively studied, notably in South-east Asia (Michon, 2005). Box 4.1 gives an example of an indigenous process of domestication of production systems in respect to the popular South-east Asian fruit tree, the durian (*Durio zibethinus*).

The locally evolved fruit tree production systems are often dynamic, as they gradually evolve in response to changing conditions. Such changes may involve a variety of ecological, socio-economic, cultural and political factors (Arnold and Dewees, 1995; Belcher *et al.*, 2005). The four most important changes which may affect the structure and species composition of fruit tree production are:

1. Changed ecological conditions, such as resource depletion or land degradation.
2. Changed technological conditions caused by the introduction of new agricultural and forestry techniques.
3. Changed economic conditions such as development of new markets and increased commercialization, changed demands for forest products and changed opportunities for off-farm employment.

Box 4.1. Co-domestication of trees and forests: the example of *Durio zibethinus*.

Durian (*Durio zibethinus* Murray) is a popular fruit in South-east Asia. Wild durian are found in Borneo and Sumatra, but as a cultivated species the tree has spread over a much larger area, ranging from Sri Lanka and South India to New Guinea. The tree is not only popular because of the fruit's unique taste, well liked by South-east-Asian people, but also because of its good production capacity. A major tree can produce 200–800 fruits a year and maintain such production over several generations. Fruits may be sold with common prices ranging from US$0.25 to US$4, making it one of the most profitable fruit trees in the region for local households (Goloubinoff and Hoshi, 2004). It has even been reported that the fruit is liked so much that rice harvesting suffers when it coincides with the durian harvest (Subhadrabandhu *et al.*, 1991). In its area of origin the fruit is still collected from the wild, and its cultivation mostly takes place in mixed cultivation systems such as forest and homegardens or on field boundaries of agricultural fields. It is only in Thailand that the tree is predominantly cultivated in orchards. Propagation is still mostly by seeds, but in Thailand clonal propagation methods are common (Subhadrabandhu *et al.*, 1991).

Because of its local popularity, durian has undergone a long process of domestication by local people, and only relatively recently has it been possible to state that '(professional) standardization of husbandry techniques is being achieved in durian orchards in Thailand' (Subhadrabandhu *et al.*, 1991). The process of domestication of durian by local people clearly reflects the notion of domestication as a coevolutionary process between trees and forests. On the one hand, trees were selected on the basis of their taste, resulting in the development of several local varieties. On the other hand, the extraction of fruits from wild plants was gradually replaced by cultivation in forest and homegardens. Within Indonesia, the various phases in this process of domestication are still represented:

1. The first step in the domestication of durian consists of the development of social measures for control over valuable tree species in the form of local rights of use of wild durian trees. Such tree tenure rights may relate either to individual trees or fruit tree groves; often they are maintained for several generations. The tenure rights are often complex (Peluso, 1996); they do not only pertain to rights to collect durian trees, but also to compensation for damaging trees, e.g. as a result of timber exploitation.
2. The next step in the process of domestication consists of the development of technical measures to stimulate *in situ* production (e.g. by removing competing vegetation) and natural regeneration (e.g. by protecting or transplanting wildlings or through seeding). These measures result in gradual enrichment of natural forests with durian trees and development of forest gardens; in such gardens other local fruit species are also maintained. As a result of the popularity of the durian, the species is not only maintained in forest gardens in its area of origin in Kalimantan and Sumatra (e.g. Aumeeruddy and Sansonnens, 1994; Salafsky, 1995; Marjokorpi and Ruokolainen, 2003), but has also been incorporated in forest gardens on other islands such as Sulawesi (Brodbeck *et al.*, 2003) and Maluku (Kaya *et al.*, 2002).
3. Further intensification in the process of domestication entails the incorporation in home gardening systems such as on the island of Java (e.g. Dury *et al.*, 1996). In rural areas such homegardens often consist of a mixture of trees, vegetables and other useful

plants. However, in case of good commercial prospects near cities, they may be gradually transferred into mixed fruit-tree plantations (Verheij, 1991).

The various types of durian exploitation systems are not just production systems, but also have important social functions. The development of access and use rights to wild trees forms a component in the territorialization strategies of local communities and households (Peluso, 1996). In areas with more intensive durian production, the guarding and collection of ripened fruits is often a joint activity of the tree owner, relatives and friends (Goloubinoff and Hoshi, 2004). In West Java, the trees serve as a form of financial security for owners who may pawn trees from homegardens (Dury *et al.*, 1996). These examples demonstrate that the process of domestication does not only involve a change in tree characteristics and production systems, but also the incorporation in socio-economic networks.

Although the incorporation of the durian in forest and homegardens illustrates its popularity in mixed production systems of multi-enterprise households, horticulturists have claimed that the trees usually have a low productivity, which keeps up the price and limits consumption. This is related to the irregular fruiting cycle; it can be improved through manipulation of the growth rhythm of the tree, thus extending the harvest season (Subhadrabandhu *et al.*, 1991). This production-oriented view on the need for further specialization in cultivation is in accordance with the view that 'in a typical home gardening situation, where everybody grows fruit but nobody is a fruit grower, the traditional expertise is dispersed in the community and much of it is latent, i.e. not put to use ... It is difficult to take stock of local knowledge as long as it has not yet been accumulated by professional growers' (Verheij, 1991).

However, several authors have recently suggested that durian production in mixed multistoreyed cropping systems is a promising multifunctional production system for meeting several newly arising concerns. Four major arguments have been brought forward for considering such production systems as being adjusted to both social, environmental and production demands rather than only forming an 'intermediate' position in the process of co-domestication of forests and trees:

1. The mixed production systems, yielding a multitude of products for use in the household and generating cash income, have an important role as a 'safety net' in times of hardship and hence play an important role in the coping strategies of rural people, contributing to livelihood security (Brodbeck *et al.*, 2003). Thus, these systems can play an important role in programmes for poverty alleviation.

2. Rather than considering that the management of forest and homegardens is based on a low degree of professional knowledge of fruit production, it can be considered that it represents a profound knowledge of maintaining multi-species production systems. For instance, Salafsky (1995) demonstrated that durian producers have accurate perceptions as to how ecological factors influence durian production in forest gardens.

3. The forest garden systems can play an important role in the conservation of biodiversity (Marjokorpi and Ruokolainen, 2003). Consequently, they could play an important role in the management of buffer zones around protection areas (Aumeeruddy and Sansonnens, 1994).

4. Establishment of multi-stratum fruit-based systems incorporating durian can contribute towards the sequestration of carbon dioxide, while providing a positive financial return (Ginoga *et al.*, 2002).

4. Changed socio-political conditions, e.g. population growth and migration, increased interaction with other (ethnic) groups, changed tenure conditions including gradual privatization or nationalization of forest lands, development of state organizations for forest management and rural development.

These changes increase pressure on forest and tree resources. In several cases this has resulted in deforestation and forest degradation, but in other cases farmers have reacted by modifying their management strategies, e.g. by intensifying or reducing the cultivation of valuable tree species (Belsky, 1993; Dove, 1994; Arnold and Dewees, 1995). Recently, it has been noted that in several forest and homegarden systems such intensification has resulted in a gradual transfer from mixed-species multistoreyed cropping systems towards more uniform systems (Belcher *et al.*, 2005). However, the process of change in these systems is not always uniform, and detailed studies have shown that whereas some smallholder cultivators gradually change their 'intermediate' fruit tree production systems towards more specialized arboricultural practices, other smallholders maintain these systems, although they modify the species composition and/or specific management practices (Peyre *et al.*, 2006).

The dynamics of the fruit exploitation systems demonstrate that many local communities have been actively engaged in domesticating tree species as well as production systems, and gradually adapting these to their household needs. Berkes *et al.* (2000) have characterized such locally evolved coevolutionary processes as an excellent example of adaptive management. It may be suggested that research to stimulate domestication of fruit tree species as a context-specific process, rather than a standard biological process, will be most successful if it is built upon such locally evolved domestication processes (Wiersum, 1996). An intriguing question then becomes what the future scope is for the 'intermediate' production systems such as forest and homegardens.

4.4 Forest Gardens as a Target for Domestication of Fruit Trees

In the past, in research programmes on tree production, two common options were normally considered: improved methods for harvesting from natural stocks or domestication in specialized plantations. Recently, it has been argued that more attention should be given to the scope for conservation and development of 'intermediate' tree production systems such as forest gardens. Rather than considering such systems as a temporary development stage in a linear process of change from exploitation of tree resources from natural vegetation to modern tree production systems, they should be considered as a true domestication of forest ecosystems (Michon and De Foresta, 1997). This opinion mirrors the increasing attention being given to innovative tree production systems that are both multifunctional and ecologically responsible, and that can contribute to biodiversity conservation (MacNeely and Scherr, 2003). Several recent concerns may have a positive impact on the social appreciation of forest gardens as a fruit tree production system that meets current human requirements (Wiersum, 2004; Michon, 2005):

- Concerns about the need to develop new approaches towards conservation of tropical forests and biodiversity. As a result of the high rates of deforestation in tropical countries, increasingly not only the need for conservation of natural forests, but also the need for conservation and sustainable management of secondary forests derived from human interaction with forests is acknowledged. Forest gardens are increasingly recognized as an important anthropogenic forest types. These land-use systems do not only offer scope for *in situ* conservation of naturally evolved biodiversity, but also for *in domo* conservation of anthropogenically derived forms of biodiversity (Attah-Krah *et al.*, 2004; MacNeely, 2004; Wiersum, 2004).
- Concerns about the need to stimulate multifunctional production systems. In the past, fruit production was scientifically conceived of as a specialized horticultural activity without connection to silvicultural practices aimed at timber production. It is now recognized, however, that fruit trees often provide important amounts of fuelwood from prunings, and increasingly, fruit trees are also being used for timber production (Durst *et al.*, 2004). Moreover, in response to environmental concerns, increased attention is also given to the possibility of fruit tree production systems contributing to carbon sequestration (Montagnini and Nair, 2004).
- Within the context of forest conservation much attention is given to the need to decentralize forest management and to stimulate greater community involvement. Often, such community-level forest management will not be based on the same degree of organizational specialization as was the case in professional forest management. It can be expected that the development of community forestry will bring with it new interests in the role of forests within the local livelihood systems, including options for provision of non-timber forest products. The forest gardens offer a good example of what such integration could entail.
- At present it is also increasingly appreciated that the approach of transfer of scientifically developed technological innovations, which dominated agricultural and forestry development in the past, has reached the limits of its applicability, and that new approaches towards land-use development are needed, based on endogenous instead of exogenous innovations. This has brought with it increased attention to the scope for the application of principles embedded in indigenous land-use management systems (MacNeely and Scherr, 2003).

As illustrated by the case of durian (Box 4.1), forest gardens are excellent demonstrations of fruit tree production systems which transcend the traditional nature–culture dichotomy as well as the disciplinary distinction between horticulture and forestry. Hence, they provide a valuable example that can be used in the search for new production systems balancing productive, ecological and social concerns. Moreover, they provide scope for identification of research questions to be considered when developing new ways to achieve a more integrated approach to domestication as an inclusive process. They also suggest new research questions in respect of the ecological principles of plant domestication. For instance, how biodiversity impacts on production conditions (Siebert, 2002; Schroth *et al.*, 2004b) or how the processes of synergy and competition between plants can be balanced (Wiersum, 2004). Another interesting question is whether and how the vegetation and species composition of the forest gardens could be further adapted to meet production requirements. This might require a new approach to selecting and breeding high-yielding tree varieties, which can be cultivated in the forest rather than in a plantation environment (Wibawa *et al.*, 2005).

4.5 Conclusions

The domestication of a tree species involves the manipulation and cultivation of trees for specific uses. Domestication of fruit trees is a multifaceted process in which a progressively closer interaction between people and trees takes place. This process consists of three dimensions:

- *Acculturation*: the incorporation of trees in an increasingly complex social environment through the formation of management entities and the formulation of objectives as well as rules for tree utilization and management.
- *Modification of the biophysical environment*: the protection and stimulation of production of useful trees in natural forests, the (partial) clearing of natural forests followed by the planting of useful wild species, as well as the manipulation of the biophysical environment to stimulate production of the valuable tree species.
- *Modification of a tree's biological characteristics*: the manipulation of its morphological and genetic characteristics.

The utilization of tree species usually starts with the exploitation of trees from natural forests. Gradually, uncontrolled utilization of the wild tree products is changed to controlled exploitation. Subsequently, wild trees may be purposefully cultivated in either resource-enriched (fallow) vegetation systems or in (indigenous) multistoreyed production systems such as forest gardens. The cultivation of selected varieties of tree in either mixed tree plantations or commercial tree-crop plantations is the last phase of this domestication process. Each of these phases is characterized by specific conditions with respect to social production conditions and management objectives as well as ecological conditions and requirements for biological characteristics of fruit trees.

The recognition of these various stages of the domestication process is of importance for understanding the scope for domestication of fruit trees. It allows better understanding of the various options for fruit tree production under different land-use conditions in the continuum between natural forests and commercial plantation with high-yielding cultivars. It illustrates that efforts to domesticate fruit trees should focus on integrating biological and ecological considerations with social aspirations, and that it should increasingly focus on addressing a variety of production systems rather than assuming one ideal typical production system. It also allows the identification of new approaches towards plant domestication in response to newly emerging social demands regarding multifunctional rather than specialized production systems. In response to several newly emerging considerations the domestication of fruit trees should be focused on more aspects than productivity only. Consequently, the traditional idea of domestication involving mainly a fundamental change towards vegetative propagation of superior trees for production in orchards needs to be adjusted, and much more attention should be given to the development of multifunctional fruit production systems appropriate to present-day social and environmental concerns. The 'intermediate' systems, ranging from wild forests modified for increased fruit production to anthropogenic forests with fruit species growing in a biodiverse multistoreyed production system, can be considered as good models of domesticated forests, combining productivity for multi-enterprise farmers and biodiversity conservation, and as providing a good basis for novel approaches towards a more integrated approach to domestication.

References

Anderson, A.B. (1990) Extraction and forest management by rural inhabitants in the Amazon estuary. In: Anderson, A.B. (ed.) *Alternatives to Deforestation: Steps Towards Sustainable Use of the Amazon Rain Forest.* Columbia University Press, New York, pp. 65–85.

Arnold, J.E.M. and Dewees, P.A. (eds) (1995) *Tree Management in Farmer Strategies: Responses to Agricultural Intensification.* Oxford University Press, Oxford, 292 pp.

Attah-Krah, K., Kindt, R., Skilton, J.N. and Amaral, W. (2004) Managing biological and genetic diversity in tropical agroforestry. *Agroforestry Systems* 61, 183–194.

Aumeeruddy, Y. and Sansonnens, B. (1994) Shifting from simple to complex agroforestry systems: an example for buffer zone management from Kerinci (Sumatra, Indonesia). *Agroforestry Systems* 28, 113–141.

Belcher, B., Michon, G., Angelsen, A., Ruiz-Perez, M. and Asbjornsen, H. (2005) The socioeconomic conditions determining the development, persistence, and decline of forest garden systems. *Economic Botany* 59, 245–253.

Belsky, J.M. (1993) Household food security, farm trees, and agroforestry: a comparative study in Indonesia and the Philippines. *Human Organization* 52, 130–141.

Berkes, F., Colding, J. and Folke, C. (2000) Rediscovery of traditional ecological knowledge as adaptive management. *Ecological Applications* 19, 1251–1262.

Brodbeck, F., Hapla, F. and Mitlöhner, R. (2003) *Traditional forest gardens as 'safety net' for rural households in Central Sulawesi, Indonesia.* Paper presented at International Conference on Rural Livelihoods, Forests and Biodiversity, Bonn, Germany, 19–23 May 2003.

Chase, A.K. (1989) Domestication and domiculture in northern Australia: a social perspective. In: Harris, D.R. and Hillman, G.C. (eds) *Foraging and Farming: The Evolution of Plant Exploitation.* Unwin Hyman, London, pp. 42–54.

Chuntanaparb, L. and MacDicken, K.G. (1991) Tree selection and improvement in agroforestry. In: Avery, M.E., Cannell, M.G.R. and Ong, C. (eds) *Biophysical Research for Asian Agroforestry.* Winrock International, USA and South Asia Books, USA, pp. 41–57.

Dove, M.R. (1994) Transition from native forest rubbers to *Hevea brasiliensis* (Euphorbiaceae) among tribal smallholders in Borneo. *Economic Botany* 48, 382–396.

Durst, P.B., Killman, W. and Brown, C. (2004)

Asia's new woods. *Journal of Forestry* 102, 46–53.
Dury, S., Vilcosqui, L. and Mary, F. (1996) Durian trees (*Durio zibethinus* Murr.) in Javanese home gardens: their importance in informal financial systems. *Agroforestry Systems* 33, 215–230.
FAO (1999) *Use and Potential of Wild Plants in Farm Households.* FAO Farm System Management Series No. 15, Food and Agriculture Organization, Rome.
Ginoga, K., Cahya Wulan, Y. and Lugina, M. (2002) *Potential of Agroforestry and Plantation Systems in Indonesia for Carbon Stocks: an Economic Perspective.* ACIAR Project ASEM 2002/066 Working Paper CC14. Center for Socio-Economic Research in Forestry, Bogor, Indonesia, 23 pp.
Goloubinoff, M. and Hoshi, R.S. (2004) Durian. In: Lopez, C. and Shanley, P. (eds) *Riches of the Forest: Food, Spices, Crafts and Resins of Asia.* Center for International Forestry Research, Bogor, Indonesia, pp. 21–24.
Harris, D.R. (1989) An evolutionary continuum of people–plant interaction. In: Harris, D.R. and Hillman, G.C. (eds) *Foraging and Farming: The Evolution of Plant Exploitation.* Unwin Hyman, London, pp. 11–26.
Kaya, M., Kammesheidt, L. and Weidelt, H.J. (2002) The forest garden system of Saparua island, Central Maluku, Indonesia, and its role in maintaining tree species diversity. *Agroforestry Systems* 54, 225–234.
Leakey, R.R.B. and Newton, A.C. (eds) (1994) *Domestication of Tropical Trees for Timber and Non-timber Products.* MAB Digest 17. UNESCO, Paris.
Leakey, R.R.B., Temu, A.B., Melnyk, M. and Vantomme, P. (eds) (1996) *Domestication and Commercialization of Non-timber Forest Products in Agroforestry Systems.* Non-wood Forest Products No. 9. Food and Agriculture Organization, Rome.
Leakey, R.R.B., Tchoundjeu, Z., Schreckenberg, S., Shackleton, S.E. and Shackleton, C.M. (2005) Agroforestry tree products (AFTPs): targeting poverty reduction and enhanced livelihoods. *International Journal of Agricultural Sustainability* 3, 1–23.
MacNeely, J.A. (2004) Nature or nurture: managing relationships between forests, agroforestry and wild biodiversity. *Agroforestry Systems* 61, 155–165.
MacNeely, J.A. and Scherr, S. (2003) *Ecoagriculture: Strategies to Feed the World and Save Biodiversity.* Island Press, Washington, DC, 323 pp.
Marjokorpi, A. and Ruokolainen, K. (2003) The role of traditional forest gardens in the conservation of tree species in West Kalimantan, Indonesia. *Agroforestry Systems* 12, 799–822.
McKey, D., Linares, O.F., Clement, C.R. and Hladik, C.M. (1993) Evolution and history of tropical forests in relation to food availability: background. In: Hladik, C.M., Hladik, A., Linares, O.F., Pagezy, H., Semple, A. and Hadley, M. (eds) *Tropical Forests, People and Food: Biocultural Interactions and Application to Development.* Man and Biosphere Series 13. UNESCO and Parthenon Publishers, New York, pp. 17–24.
Michon, G. (2005) *Domesticating Forests: How Farmers Manage Forest Resources.* Center for International Forestry Research, Bogor, Indonesia, and World Agroforestry Centre, Nairobi, Kenya, 187 pp.
Michon, G. and De Foresta, H. (1996) Agroforests as alternative to pure plantations for the domestication and commercialization of NTFPs. In: Leakey, R.R.B., Temu, A.B., Melnyk, M. and Vantomme, P. (eds) *Domestication and Commercialization of Non-timber Forest Products in Agroforestry Systems.* Non-wood Forest Products No. 9, Food and Agriculture Organization, Rome, pp. 160–175.
Michon, G. and De Foresta, H. (1997) Agroforests: pre-domestication of forest trees or true domestication of forest ecosystems? *Netherlands Journal of Agricultural Science* 45, 451–462.
Montagnini, F. and Nair, P.K.R. (2004) Carbon sequestration: an underexploited environmental benefit in agroforestry systems. *Agroforestry Systems* 61, 281–295.
Peluso, N. (1996) Fruit trees and family trees in an anthropogenic forest: ethics of access, property zones and environmental change

in Indonesia. *Comparative Studies in Society and History* 38, 510–548.

Peyre, A., Guidal, A., Wiersum, K.F. and Bongers, F. (2006) Dynamics of homegarden structure and function in Kerala, India. *Agroforestry Systems* 66, 101–115.

Raintree, J.B. (1991) *Socioeconomic Attributes of Trees and Tree Planting Practices.* Community Forestry Note No. 9. Food and Agriculture Organization, Rome.

Roshetko, J.M. and Evans, D.O. (eds) (1999) *Domestication of Agroforestry Trees in Southeast Asia: Proceedings of a Regional Workshop held in Yogyakarta, Indonesia.* Forest, Farm, and Community Tree Research Reports, Special Issue. Taiwan Forestry Research Institute, Taiwan, and Winrock International, Morrilton, Arkansas.

Salafsky, N. (1995) Ecological factors affecting durian production in the forest gardens of West Kalimantan, Indonesia. *Agroforestry Systems* 32, 63–79.

Schroth, G., da Fonseca, G.A.B., Harvey, C.A., Gascon, C., Vasconcelos, H.L. and Izac, A.-M.N. (2004a) *Agroforestry and Biodiversity Conservation in Tropical Landscapes.* Island Press, Washington, DC, 523 pp.

Schroth, G., Harvey, C.A. and Vincent, G. (2004b) Complex agroforests: their structure, diversity, and potential role in landscape conservation. In: Schroth, G., da Fonseca, G.A.B., Harvey, C.A., Gascon, C., Vasconcelos, H.L. and Izac, A.-M.N. (eds) *Agroforestry and Biodiversity Conservation in Tropical Landscapes.* Island Press, Washington, DC, pp. 227–260.

Siebert, S.F. (2002) From shade- to sun-grown perennial crops in Sulawesi, Indonesia: implications for biodiversity conservation and soil fertility. *Biodiversity and Conservation* 11, 1889–1902.

Subhadrabandhu, S., Schneeman, J.M.P. and Verheij, E.W.M. (1991) *Durio zibethinus* Murray. In: Verheij, E.W.M. and Coronel, R.E. (eds) *Edible Fruits and Nuts.* Plant Resources of South-East Asia No. 2. Pudoc, Wageningen, The Netherlands, pp. 157–161.

Verheij, E.W.M. (1991) Introduction. In: Verheij, E.W.M. and Coronel, R.E. (eds) *Edible Fruits and Nuts.* Plant Resources of South-East Asia No. 2. Pudoc, Wageningen, The Netherlands, pp. 15–56.

Warner, K. (1995) *Selecting Tree Species on the Basis of Community Needs.* Community Forestry Field Manual No. 5. Food and Agriculture Organization, Rome.

Wibawa, G., Hendrato, S. and van Noordwijk, M. (2005) Permanent smallholder rubber agroforestry systems in Sumatra. In: Palm, C.A., Vost, S.A., Sanchez, P.A. and Ericksen, P.J. (eds) *Slash-and-burn Agriculture: The Search for Alternatives.* Columbia University Press, New York, pp. 222–232.

Wiersum, K.F. (1996) Domestication of valuable tree species in agroforestry systems: evolutionary stages from gathering to breeding. In: Leakey, R.R.B., Temu, A.B., Melnyk, M. and Vantomme, P. (eds) *Domestication and Commercialization of Non-timber Forest Products in Agroforestry Systems.* Non-wood Forest Products No. 9. Food and Agriculture Organization, Rome, pp. 147–158.

Wiersum, K.F. (1997a) From natural forest to tree crops: co-domestication of forests and tree species – an overview. *Netherlands Journal of Agricultural Science* 45, 425–438.

Wiersum, K.F. (1997b) Indigenous exploitation and management of tropical forest resources: an evolutionary continuum in forest–people interaction. *Agriculture, Ecosystems and Environment* 63, 1–16.

Wiersum, K.F. (2004) Forest gardens as an 'intermediate' land-use system in the nature–culture continuum: characteristics and future potential. *Agroforestry Systems* 61, 123–134.

5 Homegarden-based Indigenous Fruit Tree Production in Peninsular India

B.M. KUMAR

College of Forestry, Kerala Agricultural University, Kerala, India

5.1 Introduction

Tropical homegardens are 'intimate, multi-storey combinations of various trees and crops, sometimes in association with domestic animals, around homesteads' (Kumar and Nair, 2004, p. 135). They are the oldest land-use activity in the world after shifting cultivation. Although as a land-use system, homegarden cultivation is most highly evolved in Java (Indonesia) and Kerala (India), its coverage is almost pan-tropical, sometimes extending into the Mediterranean and temperate regions (Nair and Kumar, 2006). The notion of homegardens as a component of integrated farming systems, which include fields for staple food production, is also found in the literature (Tesfaye *et al.*, 2006). In this respect, homegarden production is mostly supplementary and mainly concentrates on vegetables, fruits and condiments, with fuelwood and small timber as a profitable sideline. Although homegarden research in the past has concentrated on certain functional groups such as medicinal plants (Rao and Rao, 2006), phytofuels (Shanavas and Kumar, 2003) and timber-yielding species (Kumar *et al.*, 1994; Shanavas and Kumar, 2006), aspects relating to the production of native fruits have not been adequately addressed, despite its economic importance. In this chapter, an attempt is made to summarize the information on indigenous fruit trees in tropical homegardens with special reference to Kerala State in Peninsular India.

5.2 Tropical Homegardens as Prehistoric Loci for Fruit Tree Domestication

Farmers in many traditional cultures have been domesticating fruit trees and other agricultural crops in areas around their dwellings for millennia, primarily to meet their subsistence needs. This early domestication process may have started as a spontaneous growth of plants from the remnants of products brought to the camps by the hunter-gatherers. Anderson (1952) described it as the 'dump-heap' or incidental route to domestication. In this route, the seeds or vegetative propagules of edible fruit trees and other useful species collected from the forest by early man were discarded near the dwellings, where they germinated and grew. The sites around habitations provided an environment conducive to the survival of such regenerants. Apparently, fruit-tree domestication in South Asia followed this pattern, although little documented information exists. The recognition and management of such 'volunteers' would have been the next step along the road to domestication. The domestication of fruit trees may also have coincided with that of root crops, as the hunter-gatherers used to collect both fruits and tubers from the wild.

Slowly, however, the unintentional dissemination of seeds became more purposeful, with important species being planted to ensure their utilization (Wiersum, 2006). The prehistoric people may also have instinctively selected trees with larger fruit size, better quality or other desirable features from the wild, besides supporting their regeneration. This, in turn, resulted in the cultivated populations becoming genetically distinct from their wild progenitors (Ladizinsky, 1998). However, only a few studies have addressed the question of genetic diversity of fruit trees in the South and South-east Asian homegardens.

5.2.1 Archaeological and literary evidence available from India

Although scientific evidence for homegarden-based tree domestication in Peninsular India is hard to find, the ancient literature and evidence from archaeological excavations provide some indications about the diversity of such plants around the homesteads. In general, they corroborate the idea that fruit-tree domestication started around the settlements in the prehistoric period. The earliest evidence on this probably dates back to the Mesolithic period (between 10,000 and 4000 years ago), when fruits of 63 plants including *Aegle marmelos*, *Buchanania lanzan*, *Emblica officinalis*, *Mangifera indica*, *Ficus* spp., *Madhuca* spp. and *Ziziphus* spp. were reportedly eaten raw, roasted or pickled by the inhabitants of Madhya Pradesh in central India (Randhawa, 1980). Subsequently, Emperor Ashoka, a great Indian ruler (273–232 BC), encouraged a system of arbori-horticulture of plantains (*Musa* spp.), mango (*Mangifera indica*), jackfruit (*Artocarpus heterophyllus*) and grapes (*Vitis* spp.). Confirming the predominance of homegardens in ancient India, Vatsyayana wrote in his great book of Hindu aesthetics, the *Kama Sutra* (composed between AD 300 and AD 400), that the housegardens were a significant source of fruits and vegetables (cf. Randhawa, 1980). The travelogue of Ibn Battuta, a Persian traveller (AD 1325–1354), provides the earliest literary evidence from Peninsular India; it mentions that in the densely populated and intensively cultivated landscapes of the Malabar Coast of Kerala, coconut (*Cocos nucifera*) and black pepper (*Piper nigrum*) were prominent around the houses (Randhawa, 1980).

5.3 Homegardens as Laboratories for Crop Evolution and Diversity

The continuing inventive spirit of the prehistoric gardeners, who nurtured the volunteers and deliberately planted trees around their homesteads, probably makes the present-day homegardeners 'perpetual experimenters' as well. These garden owners constantly try out and test new species and varieties, including methods to improve their management. Coincidentally, both naturally occurring wild plants and deliberately introduced plants abound in these gardens (Kumar *et al.*, 1994). A new species may be chosen because of its properties (i.e. food, wood, medicinal, religious or ornamental), based on intuition or information passed on by neighbours and relatives. The suite of homegarden species therefore contains diverse functional groups such as fruit trees, vegetables, medicinal plants, ornamental plants, and tall trees for shade, timber, fruits, nuts and resins.

Women play a significant and special role in the acquisition of new species in the Kerala homegardens, in common with some other regions (e.g. Yamada and Osaqui, 2006). The seeds may be gathered from other homegardens, from wild populations, or from cultivated areas. Friends, relatives, neighbours, workers and visitors also contribute plant species as gifts, or give them in exchange for other species. A more recent phenomenon is the procurement of planting materials of medicinal/ornamental plants, fruits and nuts from government farms and/or commercial nurseries.

The choice of species and planting techniques reflects the accrued wisdom and insights of people who have interacted with the environment for generations and have made homegardens a principal hub of crop evolution and diversity. It is also reasonable to assume that indigenous cultivators use rational ecological approaches to manipulate the plants, and this may confer sustainability on the system. Homegardening thus constitutes a unique land-use activity that not only integrates the key processes of tree domestication, such as identification, production, management and adoption of agroforestry tree genetic resources (*sensu* Leakey and Tchoundjeu, 2001), but also mirrors a substantial indigenous knowledge base. None the less, agroforestry tree domestication is regarded as a farmer-driven, market-led process, focusing on conventional timber-tree improvement, which emerged in the early 1990s (Simons and Leakey, 2004). The result is that the remarkable breakthroughs attained by conventional gardeners over generations, largely through uncoordinated activities, are often overlooked. Furthermore, agricultural transformations brought about by market economies in the recent past, especially the incorporation of exotic commercial crops (e.g. the rubber tree, *Hevea brasiliensis*) has led to the destruction of some of these traditional gardens (Kumar and Nair, 2004).

5.4 Fruit-tree Diversity in Kerala Homegardens

While progress in modern agriculture has been through monospecific stands, sometimes described as 'biological deserts' of low species diversity, the tropical homegardens are spectacular examples of fruit-tree richness and diversity. Almost all types of fruit crops, ranging from tropical evergreen to temperate deciduous types, are grown. Overall, 69 fruit- and nut-yielding trees accounted for more than 30% of all tree species observed in field surveys conducted in Kerala (Kumar *et al.*, 1994; Patil, 2005). Prominent examples include mango (*Mangifera indica*) – the 'king of fruits' – and many underutilized fruits such as jackfruit (*Artocarpus heterophyllus*), aonla or Indian gooseberry (*Emblica officinalis*), bael (*Aegle marmelos*), custard apple (*Annona squamosa*), jamun (*Syzygium cuminii*), karonda (*Carissa congesta*) and tamarind (*Tamarindus indica*). Many of these have good nutritive value and attract considerable local demand (Pareek and Sharma, 1993). Exotic fruits such as papaya (*Carica papaya*), guava (*Psidium guajava*), pomegranate (*Punica granatum*) and cashew (*Anacardium occidentale*) are also important. The relative proportions of individual species found within the homegardens, however, are variable (Table 5.1).

Each homegarden owes its unique appearance to its history, species diversity and physical arrangements. None the less, most of the gardeners included in the surveys showed a propensity to grow fruit and nut trees. Therefore it is not surprising to find each garden characterized by the presence of tall native and exotic fruit trees such as coconut, mango, jackfruit, guava and plantains. This, however, does not imply that the fruit-tree composition of all homegardens within a region is alike or static. Indeed, the species assemblages are strongly influenced by the specific needs and preferences of each household, including the tastes of household members, food culture and local customs, nutritional complementarity with other major food sources, and ecological and socio-economic factors such as market forces, policies and local development projects (Kumar and Nair, 2004). These factors probably make the fruit-tree composition of the homegarden a transient phenomenon.

5.4.1 Temporal dynamics in major homegarden components

Table 5.1 summarizes the macro-level changes in major fruit trees (both indigenous and exotic) of Kerala homegardens over a 10-year period between 1992–1995 and 2002–2004. Although the data presented are not strictly comparable because of the disparate nature of the sampling units and intensities, the number of fruit-tree taxa in the homegardens has apparently increased over time. For example, 29 additional species were encountered in the 2002–2004 survey over the previous period, and another 24 species increased their presence. None the less, for as many as 41 of the 69 fruit and nut trees listed in Table 5.1, the increases in relative proportion of homegardens were rather modest, while one decreased, and data on three species are not available. The implicit temporal variations in garden composition probably reflect the processes of rural transformations that influence homegarden structure, composition and dynamics (Peyre *et al.*, 2006; Wiersum, 2006). Kumar and Nair (2004) summarized the work on homegarden variability and reported that both temporal and spatial variations in species diversity and plant density are probable, even within the same geographical/eco-climatic region.

Table 5.1. Frequency of occurrence of fruit and nut trees in the homegardens of Kerala.

Species		Frequency (%)			
		Mid-1990s[a]			
Botanical name	Common name	Mean	Range[d]	Mid-2000s[b]	Change[c]
Indigenous					
Aegle marmelos (L.) Corr	Bael fruit	7.5	3.7–26.7	5	0
Alangium salviifolium (L. f.) Wang.	Sage-leaved alangium	0	0	2	0
Areca catechu L.	Arecanut	55.3	16.7–93.3	79	+
Artocarpus gomezianus Wall. ex Trecul	Monkey jack	0	0	6	0
Artocarpus heterophyllus Lam.	Jackfruit	69.8	30.4–100	95	+
Artocarpus hirsutus Lam.	Wild jack	30.6	3.8–79.5	44	+
Azadirachta indica A. Juss.	Neem	0	0	22	+
Borassus flabellifer L.	Palmyra palm	12.5	2.5–61.5	2	–
Calophyllum inophyllum L.	Alexandrian laurel	9.2	2.5–13.6	4	0
Carissa carandas L.	Karaunda	0	0	3	0
Chrysophyllum lanceolatum (Blume) A. DC.	Star apple	0	0	6	0
Citrus spp.		14.4	1.7–33.3	47	+
Cocos nucifera L.	Coconut	92.2	46.2–100	98	0
Emblica officinalis Gaertn.	Indian gooseberry	15.9	2.3–83.9	30	+
Ficus racemosa L. Guler.	Cluster fig	0	0	9	0
Flacourtia montana Grah.	Charalpazham	0	0	9	0
Garcinia gummi-gutta (L.) Robs.	Kudampuli	15.6	3.2–37	37	+
Garcinia indica (Thouars) Choisy	Kokkam butter	1.8	1.7–1.9	0	0
Grewia tiliifolia Vahl.	Chadachi	9.1	6.7–13.3	7	0
Hydnocarpus pentandra (Buch.-Ham.) Oken	Marotti	11.3	1.7–40	4	0
Madhuca longifolia (Koenig) J.F. Macbr.	Mahua tree	0	0	1	0
Mangifera indica L.	Mango	75.6	35.3–100	90	+
Mimusops elengi L.	Spanish cherry	0	0	11	+
Moringa oleifera Bedd.	Drumstick	15.3	3.3–36.2	73	+
Morus alba L.	Mulberry	10.2	1.7–33.3	8	0
Myristica malabarica Lam.	False nutmeg	0	0	4	0

Musa spp.	Plantains/banana	55	13.3–100		*
Prunus ceylanica (Wight) Miq.	Saberjelli	0	0	1	0
Sapindus emarginatus Vahl.	Soapnut tree	7.5	2.5–20	2	0
Semecarpus anacardium L.f.	Marking-nut tree	6.7	0	0	0
Spondias pinnata (L.f.) Kurz.	Indian hog-plum	12	3.2–40	19	0
Syzygium cuminii (L.) Skeels.	Malabar plum	7.7	2.5–21.4	17	0
Syzygium jambos (L.) Alston.	Malabar plum	15	12.1–20	30	+
Terminalia bellirica (Gaertn.) Roxb.	Belliric myrobalan	8.2	2.5–13.3	8	0
Exotic					
Anacardium occidentale L.	Cashew	35.8	6.7–93.3	59	+
Annona muricata L.	Soursop	5	2.3–7.5	4	0
Annona reticulata L.	Bullock's heart	4.7	2.3–6.7	26	+
Annona squamosa L.	Custard apple	16	6.7–27.5	14	0
Artocarpus communis J.R. & G. Forst.	Breadfruit	15.2	6.5–37	41	+
Averrhoa bilimbi L.	Bilimbi	13.1	1.7–31.3	26	+
Averrhoa carambola L.	Carambola	0	0	1	0
Bixa orellana L.	Annatto tree	7.5	1.7–13.3	1	0
Carica papaya L.	Papaya	19.8	5.2–66.7	71	+
Chrysophyllum cainito L.	Star apple	0	0	6	0
Cinnamomum camphora (L.) Presl.	Camphor tree	0	0	16	+
Citrus grandis (L.) Osbeck.	Pummelo	0	0	9	0
Citrus reticulata Blanco	Mandarin orange	0	0	1	0
Citrus sinensis (L.) Osbeck.	Thick-skinned orange	0	0	5	0
Coffea arabica	Coffee	8.2	2.5–20	–	*
Elaeis guineensis Jacq.	African oil palm	0	0	1	0
Eugenia uniflora L.	Surinam cherry	0	0	2	0
Ficus carica L.	Anjur	0	0	1	0
Garcinia mangostana L.	Mangosteen	0	0	7	0
Malpighia punicifolia L.	West Indian cherry	0	0	2	0
Manilkara zapota (L.) P. Royan	Sapota	12	2.5–21.4	24	+
Myristica fragrans Houtt.	Nutmeg	12.7	1.9–33.3	27	+

Continued

Table 5.1. *Continued*

Species		Frequency (%)			
		Mid-1990s[a]			
Botanical name	Common name	Mean	Range[d]	Mid-2000s[b]	Change[c]
Persea americana Mill.	Avocado	0	0	3	0
Phyllanthus acidus (L.) Skeels	Star gooseberry	13.9	3.4–25	14	0
Pimenta officinalis Lindl.	Allspice	0	0	14	+
Pithecellobium dulce (Roxb.) Benth.	Bread and cheese tree	0	0	1	0
Pouteria campechiana (Kunth.) Baehni	Egg fruit	0	0	17	+
Psidium guajava L.	Guava	31.1	6.7–70	84	+
Psidium cattleyanum Weinw	Red strawberry guava	0	0	1	0
Punica granatum L.	Pomegranate	0	0	5	0
Syzygium aromaticum (L.) Merr. & Perry	Clove	8.5	2.3–17.4	12	0
Syzygium malaccensis (L.) Merr. & Perry	Malay apple	0	0	30	
Tamarindus indica L.	Tamarind	32.9	6.7–48.1	66	+
Terminalia catappa L.	Barbados almond	8.9	2.3–21.4	18	+
Theobroma cacao L.	Cacao	8.1	1.7–20	–	*

[a] Survey conducted during 1992–1995, includes data on the 252 farms reported by Kumar *et al.* (1994) supplemented with 333 additional farms (B.M. Kumar, 1995, unpublished results; total respondents = 585).
[b] Survey of 100 farms in Kerala, which constitutes a subset of the 544 samples enumerated by Patil (2005) during 2002–2004 from three South Indian States.
[c] Shift in magnitude less than 10% in either direction is ignored ('0'), more than 10% increase in frequency is '+', and more than 10% decrease is '–'. * Indicates species such as cacao, plantains and coffee, which were not enumerated in the 2002–2004 survey.
[d] Range of values for different *thaluks* (strata) during the first sampling.

5.4.2 Genetic diversity of indigenous fruit trees

Aside from floristic richness, several of the indigenous fruit trees exhibit considerable intraspecific variations in tree growth, phenology of flowering, and fruit characters (Table 5.2). This is consistent with the observations of Watson and Eyzaguirre (2002), who reported that several landraces and cultivars, as well as rare and endangered species, are preserved in homegardens. Contemporary patterns of genetic variation in homegarden components, however, reflect the historical processes associated with domestication, such as the geographical origin(s) of the cultivated populations, interpopulation genetic exchange and ancestry, which influence the genetic structuring of such populations (Schaal *et al.*, 1998). However, in view of the complex patterns of germplasm exchange prevailing in many indigenous cultures and the multiple origins of cultivated plant populations, it is difficult to make firm generalizations. Furthermore, only limited research on population genetics and systematics, in order to characterize this diversity and understand the mechanisms through which it arises, has been done in Peninsular India. The high degree of genetic diversity observed in the Kerala homegardens (Table 5.2), nevertheless, ensures compliance with the Convention on Biological Diversity (CBD) on aspects such as *ex situ* biodiversity conservation.

5.4.3 Utilizing and conserving homegarden genetic resources

In certain cases, the rich genetic diversity of fruit trees has been utilized in clonal selection programmes, which have led to the development of many commercial varieties. For example, in mango about 30 varieties have been developed in India (Ghosh, 1998). However, traditional mango varieties in the homegardens of Kerala, which constitute an important segment of the genetic diversity of this crop (Anila and Radha, 2003), have been largely ignored by organized research. Due to commercialization, many of these cultivars are also disappearing (Kumar and Nair, 2004), which calls for urgent steps to be taken to conserve the indigenous germplasm. Jackfruit, another important product of Kerala homegardens, grows wild in the forests of the Western and Eastern Ghats of India, besides being cultivated as a horticultural crop (Ghosh, 1998). Being cross-pollinated and mostly seed-propagated, it exhibits great variations in fruit characteristics such as density of spikes on the rind, periodicity of bearing, size and shape, quality, and period of maturity (Melantha, 1998). Here too, selections involving culinary/table types with superior traits have been attempted (Ghosh, 1998), but no major conservation efforts to preserve the native populations are in the pipeline.

5.5 Utilization of Homegarden Products

While a large proportion of the homegarden production is consumed domestically (30–72%), products such as fruits, vegetables and medicinal/ornamental plants are also generously shared within the local communities (Kumar and Nair, 2004). In addition, there are many tree species producing edible fruits and other products,

Table 5.2. Variability in fruit characters of certain under-exploited indigenous fruit trees of Kerala.

Species/parameter studied	Mean	SE	Range
Emblica officinalis[a]			
Fruit girth (cm)	6.74	0.41	2.4–9.1
Fruit weight (g)	7.14	0.53	2.9–13.9
Acidity (%)	0.49	0.03	0.38–0.82
Total soluble solids (TSS) (%)	10.97	0.22	8.6–12.3
Vitamin C (mg/100 g)	329.09	16.90	201.0–420.4
Fibre (%)	2.56	0.10	2.0–3.2
Garcinia gummi-gutta fruit types[b]			
Round mammiform			
Fruit weight (g)	130.0	16.40	75–166
Fruit diameter (cm)	6.6	0.31	5.5–7.3
Rind ratio	75.4	3.10	62–82
Rind–seed (%)	3.3	0.45	1.6–4.6
Round			
Fruit weight (g)	95.0	15.8	63–128
Fruit diameter (cm)	6.0	0.4	5.4–7.0
Rind ratio	76.3	4.2	66–83
Rind–seed (%)	3.6	0.7	1.9–4.8
Oval mammiform			
Fruit weight (g)	51.0	8.8	30–70
Fruit diameter (cm)	4.7	0.3	4.0–5.4
Rind ratio	74.5	5.2	67–89
Rind–seed (%)	3.8	1.4	2.0–8.0
Oval			
Fruit weight (g)	67.0	9.6	50–83
Fruit diameter (cm)	5.0	0.2	4.5–5.4
Rind ratio	73.0	2.9	68–78
Rind–seed (%)	2.8	0.4	2.8–3.6
Cylindrical			
Fruit weight (g)	82.0	5.3	71–90
Fruit diameter (cm)	5.4	0.1	5.3–5.5
Rind ratio	76.7	1.3	74–78
Rind–seed (%)	3.3	0.2	2.9–3.5
Moringa oleifera[c]			
Fruit characters			
Length (cm)	51.6	3.62	32.3–100.0
Girth (cm)	5.7	0.22	4.22–8.36
Number per plant	326.0	23.62	174–612
Weight (g)	75.3	10.75	25.3–227.3
Yield (kg/plant)	27.4	3.32	8.3–70.46
Vitamin A (IU)	141.2	6.14	94.6–184.7
Vitamin C (mg/100 g)	114.8	2.25	87.5–129.2
Seeds (no./fruit)	20.5	0.60	14.6–25.6
Syzygium cuminii[d]			
Fruit weight (g)	5.92	0.48	2.2–13.8
Fruit length (cm)	23.10	0.91	11–34
Fruit diameter (cm)	18.30	0.66	11.4–27
Pulp contents (%)	77.68	1.85	58.19–97.71
Total soluble solids (TSS) (%)	17.09	0.58	10–23
Acidity (%)	0.50	0.12	0.32–0.77
Total sugars (%)	8.74	0.29	5.68–12.5
Ascorbic acid (mg/100 g)	48.23	1.95	34.38–76.25

[a] Aravindakshan *et al.* (1986) on trees situated at 19 sites including forest areas and homegardens.
[b] Manomohandas *et al.* (2002) on 21 trees in farmers' fields in Mananthavady, Kerala.
[c] Resmi *et al.* (2005) on 28 accessions from Kerala.
[d] Patel *et al.* (2005) in northern India.

which are grown and marketed on a small scale (Table 5.1). The range of species available in homegardens producing saleable products is seemingly large and accommodates choice by those who are able to fulfil the conditions of labour availability, market demands, systems of tenure and variations in soils and climate. The net income generated from homegardens is correspondingly variable. Although data on financial aspects are not readily available, income from homegardens broadly ranges between 21.1 and 29.5% of the total income for Indonesian and Vietnamese gardens (Soemarwoto, 1987; Trinh *et al.*, 2003). The number of individuals per species of fruit trees in a homegarden also varies depending on each individual farmer's need for cash income generation *vis-à-vis* consumption requirements. In general, when the fruits are used for home consumption, the number of plants per garden is low. Conversely, with market-orientated production, a relatively large number of plants is cultivated with increasing levels of input. Size of the gardens, family needs, geographical location and species composition are other important factors.

5.6 Underutilized Tropical Fruit Trees

Although considerable research has been done in India on a few selected fruit- and nut-yielding tree species such as coconut, arecanut (*Areca catechu*) and mango (Arora and Rao, 1998; Parthasarathy *et al.*, 2006), little or no formal research has been carried out on many of the so-called 'Cinderella trees' – the hitherto wild species – to assess their potential for genetic improvement, reproductive biology or suitability for cultivation. In a few cases, e.g. the earlier domesticates such as *Emblica officinalis*, *Garcinia gummi-gutta* (Abraham *et al.*, 2006), *Tamarindus indica* (Hanamashetti *et al.*, 2000), *Moringa oleifera* (Ramachandran *et al.*, 1980), *Syzygium cuminii* (Patel *et al.*, 2005), *Aegle marmelos* (Misra *et al.*, 2000; Gupta and Misra, 2002) and *Azadirachta indica* (Singh *et al.*, 1999), a limited amount of research has been carried out. None the less, this is mostly anecdotal and deals with the evaluation of morphological and phenological variations of certain managed and natural populations. Scientific and managerial research for genetic improvement for an array of species including *Acacia sinuata* (soapnut), *Sapindus emarginatus* (soapnut tree), *Terminalia bellirica* and *Terminalia chebula*, which are normally harvested by the tribal people, are lacking. Therefore, more attention should be focused on the under-researched species for which considerable genetic diversity exists in the homegardens. This is of special significance if the indigenous fruit-tree production programmes are to move forward.

5.7 Wild Fruits for Maintaining Food, Nutritional and Livelihood Security

In addition to the homegarden-grown fruits and vegetables, the local people also collect an array of wild fruits for food and medicinal purposes (edible fruits, spices, condiments) from the nearby forests, especially during food shortages. Muraleedharan *et al.* (2005) documented 229 non-timber forest products (NTFP) from 73 plant families, including 68 trees and 35 shrubs, that are collected from three sites in the Kerala region of the Western Ghats, which included many significant wild fruits. Such a diversity of products, available year-round either from the homegardens or from the adjacent forests, is expected to contribute to food security, especially during 'lean' seasons. In addition, a substantial quantity may be also sold in the local markets, which provides a major source of income for the cultivators.

As the forest-derived NTFPs become commercially valuable, however, market forces start to control their extraction, leading to over-exploitation in the wild – thus driving them closer to extinction. Field observations suggest that tribal peoples and others selectively harvest fruits from trees that produce bumper yields, thereby ensuring relatively high economic returns per unit effort (Muraleedharan *et al.*, 2005). That is, the extraction of NTFPs is generally dependent on the opportunity cost of collection – with a higher cost for a less frequent commodity than a more common product. To maximize returns, therefore, destructive harvesting practices are resorted to: e.g. branches of *Emblica officinalis*, *Myristica dactyloides*, *Mangifera indica* and *Hydnocarpus pentandra* are lopped, and *Alnus sinuata*, a woody climber, is cut off at the base to avoid the effort of climbing up to harvest the fruits. A few days later, the fallen fruits are gathered from the ground; a process that sometimes results in the death of an entire population of plants. Extraction patterns of *Sapindus emarginatus* and *Emblica officinalis* have also changed from the subsistence mode to large-scale commercial removals (Muraleedharan *et al.*, 2005). The domestication of such species in homegardens may go a long way towards safeguarding these genetic resources and their diversity in the wild. Homegardens can thus serve as loci for conserving the shrinking NTFP resources in the wild.

Wild fruits also bring diversity to the diet, serve as a source of vitamins and minerals, and are valuable sources of indigenous medicines (Leakey, 1999; Muraleedharan *et al.*, 2005). In certain cases where homegardens already contained such species, the consumption of fruits and vegetables alleviated deficiencies in iodine, vitamin A and iron (Molina *et al.*, 1993) and made the children of the garden owners less prone to xerophthalmia (Shankar *et al.*, 1998). In experimental studies conducted in South Africa, the target families significantly increased their year-round production and consumption of vitamin-rich fruits and vegetables compared with the control group without gardens (Faber *et al.*, 2002). Wild fruits also confer important health benefits against malnutrition and possibly improve resilience against epidemics such as HIV/AIDS (Barany *et al.*, 2001). Consequently, there is now a growing awareness that homegardening combined with nutrition education can be a viable strategy for improving household nutritional security for at-risk populations, particularly women and children. However, few attempts have been made to investigate the indigenous species in terms of their traditional characteristics and nutritional value; neither have significant efforts been made to build on the potential of traditional food supply systems, integrating them into homegarden projects (Cleveland and Soleri, 1987).

5.8 Tree Domestication Programme for Homegardens

Historically, tree domestication was mostly intuitive. Local people learned to determine visually which trees produced better fruit quality, size and other desirable characteristics. Modern domestication strategies, however, may involve a range of activities such as 'nurturing wild plants through to plant breeding through to genetic modification *in vitro*' (Simons and Leakey, 2004). This may be a scale-neutral technology, which benefits resource-poor and resource-rich farmers alike; whereas, with most other new technologies, farmers with the larger and better-endowed lands tend to gain the most, and income disparities are often accentuated. New initiatives in agroforestry everywhere are seeking to integrate more and more indigenous trees, whose products have traditionally been gathered from natural forests (Leakey and Tchoundjeu, 2001).

There are many tree species in the Kerala homegardens which have commercial potential in the local, regional or even international markets (Table 5.1). Through selection, it is possible to achieve rapid and substantial genetic improvements of these trees. Additionally, many potentially useful species exist in nearby forests (Muraleedharan *et al.*, 2005). Domestication is a dynamic process in which genetic and cultivation aspects are continuously refined. This would probably require the sustained exploration and collection of natural or anthropogenic populations and the evaluation and selection of suitable species and provenances. Such domestication processes may conserve the genetic diversity of indigenous species (e.g. *Mangifera indica*) and ease pressure on wild populations (e.g. *Sapindus emarginatus*, *Emblica officinalis*, *Hydnocarpus pentandra*). Although participatory domestication, multiplication and dissemination of planting materials by key farmers has been advocated (Weber *et al.*, 2001), not much headway has been made in India. Strategic approaches to domestication may help avoid the potential pitfalls of traditional methods, such as introducing narrow genetic bottlenecks, using maladapted materials, or focusing on exotics.

5.8.1 Plant propagation

Once an appropriate species/population has been identified, providing farmers with high-quality propagules in a timely manner is a key challenge in tree domestication. In several cases, the best types are identified, multiplied by vegetative means, and distributed among farmers. Examples include the cross-pollinated species such as jack (Nazeem *et al.*, 1984; Kelaskar *et al.*, 1993), mango (Arora and Rao, 1998), Indian gooseberry (Tewari and Bajpai, 2005) and *Garcinia gummi-gutta* (Muthulakshmi *et al.*, 1999; Nair *et al.*, 2005). In some cases, tissue-culture protocols have also been standardized (e.g. Rajmohan and Mohanakumaran, 1988; Puri and Swamy, 1999; Mishra *et al.*, 2005). Problems of seed storability, dormancy and aspects of nursery stock production, and also those relating to protection and product utilization, have been addressed (Mathew and George, 1995; Yadav, 1998), albeit in a limited way. Despite the studies reported, standardization of the propagation techniques has been one of the greatest challenges of fruit-tree domestication.

5.9 Conclusions

Over long periods in the history of tropical land use, homegarden systems have remained as repositories of indigenous fruits, nuts and spices. In certain cases, demand for indigenous fruits is partly satisfied by gathering from the wild. However, wild populations are often threatened, and tree domestication in homegardens and other cultivated fields offers scope for further improvements in production. Homegardens are also endowed with considerable genetic diversity, which is an important aspect of biodiversity conservation. While extensive infraspecific classifications based on morphological, geographical or ecological items, or a combination of these, have been attempted for important crop plants, such studies on the genetic diversity of the native fruit-tree components of Kerala homegardens are rare. Likewise, tropical homegardens receive far less attention from researchers, land managers and extensionists than monospecific production systems. Despite this, the importance of homegarden products, including under-exploited native fruit trees, in supporting regional economies and food and nutritional security cannot be underestimated. On a final note, participatory approaches for locating promising genotypes and their domestication are suggested as corrective measures to augment fruit-tree production in tropical homegardens.

References

Abraham, Z., Malik, S.K., Rao, G.E., Narayanan, S.L. and Biju, S. (2006) Collection and characterisation of Malabar tamarind [*Garcinia cambogia* (Gaertn.) Desr.]. *Genetic Resources and Crop Evolution* 53, 401–406.

Anderson, E. (1952) *Plants, Man, and Life.* Little, Brown and Company, Boston, Massachusetts, 245 pp.

Anila, R. and Radha, T. (2003) Physico-chemical analysis of mango varieties under Kerala conditions. *Journal of Tropical Agriculture* 41, 20–22.

Aravindakshan, M., Gopikumar, K., Sreekumar, K. and Nair, C.S.J. (1986) Natural variability in economic characters of *Phyllanthus emblica* (Indian gooseberry), a tree species suitable for restoration of some degraded areas of Western Ghats. In: Nair, K.S.S., Gnanaharan, R. and Kedharnath, S. (eds) *Ecodevelopment of Western Ghats.* Kerala Forest Research Institute, Peechi, India, pp. 130–132.

Arora, R.K. and Rao, V.R. (eds) (1998) *Tropical Fruits in Asia: Diversity, Maintainance, Conservation and Use.* Proceedings of the IPGRI-ICAR-UTFANET Regional Training Course on the Conservation and Use of Germplasm of Tropical Fruits in Asia. IPGRI Office for South Asia, New Delhi, India, 270 pp.

Barany, M., Hammett, A.L., Sene, A. and Amichev, B. (2001) Non-timber forest benefits and HIV/AIDS in sub-Saharan Africa. *Journal of Forestry* 99, 36–42.

Cleveland, D.A. and Soleri, D. (1987) Household gardens as a development strategy. *Human Organization* 46, 259–270.

Faber, M., Phungula, M.A.S., Venter, S.L., Dhansay, M.A. and Benadé, A.J.S. (2002) Home gardens focusing on the production of yellow and dark-green leafy vegetables increase the serum retinol concentrations of 2–5-y-old children in South Africa. *American Journal of Clinical Nutrition* 76, 1048–1054.

Ghosh, S.P. (1998) Fruit wealth of India. In: Arora, R.K. and Rao, V.R. (eds) *Tropical Fruits in Asia: Diversity, Maintenance, Conservation and Use.* Proceedings of the

IPGRI-ICAR-UTFANET Regional Training Course on the Conservation and Use of Germplasm of Tropical Fruits in Asia. IPGRI Office for South Asia, New Delhi, India, pp. 3–14.

Gupta, N.K. and Misra, K.K. (2002) Growth, yield and photosynthetic efficiency of bael (*Aegle marmelos*) genotypes in foothills region of Uttaranchal. *Indian Journal of Agricultural Sciences* 72, 220–222.

Hanamashetti, S.I., Sulikeri, G.S. and Salimath, P.M. (2000) Multivariate analysis in clonal progenies of tamarind (*Tamarindus indicus* L). In: Solanki, K.R., Bisaria, A.K. and Handa, A.K. (eds) *Multipurpose Tree Species Research: Retrospect and Prospect.* Agrobios (India). Jodhpur, India, pp. 105–108.

Kelaskar, A.J., Desai, A.G. and Salvi, M.J. (1993) Effect of season on the success and growth of bud grafts of jack (*Artocarpus heterophyllus* Lam.). *Journal of Tropical Agriculture* 31, 112–115.

Kumar, B.M., George, S.J. and Chinnamani, S. (1994) Diversity, structure and standing stock of wood in the homegardens of Kerala in peninsular India. *Agroforestry Systems* 25, 243–262.

Kumar, B.M. and Nair, P.K.R. (2004) The enigma of tropical homegardens. *Agroforestry Systems* 61, 135–152.

Ladizinsky, G. (1998) *Plant Evolution under Domestication.* Kluwer, Dordrecht, The Netherlands, pp. 4–9.

Leakey, R.R.B. (1999) Potential for novel food products from agroforestry trees. *Food Chemistry* 64, 1–14.

Leakey, R.R.B. and Tchoundjeu, Z. (2001) Diversification of tree crops: domestication of companion crops for poverty reduction and environmental services. *Experimental Agriculture* 37, 279–296.

Manomohandas, T.P., Anith, K.N., Gopakumar, S. and Jayaraja, M. (2002) Kodampuli: a fruit for all reasons. *Agroforestry Today* 13(1–2), 7–8.

Mathew, K.L. and George, S.T. (1995) Dormancy and storage of seeds in *Garcinia cambogia* Desr. (kodampuli). *Journal of Tropical Agriculture* 33, 77–79.

Melantha, K.R. (1998) Genetic variability, characterization and utilization of germplasm in jackfruit. In: Arora, R.K. and Rao, V.R. (eds) *Tropical Fruits in Asia: Diversity, Maintainance, Conservation and Use.* Proceedings of the IPGRI-ICAR-UTFANET Regional Training Course on the Conservation and Use of Germplasm of Tropical Fruits in Asia. IPGRI Office for South Asia, New Delhi, India, pp. 207–208.

Misra, K.K., Singh, R. and Jaiswal, H.R. (2000) Performance of bael (*Aegle marmelos*) genotypes under foothills region of Uttar Pradesh. *Indian Journal of Agricultural Sciences* 70, 682–683.

Mishra, M., Chandra, R., Tiwari, R.K., Pati, R. and Pathak, R.K. (2005) Micropropagation of certain underutilized fruit crops: a review. *Small Fruits Review* 4, 7–18.

Molina, M.R., Noguera, A., Dary, O., Chew, F. and Valverde, C. (1993) Principal micronutrient deficiencies in Central America. *Food, Nutrition and Agriculture* 7, 26–33.

Muraleedharan, P.K., Sasidharan, N., Kumar, B.M., Sreenivasan, M.A. and Seethalakshmi, K.K. (2005) Non-wood forest products in the Western Ghats of Kerala, India: floristic attributes, extraction, regeneration and prospects for sustainable use. *Journal of Tropical Forest Science* 17, 243–257.

Muthulakshmi, P., George, S.T. and Mathew, K.L. (1999) Morphological and biochemical variations in different sex forms of kodampuli (*Garcinia gummi-gutta* L.). *Journal of Tropical Agriculture* 37, 28–31.

Nair, K.K.N., Mohanan, C. and Mathew, G. (2005) Preliminary observations on seed and plantation technology for *Garcinia gummi-gutta* (Guttiferae): an indigenous and economic tree crop of the Indian Peninsula. *Forests, Trees and Livelihoods* 15, 75–87.

Nair, P.K.R. and Kumar, B.M. (2006) Introduction. In: Kumar, B.M. and Nair, P.K.R. (eds) *Tropical Homegardens: A Time-Tested Example of Sustainable Agroforestry.* Springer Science, Dordrecht, The Netherlands, pp. 3– 11.

Nazeem, P.A., Gopikumar, K. and Kumaran, K. (1984) Vegetative propagation in jack (*Artocarpus heterophyllus* Lam.).

Agricultural Research Journal of Kerala 22, 149–154.

Pareek, O.P. and Sharma, S. (1993) Genetic resources of underexploited fruits. In: Chadha, K.L. and Pareek, O.P. (eds) *Advances in Horticulture: Fruit Crops.* Malhotra Publishing House, New Delhi, India, pp. 189–241.

Parthasarathy, V.A., Chattopadhyay, P.K. and Bose, T.K. (eds) (2006) *Plantation Crops, Vol. 2.* Naya Udyog, Kolkata, India, 519 pp.

Patel, V.B., Pandey, S.N., Singh, S.K. and Das, B. (2005) Variability in jamun (*Syzygium cuminii* Skeels) accessions from Uttar Pradesh and Jharkhand. *Indian Journal of Horticulture* 62, 244–247.

Patil, V.S. (2005) *MARAM: Multipurpose Tree Database for Agroecosystem Research and Appropriate Management* (CD-Rom). The French Institute of Pondicherry, Pondicherry, India.

Peyre, A., Guidal, A., Wiersum, K.F. and Bongers, F. (2006) Dynamics of homegarden structure and function in Kerala, India. *Agroforestry Systems* 66, 101–115.

Puri, S. and Swamy, S.L. (1999) Geographical variation in rooting ability of stem cuttings of *Azadirachta indica* and *Dalbergia sissoo. Genetic Resources and Crop Evolution* 46, 29–36.

Rajmohan, K. and Mohanakumaran, N. (1988) Effect of plant growth substances on the *in vitro* propagation of jack (*Artocarpus heterophyllus* Lam.). *Agricultural Research Journal of Kerala* 26, 29–38.

Ramachandran, C., Peter, K.V. and Gopalakrishnan, P.K. (1980) Drumstick (*Moringa oleifera*): a multipurpose Indian vegetable. *Economic Botany* 34, 276–283.

Randhawa, M.S. (1980) *The History of Indian Agriculture, Vol. 2.* Indian Council of Agricultural Research, New Delhi, India, pp. 67–68, 98–99, 414–415.

Rao, M.R. and Rao, B.R.R. (2006) Medicinal plants in tropical homegardens. In: Kumar, B.M. and Nair, P.K.R. (eds) *Tropical Homegardens: A Time-Tested Example of Sustainable Agroforestry.* Springer Science, Dordrecht, The Netherlands, pp. 205–231.

Resmi, D.S., Celine, V.A. and Rajamony, L. (2005) Variability among drumstick (*Moringa oleifera* Lam.) accessions from central and southern Kerala. *Journal of Tropical Agriculture* 43, 83–85.

Schaal, B.A., Hayworth, D.A., Olsen, K.M., Rauscher, J.T. and Smith, W.A. (1998) Phylogeographic studies in plants: problems and prospects. *Molecular Ecology* 7, 465–474.

Shanavas, A. and Kumar, B.M. (2003) Fuelwood characteristics of tree species in homegardens of Kerala, India. *Agroforestry Systems* 58, 11–24.

Shanavas, A. and Kumar, B.M. (2006). Physical and mechanical properties of three multipurpose trees grown in the agricultural fields of Kerala. *Journal of Tropical Agriculture* 44, 23–30.

Shankar, A.V., Gittelsohn, J., Pradhan, E.K., Dhungel, C. and West, K.P., Jr (1998) Homegardening and access to animals in households with xerophthalmic children in rural Nepal. *Food and Nutrition Bulletin* 19, 34– 41.

Simons, A.J. and Leakey, R.R.B. (2004) Tree domestication in tropical agroforestry. *Agroforestry Systems* 61, 167–181.

Singh, A., Negi, M.S., Rajagopal, J., Bhatia, S., Lakshmikumaran, M., Tomar, U.K. and Srivastava, P.S. (1999) Assessment of genetic diversity in *Azadirachta indica* using AFLP markers. *Theoretical and Applied Genetics* 99, 272–279.

Soemarwoto, O. (1987) Homegardens: a traditional agroforestry system with a promising future. In: Steppler, H.A. and Nair, P.K.R. (eds) *Agroforestry: A Decade of Development.* ICRAF, Nairobi, pp. 157–170.

Tesfaye, A., Wiersum, K.F., Bongers, F. and Sterck, F. (2006) Diversity and dynamics in homegardens of southern Ethiopia. In: Kumar, B.M. and Nair, P.K.R. (eds) *Tropical Homegardens: A Time-Tested Example of Sustainable Agroforestry.* Springer Science, Dordrecht, The Netherlands, pp. 123–142.

Tewari, R.K. and Bajpai, C.K. (2005) Bench grafting in aonla (*Emblica officinalis* Gaertn.) for easy transport and establishment. *Indian Journal of Agroforestry* 7, 65–67.

Trinh, L.N., Watson, J.W., Hue, N.N., De, N.N., Minh, N.V., Chu, P., Sthapit, B.R.

and Eyzaguirre, P.B. (2003) Agrobiodiversity conservation and development in Vietnamese homegardens. *Agriculture, Ecosystems and Environment* 97, 317–344.

Watson, J.W. and Eyzaguirre, P.B. (eds) (2002) *Home Gardens and In Situ Conservation of Plant Genetic Resources in Farming Systems: Proceedings of the Second International Home Gardens Workshop, 17–19 July 2001, Witzenhausen, Germany.* International Plant Genetic Resources Institute (IPGRI), Rome, 184 pp.

Weber, J.C., Montes, C.S., Vidaurre, H., Dawson, I.K. and Simons, A.J. (2001) Participatory domestication of agroforestry trees: an example from the Peruvian Amazon. *Development in Practice* 11, 425–433.

Wiersum, K.F. (2006) Diversity and change in homegarden cultivation in Indonesia. In: Kumar, B.M. and Nair, P.K.R. (eds) *Tropical Homegardens: A Time-Tested Example of Sustainable Agroforestry*. Springer Science, Dordrecht, The Netherlands, pp. 13–24.

Yadav, I.S. (1998) Indian Institute of Horticultural Research and its contribution to Indian horticulture. In: Arora, R.K. and Rao, V.R. (eds) *Tropical Fruits in Asia: Diversity, Maintenance, Conservation and Use.* Proceedings of the IPGRI-ICAR-UTFANET Regional Training Course on the Conservation and Use of Germplasm of Tropical Fruits in Asia. IPGRI Office for South Asia, New Delhi, India, pp. 16–28.

Yamada, M. and Osaqui, H.M.L (2006) The role of homegardens for agroforestry development: lessons from Tomé-Açu, a Japanese-Brazilian settlement in the Amazon. In: Kumar, B.M. and Nair, P.K.R. (eds) *Tropical Homegardens: A Time-Tested Example of Sustainable Agroforestry*, Springer Science, Dordrecht, The Netherlands, pp. 299–316.

6 Native Fruit Tree Improvement in Amazonia: An Overview

C.R. Clement,[1] J.P. Cornelius,[2] M.H. Pinedo-Panduro[3] and K. Yuyama[1]

[1] *Instituto Nacional de Pesquisas da Amazônia (INPA), Manaus, Amazonas, Brazil*; [2] *World Agroforestry Centre (ICRAF), CIP, Lima, Peru*; [3] *Instituto de Investigaciones de la Amazonia Peruana, Iquitos, Peru*

6.1 Introduction

6.1.1 Background

Amazonia contains the largest remaining area of humid tropical forest on the planet and is undergoing rapid and dramatic human-driven environmental change. The region is an extremely heterogeneous biome, extending from the cool Andean foothills in the west to the Atlantic Ocean in the east with seasonally dry savannah plateaus in both the north and the south. The Amazon River Basin's landforms are composed of sediments from three sources: the Guyana (north) and Brazilian (south) shields (the remaining highly weathered surfaces of the palaeo-continent) and the Andes mountains, derived from recent tectonic uplift. The rivers that drain the ancient shields are nutrient-poor, while those that drain the Andes are nutrient-rich, which creates a mosaic of river types: black water (organic acid-rich) rivers from the Guyana shield and consolidated sedimentary plateaus within the basin, clear water (little organic acid) rivers from the Brazilian shield and white water (sediment-rich) rivers from the Andes. Generally low relief means that all rivers meander and the Amazon River itself has wide flood plains that rival the Nile and Mekong Rivers in potential primary productivity. Rainfall varies from 1200 mm along the south and south-eastern limits to more than 3000 mm in the north-western sector, with correspondingly variable dry seasons (6–8 months in the south-east to no dry season in the north-west). This complex physical mosaic supports and has helped create an equally complex biological mosaic as its mantle, making it one of the planet's centres of mega-biodiversity, with an estimated 15–20,000 species of higher plants, including hundreds of species with edible fruit.

With increased worldwide awareness of the importance of the biosphere to human society, the concept of sustainable development has become central to all discussions about Amazonia. Agroforestry and fruit crops are always included among the environmentally friendly and economically viable options for the region. Smith *et al.* (1998) analysed the numerous constraints on the adoption of agroforestry, while Clement (1997) analysed those on development of underutilized fruit tree species, both of which help to explain the expansion of pasture and grains. These constraints notwithstanding, fruit crops in orchards or agroforestry systems are important in the region; as this importance increases, area, labour involvement, productivity and income generated also increase. The most important species, however, are not native to Amazonia, which is apparently an anomaly when compared with South-east Asia, the other world centre of origin of tropical fruit crops (Clement, 2004).

The majority of the fruit species used, cultivated and domesticated by Native Amazonians are indigenous to Amazonia and include at least eight domesticates, 18 semi-domesticates and 21 incipient domesticates, as well as another 33 fruit crops from other parts of Central and South America that have been domesticated to varying degrees (see Tables A6.1–A6.3). Hundreds of other species remain wild, either because they never attracted enough Native Amazonian attention or because they naturally occur in sufficient abundance to make cultivation unnecessary. Two of the Amazonian natives that are important around the world today are cocoa (*Theobroma cacao* L.) and pineapple (*Ananas comosus* Merr.). Others are moderately familiar as minor fruits and agroforestry fruit tree species: peach palm (*Bactris gasipaes* Kunth), abiu or caimito (*Pouteria caimito* Radlk.), inga or ice cream bean (*Inga edulis* Mart.) and Brazil nut (*Bertholletia excelsa* Kunth). All of the other native fruits are of minor importance regionally, although they may be important locally. At the same time, the most important fruits in Amazonia are banana and plantains (*Musa* spp.), orange (*Citrus sinensis* Osbeck) and other citrus species, coconut (*Cocos nucifera* L.), mango (*Mangifera indica* L.) and several American species from outside Amazonia that had been introduced before European conquest, such as papaya (*Carica papaya* L.), passion fruit (*Passiflora edulis* Sims.) and avocado (*Persea americana* Mill.).

The small percentage of native Amazonian fruits known outside the region raises a question. Why are Amazonian fruits so often unimportant in the market? Are the fruits to blame? Many cannot be consumed *in natura*. Patiño (2002) divides neotropical fruits into those that require processing and those that do not, with many more in the first category. Most have unacceptably short shelf lives. Many have strange flavours, aromas or textures that are acquired tastes. Most are extremely variable in quality because they are seed-propagated, and only the most intensively selected come relatively true from seed. This latter point is the major problem for most fruits around the world and can be solved through appropriate improvement strategies, which is the topic of this book. Additionally, many neotropical fruits are more like staple foods than desserts (see Table 6.1; most of the palm and starchy oily fruits), which places them at a competitive disadvantage in extra-Amazonian markets that concentrate on dessert or fresh out-of-hand fruits, although this characteristic may make them more important for food security and sovereignty at the subsistence level.

Over the last half-century there has been considerable research and development (R&D) on native American fruits implemented principally by

Table 6.1. Comparison of the mean chemical compositions (g/100 g)[a] and energy values (kcal) of contrasting native Amazonian fruit groups. The mesocarp is generally the most important part of the palm, starchy/oily and juicy fruit groups (Clement, 2006).

Group (*n* species)	Water	Protein	Fats	Carbohydrates	Fibre	Energy
Nuts and seeds (8)	3.9	14.1	57.4	18.1	4.8	62.1
Palm fruits (8)	45.3	3.5	21.8	16.0	12.2	31.0
Starchy/oily fruits (8)	51.1	2.5	8.3	32.4	9.0	23.1
Juicy fruits (21)	82.8	0.9	0.8	11.9	2.9	6.3

[a] Fresh weights; the difference between the sum of these means and 100 is due to ash content.

national agricultural research services. The primary objective has generally been to get fruit into national and international markets, rather than using it as an additional item in food security. Unfortunately, this objective has not been achieved very often. In part, this may have been because most fruits in Amazonia, especially the native fruits, are produced by family farmers, whereas the clients for R&D have been considered to be commercial farmers and agroindustry.

Additionally, international fruit markets are highly competitive and any new fruit takes market share from existing fruits because overall growth of the international fruit market is small. According to Alonso González (CIAT, 2006, personal communication) growth is currently 3% per year. However, economic growth currently averages 4–5%. Hence, new fruits must have high quality, uniformity and good price, which are difficult objectives to attain quickly, especially for family farmers. Because of globalization, national urban supermarkets follow international markets in their demands for quality, uniformity and price. The result is that rural Amazonian fruit producers compete with fruits of international standard, both imported and nationally produced, while at the same time suffering from developing-world limitations (Clement, 1997; Smith *et al.*, 1998). In Manaus, Amazonas, Brazil, for example, apples from Argentina and Rio Grande de Sul, Brazil, are cheaper and more uniform in quality than locally produced native fruits. Consequently (and, possibly, because of local preferences), demand in supermarkets is higher for the imported fruits.

The need for improvement of native species, whether already domesticated or still wild, is clear when the objectives are the regional, national or international markets. Is the same true for species used exclusively for food security? We argue that it is, as any family farmer will market excess yields when possible, and the better the quality of fruit sent to market the more likely the family will be to receive a good price. Sale of farm produce also contributes to food security, as a considerable number of off-farm products are now part of most people's diets even when the family is far from an urban centre.

In this chapter, we list some of the current native fruit improvement activities in Amazonia, examine two cases that contrast conventional and participatory approaches, and consider the challenges for the next decade. We concentrate on Brazilian and Peruvian Amazonia, which constitute most of the Basin and have the greatest concentration of research and development activity.

6.1.2 History of Amazonian agriculture

It is thought that between 12,000 and 25,000 years ago the ancestors of the Native Amazonians arrived from the north, although the exact period of immigration is not certain. Initially, these people were hunter-gatherers, but as climates changed and forests expanded during the early Holocene, the people gradually developed food production systems (Piperno and Pearsall, 1998), domesticating both landscapes and plant populations in the process (Clement, 1999). During the last 5000 years, in particular, Amazonia became a biome strongly influenced by humans, with at least 15% modified by human activities – to such an extent that these modifications are still visible today, 400 years after the demographic collapse caused by European conquest. It is probable that most of the biome was modified to some extent (Mann, 2005). Before the conquest, there were probably between 5 and 25 million people in the Amazonian biome and these people depended principally upon horticulture for their subsistence, with a significant number of fruit crops in their diet (Clement, 1999).

After the demographic collapse (AD 1600–1700), Amazonia started to be repopulated, a process that accelerated during the rubber boom (1880–1915), which brought an influx of peoples from north-eastern Brazil, the Andes, Africa and Europe. After the rubber crash (1915–1916), the immigration rate slowed until the 1970s, when national efforts to integrate Amazonia into its various Nation States and their economies accelerated, financed by cheap international credit. This immigration soon led to the current worldwide concern about environmental change in Amazonia, as all of the immigrants are agricultural peoples with varying interests, knowledge and access to capital. In Brazil alone, nearly 20% of the original forest cover has been removed since the 1970s. In general, this has contributed little to regional development, and enormous areas are now in secondary forests of varying ages. The lack of contribution to regional development is mostly because the great majority of these immigrants discovered that agriculture in Amazonia is more of a challenge than expected, especially to those who have neither much knowledge (traditional or other) about the region, nor the capital to obtain knowledge and inputs readily. This discovery, in turn, caused migration within Amazonia, this time from the countryside to the cities, although other factors obviously influenced personal decisions about migration. In Brazilian Amazonia, 80% of the population is now urban, including a considerable proportion of the Native American population and the peasants who had learned horticulture from the natives, thus leaving a large proportion of recent immigrants in the countryside – precisely those with least traditional knowledge about Amazonia. This strongly influenced decisions about which fruit trees to plant, and recent R&D on fruit trees reflects these decisions.

6.2 Partial Inventory of Amazonian Fruit Improvement Activities

Research and development activities on native Amazonian fruit started in the 1930s in Brazil, with the creation of the Instituto Agronômico do Norte, Belém,

Pará (now Embrapa Amazônia Oriental). Work initially concentrated on cacao and gradually expanded to include minor efforts on numerous native fruits, including abiu (*Pouteria caimito*), bacuri (*Platonia insignis* Mart.), biribá (*Rollinia mucosa* Bail.) and cupuaçu (*Theobroma grandiflorum* Schumm.) (Calzavara, 1970), without, however, creating much impact in regional or national markets. The same institute pioneered studies on açaí-do-Pará (*Euterpe oleracea* Mart.) and peach palm (*Bactris gasipaes* Kunth), but these studies also had little impact and the institution suffered from lack of continuity during the decade before the creation of Embrapa in 1973.

In the 1960s, the Instituto Agronômico da Amazônia Ocidental, Manaus, Amazonas (now Embrapa Amazônia Ocidental), started working on guaraná (*Paullinia cupana* Mart.) as demand grew for its caffeine-rich seed for soft drinks. This work continued both in Manaus and in Maués, Amazonas, where the multinational American Beverage Company (AmBev) has its major Amazonian plantation of this fruit crop. Field trials were organized in Bahia by the Comissão Executiva do Plano da Lavoura Cacaueira (CEPLAC). Bahia now produces nearly 80% of the guaraná used in Brazil, although R&D has increased in Amazonia. Guaraná was chosen as the target for the first Brazilian Amazonian regional genome network. This project has already determined that guaraná is a high-level polyploid and is identifying the genes involved in disease susceptibility and resistance.

In 1975, the Instituto Nacional de Pesquisas da Amazônia (INPA: National Research Institute for Amazonia) created a fruit studies group that initially concentrated on peach palm, cupuaçu, sapota (*Quararibea cordata* Vischer), graviola (*Annona muricata* L.), araçá-boi (*Eugenia stipitata* McVaugh), camu-camu (*Myrciaria dubia* McVaugh) and cubiu (*Solanum sessiliflorum* Dunal) (Clement *et al.*, 1997). This group pioneered the idea of creating agroforestry systems with fruit trees of different stature and shade tolerance, without, however, making much impact on regional or national markets (van Leeuwen *et al.*, 1997). In the mid-1980s, the INPA group collaborated with Embrapa Genetic Resources and Biotechnology (then the National Centre for Genetic Resources) and helped create the peach palm germplasm collections in Brazil, Colombia, Costa Rica, Ecuador and Peru (Clement and Coradin, 1988). Although the group made a major impact when it identified and imported spineless peach palm from Peru for heart-of-palm production, the work with the fruit failed to increase demand in local and regional markets (Clement *et al.*, 2004). This group is now working principally on peach palm and camu-camu and slowly expanding its activities with tucumã (*Astrocaryum tucuma* Mart.).

In the late 1970s, Embrapa Amazonia Oriental expanded its work on fruits again, concentrating on Brazil nut, cupuaçu, guaraná, bacuri, camu-camu, peach palm and açaí-do-Pará. During the early 1980s, they collected patauá (*Oenocarpus bataua* Mart.) and bacaba (*Oenocarpus bacaba* Mart.) with international support, but the project did not advance beyond the germplasm collection. More recently they have added uxi (*Endopleura uchi* Cuatrecasas). Embrapa Amazonia Ocidental also expanded their work on fruits beyond guaraná, concentrating on cupuaçu and camu-camu. Embrapa Acre (Rio

Branco, Acre) initiated work on pineapple and peach palm, although the latter was more for heart-of-palm than for fruit.

During the last three decades, these institutions have contributed to the expansion of peach palm production for heart-of-palm, cupuaçu and açaí-do-Pará, as well as the expansion of interest in camu-camu and cubiu, principally in other parts of Brazil. The lack of greater impact is partly due to an excessive emphasis on germplasm collections rather than progeny trials designed to meet consumer demands as rapidly as possible (van Leeuwen *et al.*, 2005) and partly to a lack of continuity in the various R&D projects. Açaí-do-Pará is a success story, however, which will be mentioned below.

While this history was unfolding in Brazil, both Peru and Colombia also invested in R&D on native Amazonian fruits. The San Roque Experiment Station, Iquitos, Peru, is a part of INIA and started collecting native fruit germplasm in 1972. Camu-camu, araçá-boi, abiu, peach palm, macambo (*Theobroma bicolor* H.B.K.), cacahuillo (*Herrania nitida* R.E. Schultes), naranjo podrido (*Parahancornia peruviana* Monach.), chope (*Gustavia longifolia* Poepp. ex. O. Berg), among others, were collected and characterized. Personnel affiliated with San Roque participated in the 1983–1984 USAID-financed prospecting for peach palm germplasm and established the two major Peruvian collections as a result, one at San Roque and the other at Yurimaguas (Clement and Coradin, 1988). The same period saw concentrated prospecting of camu-camu as commercial interest expanded. In the mid-1990s, INIA invited ICRAF to collaborate on fruit crop development and considerable innovative R&D was initiated, including surveys of family farmer interests and consumer preferences for native fruit (Sotelo Montes and Weber, 1997), studies of family farmer management of fruit germplasm (Brodie *et al.*, 1997), and the beginnings of several participatory improvement projects with fruit and timber species that had been selected by family farmers (Weber *et al.*, 2001). The genetic consequences of the participatory project with peach palm were examined in terms of conservation and improvement (Cornelius *et al.*, 2006).

In the 1980s, the Peruvian government created the Instituto de Investigaciones de la Amazonía Peruana (IIAP), and fruit crops were included in its mandate. Their first major project was with annatto (*Bixa orellana* L.), a domesticated shrub that produces a vibrant yellow-orange to red food colouring in the scarce pulp around the seeds. Between 1993 and 2005, the Amazonian Cooperation Treaty financed a wide-ranging collection of fruit species that is still being evaluated. This allowed priority-setting and interest focused on camu-camu, meto huayo (*Caryodendron orinocense* H. Karst.), uvilla (*Pourouma cecropiifolia* Mart.), aguaje (*Mauritia flexuosa* L.) and macambo (A. González, IIAP, 2006, personal communication). Today, IIAP leads the Peruvian R&D effort on camu-camu; this project will be examined in detail below.

Work in Colombia has also suffered from inconsistent investment, but much has been done during the past few decades. CorpoICA is the major Colombian agricultural research institution and has worked on peach palm and meta huayo (called 'inchi' in Colombia), among other native fruits. The Corporación Araracuara was created in the 1980s with international funding to work on agroforestry systems and native fruit crops in Amazonia. They participated in the peach palm expeditions of 1983–1984 and created a collection at Araracuara on the Caquetá River. They also did considerable work with inchi as a nut and oil crop, but it has not been adopted by Colombian fruit growers.

In the 1990s, the Bolivian government, with support from the USA and the European Community, started a programme called Alternative Tropical Development, designed to identify and develop economically attractive alternatives to illicit coca production (*Erythroxylum coca* Lam.). Among the native Amazonian fruit species, they worked with camu-camu, pineapple, peach palm (for heart-of-palm) and cocoa in both monoculture orchards and agroforestry systems (F. Alemán, Bolivia, 2006, personal communication). Although they have had some success, their impact has been limited due to the high returns available from illicit coca production.

6.3 Participatory versus Conventional Improvement

Participatory improvement is defined as genetic improvement that includes close farmer–researcher collaboration (Christinck *et al.*, 2005). The approach is essentially a reaction to some shortcomings of conventional plant breeding. The basic premise of participatory plant breeding is that by involving farmers in such a way that they can express their preferences and their local knowledge can be effectively used, the relevance of research products to the needs of small farmers will be increased. Thus, the fundamental objective is to ensure that research responds to farmers' needs (Vernooy, 2003). Increased relevance can be achieved at various levels. For example, selection of species and identification of traits for improvement, and production of germplasm for 'real' conditions, including potentially marginal sites, are feasible.

The increased relevance of participatory research products, coupled with the nature of the process itself, also promises to facilitate adoption, especially by avoiding delays in transfer of technology from experimental stations to farmers' fields (Simons and Leakey, 2004; Christinck *et al.*, 2005). However, careful planning and design will still be necessary to avoid delays in transferring technology from participating farmers to wider target groups. In addition, the approach promises to correct another problem associated with conventional plant breeding, i.e. the replacement of landraces with 'modern' varieties, which is widely considered to be the principal cause of erosion of crop genetic resources. Participatory approaches are likely to maintain and even increase diversity, as different groups of farmers in different places are unlikely to have the same priorities and needs and, even in one location, farmers may well prefer retention of a 'bundle' of landraces for different needs and conditions. Rather than being abandoned, in many cases landraces may be preserved as sources of material for breeding programmes (Christinck *et al.*, 2005). Also, the presence in farmers' fields of diverse experimental plantings and, in some cases, of new seed sources, may directly increase within-species diversity.

Finally, participatory improvement has the potential to serve as a mechanism for sharing the benefits from the sustainable use of biodiversity, as mandated implicitly or explicitly in international treaties, serving as a best practice alternative to biopiracy (Simons and Leakey, 2004). These benefits include, but are not restricted to, access to germplasm for own use or for sale. Other potential benefits include opportunities for joint and mutual learning, capacity and skill-building, institutional development (formal and farmer institutions) and policy development (Christinck *et al.*, 2005).

Participatory improvement also has some possible disadvantages with respect to the conventional approach. The logistical difficulties and, in some cases, the costs of maintaining participatory field experiments may be considerable. Although farmers may themselves meet part of the cost through their own labour, distances between experiments can be considerable, and the difficulty of regular monitoring can lead to problems in trial maintenance and protection. Second, close involvement of farmers in research planning and execution requires the effective deployment of skills and techniques that may be unfamiliar to many researchers. Thus, the approach is difficult to implement in situations where professionals with the necessary skills in participatory research are absent, both in the research community and in the extension services. Conventional improvement done well may be better than participatory research done poorly or in a 'token' way.

6.3.1 An example of participatory and conventional improvement: camu-camu

Myrciaria dubia (H.B.K.) McVaugh, Myrtaceae, called camu-camu in Peru and caçari or river guava in Brazil, is a shrub that may attain 4 m in height and be generally abundantly branched from the base, forming an open vase-shaped crown (Pinedo *et al.*, 2001). The fruit is a round berry with a smooth, somewhat shiny skin, dark red to almost black purple when ripe; the colour is due to a mixture of anthocyanins. Fruit weight varies from 3 to 20 g, with an average of 7 g; diameter varies from 2 to 4 cm. The fruit contains 1–4 flattened kidney-shaped seeds that are 8–11 mm wide. The average fruit has 51% white to nearly translucent, slightly fibrous, juicy pulp, although this percentage varies depending upon fruit size and number of seeds; the average fruit also has 20% skin and 29% seeds (Pinedo *et al.*, 2001); the use of mechanical separators leaves only 30% skin and seeds. Because the fruit skin is thin and coloured, this is generally processed with the pulp to obtain a pleasant reddish purple colour. The chemical composition of the skin–pulp mixture has been reported (Calzada, 1980). Recent prospecting has identified fruit with enormous variation in ascorbic acid content: 0.8–6.1 g in 100 g skin–pulp (Yuyama *et al.*, 2002).

Camu-camu has attracted considerable attention because of its high ascorbic acid content, especially as this is an important antioxidant (Rodrigues *et al.*, 2006). Initially, numerous food products based on the fruit were developed, such as juices, nectars and yogurts, and frozen pulp is also commercialized. Currently numerous healthcare and personal hygiene products have also been made available, including skin creams, shampoos and mascara, as well as capsules and powders made from dehydrated skin (as this is richest in ascorbic acid and anthocyanins).

Natural populations of camu-camu occur most often along the edges of black-water rivers throughout the Amazon Basin and into the Orinoco Basin. Sometimes it is found along white-water rivers, but in this case generally in oxbow lakes and abandoned channels where sediments are not so abundant. In these riverside environments, plants are subject to inundation during the annual cycle of flooding, and will be partially or totally immersed for 4–6 months. In the upper reaches of the tributaries, flooding may occur several times during the rainy season and appears to have a favourable effect on ascorbic acid production. Fruit ripening occurs during the period of rising waters, and ripe fruit are generally harvested from canoes or left for the fish, many of which depend on camu-camu. Until the high ascorbic acid content was identified, camu-camu was seldom used as human food. Roraima, Brazil, is an exception, as the peasants along the rivers consider caçarí a popular juice.

Once the high ascorbic acid content was identified in Peru, a development programme was designed and put into practice, with ecological, agronomic, genetic and processing studies to add value (Pinedo, 2004). One of the first initiatives was to cultivate camu-camu on the 'restingas', the high beaches formed by deposition of sediments during flooding, because this was found to reduce the seasonality of the crop, making fruit available nearly year-round. The first 'restinga' trials were established in 1980 near Iquitos, Peru, through collaboration between the Instituto Nacional de Investigaciones Agrarias (INIA: National Agrarian Research Institute) and the Instituto Veterinario de Investigación de Trópicos y de Altura (IVITA: Tropical and Mountain Veterinary Research Institute). In 1991, inviting 'restinga' farmers to participate actively in the agronomic and genetic trials enriched the research process. Initially, seven producers in the Santa Ana Community (Amazonas River) participated. This number expanded to 28 producers in six communities in 1994, and to 4000 producers in 150 communities in the Departments of Loreto and Ucayali in 1997, as a political decision by the Secretariats of Agriculture to encourage camu-camu production. These 4000 producers manage natural populations and have started planting orchards on 'restingas'.

Peru started to export frozen pulp to Japan in 1995, with most of the production obtained from natural populations. From 1995 to 2000, exports expanded gradually, but were interrupted for 3 years (Pinedo and Jong, 2004). In 2004, exports were reinitiated and in the 2005–2006 harvest season attained a FOB value of US$1.1 million, of which 93% went to Japan. At present, the following export products have low added value: frozen, clarified and concentrated pulp. The internal market is relatively small, although numerous products are now available, including yogurt, nectar and juices that mix camu-camu with other fruits.

The Peruvian improvement programme

The Peruvian development programme was created to transform camu-camu's ascorbic acid potential into a lucrative market for small farmers given preliminary demand from developed countries. The first decision was to concentrate on *Myrciaria dubia* rather than *Myrciaria floribunda* O. Berg, as the former has five times as much ascorbic acid as the latter. None the less, *M. floribunda* may have a role in the future and has been included in germplasm collections.

The improvement programme planners decided to concentrate on an ideotype with four components (Pinedo *et al.*, 2004): high ascorbic acid content (at least 2 g/100 g), high yield, precocity (at least 0.5 kg at 3 years from seed) and fruit size (at least 10 g). The Japanese importers demand at least 1.8 g/100 g of ascorbic acid, which can generally be obtained from wild populations, but collecting from the wild makes quality control more difficult. Germplasm with the desired characteristics existed in the *ex situ* collections and had been evaluated, but vegetative propagation techniques were inadequate and had to be developed in order to get the programme started.

Initially, farmers' perceptions and preferences were not considered, principally because most of them had no experience with the crop. Recently farmers have started to identify plants that do not grow too tall, have large fruits and good yields that are stable year-to-year, as well as plants that yield out of season. The research team from IIAP now does participatory plant evaluation with farmers in numerous communities along the Ucayali River, as well as a few others elsewhere.

A collaborative participatory improvement plan designed by IIAP and INIA started in 2000 to build on previous on-station and participatory work. The plan was designed principally to meet the criteria of the Japanese market, as local demand does not have stringent quality requirements. The participatory aspects include:

- The farmer identifies plants with elite characteristics (see above) and a research team collects samples (both seeds and cuttings).
- After propagation on-station, the research team returns some of the plants to the farmer and the remainder is incorporated into on-station clonal and progeny trials (there are no on-farm progeny or clonal trials).
- The farmer also propagates his best plants from seed to expand his orchard, and he is encouraged to trade seed and seedlings with other farmers as well.
- Seed from F_2 INIA selections (taken from long-term progeny evaluation trials) are also distributed to interested farmers, who plant them in the same plot as their own selections where abundant cross-pollination will be expected.
- Farmers participate in the evaluation of all progenies on their farms, both their own and INIA's selections.

Technologies are being transferred with the improved seed and cuttings. Each time a research team visits farmers they discuss pest control practices and the use of legume ground covers, seeds of which were provided free of charge and with appropriate management instruction.

EXPECTED IMPACTS IN PERU At present, the supply of fruits is insufficient to meet the demand of the Japanese importers. During the 2005–2006 harvest, only 500 t of frozen pulp was exported, which was less than 10% of the stated demand

(Farronay, 2005). Hence, there is strong demand for the fruit coming from new plantations and from managed populations. However, it is not yet clear if demand will expand as supply expands, once all the plantations come into production.

The rapid expansion of the camu-camu project is already benefiting the farmers who joined the effort early, while those who joined later are expecting benefits in the near future. Most of the farmers who joined the project have included camu-camu as an additional component of their diversified traditional production systems. With current high prices, they are enjoying considerable additional income. Near Pucallpa, on the Ucayali River, a community commercialized 25 t of fruit from their plantations and adjacent wild populations, earning about US$11,500 in the 2005/2006 season.

Most families have also started using the fruit for subsistence, which is quite probably improving family health. Many families have started to process fruit for local consumption and markets; in general, home processing is done by the women and older girls, who also benefit directly from sales. The men and women will often take fruit and fruit products to market in alternate weeks, with the benefits managed by whoever goes to market. One group has used family contacts to market directly to Lima, the capital, and is receiving between US$1 and US$1.50 per kilogram of pulp, considerably more than if they sold locally. Although there are health benefits, camu-camu contributes to food security more via income than as a food.

In terms of long-term conservation of genetic resources, the programme has enhanced local genetic diversity by introducing germplasm from other communities. Each farmer now has control over his germplasm and protects it from loss. Additionally, before the current improvement programme was designed and implemented, INIA had encouraged plantings with non-selected seed from numerous natural populations in different river basins (Ucayali, Tigre, Curaray, Yavari, Putumayo, etc.). This distribution enhanced local diversity in the project area and provided an ample genetic base for the farmers to start selecting from.

The Brazilian improvement programme

The INPA group introduced camu-camu in the late 1970s and distributed seed nationwide, but sufficient commercial interest in the fruit only appeared in the early 1990s. With this new demand, the INPA group expanded its prospecting efforts to capture variability throughout the basin, especially along tributaries rather than the main river (Yuyama, 2001). This is the effort that identified camu-camu with 6.1 g of ascorbic acid in 100 g of skin-pulp (Yuyama *et al.*, 2002). The germplasm collected (150 accessions to date) is planted in replicated progeny trials or in collections, where phenotypic characterization and evaluation is executed (Gomes *et al.*, 2004). As soon as plants fruit, their proximate and ascorbic acid compositions are analysed. The group has also worked on vegetative propagation techniques with some success (Pereira and Yuyama, 2002), but a commercially viable protocol remains to be perfected.

In 2003, a collaborative effort among INPA, Embrapa Genetic Resources and Biotechnology and the Federal University of Amazonas was funded to develop an expressed sequence tag (EST) database that could be used to identify the genes involved in ascorbic acid synthesis, as well as other biotechnologically interesting compounds. Numerous ESTs associated with ascorbic acid synthesis and degradation were identified (Silva, 2006), although the full synthesis pathway is still incomplete. ESTs associated with anthocyanin synthesis, oxidative stress and transcription factors were also identified. The same study identified numerous EST microsatellites (SSRs) that are currently being used to examine genetic diversity in the INPA collections and progeny trials. A set of nuclear SSRs is also being developed, since these tend to be more variable than EST-SSRs. The INPA group expects to have a full analysis of genetic diversity throughout the Amazon Basin by 2008.

An ideotype similar to the Peruvian ideotype is used to identify elite plants in the progeny trials and collections. These are now being hybridized in a diallel design to examine the general and specific combining abilities for ascorbic acid production, fruit yield, precocity and plant architecture. Hybrids are also being offered to local farmers, but few are yet convinced that camu-camu will be economically viable in orchards on the non-flooding plateaus of Brazilian Amazonia.

EXPECTED IMPACTS IN BRAZIL Interest in camu-camu as a functional food continues to expand in Europe, Japan and the USA (Yuyama *et al*., 2002, 2003; Rodrigues *et al*., 2006). None the less, interest in cultivating camu-camu in central Amazonia has been minimal to date. An attempt to develop a participatory improvement programme near Manaus was not funded because of lack of demonstrable farmer interest. As the information on the nutritional qualities of camu-camu accumulate, however, increasing interest is being expressed in São Paulo State and camu-camu may be the next Amazonian fruit to migrate out of the region. Although some plantings already exist, at Iguape, São Paulo, for example, they are based on an extremely narrow genetic base and a new pest has already appeared to exploit this. Without expanding the genetic base via the Amazonian collections and improvement programmes, expansion outside of Amazonia may be slower than with other species that are more pest-resistant or have more ample genetic bases to start from.

6.4 Final Thoughts

Even though there is considerable R&D, new and underutilized fruits are seldom institutional priorities. This is partly because public institutions are continuously and seriously underfunded in Latin America and partly because the underfunding causes priorities to be established around export crops. Perhaps worse than the lack of R&D is the fact that underutilized fruits are often socially and cultural marginalized; a tendency that seems to be increasing. Media attention usually focuses on exotics, both as examples of successful exports and as raw materials for international cuisine. Furthermore, schools rarely teach about native species, even in rural areas. Authors who have never seen a native Amazonian cupuaçu or abiu, but who are quite familiar with apples, pears and oranges, write the textbooks used in Brazilian Amazonia. Promotion of underutilized native species is an uphill struggle and the slope is increasing.

With all the difficulties mentioned, there are new things happening in underutilized fruit development in Latin America. These developments can be considered as 'push' and 'pull' vectors. The 'push' vector is the traditional R&D work to provide information, ideas and training to local producers and agro-industries. These actors are more or less competent to use the information, ideas and training to improve productivity, quality and final product. The 'pull' vector is new demand from local, national and international entrepreneurs. These actors may have known native fruits since childhood or have learned about them during a visit to the region, or even seen them on television.

A new 'pull' is that European entrepreneurial demands are starting to filter into Latin America. There are also demands from the USA, Japan and a few other developed countries, but they are less numerous and directed than the European demands. The Japanese demand for camu-camu is an exception. These entrepreneurs are seeking exotic flavours, new colours, different appearances, but they, especially the Europeans, are demanding organic and fair trade certification and high quality in exchange for good prices.

An example of these new times and actions is the fruit of the açaí-do-Pará. This palm has been part of daily subsistence in the Amazon River estuary for millennia and is consumed in enormous quantities. The small (1–2 g) fruits have a thin (1 mm), oily, fibrous, starchy, purple pulp around the seed. The pulp is softened in warm water for a couple of hours and removed from the seed by scraping on metal, wood or fibre screens, resulting in a thick gruel popularly called açaí wine (although it is not fermented). A medium-thick wine contains 12.5% dry matter, of which 52% is fats, 25% is fibres, 10% is proteins, 3% is ash, and 2% is sugars; the wine is also relatively rich in the antioxidants anthocyanins and alpha-tocopherol (Rogez, 2000). The cosmopolitan area of Belém, Pará, Brazil, has a population of nearly 1.2 million, who consume 400 t of açaí fruit per day in the form of açaí wine of various consistencies, depending upon how much water is used during processing. Among the poorer social classes, açaí wine is consumed at every meal, mixed with tapioca or manioc flour.

At first glance, açaí would seem to be an unlikely candidate for development outside its traditional area of consumption. Açaí wine is an acquired taste, somewhat nutty, often somewhat metallic, a little acidic (pH = 5.2; easily corrected with sugar), especially if not truly fresh, so it is actually surprising that it became a fad. There are half a dozen similar wines in Amazonia, none of which is as important as açaí, but all of which are locally popular, just as açaí wine was until some entrepreneurs came into the picture. Since the mid-1990s the popularity of açaí wine has expanded throughout Brazil and has caught the attention of American, European and Japanese entrepreneurs. Brondizio (2004) discusses the local, national and international history of this development.

As an underutilized product from the Brazilian periphery, açaí wine has quality problems, due to fruit quality variation, harvesting and postharvesting practices, shelf life (fruit must be processed within 48 h after harvesting if properly handled), processing practices and storage, all of which affect food quality and safety (Rogez, 2000). As demand started to pick up in the mid-1990s, these problems became limiting very quickly. The Federal University of Pará and the Embrapa Amazônia Oriental both expanded their work on açaí to address these quality issues. Local businesses have adopted the new technologies and best practices and are investing to meet both national and international demands for quality. Embrapa Amazônia Oriental is actively prospecting for açaí that fruits at various times during the year, in an attempt to make açaí available year-round. Initial results are promising and improved seed will start returning to the production areas within the next couple of years (J.T. Farias Neto, Embrapa Amazônia Oriental, 2006, personal communication).

Açaí wine is certainly the major success story in Amazonia today. What can be learned from this success? Chance aside, in our opinion the principal element was entrepreneurial involvement. Açaí wine has existed for millennia in the estuary, but never attracted interest outside Amazonia, although Amazonian researchers have touted its charms for decades. The entrepreneurial motor could only accelerate, however, because there was sufficient product available near Belém, a primitive agro-industry was already processing the wine and the logistics were in place to move it from Belém to south-eastern Brazil and the world. The third, and perhaps decisive, factor was the agility of the R&D institutions in Belém. As soon as demand met local limitations, these institutions moved to solve the quality problems that were causing uneasiness among the budding entrepreneurs and new consumers. Although work and investments are ongoing, the directions are correct and solutions are coming online as soon as they are available.

This example of the synergy between pull and push vectors offers a lesson for R&D institutions interested in the development of underutilized fruits. These institutions must be prepared to work with entrepreneurs to make use of the information that has been accumulating for decades, but which is often stored on library shelves. The best information and product are essential, but it is only an entrepreneur who can transform potential into profit in the highly competitive world fruit market.

While this success certainly stimulates native fruit researchers, it is pertinent to ask who benefits. In the case of açaí, traditional producers are benefiting at present, but once plantations on non-flooding plateaus come into production the traditional producers may lose market share and eventually be pushed out altogether. This projection is similar to Homma's (1993) analysis of non-timber forest products and simply reflects the logic of a capitalist economic system. Numerous Amazonian fruits have already shown that this logic is inexorable; guaraná is now produced principally in Bahia; heart-of-palm from the Amazonian peach palm is now produced principally in São Paulo; cocoa is now produced principally in Bahia, Africa and Asia.

None the less, continued development of Amazonian fruit crops does contribute to food security in the region. If a fruit crop is successful and leaves the region, it will still be produced locally for consumption and local markets. Hence, it will continue to contribute to local diets and nutrition.

What about the conservation of genetic resources? Most institutional programmes suffer from lack of funds for conservation (van Leeuwen *et al.*, 2005) and successful development of a native fruit is no guarantee that this will change – even though the idea of conservation-through-use is widespread. In contrast, however, Cornelius *et al.* (2006) showed that participatory improvement can successfully conserve genetic resources, not only because their use encourages their conservation, but also because of the large numbers involved – numbers that most institutions working with native fruits are unable to manage. We feel that participatory improvement offers other advantages also and are glad to see this practice expanding in Amazonia.

Acknowledgements

We are grateful to Alindo Chuquipoma Diaz, INIA, Tarapoto, Peru; Carlos Oliva, IIAP, Pucallpa, Peru; Fimo Alemán, BASFOR, Cochabamba, Bolivia; and Blas Garcia, CIAT Bolivia, Cochabamba, Bolivia, for information on fruit improvement activities at their respective institutions. We also thank the editors for their encouragement and patience.

References

Brodie, A.W., Labarta-Chavarri, R.A. and Weber, J.C. (1997) *Tree Germplasm Management and Use On-farm in the Peruvian Amazon: A Case Study from the Ucayali Region, Peru.* Research Report, Overseas Development Institute, London, and International Centre for Research in Agroforestry (ICRAF), Nairobi.

Brondizio, E. (2004) From staple to fashion food: shifting cycles and shifting opportunities in the development of the açaí palm fruit economy in the Amazon estuary. In: Zarin, D.J., Alavalapati, J.R.R., Putz, F. and Schmink, M. (eds) *Working Forest in the Neotropics.* Columbia University Press, New York, pp. 339–365.

Calzada, B.J. (1980) *143 Frutales Nativos.* Universidad Nacional Agraria La Molina, Facultad de Agronomia, El Estudiante, Lima.

Calzavara, B.B.G. (1970) Fruteiras: abiu, mamey, bacuri, biribá, cupuaçu. *Série Culturas da Amazônia* 1(2), 45–84, Instituto de Pesquisas e Estudos Agronômicos do Norte, Belém, Pará.

Christinck, A., Weltzien, E. and Hoffmann, V. (2005) Introduction. In: Christinck, A., Weltzien, E. and Hoffmann, V. (eds) *Setting Breeding Objectives and Developing Seed Systems with Farmers.* Margraf Publishers, Wageningen, The Netherlands, pp. 11–24.

Clement, C.R. (1997) Environmental impacts of, and biological and socio-economic limitations on new crop development in Brazilian Amazonia. In: Smartt, J. and Haq, N. (eds) *Domestication, Production and Utilization of New Crops.* International Centre for Underutilised Crops, Southampton, UK, pp. 134–46.

Clement, C.R. (1999) 1492 and the loss of Amazonian crop genetic resources. I. The relation between domestication and human population decline. *Economic Botany* 53, 188–202.

Clement, C.R. (2004) Fruits. In: Prance, G.T. and Nesbitt, M. (eds) *The Cultural History of Plants.* Routledge, London, pp. 77–95.

Clement, C.R. (2006) Fruit trees and the transition to food production in Amazonia. In: Balée, W. and Erickson, C.L. (eds) *Time and*

Complexity in the Neotropical Lowlands: Studies in Historical Ecology. Columbia University Press, New York, pp. 165–185.

Clement, C.R. and Coradin, L. (1988) *Final Report (Revised): Peach Palm (Bactris gasipaes H.B.K.) Germplasm Bank*. US AID Project Report. INPA/Embrapa Cenargen, Manaus, Amazonas, Brazil.

Clement, C.R., Alfaia, S.S., Iriarte-Martel, J.H., Yuyama, K., Moreira Gomes, J.B., van Leeuwen, J., Souza, L.A.G. and Chávez Flores, W.B. (1997) Fruteiras nativas e exóticas. In: Noda, H., Souza, L.A.G. and Fonseca, O.J.M. (eds) *Duas Décadas de Contribuições do INPA à Pesquisa Agronômica no Trópico Úmido*. Instituto Nacional de Pesquisas da Amazônia, Manaus, Amazonas, Brazil, pp. 111–129.

Clement, C.R., Weber, J.C., van Leeuwen, J., Domian, C.A., Cole, D.M., Arévalo Lopez, L.A. and Argüello, H. (2004) Why extensive research and development did not promote use of peach palm fruit in Latin America. *Agroforestry Systems* 61, 195–206.

Cornelius, J.P., Clement, C.R., Weber, J.C., Sotelo-Montes, C., van Leeuwen, J., Ugarte-Guerra, L.J., Rices-Tembladera, A. and Arévalo-López, L. (2006) The trade-off between genetic gain and conservation in a participatory improvement programme: the case of peach palm (*Bactris gasipaes* Kunth). *Forests, Trees and Livelihoods* 16, 17–34.

Farronay, P.R. (2005) *Actualización sobre logros económicos de las plantaciones del camu-camu en el distrito de Sapuena, rio Ucayali*. Instituto de Investigaciones de la Amazonia Peruana (IIAP), Iquitos, Peru.

Gomes, J.C.B., Costa, S.S., Soares, C.P. and Yuyama, K. (2004) Caracterização fenológico do Banco de Germoplasma do camu-camu abordado por meio de análise estatistica multivariadas. In: *Proceedings of the Congresso Brasileiro de Fruticultura*. Sociedade Brasileira de Fruticultura, Jaboticabal, SP, Brazil, v. 18, pp. 750–755.

Homma, A.K.O. (1993) *Extrativismo vegetal na Amazônia: Limites e oportunidades*. Embrapa SPI, Brasília.

Mann, C.C. (2005) *1491: New Revelations of the Americas before Columbus*. Alfred A. Knopf, New York.

Patiño, V.M. (2002) *Historia y Dispersión de los Frutales Nativos del Neotrópico*. CIAT Publication No. 326. CIAT, Cali, Colombia.

Pereira, B.G. and Yuyama, K. (2002) Produção de mudas de camu-camu (*Myrciaria dubia* (H.B.K.) McVaugh) por estaquia utilizando ramos provenientes de diferentes tipos e posição da planta. In: *Proceedings of the Congresso Brasileiro de Fruticultura*. Sociedade Brasileira de Fruticultura, Jaboticabal, SP. Brazil, v. 17, pp. 161–165.

Pinedo, P.M. (2004) Reflexiones criticas acerca de un programa de promoción del camu-camu (*Myrciaria dubia* McVaugh, Mirtaceae) en zonas riberenas de la Amazonia Peruana. In: *Critica de Proyectos y Proyectos Criticos de Desarrollo*. Instituto de Investigaciones de la Amazonia Peruana (IIAP), Iquitos, Peru, pp. 61–76.

Pinedo, P.M. and Jong, W. (2004) Camu-camu (*Myrciaria dubia* McVaugh, Myrtaceae), un arbusto amazonico de areas inundables con alto contenido de vitamina C en Loreto, Peru. In: Alexiades, M. and Shanley, P. (eds) *Productos Forestales, Medios de Subsistencia y Conservacion*; Estudios de Caso sobre Sistemas de Manejo de Productos Forestales No Maderables. Volumen 3 – America Latina. CIFOR, Bogor, Indonesia, pp. 275–294.

Pinedo, P.M., Riva, R.R., Rengifo, S.E., Delcado, V.C., Villacrés, V.J., Gonzales, C.A., Inga, S.H., Lopez, U.A., Farronay, P.R., Vega, V.R. and Linares, B.C. (2001) *Sistema de Producción de Camu-camu en Restinga*. Instituto de Investigaciones de la Amazonia Peruana (IIAP), Iquitos, Peru.

Pinedo, P.M., Linares, B.C., Mendoza, H. and Anguiz, R. (2004) *Plan de mejoramiento genetico de camu-camu*. Instituto de Investigaciones de la Amazonia Peruana (IIAP), Iquitos, Peru.

Piperno, D.R. and Pearsall, D.M. (1998) *The Origins of Agriculture in the Lowland Neotropics*. Academic Press, San Diego, California.

Rodrigues, R.B., Papagiannopoulos, M., Maia, J.G.S., Yuyama, K. and Marx, F. (2006) Antioxidant capacity of camu camu [*Myrciaria dubia* (H.B.K.) McVaugh] pulp. *Ernährung/Nutrition* 30, 357–362.

Rogez, H. (2000) *Açaí: Preparo, composição e melhoramento da conservação.* EDUFPA, Belém, Brazil, 313 pp.

Silva, M.L. (2006) Estudo de genes expressos em frutos de camu-camu: sequenciamento de ESTs. Doctoral thesis, Biotechnology Post-Graduate Program, Universidade Federal de Amazonas, Manaus, Amazonas, Brazil, 117 pp.

Simons, A.J. and Leakey, R.R.B. (2004) Tree domestication in tropical agroforestry. *Agroforestry Systems* 61, 167–181.

Smith, N., Dubois, J., Current, D., Lutz, E. and Clement, C. (1998) *Agroforestry Experiences in the Brazilian Amazon: Constraints and Opportunities.* Pilot Program to Conserve the Brazilian Rain Forest, Ministério do Meio Ambiente, Recursos Hídricos e Amazônia Legal & World Bank, Brasília.

Sotelo Montes, C. and Weber, J.C. (1997) Priorización de especies arbóreas para sistemas agroforestales en la selva baja del Perú. *Agroforestería en las Américas* 4, 12–17.

van Leeuwen, J., Menezes, J.M.T., Moreira Gomes, J.B., Iriarte-Martel, J.H. and Clement, C.R. (1997) Sistemas agroflorestais para a Amazônia: importância e pesquisas realizadas. In: Noda, H., Souza, L.A.G. and Fonseca, O.J.M. (eds) *Duas Décadas de Contribuições do INPA à Pesquisa Agronômica no Trópico Úmido.* Instituto Nacional de Pesquisas da Amazônia, Manaus, Amazonas, Brazil, pp. 131–146.

van Leeuwen, J., Lleras Pérez, E. and Clement, C.R. (2005) Field genebanks may impede instead of promote crop development: lessons of failed genebanks of 'promising' Brazilian palms. *Agrociencia* 9, 61–66.

Vernooy, R. (2003) *Semillas Generosas. Mejoramiento Participativo de Plantas.* IDRC, Ottawa.

Weber, J.C., Sotelo Montes, C., Vidaurre, H., Dawson, I.K. and Simons, A.J. (2001) Participatory domestication of agroforestry trees: an example from the Peruvian Amazon. *Development in Practice* 11, 425–433.

Yuyama, K. (2001) Domesticação de germoplasma de camu-camu [*Myrcyaria dubia* (H.B.K.) McVaugh] para o uso em agroindústria na Amazônia. In: *Relatório do Programa Piloto para a Proteção das Florestas Tropicais do Brasil.* PPG/MCT, 2001. Ministério de Ciência e Tecnologia, Brasília, pp. 149–153.

Yuyama, K., Aguiar, J.P.L. and Yuyama, L.K.O. (2002) Camu-camu: um fruto fantástico como fonte de vitamina C. *Acta Amazonica* 36, 169–174.

Yuyama, L.K.O., Aguiar, J.P.L., Yuyama, K., Lopes, T.M., Fávaro, D.I.T., Berge, P.C.P. and Vasconcellos, M.B.A. (2003) Teores de elementos minerais em algumas das populações de camu-camu. *Acta Amazonica* 33, 949–954.

Appendix

Table A6.1. Fruit crops with domesticated populations that were present in Amazonia at the time of European conquest (see Clement, 1999, for bibliography).

Species	Family	Probable origin	Uses
Annona muricata L.	Anonaceae	N.–S. America	Fruit
Rollinia mucosa (Jacq.) Baillón	Anonaceae	Amazonia	Fruit
Crescentia cujete L.	Bignoniaceae	N.–S. America	Tree gourd
Bixa orellana L.	Bixaceae	SW. Amazonia	Colourant
Ananas comosus (L.) Merrill.	Bromeliaceae	Brazil/Paraguay	Fruit
Carica papaya L.	Caricaceae	MesoAmerica	Fruit
Poraqueiba paraensis Ducke	Icacinaceae	E. Amazonia	Fruit, oil
Poraqueiba sericea Tul.	Icacinaceae	W. Amazonia	Fruit, oil
Persea americana Mill.	Lauraceae	MesoAmerica	Fruit
Bactris gasipaes Kunth	Palmae	SW. Amazonia	Fruit
Passiflora edulis Sims	Passifloraceae	N.–S. America	Fruit
Passiflora quadrangularis L.	Passifloraceae	N.–S. America	Fruit
Genipa americana L.	Rubiaceae	N.–S. America	Colourant
Paullinia cupana Kunth	Sapindaceae	C. Amazonia	Stimulant
Pouteria caimito Radlk.	Sapotaceae	Amazonia	Fruit
Solanum sessiliflorum Dunal	Solanaceae	W. Amazonia	Fruit

Table A6.2. Fruit crops with semi-domesticated populations that were present in Amazonia at the time of European conquest (see Clement, 1999, for bibliography).

Species	Family	Probable origin	Uses
Anacardium occidentale L.	Anacardiaceae	NE. Brazil?	Fruit, nut
Spondias mombin L.	Anacardiaceae	N.–S. America	Fruit
Annona montana Macf.	Anonaceae	Amazonia	Fruit
Annona reticulata L.	Anonaceae	MesoAmerica	Fruit
Macoubea witotorum Schultes	Apocynaceae	W. Amazonia	Fruit juice
Thevetia peruvianum Merr.	Apocynaceae	C. Andes	Poison
Quararibea cordata Vischer	Bombacaceae	W. Amazonia	Fruit
Couepia subcordata Benth.	Chrysobalanaceae	Amazonia	Fruit
Mammea americana L.	Guttiferae	Antilles	Fruit
Platonia insignis Mart.	Guttiferae	E. Amazonia	Fruit, seed?
Cassia leiandra Benth.	Leg. Caesalpinoidae	Amazonia	Fruit
Inga cinnamomea Benth.	Leg. Caesalpinoidae	Amazonia	Fruit
Inga edulis Mart.	Leg. Caesalpinoidae	W. Amazonia	Fruit
Inga feuillei DC	Leg. Caesalpinoidae	W. Amazonia	Fruit
Inga macrophylla H.B.K.	Leg. Caesalpinoidae	W. Amazonia	Fruit
Bunchosia armeniaca DC	Malpigiaceae	Amazonia	Fruit
Byrsonima crassifolia H.B.K.	Malpigiaceae	MesoAmerica	Fruit
Pourouma cecropiifolia Mart.	Moraceae	W. Amazonia	Fruit
Eugenia stipitata McVaugh	Myrtaceae	W. Amazonia	Fruit
Myrciaria cauliflora McVaugh	Myrtaceae	S. Brazil	Fruit
Psidium guajava L.	Myrtaceae	NE. Brazil	Fruit
Astrocaryum aculeatum Meyer	Palmae	W. Amazonia	Fruit
Talinum triangulare Willd.	Portulacaceae	N.–S. America	Vegetable
Borojoa sorbilis Cuatr.	Rubiaceae	Amazonia	Fruit
Paullinia yoco Schult. & Killip	Sapindaceae	W. Amazonia	Stimulant
Pouteria macrocarpa Baehni	Sapotaceae	Amazonia	Fruit
Pouteria macrophylla (Lam.) Eyma.	Sapotaceae	Amazonia	Fruit
Pouteria obovata H.B.K.	Sapotaceae	C. Andes	Fruit
Theobroma bicolor H. & B.	Sterculiaceae	W. Amazonia	Fruit, seed
Theobroma cacao L.	Sterculiaceae	W. Amazonia	Stimulant

Table A6.3. Fruit crops with incipiently domesticated populations that were present in Amazonia at the time of European conquest (see Clement, 1999, for bibliography).

Species	Family	Probable origin	Uses
Couma utilis Muell.	Apocynaceae	Amazonia	Fruit, latex
Hancornia speciosa Gomes	Apocynaceae	NE. Brazil	Fruit, latex
Caryocar glabrum (Aubl.) Pers.	Caryocaraceae	W. Amazonia	Nut
Caryoca nuciferum L.	Caryocaraceae	N.–S. America	Nut
Caryoca villosum (Aubl.) Pers.	Caryocaraceae	C. Amazonia	Fruit
Chrysobalanus icaco L.	Chrysobalanaceae	N.–S. America	Fruit
Couepia bracteosa Benth.	Chrysobalanaceae	C. Amazonia	Fruit
Couepia edulis Prance	Chrysobalanaceae	Amazonia	Nut
Couepia longipendula Pilger	Chrysobalanaceae	Amazonia	Nut
Caryodendron orinocense Karst.	Euphorbiaceae	W. Amazonia	Nut
Hevea spp. (various)	Euphorbiaceae	Amazonia	Seed, latex
Rheedia brasiliensis Pl. & Tr.	Guttiferae	Amazonia	Fruit
Rheedia macrophylla (Mart.) P. & Tr.	Guttiferae	Amazonia	Fruit
Bertholletia excelsa H. & B.	Lecythidaceae	E. Amazonia	Nut
Lecythis pisonis Camb.	Lecythidaceae	Amazonia	Nut
Grias neubertii MacBride	Lecythidaceae	W. Amazonia	Fruit
Grias peruviana Miers	Lecythidaceae	W. Amazonia	Fruit
Hymenaea courbaril L.	Leg. Caesalpinioidae	Amazonia	Starchy fruit
Campsiandra comosa Cowan	Leg. Mimosoidae	NW. Amazonia	Fruit
Inga spp. (numerous)	Leg. Mimosoidae	Amazonia	Fruit
Eugenia uniflora L.	Myrtaceae	S. America	Fruit
Psidium acutangulum DC	Myrtaceae	Amazonia	Fruit
Psidium guineensis Sw.	Myrtaceae	N.–S. America	Fruit
Acrocomia aculeata (Jacq.) Lood	Palmae	E. Amazonia	Oily fruit
Astrocaryum murumuru Mart.	Palmae	E. Amazonia	Oily fruit
Elaeis oleifera (H.B.K.) Cortés	Palmae	N.–S. America	Oily fruit
Euterpe oleracea Mart.	Palmae	E. Amazonia	Oily fruit
Jessenia bataua (Mart.) Burret	Palmae	N.–S. America	Oily fruit
Mauritia flexuosa L. f.	Palmae	N.–S. America	Oily fruit
Maximiliana maripa Drude	Palmae	E. Amazonia	Oily fruit
Oenocarpus bacaba Mart.	Palmae	Amazonia	Oily fruit
Oenocarpus distichus Mart.	Palmae	E. Amazonia	Oily fruit
Alibertia edulis A. Rich ex DC	Rubiaceae	Amazonia	Fruit
Melicoccus bijugatus Jacq.	Sapindaceae	C. & N.–S. America	Fruit
Talisia esculenta Radlk.	Sapindaceae	W. Amazonia	Fruit
Manilkara huberi (Huber) Standl.	Sapotaceae	Amazonia	Fruit, latex
Pouteria spp. (numerous)	Sapotaceae	Amazonia	Fruit
Sterculia speciosa K. Sch.	Sterculiaceae	Amazonia	Fruit
Theobroma grandiflorum Schum.	Sterculiaceae	E. Amazonia	Fruit
Theobroma speciosum Willd.	Sterculiaceae	Amazonia	Fruit
Theobroma subincanum Mart.	Sterculiaceae	Amazonia	Fruit
Erisma japura Spruce	Vochysiaceae	NW. Amazonia	Fruit

7 The Domestication of Fruit and Nut Tree Species in Vanuatu, Oceania

V. Lebot,[1] A. Walter[2] and C. Sam[3]

[1]*CIRAD, Port-Vila, Vanuatu*; [2]*IRD, Agropolis, Montpellier, France*; [3]*MQAFF, National Herbarium, Port-Vila, Vanuatu*

7.1 Introduction

The population of the Pacific Islands, comprising Melanesia, Polynesia and Micronesia (hereafter named Oceania), is increasing very rapidly (SPC Demography/Population Programme, 2000). In some Melanesian countries, such as Vanuatu, it is doubling every 20 years. There is obviously an urgent need to develop smallholder production and plant improvement. Crop improvement and protection and the maintenance of soil fertility are the key areas. Local species of fruit and nut trees can play a major role because they are well adapted to traditional cropping systems. In Oceania arboriculture is the necessary complement of the traditional cropping system (Pollock, 2002). Most of the traditional staple crops are propagated vegetatively and are established asexually without ploughing. Their root systems are very superficial and their cultivation is safer when conducted in very small plots (0.1–0.3 ha) within agroforestry systems where shade and windbreaks are well established. Fruit and nut species are therefore omnipresent, not only because their production is appreciated but also because their protective role is essential for sustainable food production (Walter and Sam, 1999; Walter and Lebot, 2003).

The fruit tree species that have been introduced since the 16th century include mango, papaya, lime, guava, custard apple, orange, grapefruit, mandarin and avocado (Yen, 1998). However, it is much harder to define the centres of origin and original distributions of other fruiting species because they were spread more widely by humans during earlier aboriginal migrations (Yen, 1985, 1993). Thus, a large number of fruit tree species are present over a vast area from the Indomalayan region to eastern Polynesia. These include the sea almond (*Terminalia catappa*), Indian mulberry (*Morinda citrifolia*), Tahitian chestnut (*Inocarpus fagifer*) and candlenut (*Aleurites moluccana*), to mention just a few of the most common species (Smith, 1985).

The richness of individual floras decreases progressively from the large island of New Guinea (French, 1986), insular Melanesia (Whitmore, 1986; Wheatley, 1992), Fiji (Smith, 1981) to Polynesia (Whistler, 1984), but the Polynesian islands also host some of the most improved morphotypes for a given species. This is the case, of course, for the famous seedless breadfruits but also for a few other species. The reduction of species diversity from west to east, New Guinea being the richest reservoir, is also responsible for bottlenecks, with drastic narrowing of the genetic base on the oceanic islands (Lebot, 1992, 1999).

The majority of societies in Melanesia cultivate fruit trees and regularly eat fruits and nuts (Barrau, 1962). In Polynesia, however, the situation is different and species introduced subsequent to contact with Europeans are nowadays becoming ever more dominant, to the detriment of indigenous species. The aim of this chapter is to review the present status of indigenous fruit and nut species exploited in Vanuatu (Fig. 7.1), Melanesia, and to discuss their stage of domestication, improvement, future breeding prospects and commercial development.

7.2 Stage of Domestication

A comprehensive list of indigenous Oceanian species of fruit trees is presented in Table 7.1. A list of nut trees is presented in Table 7.2. These species are not equally exploited. Some are only foraged while others are protected; only a few are truly cultivated.

7.2.1 Foraged species

Numerous species are simply foraged from the wild. They are naturally disseminated by bats, birds and/or ocean currents when they can reproduce spontaneously. This is the case with *Aceratium*, *Burckella*, *Corynocarpus*, *Ficus*, *Garcinia*, *Garuga*, *Haplolobus*, *Horsfieldia*, *Maesa*, *Mangifera*, *Myristica*, *Parartocarpus*, *Parinari*, *Phyllocladus*, *Pipturus*, *Planchonella*, *Pleiogynium*, *Pouteria*, *Semecarpus*, *Syzygium* and *Terminalia* species. Quite often their growth is protected by humans who avoid weeding out or destroying the seedlings and young trees. These species are well identified by local communities and are designated in vernacular languages by a name and often by a qualifier to distinguish the diverse morphotypes existing within species. In most cases, villagers know the exact location of all the useful trees and are thus in a position to harvest them according to the season and to their needs. The harvest is often opportunistic and occurs when people walk through their territory to reach their food gardens or to exploit forest resources. However, when the frequent cyclones destroy their gardens foraged species become essential to secure subsistence. Nuts are collected from *Adenanthera*, *Agathis*, *Barringtonia*, *Canarium*, *Castanopsis*, *Cordia*, *Elaeocarpus*, *Finschia*, *Heritieria*, *Pandanus*, *Pangium*, *Pouteria*, *Sterculia* and *Terminalia* species. For these foraged species, most of the fruits and nuts are eaten raw or cooked on the spot as people move through their territory or work in their plots and they are rarely picked up to be taken back to the village.

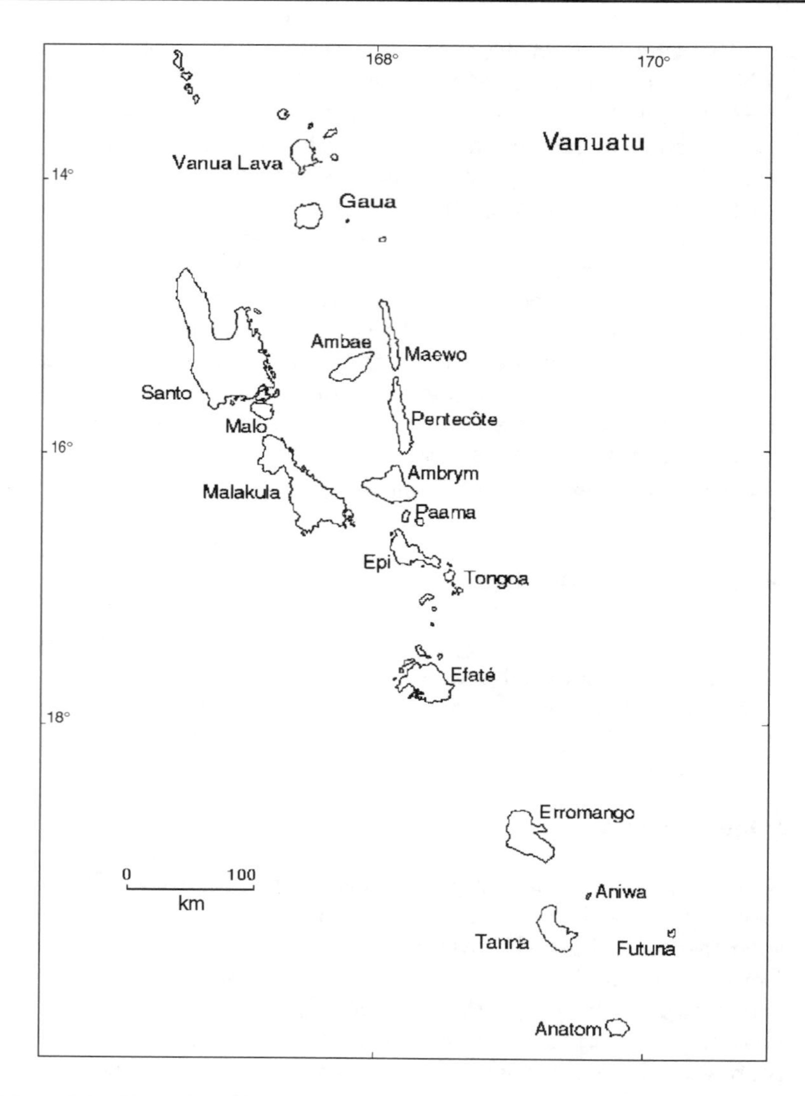

Fig. 7.1. Map of the Vanuatu archipelago.

7.2.2 Protected species

These species are not truly cultivated but are gathered in abundance and eaten regularly. This is the case for the fruits collected from *Bruguiera gymnorrhiza*, *Burckella fijiensis*, *B. obovata*, *Dracontomelon* spp., *Ficus scabra*, *Pandanus* spp., *Pometia pinnata* and *Syzygium clusiifolium* and the nuts collected from *Gnetum costatum* and *G. latifolium*. They are often represented by important groves established by currents on the shore close to villages. They reproduce well spontaneously and provide abundant fruits and nuts that are eaten regularly. Villagers, however, can collect seedlings and attempt to establish

Table 7.1. Fruit tree species of Oceania.

Species	Family	Foraged	Protected	Cultivated
Aceratium insulare	Elaeocarpaceae	✓		
Aceratium oppositifolium D.C.	Elaeocarpaceae	✓		
Antidesma bunius	Euphorbiaceae	✓		
Artocarpus altilis (Parkinson) Fosberg	Moraceae			✓
Artocarpus vriesianus Miquel	Moraceae	✓		
Bacaura papuana F.M. Bailey	Euphorbiaceae	✓		
Bruguiera gymnorrhiza H.J. Lam	Rhizophoraceae		✓	
Bruguiera sexangula Poiret	Rhizophoraceae	✓		
Burckella fijiensis (Hemsley) A.C. Smith & S. Darwin	Sapotaceae		✓	
Burckella obovata (G. Forster) Pierre	Sapotaceae		✓	
Burckella richii (A. Gray) Lam	Sapotaceae	✓		
Burckella sorei Van Royen	Sapotaceae	✓		
Burckella sp. (*Cassidispermum megahilum* Hemsley)	Sapotaceae	✓		
Citrus macroptera Montrouzier	Rutaceae	✓		
Clymenia polyandra (Tanaka) Swingle	Rutaceae	✓		
Corynocarpus cribbianus (F.M. Bail) L.S. Sm.	Corynocarpaceae	✓		
Corynocarpus similis Hemsley	Corynocarpaceae	✓		
Diospyros elliptica (Forster) Green	Ebenaceae	✓		
Diospyros major (Forster) Bakh.	Ebenaceae	✓		
Dracontomelon dao (Blanco) Merr. & Rolfe	Anacardiaceae		✓	
Dracontomelon lenticulatum Wilkinson	Anacardiaceae		✓	
Dracontomelon vitiense Engl. ex Guillaumin	Anacardiaceae		✓	
Eriandra fragrans Van Royen & Van Steenis	Polygalaceae			
Ficus adenosperma Miquel	Moraceae	✓		
Ficus arbuscula K. Schum. & Lauterb.	Moraceae	✓		
Ficus aspera Forster f.	Moraceae		✓	
Ficus austro-caledonica Bur.	Moraceae	✓		
Ficus bambusaefolia Seemann	Moraceae	✓		
Ficus barclayana (Miq.) Summerhayes	Moraceae	✓		
Ficus barraui Guillaumin	Moraceae	✓		
Ficus botryocarpa Miquel	Moraceae	✓		
Ficus calopilina Diels	Moraceae	✓		
Ficus copiosa Steudel	Moraceae	✓		
Ficus dammaropsis Diels	Moraceae	✓		
Ficus granatum Forster f.	Moraceae		✓	
Ficus gymnocrygma Summerhayes	Moraceae	✓		
Ficus itoana Diels	Moraceae	✓		
Ficus masonii Horne ex Baker	Moraceae	✓		
Ficus obliqua Forst. f.	Moraceae	✓		
Ficus pritchardii Seemann	Moraceae	✓		
Ficus pungens Reinw. ex Blume	Moraceae	✓		
Ficus scabra Forster f.	Moraceae		✓	
Ficus tinctoria Forster f.	Moraceae	✓		
Ficus virgata Reinw. ex Blume	Moraceae	✓		
Ficus vitiensis Seemann	Moraceae	✓		
Ficus wassa Roxburgh	Moraceae		✓	

Continued

Table 7.1. *Continued*

Species	Family	Foraged	Protected	Cultivated
Flacourtia jangomas (Lour.) Raeusch	Flacourtiaceae	✓		
Flacourtia rukam Zollinger & Morritzi	Flacourtiaceae	✓		
Galbulimima belgrayeana Sprague	Magnoliaceae	✓		
Garcinia hollrungii Lauterb.	Clusiaceae	✓		
Garcinia jaweri Lauterb.	Clusiaceae	✓		
Garcinia pseudoguttifera Seemann	Clusiaceae	✓		
Garuga floribunda M.J. Decaisne	Burseraceae	✓		
Gnetum gnemon L.	Gnetaceae			✓
Grewia crenata Hochst ex Mast	Tiliaceae	✓		
Haplolobus floribundus (K. Schum.) H.J. Lam	Burseraceae	✓		
Horsfieldia spicata (Roxb.) J. Sinclair	Myristicaceae	✓		
Horsfieldia sylvestris Warburg	Myrsinaceae	✓		
Maesa ambrymensis Guillaumin	Myrsinaceae	✓		
Mangifera minor Blume	Anacardiaceae	✓		
Mangifera mucronulata Blume	Anacardiaceae	✓		
Morinda citrifolia L.	Rubiaceae			✓
Myristica hollrungii Warburg	Myristicaceae	✓		
Pandanus conoideus Lam	Pandanaceae		✓	
Pandanus englerianus Martelli	Pandanaceae		✓	
Pandanus magnificus Martelli	Pandanaceae		✓	
Parartocarpus venenosus Beccari	Urticaceae	✓		
Parinari solomonensis	Chrysobalanaceae	✓		
Phyllocladus hypophyllus Hook. f.	Coniferae	✓		
Pipturus argenteus (Forster f.) Weddell	Urticaceae	✓		
Planchonella grayana St John	Sapotaceae	✓		
Pleiogynium timoriense (D.C.) Leenhouts	Anacardiaceae	✓		
Pometia pinnata J.R. & G. Forster	Sapindaceae			✓
Pouteria campechiana (H.B.K.) Baehni	Sapotaceae	✓		
Semecarpus cassuvium Roxburgh	Anacardiaceae	✓		
Spondias cytherea Sonnerat	Anacardiaceae			✓
Spondias novoguineensis A.J.G.H. Kostermans	Anacardiaceae	✓		
Syzygium aquaeum (Burm.) Alston	Myrtaceae	✓		
Syzygium clusiifolium (A. Gray) Mueller	Myrtaceae		✓	
Syzygium corynocarpum (A. Gray) Mueller	Myrtaceae	✓		
Syzygium malaccense (L.) Merrill & Perry	Myrtaceae			✓
Syzygium nutans (K. Schum.) Merill & Perry	Myrtaceae	✓		
Syzygium richii (A. Gray) Merrill & Perry	Myrtaceae	✓		
Syzygium samarangense (Blume) Merril & Perry	Myrtaceae	✓		
Terminalia microcarpa Decne	Combretaceae	✓		
Ximenia americana L.		✓		

Sources: Smith, 1981; Whistler, 1984; French, 1986; Whitmore, 1986; Wheatley, 1992; Walter and Sam, 1999.

Table 7.2. Nut trees species of Oceania.

Species	Family	Foraged	Protected	Cultivated
Adenanthera pavonina L.	Fabaceae	✓		
Agathis hypoleuca Warburg	Araucariaceae	✓		
Agathis lanceolata Warburg	Araucariaceae	✓		
Agathis macrophylla (Lindley) Masters	Araucariaceae	✓		
Agathis moorei Masters	Araucariaceae	✓		
Agathis ovata Warburg	Araucariaceae	✓		
Agathis spinulosa	Araucariaceae	✓		
Aleurites moluccana Willd.	Euphorbiaceae	✓		
Artocarpus fretessii Teysm. & Binn. ex Hassk	Moraceae	✓		
Barringtonia edulis Seemann	Lecythidaceae			✓
Barringtonia niedenzuana (K. Schum.) Knuth	Lecythidaceae	✓		
Barringtonia novae-hiberniae Lauterbach	Lecythidaceae			✓
Barringtonia procera (Miers) Knuth	Lecythidaceae			✓
Barringtonia seaturae H.B. Guppy	Lecythidaceae	✓		
Bruguiera cylindrical Blume	Rhizophoraceae	✓		
Canarium decumamum Gaertner	Burseraceae	✓		
Canarium harveyi Seemann	Burseraceae			✓
Canarium hirstum	Burseraceae	✓		
Canarium indicum L.	Burseraceae			✓
Canarium kaniense Lauterbach	Burseraceae	✓		
Canarium oleiferum Baillon	Burseraceae	✓		
Canarium schlechteri Lauterbach	Burseraceae	✓		
Canarium solomonense	Burseraceae	✓		
Canarium vanikoroense Leenhouts	Burseraceae	✓		
Canarium vulgare Leenhouts	Burseraceae	✓		
Castanopsis acuminatissima A. DC. ex Hance	Cupuliferae	✓		
Castanospermum australe A. Cunn. & Fraser	Papillionaceae	✓		
Cordia dichotoma Forst. f.	Boraginaceae	✓		
Cordia subcordata H.J. Lam	Boraginaceae	✓		
Cryptocarya wilsonii Guillaumin	Lauraceae	✓		
Elaeocarpus chelonimorphus Gillespie	Elaeocarpaceae	✓		
Elaeocarpus kambi L.S. Gibbs	Elaeocarpaceae	✓		
Elaeocarpus polydactylus Schlecther	Elaeocarpaceae	✓		
Elaeocarpus pullenii Weibel	Elaeocarpaceae	✓		
Elaeocarpus womersleyi Weibel	Elaeocarpaceae		✓	
Finschia chloroxantha Diels	Proteaceae	✓		
Finschia ferruginiflora C.T. White	Proteaceae	✓		
Gnetum costatum K. Schum.	Gnetaceae		✓	
Gnetum latifolium Blume	Gnetaceae		✓	
Heritiera trifoliate (F. Muell.) Kosterman	Sterculiaceae	✓		
Inocarpus fagifer (Parkinson ex. Zollinger) Fosberg	Fabaceae			✓
Kermadecia sinuate Brongn. & Gris.	Proteaceae	✓		
Macadamia leptophylla (Guillaumin) Virot	Proteaceae			✓
Neisosperma oppositifolium (Lam.) Fosberg & Sachet	Apocynaceae	✓		
Pandanus antaresensis St John	Pandanaceae	✓		
Pandanus brosimos Merrill & Perry	Pandanaceae	✓		
Pandanus dubius Sprengel	Pandanaceae		✓	

Continued

Table 7.2. *Continued*

Species	Family	Foraged	Protected	Cultivated
Pandanus foveolatus Kanehira	Pandanaceae	✓		
Pandanus julianettii Martelli	Pandanaceae	✓		
Pandanus tectorius Soland. ex Balf. f.	Pandanaceae	✓		
Pangium edule Warburg	Flacourtiaceae		✓	
Pittosporum pullifolium Burkill	Pittosporaceae	✓		
Pouteria endlicheri (Montrouz) Baehni	Sapotaceae	✓		
Pouteria linggensis (Burck) Baehni	Sapotaceae	✓		
Pouteria maclayana (F. Muell.) Baehni	Sapotaceae	✓		
Scleropyrum aurantiacum Pilge	Santalaceae	✓		
Semecarpus atra Vieillard	Anacardiaceae	✓		
Sleumerodendron austro-caledonicum (Brongn. & Gris) Virot	Proteceae	✓		
Sloanea tieghemmii (F. Muell.) A.C. Smith	Asclepiadaceae	✓		
Sterculia foetida L.	Sterculiaceae	✓		
Sterculia schumanniana Milbraed	Sterculiacea	✓		
Sterculia vitiensis Seemann	Sterculiaceae			✓
Terminalia catappa L.	Combretaceae			✓
Terminalia copelandii Elmer	Combretaceae	✓		
Terminalia impediens Coode	Combretaceae	✓		
Terminalia kaernbachii Warburg	Combretaceae			✓
Terminalia litoralis Panch. ex Guillaumin	Combretaceae	✓		
Terminalia samoensis Rechinger	Combretaceae	✓		
Terminalia sepicana Diels	Combretaceae	✓		

Sources: Smith, 1981; Whistler, 1984; French, 1986; Whitmore, 1986; Wheatley, 1992; Walter and Sam, 1999.

them in areas where they do not exist in order to ease collection, for example near gardens, thus extending their area of distribution and increasing the number of selected individuals within that area.

7.2.3 Cultivated species

Certain species are truly and regularly cultivated and they have a wide geographical distribution. These are often staple foods for Oceanians and their distribution is the result of early migrations on voyaging canoes. They are mostly coastal species and are found around homesteads. Through millennia of cultivation, genotypes have been improved by following a continuous selection process. This is the case for the breadfruit (*Artocarpus altilis*), well known for its fruits, but also foraged, protected and cultivated for its nuts, the seeds of the fruits, in northern Melanesia and Papua New Guinea. It is also the case for the fruits of *Gnetum gnemon*, *Morinda citrifolia*, *Pometia pinnata*, *Spondias cytherea* and *Syzygium malaccense* and the nuts of *Barringtonia edulis*, *B. novae-hiberniae*, *B. procera*, *Canarium harveyi*, *C. indicum*, *Inocarpus fagifer*, *Sterculia vitiensis* and *Terminalia catappa*. Some of these species are foraged in the wild, protected along pathways or around gardens and, at the same time, truly cultivated around villages, so there is a continual gene flow from space to space.

7.3 Indigenous Species with Economic Potential

Many indigenous fruits and nut species have been identified in Oceania. However, only a few have an economic potential which can be readily exploited. These include *Artocarpus altilis*, *Barringtonia edulis*, *Barringtonia novae-hiberniae*, *Barringtonia procera*, *Burckella fijiensis*, *Burckella obovata*, *Burckella* spp. (*Cassidispermum megahilum*), *Canarium harveyi*, *Canarium indicum*, *Dracontomelon vitiense*, *Inocarpus fagifer*, *Morinda citrifolia*, *Spondias cytherea*, *Syzygium malaccense* and *Terminalia catappa*. In the following sections we will discuss the major species.

7.3.1 *Artocarpus altilis*

Artocarpus altilis (Parkinson) Fosberg, commonly called breadfruit, is cultivated throughout the Pacific generally by vegetative means except in Melanesia, where sexual propagation is often used. The tree grows abundantly up to 600 m above sea level (m.a.s.l.). Mature trees may be found in secondary forests, where they usually indicate a site of earlier settlement. The fruits are cooked in various ways: grilled, braised in an oven, sometimes cut up and boiled, often grated and cooked into a *laplap* (the national dish of Vanuatu, consisting of an amylaceous pudding made of a graded root crop, bananas or breadfruit). The mode of cooking of each cultivar depends on the texture of its fruits (Walter, 1989). The seeds are sometimes eaten grilled or boiled. Young, unrolled, leaves can also be eaten after cooking by steaming. More than 100 cultivars are known on the island of Malo in Vanuatu alone, whereas only ten or so occur in Tonga.

In Melanesia the seeded forms are very abundant whereas the parthenocarpic, seedless clones are rare; in Polynesia the situation is reversed. Seeded forms are mostly diploids while seedless cultivars are triploids. There is tremendous genetic diversity within diploids and the genetic base of triploids is extremely narrow (Ragone, 1991, 1997). Using molecular markers, it has now been demonstrated that two species (*A. camansi* Blanco and *A. mariennensis* Trécul) and two different events (vegetative propagation and introgressive hybridization) were involved in the origin of breadfruit (Zerega *et al.*, 2006). In Vanuatu, each community recognizes different cultivars according to the size of the tree, the shape of the leaves, the size and the shape of the fruit, the presence or the absence of spines on the fruit and its colour, the colour, texture and taste of the flesh, the number of seeds, and the fruiting season. A few seedless forms (triploids) have been found in Vanuatu but the country appears to be an important centre for diversity of breadfruit and a key locality for its domestication. Farmers select genotypes found in the forest and plant them around their houses using root cuttings.

At present, only the Polynesian triploids have some export potential for the ethnic markets of the US West Coast, New Zealand, Australia and New Caledonia. More research is needed to produce artificial triploids. This could be achieved either through doubling artificially existing diploids to obtain fertile tetraploids, which could be crossed again with diploids to generate triploids, or by developing techniques to induce unreduced gametes. Also, additional research on pollination biology, the development of breadfruit food products with a long shelf life and the collection of cultivars from underrepresented areas, such as Vanuatu, for deposit in *ex situ* collections will all contribute to the future use and conservation of breadfruit (Zerega *et al.*, 2006).

7.3.2 *Barringtonia* species

The English common name for *Barringtonia* species is cut nut, including *B. edulis*, *B. novae-hiberniae* and *B. procera*. Some species are toxic. For example, *B. asiatica* is used as a fish poison. *B. edulis* is found cultivated in villages, gardens and along tracks and roads up to 600 m.a.s.l. The tree germinates in the wild from fruits dropped on the ground or spread by bats. Humans transplant young saplings close to dwellings, where they are carefully protected. In Vanuatu, each community possesses its own germplasm collection of cut nut trees and each morphotype has a distinct name in the vernacular language. When exceptional genotypes are found, farmers sometimes obtain planting material by marcotting. Although most varieties have dark green foliage, it is possible to find some with red foliage and red inflorescences with red fruits. Flowering occurs several times a year, even continuously, with some variation according to the island and individual trees. Production varies greatly from tree to tree and fruits reach maturity 2 months after flowering. Flowers are, however, fragile, and heavy rains can cause great losses before fruit set. The kernels have a very fine, distinct, taste and are eaten raw, grilled or boiled. Sometimes they may be powdered over a *laplap*. The shelf life of these kernels varies from 1 to 2 weeks. The wood is of poor quality and is not used for manufacturing purposes.

The morphological variability of *B. novae-hiberniae* is less than that of *B. edulis* but several types may be easily distinguished. There are trees with red foliage, although the foliage of this species is usually glossy green. Some trees have green or red fruits; others have very small fruits (approximately 4 cm in diameter). The tree is tolerant of shade, and flowering tends to be continuous; fruiting occurs 2 months after flowering. The tree is usually very productive. *B. procera* is almost always cultivated and does not tolerate shade. It requires much care and is rarely found in the wild. Flowering and fruiting occur once a year, usually during the warm and wet season (December to April). Being slender, frail and bearing few infructescences, the trees are not very productive. Nevertheless, their fruits and kernels are larger than those of other *Barringtonia* species. The fruits vary in size, shape and colour. Some dwarf types are found with a height which does not exceed 2 m. These are quite rare but are present in most islands. Unfortunately, they bear very poorly, although an improvement in their productivity would lead to orchards with small trees that are easy to harvest. Some trees have cylindrical fruits up to 8 cm long, which can be green or purple.

7.3.3 *Burckella* species

Three species of *Burckella*, namely *B. fijiensis*, *B. obovata* and *Burckella* sp. (= *Cassidispermum megahilum*), will be discussed here. In Vanuatu, *B. fijiensis* is found only on the island of Futuna and is probably a Polynesian introduction. The species is cultivated but numerous volunteer plants are found under the groves. It appears to be very variable in the size and taste of the fruits and their degree of sweetness. Vegetative propagation is very difficult to achieve and farmers use very ripe fruits. The flesh of the fruit has a fragrant, distinct and delicate taste and can be eaten raw or roasted. The tree is propagated from germinated seedlings, which are transplanted closer to settlements. Two main types of *B. obovata* are found in Vanuatu: one with elongated fruits and one with rounded fruits; the latter may sometimes be very large. The elongated fruit is the commoner and sweeter of the two types of fruit, but the taste can vary from tree to tree and villagers select and transplant plants with sweet and juicy fruits. To beat the fruit bats, the fruits are harvested just before maturity by picking them directly from the tree, and they are ripened in baskets.

Burckella sp. grows at the sea's edge. It is protected by humans and spread by fruit bats that feed on the flesh of the fruits. The production of this tree is significantly more important than that of *B. obovata*. However, this species seems to be heading towards extinction and the very good quality of its fruits justifies urgent conservation measures. Its geographical distribution in Vanuatu is very narrow since it is known only from a single village in South West Bay on Malekula. The fruits may be cooked according to need and baked on embers just before they reach maturity, or they may be eaten raw when they are fully ripe.

7.3.4 *Canarium harveyi* and *Canarium indicum*

The English common name of these species is pili nut. There are several botanical varieties of *C. harveyi* Seemann and two are found in Vanuatu: var. *nova hebridense* in the north (Banks archipelago) and var. *harveyi* in Erromango. Distinctions are made mostly on the shape of the nut (var. *harveyi* has a triangular section while var. *nova hebridense* has three dorsal crests and one ventral crest) but there are no molecular data to support this morphological distinction. The vegetative parts do not differ from variety to variety, and the taxonomic separation of the two species is therefore debatable. Propagation is done by planting the endocarp of very ripe fruits. The uses are the same as those of *C. indicum* L. but the kernels are bigger and oilier and are consequently preferred. However, there is tremendous variation, and some trees produce kernels with unwanted aroma. Quality control is thus a complex endeavour.

Canarium indicum also shows great variability in Vanuatu. The variation includes the shape of the fruits but also the number of kernels in a shell, their colour, the rhythm of flowering, productivity and the ease of cracking the nuts. The tree is found in secondary forest up to 400 m.a.s.l. The species is abundant in the northern part of Vanuatu but rarer in the southern part. Trees are protected in forests and cultivated around villages. Wildings or fruits that have reached full ripeness are planted in shallow holes. Mature trees are regularly pruned to make the fruits easier to pick. Apparently, human selection has produced some trees that continue to fruit out of season and some very productive ones.

The kernels are boiled, roasted or crushed and sprinkled on the *laplap* (Walter *et al.*, 1994). The nutritional value of the nuts and their potential for commercialization, for example in the muesli industry in Australia and New Zealand, has led the Solomon Islands and Vanuatu to undertake some research on their agronomy and processing (Stevens *et al.*, 1996). Vanuatu does not have the same resources as the Solomon Islands (Evans, 1999) but has the important advantage of commercial development driven by the private sector. This, combined with the existence of superior morphotypes, could be developed into a significant industry.

7.3.5 *Dracontomelon vitiense*

This large tree, *Dracontomelon vitiense* Engler, commonly called the dragon plum, is found in stands of primary forests. In the wild it is spread by bats but it is often found around villages or in gardens, and seedlings are transplanted close to the homestead. The Melanesians appreciate the acid taste of the small fruits, which are eaten raw as a snack or cooked in coconut milk. Harvesting is done with the help of a bamboo pole and the fruits are then transported to the village or to the urban market. There are only two types of dragon plum according to the size of the fruit: those with large and small fruits. There is continuous selection pressure to produce trees that produce larger, fleshier and sweeter fruits. Productivity varies greatly from one individual to another. This species is also exploited for its high-quality timber, which is used to make the frameworks of houses.

7.3.6 *Inocarpus fagifer*

Inocarpus fagifer (Parkinson ex. Zollinger) Fosberg is commonly known as the Tahitian chestnut. In Vanuatu this species is found surrounding villages in the coastal zone up to 400 m.a.s.l. It is more abundant in the south than in the north. The species displays a great variety of forms. The size, colour and shape of the fruit are very variable. The cooked kernel has a flavour that varies greatly from one tree to another and villagers choose trees according to the aroma of the kernels that they produce. When an isolated tree that bears fruits that are distinct for their size, shape or colour is found, the villagers collect the volunteer seedlings and plant them closer to their village. Fruit bats feed on the pulp and spread the seeds.

Tahitian chestnut can be propagated by cuttings and marcotting. It is available during the break season between two harvests of yams, and it is one of the main alternative foods of the Melanesian and Polynesian peoples. The fruit is roasted in its skin or boiled. It may also be reduced to a puree, which is wrapped in leaves of *Heliconia indica* and braised. Often harvested in large quantities, the fruits can keep for several months on bamboo drying racks placed in dark places in the hut. The seeds are removed as they become ready.

7.3.7 *Morinda citrifolia*

The English common name of *Morinda citrifolia* L. is Indian mulberry. It is rarely planted in the villages because of the strong odour of the fruits rotting on the ground. It is a wild species that is carefully tended. The cultivated form, producing edible fruits with large regular (with smooth skin) fruits, is rather rare in Vanuatu but the wild form is widespread. Wildings are often transplanted closer to the homestead. Flowering and fruiting are continuous. The fruits are rich in vitamin C and they are regularly consumed in Polynesia, where superior morphotypes have been selected.

This species has recently attracted significant interest from the Western nutraceutical industry and the so-called *noni* juice is now a very popular product in the USA, Australia and New Zealand. Scientific literature regarding the physiological properties of the compounds found in the fruits is sparse and it is therefore difficult to say whether this industry is sustainable. However, the Indian mulberry is also locally important as a medicinal plant. The raw fruit is crunched and eaten for the treatment of boils and numerous types of infection.

7.3.8 *Pometia pinnata*

Pometia pinnata J.R. & G. Forster is extremely variable and the variability appears to be greater in New Guinea than in Vanuatu. There are forms with round or oval fruits, and with orange or brown fruits. The species grows in secondary forest up to 300 m.a.s.l. The tree is wild and its better forms are transplanted close to areas of habitation, where they are carefully and regularly weeded, in some cases to encourage the growth of young plants. They are pruned to ease the harvest. Unfortunately, the fruits are often infested with maggots. They are picked with the aid of a long bamboo pole and placed carefully in baskets. However, they do not keep and have to be eaten rapidly, raw. The seeds may also be eaten once they are roasted and are sometimes dried on racks and stored.

7.3.9 *Spondias cytherea*

Spondias cytherea Sonnerat (golden apple) is found up to 300 m.a.s.l. Farmers prefer to plant it around their gardens rather than in villages. Fruit bats feed on the sweet fruits and spread them in the forest. Numerous morphotypes are distinguished by farmers according to the colour, size and taste of their fruits. In Vanuatu, a few morphotypes have very large fruits. In general, the species in Vanuatu exhibits greater morphological variation in morphotypes than appears to be found elsewhere in the region.

The tree can be propagated from a germinated seedling or a cutting of a branch planted in rich and moist soil. It is rarely cultivated in Papua New Guinea and the Solomon Islands, where it is found wild with smaller, acid, fibrous fruits. From Vanuatu and further eastwards in Polynesia the golden apple is cultivated and therefore Vanuatu might be the centre of origin of the cultivated forms. The fruits are picked up while they are still green and are eaten raw several days later when they have ripened in the houses. Fruits have a pleasant, slightly acid taste and can be processed into jams, compotes, chutneys and condiments.

7.3.10 *Syzygium malaccense*

Syzygium malaccense (L.) Merrill & Perry (Malay apple or mountain apple) grows in most villages up to 500 m.a.s.l. The fruits are spread by bats and the plant is very rare in dense forests. In most of the villages of Vanuatu, farmers can distinguish up to six different morphotypes according to the colour, size and taste of the fruits. Surprisingly, some morphotypes have flowers that are completely white while most of the forms present characteristic bright purple or fuchsia pink flowers. The cultivar with white flowers is very popular in Vanuatu because its fruits are sweeter and less infested with maggots when ripe. Wildings of preferred morphotypes are transplanted close to homesteads or are planted as ripe fruits. The fruits do not keep well but they can be traded when picked before full maturity. In most cases, they are eaten raw as soon as they are picked. Their taste and flavour are highly variable; some may be completely insipid while others are sweet and scented. They are, however, always refreshing.

7.3.11 *Terminalia catappa*

Terminalia catappa L. (the sea almond) is particularly abundant along the shore line up to 400 m.a.s.l., where it is largely spread by birds and bats. Volunteer plants are transplanted close to the homesteads. Sometimes they are planted in the centre of the village because of the quality of the shade they provide. The morphology of the leaves and fruits shows great variability in Vanuatu. The trees are very productive but the quality of the kernels is very variable. Harvesting starts when the skin of the fruit turns yellow or red and continues according to need. Fruits are harvested by knocking them down with a bamboo pole. The fruit has a delicate flavour and can be eaten raw or roasted.

The nuts are sold in Vanuatu in groceries as dried products that are convenient to store. Kernels are sold in the market threaded on to the midrib of a coconut palm leaflet. They are very popular and this trade could be further developed, just like the trade in *Canarium*, as an export industry. In Papua New Guinea the fruits of *T. kaernbachii* are used in a similar way and it might be interesting to introduce this species elsewhere in Melanesia and in Polynesia. The fruit of this species is the largest among the Combretaceae.

7.4 Discussion

There are numerous indigenous fruit and nut trees species in Oceania and most of them are still in the process of being domesticated. The domestication process can be summarized as follows:

1. Selection of a wild genotype or a seedling from a cultivated form. The domesticator identifies a productive morphotype and tests its fruits or nuts. If the quality is acceptable, a volunteer plant (a seedling found under the mother tree) or a fruit is collected. For very few species, such as breadfruit, clonal propagation is possible.
2. Improvement of the environment. The soil where the young plant is planted is well prepared. Unlike the mother plant, the seedling is planted in a considerably modified environment. The soil is loosened and weeds and bush around it are cleared off. This improved environment contributes directly to the ennobled development of the selected genotype, its canopy development and fruiting aptitude.
3. Improvement of the population composed of well-established selected seedlings which are going to intercross in a modified environment. In Melanesia, many cultivars of breadfruit, for example, are simply clones of edible wild forms and a few putative wild forms are probably feral plants that escaped from cultivation, or survivors found in secondary forests in locations where villages existed in the past. Some cultivars are also clones of hybrids between wild forms and feral or cultivated plants. The most prolific and vigorous trees are always selected, and vigour is often associated with heterozygosity and/or a heterotic effect. Over the long term, the continuous selection of the largest, sweetest, least fibrous fruits, the largest nuts or the softest shells, trees with prolific production, or those that are early- or late-maturing has led to *in situ* improvement of the species. This is a form of recurrent selection with very long cycles. Very often the person who starts the selection process may not be the one who will reap the benefits of the genetic improvement.

All cultivated species now present several variable morphotypes, to which farmers assign particular names in vernacular languages. The number of distinct cultivars varies according to species and location. For example, some communities in northern Vanuatu can distinguish up to 100 different morphotypes of breadfruit (Walter, 1989) and 20 of *Canarium* spp. and/or Tahitian chestnut (*Inocarpus fagifer*). When reproduction includes sexual recombination, the plant's genes are systematically redistributed in each generation cycle, which makes it difficult to achieve identical multiplication of these morphotypes because of cross-pollination. Only breadfruit and golden apple (*Spondias cytherea*), which may also be propagated by asexual means, provide morphotypes with stable names. Certain species with particular fruit characteristics, for example *Syzygium malaccense*, will, however, produce the same morphotype in the absence of vegetative propagation. Continuous selection of the same morphotype, one generation after another, has produced cultivars that have particular traits and are easily distinguishable by their fruit colour, nut shape and other morphological characters.

Farmers observe slight variations in the form, taste, colour and size of the fruits that they are consuming and have a marked tendency to conserve each distinct morphotype for preference, necessity or prudence purposes, and sometimes just for the sake of having orchards exhibiting variation. Some farmers have developed their own collections that include morphotypes not known to other people and that they will use to exchange with new ones, thereby increasing diversity. They also preserve them by necessity because some are early-maturing while others are late-maturing, thereby extending the consumption period. Certain morphotypes of species such as *Artocarpus altilis* and *Inocarpus fagifer* also cook more rapidly than others. Since all do not have the same taste, each one is therefore a slightly different food. Some morphotypes are also more resistant than others to diseases, and risk management is important when one considers the frequency of natural disasters in this part of the world, such as landslides, cyclones, El Niño and tsunamis.

The practice of selecting and assembling the best morphotypes has resulted in a major transformation of the landscape. The fruiting trees are found mainly in or near villages, gathered into small plantations, near the subsistence gardens whose boundaries they mark. They are often assembled in orchards, where they represent an artificial population composed of individuals differing in provenance. The flow of genes is not controlled but the system is obviously efficient in generating diversity.

Obviously, the most appreciated morphotypes are propagated more frequently than the others and are therefore more abundant. The least utilized are sometimes, but not always, cut out. There is therefore an ongoing erosion of genetic stock in favour of genotypes that correspond to local tastes and needs. As molecular data are not available it is difficult to assess the extent to which inbreeding depression may narrow the genetic base in this insular environment. It is also difficult to correlate morphological variation with geographical distribution and centres of origin. Since these species have a long lifespan, generally exceeding that of humans, it is difficult to determine the genealogy of the morphotypes and their names will disappear with them. Because in most cases these species have been introduced on these Oceanic islands from voyaging canoes, studies using molecular markers might indicate a significant bottleneck on each island (Lebot, 1999).

In the long term, it might be interesting to develop *in situ* management of these resources with the participation of farmers. After all, they have already demonstrated their ability to generate diversity and to preserve it. Their traditional management system could be greatly improved by introducing into their orchards new alleles originating from exotic germplasm, which will recombine and allow further selection of recombinants for local adaptation. The most appropriate approach to the conservation of the genetic resources of indigenous fruit and nut tree species in Oceania is to increase the farmers' long-term access to useful genes. In many islands of the Pacific, the genetic basis of most of these species is probably narrow because of the insular environment and could be broadened if the species were to be able to respond to their rapidly changing environmental as well as to patterns of human use. However, in order to be acceptable to farmers and to be kept as part of their varietal portfolio, any new genotypes must perform better than those presently cultivated, as judged by the farmers' own criteria and perceptions.

Diverse provenances could be introduced and recombined with the locally adapted varieties. The requirements of village orchards can be satisfied by recurrent selection of the best individuals in the best provenances. This directed, controlled mixing of the gene pool is the only way of ensuring the long-term conservation and sustainable use of germplasm before it is lost and before its economic potential can be assessed and exploited.

Another appropriate way of preserving this remarkable diversity is to make sure that it does not become obsolete. Because of urbanization and the ease of obtaining and preparing introduced food, thanks to globalization, the importance of traditional foods, such as those derived from indigenous species, is diminishing rapidly. A sensible market development strategy could be directed first at meeting domestic demand for fresh fruits or roasted nuts. Some of the nuts may also be packaged attractively as niche market products. Overseas markets are available for nuts in the shell as well as for oil, particularly for *Canarium harveyi*. For this species there is also a need for new on-farm storage systems. For all cultivated nut tree species, cultivar selection and improved processing are needed to reduce the unacceptable level of bitterness of the nuts. For *Terminalia*, for example, the major constraints are the low kernel-to-nut ratio and the high moisture content.

References

Barrau, J. (1962) *Les plantes alimentaires de l'Océanie, origines, distribution et usages.* Annales du Musée Colonial de Marseille, Marseille, France, 276 pp.

Evans, B. (1999) Edible nut trees in Solomon islands: a variety collection of *Canarium*, *Terminalia* and *Barringtonia*. ACIAR Technical Report No. 44, 96 pp.

French, B.R. (1986) *Food Plants of Papua New Guinea: a Compendium.* Private, 20 Main Street, Sheffield, Tasmania, Australia.

Lebot, V. (1992) Genetic vulnerability of Oceania's traditional crops. *Experimental Agriculture* 28, 309–323.

Lebot, V. (1999) Biomolecular evidence for plant domestication in Sahul. *Genetic Resources and Crop Evolution* 46, 619–628.

Pollock, N. (2002) Vegeculture as food security for Pacific communities. In: Yoshida, S. and Matthews, P.J. (eds) *Vegeculture in Eastern Asia and Oceania.* JCAS Series No. 16. JCAS, Osaka, Japan, pp. 277–292.

Ragone, D. (1991) Ethnobotany of breadfruit in Polynesia. In: Cox, P.A. and Banack, S.A. (eds) *Islands, Plants and Polynesians: An Introduction to Polynesian Ethnobotany.* Diocorides Press, Portland, Oregon, pp. 203–220.

Ragone, D. (1997) Breadfruit, *Artocarpus altilis*

(Parkinson) Fosberg. IPGRI (International Plant Genetic Resources Institute), Rome, 78 pp.

Smith, A.C. (1981) *Flora Vitiensis Nova: A New Flora of Fiji, Volume 2.* National Tropical Botanical Garden, Lawai, Kauai, Hawaii, 810 pp.

Smith, A.C. (1985) *Flora Vitiensis Nova: A New Flora of Fiji, Volume 3.* National Tropical Botanical Garden, Lawai, Kauai, Hawaii, 758 pp.

SPC Demography/Population Programme (2000) [see http//:www.spc.int/demog/ Accessed 10 July 2007].

Stevens, M.L., Bourke, R.M. and Evans, B. (1996) *South Pacific Indigenous Nuts.* Proceedings of a Workshop held from 31 October to 4 November 1994, at Le Lagon Resort, Port Vila, Vanuatu. ACIAR Proceedings No. 69. ACIAR (Australian Centre for International Agricultural Research), Canberra, 176 pp.

Walter, A. (1989) Notes sur les cultivars d'arbre à pain dans le nord du Vanuatu. *Journal de la Société des Océanistes* 88/89, 3–18.

Walter, A. and Lebot, V. (2003) *Jardins d'Océanie, Les Plantes Alimentaires du Vanuatu.* Collection 'Didactiques'. Presses de l'IRD, Montpellier, France, 320 pp.

Walter, A. and Sam, C. (1999) *Fruits d'Océanie.* Collection 'Didactiques'. Presses de l'IRD, Montpellier, France, 310 pp.

Walter, A., Sam, C. and Bourdy, G. (1994) Etude ethnobotanique d'une noix comestible: les Canarium du Vanuatu. *Journal de la Société des Océanistes* 98, 81–98.

Wheatley, J.L. (1992) *A Guide to the Common Trees of Vanuatu, with Lists of their Traditional Uses and Ni-vanuatu Names.* Department of Forestry, Port-Vila, Vanuatu, 308 pp.

Whistler, A. (1984) Annotated list of Samoan plant names. *Economic Botany* 38, 464–489.

Whitmore, T.C. (1986) *Guide to the Forests of the British Solomon Islands.* Oxford University Press, Oxford, UK, 208 pp.

Yen, D.E. (1985) Wild plants and domestication in Pacific islands. In: Misra, V.N. and Bellwood, P. (eds) *Recent Advances in Indo-Pacific Prehistory.* Oxford and IBH, New Delhi, pp. 315–326.

Yen, D.E. (1993) The origins of subsistence agriculture in Oceania and the potentials for future tropical food crops. *Economic Botany* 47, 3–14.

Yen, D.E. (1998) Subsistence to commerce in Pacific agriculture: some four thousand years of plant exchange. In: Prendergast, H.V.D., Etkin, N.L., Harris, D.R. and Houghton, P.J. (eds) *Plants for Food and Medicine.* Royal Botanic Gardens, Kew, UK, pp. 161–183.

Zerega, N., Ragone, D. and Motley, T.J. (2006) Breadfruit origins, diversity and human facilitated distribution. In: Motley, T.J., Zerega, N. and Cross, H. (eds) *Darwin's Harvest: New Approaches to the Origins, Evolution and Conservation of Crops.* Columbia University Press, New York, pp. 213–238.

8 Creating Opportunities for Domesticating and Commercializing Miombo Indigenous Fruit Trees in Southern Africa

F.K. Akinnifesi,[1] O.C. Ajayi,[1] G. Sileshi,[1] P. Matakala,[2] F.R. Kwesiga,[3] C. Ham,[4] I. Kadzere,[5] J. Mhango,[6] S.A. Mng'omba,[7] T. Chilanga[8] and A. Mkonda[9]

[1]*World Agroforestry Centre, ICRAF, Lilongwe, Malawi*; [2]*World Agroforestry Centre, ICRAF, Maputo, Mozambique*; [3]*Forum for Agricultural Research in Africa, Accra, Ghana*; [4]*Department of Forest and Wood Science, University of Stellenbosch, Matieland, South Africa*; [5]*Department of Agricultural Research and Extension, Ministry of Agriculture, Harare, Zimbabwe*; [6]*Mzuzu University, Mzuzu, Malawi*; [7]*Department of Plant Production and Soil Science, Faculty of Natural and Agricultural Sciences, University of Pretoria, Pretoria, South Africa*; [8]*Bvumbwe Agricultural Research Station, Limbe, Malawi*; [9]*Zambia-ICRAF Agroforestry Project, Chipata, Zambia*

8.1 Introduction

An estimated 500 million to 1 billion smallholder farmers grow trees on farms or manage remnant forests for subsistence and income (Scherr, 2004). The development of rural economic activities is pivotal to national well-being for rural communities in developing countries. The prospects for smallholder agriculture in sub-Saharan Africa are likely to remain bleak unless major efforts are made soon to address the limiting conditions that affect farmers. Rural community dwellers have access to a wealth of natural resources, including arable lands and forests, yet they face the highest levels of poverty, ill health and malnutrition and score low on other development indicators (de Ferranti *et al*., 2005).

Through the ages, millions of people dwelling in rural areas in tropical countries have depended on forests for income, food and other livelihoods and security options, through the gathering and processing of tree products. Indigenous fruits from miombo woodlands are central to the livelihood systems of both rural and urban dwellers in southern Africa, especially during periods of famine and food scarcity (Campbell, 1987; Campbell *et al.*, 1997; Mithöfer and Waibel, 2003; Akinnifesi *et al.*, 2006a; Mithöfer *et al.*, 2006). Miombo woodlands represent an important food supplement and cash income in better times for rural people living around forests (Mithöfer, 2005). As the rate of deforestation increases in the region, wild fruit trees become prone to overexploitation and extinction. Therefore, the livelihood of rural people who largely depend on this natural resource is seriously threatened. The food production capacity of the region is being pushed to the limit, resulting in overcultivation of fragile soils and loss of soil quality.

A study conducted in 2002 to assess the contribution of miombo fruits to the livelihoods of the rural communities in the region has shown that 65–80% of rural households in the 'Chinyanja Triangle', i.e. Malawi, Mozambique and Zambia, lack access to food for as much as 3 or 4 months per year, and 26–50% of the respondents relied on indigenous fruits for sustenance during this critical period (Akinnifesi *et al.*, 2004a). Those most at risk – principally women and children – experience high rates of malnutrition and suffer disproportionately from poor health. In an earlier survey, 97% of households did not have enough food, and 60% and 21% had relied on *Uapaca kirkiana* and *Parinari curatellifolia* fruits, respectively, as coping mechanisms. In terms of scaled scores of products derived from all semi-wild trees in Malawi, households ranked them in the following order: firewood, timber, fruit, medicine, woodwork and manure (Kruise, 2006). Many of these households (79–83%) apply firebreaks in the crop area and around homesteads.

Indigenous fruits remain one of the major options for coping with hunger and nutritional deficiency in diets and with poverty in this region. Studies have shown that harvesting fruits from the wild and also from the semi-domesticated trees growing on farms can boost rural employment and generate substantial income (Ruiz-Perez *et al.*, 2004; Leakey *et al.*, 2005; Mithöfer, 2005), especially from processing and adding value (Saka *et al.*, 2004). Moreover, 94% of rural households in four villages in South Africa were reported to have been making use of *Sclerocarya birrea* fruits (Shackleton, 2004). Miombo indigenous fruits such as *Uapaca kirkiana*, *Sclerocarya birrea*, *Strychnos cocculoides*, *Adansonia digitata* and *Parinari curatellifolia* are rich in sugars, essential vitamins, minerals, oils and proteins necessary for human nutrition (Saka *et al.*, 2004, 2006, Chapter 16, this volume; Tiisekwa *et al.*, 2004). The vitamin C level of dry baobab pulp can be as high as 5127 mg/kg (Sidibe *et al.*, 1998). The efficiency of the fruit processing industry is, however, thwarted by lack of a sustained supply of primary products (raw materials) or intermediate products, and medium-scale entrepreneurs complain about their dependence on imports. This is because supplies are based on the importation of exotic fruit concentrates only.

Fresh fruits are highly perishable and incur direct or indirect nutrient and quality losses along the market chain from production to consumption. Indigenous fruits, mostly harvested from the wild, are generally sold or consumed fresh, and large proportions (up to 55%) of the collected fruits are wasted (Kwesiga *et al.*, 2000; Ramadhani, 2002). Postharvest decay can be a serious problem in many fruits undergoing storage. Most instances of postharvest decay are the result of mechanical injury during harvesting and transport (Kadzere *et al.*, 2006c). This is due to high pH (5–8), moisture content (60–90%) and limited knowledge of fruit handling and marketing. Reducing postharvest losses of indigenous fruits through appropriate storage and conservation techniques will ensure sustainable supplies of quality fruits and the provision of a wide range of products (Kadzere *et al.*, 2001). The major causes of fruit losses are mechanical damage due to cracking, compression and bruising during harvesting and transport, insect and pest damage, and overripening (Saka *et al.*, 2002; Kadzere *et al.*, 2006a, b).

The overall objectives of domesticating and commercializing fruits from indigenous trees are to improve rural livelihoods – nutritional status, household income, entrepreneurial opportunities and economic empowerment – and to promote the conservation of biodiversity and the sustainable use of natural resources. This synthesis provides an overview of efforts in domesticating indigenous fruit trees as tree crops, and commercializing their products. This chapter is a product of several years of research by a large team of specialists in the southern Africa region and beyond.

8.2 Strategy for Commoditizing Indigenous Fruit Trees

Despite the fact that indigenous fruits are well known, consumed and traded among local consumers, very little success has been achieved with their cultivation as tree crops or in their commercialization as products. Commoditizing fruits from the miombo woodlands in southern Africa as tree crops involves a long-term iterative and integrated strategy for tree selection and improvement, for the promotion, use and marketing of selected products, and for their integration into agroforestry practices (Akinnifesi *et al.*, 2006a). The following are the key components of this strategy:

- Verification of the importance and potential of indigenous fruits in the rural economy.
- Initiation of a tree domestication programme to select and improve germplasm.
- Development and promotion of indigenous fruit production using new cultivars.
- Commercialization of new products through a functional supply chain (fruit storage and processing, product quality assurance, adding value, marketing research, rural revenue generation and enterprise development).

8.2.1 Importance and prioritization of indigenous fruit trees

To understand the importance and role of miombo indigenous fruit trees in the overall livelihood security of communities, several assessments were undertaken.

These included: (i) ethnobotanical surveys; (ii) species and priority ranking and *ex ante* impact assessment; (iii) the economics of production; and (iv) market assessments.

A region-wide ethnobotanical survey was carried out in Malawi, Tanzania, Zambia, Zimbabwe and Mozambique during 1989–1991, as the first step in understanding species diversity and the role of trees on farms, with respect to their establishment and management, location and arrangement, market opportunities, uses and functions in farmer fields (Karachi *et al.*, 1991; Maghembe and Seyani, 1991; Kwesiga and Chisumpa, 1992). These surveys have identified more than 75 indigenous fruit trees that are an important resource for rural communities as sources of food and income. The fruits were collected from the forests and consumed locally and also traded in local and roadside markets. Fox and Young (1982) indicated that the fruits of more than 200 of the 1000 indigenous trees growing in southern Africa are edible.

Using the general principle developed by Franzel *et al.* (1996) for species prioritization, several other species prioritization surveys were triangulated to identify the species, their dynamics and ecological niches. In 1996–1997, farmers' species preference was assessed in Malawi, Zambia, Tanzania and Zimbabwe (Kadzere *et al.*, 1998). A total of 451 households were interviewed in four countries from 20 districts, comprising 128 households in Malawi, 70 in Tanzania, 135 in Zambia and 118 in Zimbabwe. The results showed that the most preferred species for domestication in the region were *Uapaca kirkiana*, *Parinari curatellifolia*, *Strychnos cocculoides*, *Anisophyllea boehmii*, *Azanza garckeana*, *Flacourtia indica*, *Syzygium guineense*, *Strychnos pungens*, *Physalis peruviana* and *Uapaca nitida*. The first three species were also identified as regional priority species based on their highest preference by users. The fourth in importance was *Anisophyllea boehmii* for Zambia, but it is limited to the northern and copperbelt provinces in Zambia and the Democratic Republic of Congo and the area adjacent to Lake Malawi in Tanzania. *Vitex mombassae* was the third most important in Tanzania, mainly popular in Tabora region. Similarly, *Azanza garckeana* was the fourth most important in Zimbabwe and *Flacourtia indica* was the fourth most important for Malawi.

For the first time, detailed setting of priorities was undertaken in Mozambique, involving a survey of 156 households (76 female) in four provinces (Maputo, Gaza, Manica, Tete). The results showed that 60 indigenous fruit trees were consumed and traded by rural communities. Community members were able to process *Adansonia digitata*, *Ziziphus mauritiana* and *Strychnos cocculoides* into different products, such as beer, called 'buadwa', and flour for making porridge, called 'bozo'. *Ziziphus mauritiana* and *Sclerocarya birrea* were also used for fodder. As regards indigenous knowledge, only 28% of the respondents had knowledge of processing. Planting of indigenous fruits as well as medicinal plants among farmers was not common because 97% of the the respondents said that they had no knowledge of how to grow them.

Based on these results, the regional spearhead species (species which drive the domestication programme) were identified as *Uapaca kirkiana*, *Strychnos cocculoides* and *Sclerocarya birrea* (Akinnifesi *et al.*, 2006). The key reasons for selection of these species are that they are the species most preferred by users, marketers and consumers, and that they have wide geographical distribution, consumption and marketing potential. As a rule it is expected that species to be selected for domestication should be widely consumed by the local population and the excess products should be marketable – selection is driven by consumption and markets. In this respect, although *Sclerocarya birrea* was not mentioned by the local communities in most countries as a priority fruit species (probably because it is rarely consumed fresh and needs processing), it was included as priority species because of its market potential. The flesh of *Sclerocarya birrea* (marula) fruit contains 180 mg vitamin C per 100 g, surpassing orange, grapefruit, mango and lemon. The kernels are rich in food energy, containing almost 3000 kJ per 100 g. Moreover, *Sclerocarya birrea* is probably the only miombo indigenous fruit species that has made it to the international market: the liqueur Amarula Cream, made from the fruit of *Sclerocarya birrea* by Distell Corporation in South Africa, is sold in 63 countries worldwide, with significant benefits to the rural communities (Ham, 2005). Emphasis on *Parinari curatellifolia* as a 'spearhead' species was reduced, despite its high preference (Franzel *et al.*, Chapter 1, this volume), because it was less amenable to asexual propagation methods. Other country-specific or less valued species were considered as 'shield' species, i.e. species that are important for the on-farm maintenance of genetic biodiversity. Recent species priority-setting studies have shown not only that farmers are interested in indigenous fruits but also that they rate exotics, such as mangoes, citrus, avocado, banana and papaya, very highly (Franzel *et al.*, Chapter 1, this volume). In most cases, mangoes and citrus ranked above indigenous fruit trees, and *Uapaca kirkiana*, *Strychnos cocculoides* and *Ziziphus mauritiana* are the only species mentioned among the top ten fruits in Malawi.

There is significant consumption in both rural and urban areas, with higher consumption in the rural areas and more trade in the urban and semi-urban areas. A similar trend in preferences by farmers was also shown by market consumers. In a market survey, Ramadhani (2002) asked consumers the following question: 'If you have enough money for buying only seven types of fruit, including indigenous fruits, apples, mangoes and oranges, which ones are you likely to buy (first to seventh)?' Although the question was already biased towards the named fruit species, it gave an indication of relative preference. The responses were ranked as follows (most preferred species first): apple (*Malus domestica*), orange (*Citrus sinensis*), mango (*Mangifera indica*), *Uapaca kirkiana*, *Ziziphus mauritiana*, *Azanza garckeana*, *Adansonia digitata* and *Strychnos cocculoides*. The trend was the same for *Uapaca kirkiana* and *Strychnos cocculoides* buyers. However, when the results were recalculated, the trend changed slightly: *U. kirkiana* ranked third instead of fourth and *M. indica* ranked fourth instead of third (Table 8.1). This shows that the method of analysis can influence the priority-setting results (Franzel *et al.*, Chapter 1, this volume). Willingness to buy indigenous fruits showed the following order (most preferred species first): *Uapaca kirkiana*, *Strychnos cocculoides*, *Azanza garckeana*, *Ziziphus mauritiana* and *Adansonia digitata*.

In a market survey conducted in the southern region of Malawi, Mmangisa (2006) found that the consumer preferences were in the following order (most

Table 8.1. Consumer preferences in Malawi and Zimbabwe for exotic and indigenous fruit trees.

	Market survey, southern Malawi (n = 315)[a]		Market survey, central region of Malawi (n = na)		Household survey, Malawi (n = 223)[b]		Market survey, Zimbabwe (n = na)[c]	
Rank	Species	Weighted score	Species	Weighted score	Species	Weighted score	Species	Weighted score
1	*Mangifera indica*	734	*Citrus sinensis*	279	*Mangifera indica*	678	*Malus domestica*	661
2	*Citrus sinensis*	656	*Mangifera indica*	177	*Persea americana*	472	*Citrus sinensis*	626
3	*Malus domestica*	548	*Psidium guajava*	136	*Citrus sinensis*	291	*Uapaca kirkiana*	537
4	*Uapaca kirkiana*	424	*Musa paradisiaca*	116	*Uapaca kirkiana*	269	*Mangifera indica*	531
5	*Ziziphus mauritiana*	288	*Carica papaya*	115	*Carica papaya*	234	*Ziziphus mauritiana*	377
6	*Adansonia digitata*	281	*Citrus reticulata*	105	*Musa paradisiaca*	199	*Azanza garckeana*	337
7	*Azanza garckeana*	220	*Citrus limon*	73	*Psidium guajava*	198	*Adansonia digitata*	278
8	*Strychnos cocculoides*	178	*Uapaca kirkiana*	51	*Casmiroa edulis*	198	*Strychnos cocculoides*	242
9	–	–	*Persea americana*	47	*Citrus reticulata*	164	–	–
10	–	–	*Strychnos cocculoides*	41	*Ziziphus mauritiana*	158	–	–

[a]Adapted and recalculated from Mmangisa (2006).
[b]J. Mhango and F.K. Akinnifesi (unpublished results, 2002).
[c]Adapted and recalculated from Ramadhani (2002).

preferred species first): *Mangifera indica*, *Citrus sinensis* (orange), *Malus domestica*, *Uapaca kirkiana*, *Ziziphus mauritiana*, *Adansonia digitata*, *Azanza garckeana* and *Strychnos cocculoides* (Table 8.1). The consumer survey results were different from earlier household or farmer surveys (Malembo *et al.*, 1998) but similar to those of Ramadhani (2002). In the Malawi surveys, *Uapaca kirkiana* was the only indigenous fruit tree that featured in the top four, and *Ziziphus mauritiana* appeared in the top five most preferred fruits. *Mangifera indica* and *Citrus sinensis* were the two exotic fruits most preferred by consumers.

Market research using conjoint analysis showed that, for the fresh market, improvement effort should focus on quality by releasing trees with bigger fruits with brown colour and round appearance for *Uapaca kirkiana* (Mmangisa, 2006). Buyer characteristics and preference differences between males and females and according to education status reflect market segmentation potential.

The farmers also identified the need for improvement of tree precocity and fruit quality attributes. Farmers wanted improvements in fruit size (amount of edible pulp), sweeter taste, higher yields, improved shelf life, larger tree size, and pest and disease resistance (Maghembe *et al.*, 1998). Fruit precocity was the major interest of most farmers, as they wanted quick returns for their crops, just as they did from exotics. The next most important trait for improvement was fruit size. Since there are characters that can be manipulated by vegetative propagation and clonal selection, this activity received priority in the domestication programme. However, because of the lack of knowledge about propagation techniques, seedling production and tree husbandry skills, uncertain markets and low prices, and because of the free availability of fruits from the forests, few farmers had planted indigenous fruits trees.

8.2.2 Economics of production, wild collection and marketing

Several studies have shown that households benefit from the consumption and sale of indigenous fruit trees in southern Africa (Kaaria, 1998; Mithöfer, 2005; Ramadhani and Schmidt, Chapter 12, this volume). Tree cultivation requires *ex ante* analysis of the farmers' needs, resource endowments, and the economic benefits to be derived. Mithöfer (2005) conducted field surveys on 303 households in 2000 to determine the collection and use of indigenous fruits and the contribution of indigenous fruit trees to household economy and poverty reduction. The study showed that fruit could provide a cushioning effect against poverty during critical periods of hunger and famine, especially during the period from August to January each year.

Results of an economic model on fruit trees show that, despite the availability of wild fruits, improvements in tree yield and earlier fruiting will create incentives for farmers to cultivate indigenous fruits (Mithöfer and Waibel, 2003). The model incorporated risk-coping strategies that assumed that households can: (i) access additional sources of cash (such as from farming, livestock, exotic fruits and remittances); and (ii) increase indigenous fruit collection from the wild. Income from farming showed negative covariance with other sources of income. The results showed that vulnerability to poverty was highest between August and March, and that indigenous fruits reduced vulnerability to poverty by 33% during the critical period (Mithöfer *et al.*, 2006). The households accessing indigenous fruit were able to live above the poverty line throughout the year. A simulation model of year-round income flows of household enterprises using indigenous fruits to diversify their income suggests that the benefits from selling indigenous fruits come at a critical time when income is generally low, and that the fruits provide nutrition and food when agricultural labour demands are high (Mithöfer, 2005).

In another study, 80% of households in Zambia, Malawi and Mozambique were shown to be prone to food shortage during August–March, and 20–50% of these households had explored the use of fruits as a coping strategy (Akinnifesi *et al.*, 2004a). The proportion of households falling below the poverty line during the critical period could be reduced by 33% in Zimbabwe in seasons when fruits are abundant (Mithöfer, 2005; Mithöfer *et al.*, 2006). The studies by Mithöfer (2005) and Ramadhani (2002) agree that indigenous fruits contribute to the household economy and that women and children are the major beneficiaries. Children are the main consumers; women make more money from sales to purchase other household needs. Mithöfer (2005) showed that although the gross margins from indigenous fruit trees are lower than those from livestock and crop production, the return to labour is higher. The reasons for wild collection of indigenous fruit trees are an efficient labour allocation strategy for poor households and a minimal cost outlay in production and collection.

However, despite the economic attractiveness of wild fruit collection, one major problem is that indigenous fruits are still being treated as a public good, and increased collection has often caused rivalry and shortages (Ramadhani, 2002). In Malawi, wild fruits are disappearing in some natural areas because of lack of stewardship. For instance, *Strychnos cocculoides* is now becoming rare in the southern region. *Uapaca kirkiana* population is becoming quite sparse in the southern region as a result of increasing population pressure. Some fruits are endemic to certain areas, such as *Adansonia digitata* and *Parinari curatellifolia*. On the other hand, in some countries, such as Zimbabwe, Zambia and Mozambique, wild populations of indigenous fruits still flourish in many areas. Since growing indigenous fruit trees is an investment, a hypothesis about whether it may be worthwhile to focus more on wild collection was tested in Malawi and Zimbabwe for *Uapaca kirkiana*.

Tree planting is an investment which involves 'sunk costs' and locks up cash for a period of time. Using a real option approach in Zimbabwe, Mithöfer (2005) subjected the tree domestication effort to intensive study to find out what technological change would induce farmers to adopt tree planting. The critical point at which farmers would be willing to invest in indigenous fruits was sought. A similar study was also undertaken by Fiedler (2005) in Malawi. The results from both studies were mixed. In Zimbabwe the collection of wild fruits from the communal lands is more profitable than growing and managing own trees (Mithöfer and Waibel, 2003; Mithöfer, 2005). Also, the effect of tree improvement must be significant for investment in farmer-led cultivation to be economically attractive. However, the study did observe that a combination of technological change, e.g. fruit traits, and decreasing resources abundance, such as the impact of deforestation, a change in policy that would discourage collection from the commons, and increased competition due to value addition, would provide scope for farmer-led tree cultivation and conservation.

Fiedler (2005) showed that the availability of *Uapaca kirkiana* is declining in Malawi, as more than half of the total forest resources disappeared between 1972 and 1992 (FAO, 2001). Only a few farmers have preserved trees in their fields and deforestation has reduced the number of trees available on communal lands. As a result, collection costs have risen and the labour productivity of collecting fruits from communal areas has fallen. For households still collecting and selling *U. kirkiana* fruits, the income share was 4.1% and farmers expressed their willingness to collect more if it were possible. The study indicates motivation to keep *U. kirkiana* as a source of food and income among the rural population indicating that the adoption of improved indigenous fruit trees could be successful. The real option approach used in the economics study suggests that planting *U. kirkiana* trees even without yield improvement is profitable if the maturity period of trees were to decrease from the current level of 11–16 years to 4 years (Fiedler, 2005). This indicates that a domestication programme using clonal selection to reduce the period of maturity of indigenous fruits will be a highly economic venture. With further yield increases and better fruit quality, improved *U. kirkiana* trees become more profitable and farmer adoption is expected to be successful and sustainable.

Households have choices and they vary in how they manage their portfolios of indigenous fruit trees. Domestication improves the contribution of fruit trees to household food security, nutrition and income and conserves the biodiversity of stock that is necessary for meeting both present and future demands (Akinnifesi *et al.*, 2004a). A study of *in situ* conservation was carried out by Kruise (2006) which involved 133 households in Dedza, central Malawi, to understand the extent of conservation of semi-wild indigenous trees and factors influencing their conservation. Results showed that, of the average of four trees conserved and/or managed per household, 36% were indigenous fruit trees. Forty-four per cent of trees were located in the homestead and 48% in the crop area.

8.2.3 Commoditization of fruits

Wild tree products form part of the diversified rural economy in Africa. Agriculture operates at the interface of two complex, hierarchically distinct systems: the socio-economic system and the ecosystem. For tree domestication, there is a trade-off between new commitment to increase the level of intensification of crop production to increase returns for farmers in order to address poverty challenges, coupled with associated stresses on the ecological systems, such as loss of biodiversity and soil erosion. Although wild harvesting has been shown to be profitable in the short term (Mithöfer, 2005) and more practical for rural populations (Akinnifesi *et al.*, 2006a), it tends to have adverse effects on the ecosystems in the long term and to reduce the opportunity to build privately owned long-term assets on-farm. The many consequences include the loss of biodiversity due to the collection of propagules for the next generation of wildings; declining productivity and population as a result of postharvest damage and inappropriate postharvest techniques (Kadzere *et al.*, 2006a, b, c); and disincentives to on-farm cultivation. These effects limit the potential for rural dwellers and farmers to benefit from emerging and expanded markets from improved products.

The transition from a subsistence to a market economy is important in better optimizing the potentials for using, managing and cultivating indigenous fruit trees. Recent assessments have confirmed the conservation and economic benefits of wild tree products. These issues were reflected upon in several conferences listed by Ruiz-Perez *et al.* (2004), including those dealing with the role of forestry in poverty alleviation (4–7 September 2001, Semproniano, Italy), forests in poverty reduction strategies (1–2 October 2002, Tuusula, Finland) and rural livelihoods, forests and biodiversity (19–23 May 2003, Bonn, Germany). In a recent global assessment of the links between the livelihoods of forest dwellers and global commoditization, Ruiz-Perez *et al.* (2004) found: (i) an increasing contribution by individual non-timber forest products to the household economy of producers as they move from low to high levels of commoditization; (ii) that households with diversified strategies fall between subsistence and specialized sets of cases (household income, market size, production per hectare); and (iii) that non-timber forest products provide guaranteed additional income for households that earn the bulk of their income from agriculture and off-farm sources. Households with a specialized strategy have higher household incomes, command higher prices and trade value for tree products, and produce more per hectare.

These results show that households that engaged in cultivation had higher returns to labour, used more intensive technology for production, and had higher productivity per hectare, associated with stable tenure and a stable resource base. Wild harvesting on the other hand, was associated with declining resource base. In Africa, wild harvesting poses increased pressure on resources and collectors are more vulnerable to climatic restrictions (Ruiz-Perez *et al.*, 2004). In southern Africa the resource base is fragile, market opportunities for farm crops are limited, and markets for indigenous fruits are limited and informal. The importance of safety-net and subsistence functions should be emphasized, but not at the expense of lack of conservation and deliberate cultivation of the few priority fruit trees.

In addition, increasing market demand for wild forest products tends to increase overexploitation (Ruiz-Perez *et al.*, 2004) and leads to harvest damage (Kadzere *et al.*, 2001, 2006a, b, c). Increased competition for access and use may also lead to village conflicts and rivalry due to shortage of fruits in communally owned lands or state forests, which are generally regarded as a public good (Ramadhani, 2002) – this is the phenomenon known as the tragedy of the commons. There is a need for policy interventions and institutional arrangements to redefine property rights to indigenous fruits in the commons.

8.3 Tree Crop Development

Tree domestication refers to how humans select, manage and propagate trees; the humans involved may be scientists, civil authorities, commercial companies, forest dwellers or farmers (Simons and Leakey, 2004). Wild harvesting (extractivism) and overextraction have led to market expansion and supply shortages in many cases (Simons and Leakey, 2004). The greatest potential for tree domestication resides in the interactions of rural communities with wild tree populations through retention on farms and homesteads and the cultivation or management of fruit trees as semi-domesticates. Focusing on wild collection and management of semi-domesticated trees on farms and homesteads has been indicated as an effective way of reducing the sunk costs, providing households with improved species diversity in terms of desired fruit and tree traits (Kruise, 2006). The role of indigenous fruit trees where cultivated and semi-domesticated trees coexist to adapt to human needs has been well documented in the concepts of homegardens (Kang and Akinnifesi, 2000; Kumar, Chapter 5, this volume), forest gardens and agroforestry (Wiersum, 2004). However, producing improved elite cultivars with superior fruit and traits will require conscious domestication (Akinnifesi *et al.*, 2006; Leakey and Akinnifesi, Chapter 2, this volume).

8.3.1 Germplasm collection, provenance and progeny testing

The initial work on domestication of indigenous fruits in southern Africa started with range-wide tree germplasm collection and testing (Kwesiga *et al.*, 2000; Akinnifesi *et al.*, 2004b). In 1995, 30 germplasm collectors from six southern African countries were trained. Forty provenances of *Sclerocarya birrea* were collected from eight countries in 1996 (Botswana, Malawi, Namibia, Swaziland, Tanzania, Zambia, Mali and Kenya) and 24 provenances of *Uapaca kirkiana* from five countries (Malawi, Mozambique, Tanzania, Zambia and Zimbabwe). The weight of the germplasm collected and exchanged among Malawi, Zambia, Zimbabwe and Mozambique exceeded 10 t of fruits. In order to determine the genotype × environment interactions to isolate superior genotypes, 16 provenances of *U. kirkiana* and 20 of *S. birrea* were evaluated in four countries within the region. The trees were established in each country as multilocational provenance and progeny trials. The preliminary results have been documented (Akinnifesi *et al.*, 2004b).

For *S. birrea*, the provenances planted in Tanzania varied in their initial performance. At 12 months, the fastest-growing provenance was Chikwawa from Malawi (1.37 m), compared with Mpandamatenga provenance from Botswana (0.6 m). Wami provenance from Tanzania was the second best (1.28 m). Tree survival 12 months after establishment (MAE) was generally high (99%) (ICRAF, 2005). The provenances of *U. kirkiana* and *S. birrea* planted in Zambia and Malawi also performed differently. Tree survival at Makoka was generally good at 30 MAE, ranging from 82 to 96% (Akinnifesi *et al.*, 2004b). At 54 MAE, Mulrelwa was the tallest provenance at Makoka (2 m), and nine out of the 16 provenances were taller than 1.5 m. The result at

Chipata, Zambia was opposite. Tree survival was low at 30 MAE (25–64%), and only two of the 12 provenances exceeded 40% at 54 MAE. Tree height in Zambia also ranged from 0.47 to 0.78 m at the Chipata site. The pooled analysis of variance showed there was significant interaction between tree age and site ($P < 0.01$). There was strong correlation of the latitude at the site of origin with survival ($r = -0.96$), root collar ($r = -0.96$) and tree age ($r = -0.93$). This implies that the nearer the germplasm is planted to the mother tree (site of origin) the better the performance at both sites. In Tanzania, Malawi provenances from Mangochi outperformed all the 16 provenances planted at Tumbi, with 1.37 m height at 12 months compared with 0.6 m for Mpandamatenga provenance from Botswana.

Tree survival showed negative correlation with rainfall level, indicating that germplasm collected from sites having low rainfall adapted better than that from wetter sites (Akinnifesi *et al.*, 2004b). Poor performance and adaptability of *U. kirkiana* provenances established at Chipata signals the possibility of a site effect on tree growth. Site conditions at Makoka may offer a better potential for orchard management than those at Chipata in Zambia or Domboshawa in Zimbabwe. Zimbabwe faced challenges of frost damage, and *U. kirkiana* trees established at the Tabora site were wiped out early due to flooding (Akinnifesi *et al.*, 2004b). In addition, trees of *Uapaca* planted at the lower position in the soil catena seem to have performed poorly compared with those on the middle or upper slopes, suggesting that *Uapaca* is better adapted to well-drained soils (Akinnifesi *et al.*, 2004b). This explains why 93% of *U. kirkiana* trees are found in the mountains and hills (Malembo *et al.*, 1998). It seems that latitude and climate may have an overriding influence on tree growth (Akinnifesi *et al.*, 2004b). Whether this would translate to improved yields or fruit quality is not yet certain.

The provenance trial showed that most of the priority indigenous fruit trees showed considerable phenotypic variation across southern Africa (Akinnifesi *et al.*, 2004b). Other workers have shown that *U. kirkiana* exhibits variation in growth, flowering and fruiting that could be manipulated by management and improvement (Mwamba, 1995a; Ngulube *et al.*, 1995, 1998). Akinnifesi *et al.* (2006) reported that a few individuals from the Mapanzure and Serenje provenances from Zimbabwe and Zambia, respectively, fruited at Makoka, Malawi for the first time after 8 years. The percentage of trees fruiting was less than 5% of the provenance, and fruiting was not observed in the next 2 years for all trees in the trial. This agrees with the earlier assertion that fruiting in *U. kirkiana* may occur more than 10 years after seedling establishment.

8.3.2 Participatory selection

Genetic gain and diversity in the domestication of miombo fruit trees

The high genetic diversity of selected *Uapaca* populations in Malawi compared with the mother source and their progenies confirmed that the extensive participatory selection used to capture superior individuals was adequate

(Hamisy, 2004). Knowledge of phenotypic and genetic variation is a prerequisite for the domestication and improvement of indigenous fruits from the wild. Within-stand genetic variation of *U. kirkiana* and *Sclerocarya birrea* trees was determined (Fig. 8.1), and implications for developing germplasm collection and conservation guidelines were suggested (Agufa, 2002). *S. birrea* had more diversity than *U. kirkiana* in the study. The role of geography and structure in subspecies status in *S. birrea* in contributing to the pattern of variation was affirmed. *S. birrea* from Tanzania contained more chloroplast variation, supporting the hypothesis that this country is the centre of diversity and origin of *S. birrea* (Agufa, 2002).

In genetic terms, human-induced domestication results in either gain or loss of genes, altered gene frequencies or modifications of the way genes are packed, i.e. gene complexes (Simons, 1996). The domestication of wild fruit trees must be seen as a continuum between the undisturbed wild state, through cultivation of semi-domesticates, to monocultural plantation or orchards of advanced

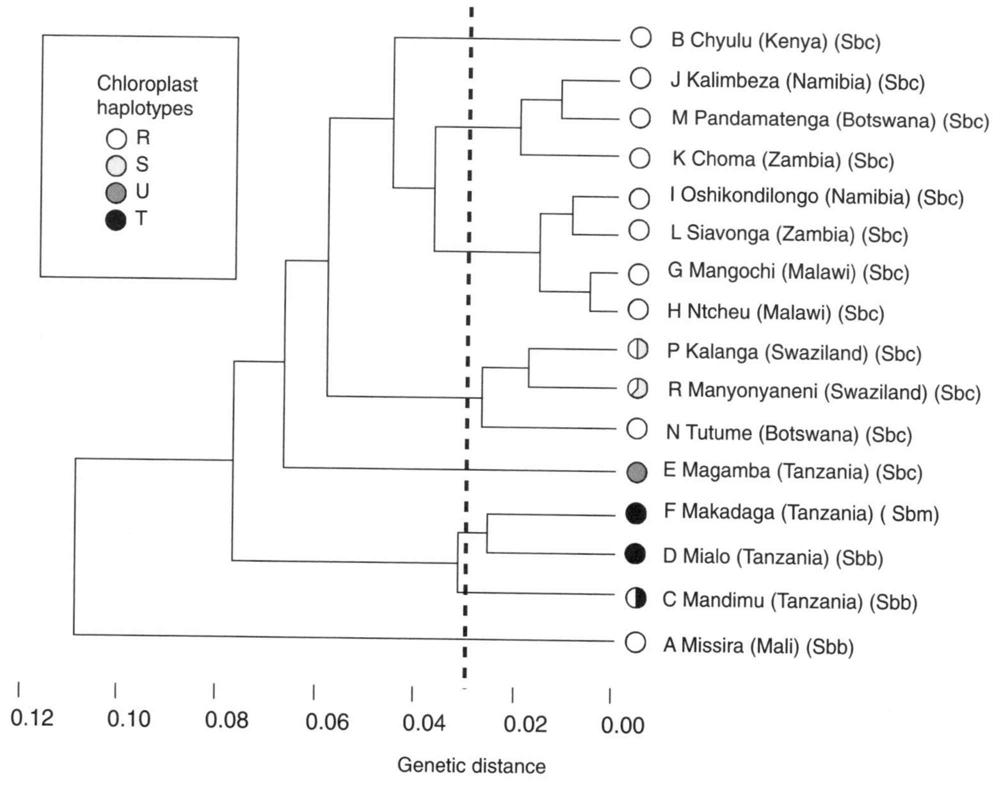

Fig. 8.1. Phenogram-based genetic distance from 80 RAPD markers for 16 populations of *Sclerocarya birrea* sampled from eight countries in sub-Saharan Africa. Codes in parentheses indicate subspecies designations made during collection. Sbc, ssp. *caffra*; Sbb, ssp. *birrea*; Sbm, ssp. *multifoliolata* (Agufa, 2002).

generation lines. The extent to which domesticated trees differ from their wild progenitors depends on population size, the heritability of the desired trait under selection, the mating system, the intensity of selection and the inherent variability of the traits (Nyland, 1996; Simons, 1996; Cornelius *et al.*, 2006). Domestication requires not just breeding, but also selection and management. Simons (1996) has asserted that qualitative traits, such as fruit size, shape and taste and tree form and precocity, are more strongly inherited from the selected mother trees, such that 60% of the progeny might be similar to their parents. Therefore cloning, after identifying and screening for a large number of superior traits, is the best way to capture such genetic variation because it eliminates the recombination or segregation of genes. This is the basis of participatory clonal selection in the domestication programme of ICRAF.

Hamisy (2004) assessed the efficiency of tree selection using the technique of random amplification of polymorphic DNA (RAPD). Two populations were used, and data on 109 RAPD loci in 181 *Uapaca kirkiana* individuals showed a lack of clear differentiation between populations and subpopulations, showing a high level of genetic identity and thus indicating close relationships and the possibility of gene flow between the populations. Farmer-mediated short-, medium- and long-distance seed movements among populations may also explain the lack of distinct genetic subdivisions. A similar, low amount of genetic variation was observed in peach palm in Brazil (Cornelius *et al.*, 2006). However, Hamisy's study confirmed that the elite trees selected from the wild using the participatory approach showed more diversity than their wild counterpart sources and their progenies. This was attributed to the extensive sampling used in the phenotypic selection from natural forests, communal lands, homesteads and farms, which enabled the capture of superior trees from wild populations (Akinnifesi *et al.*, 2006a) (Fig. 8.1). Large amounts of diversity between mother trees and their progenies indicate the potential for using vegetative propagation techniques for improvement to ensure genetic similarity as well as genetic diversity.

In practice, the use of seeds collected from isolated trees on the farm may cause inbreeding (through parent–offspring or half-sib mating), hence the need for proper selection guidelines for participatory domestication. A comprehensive framework for breeding programmes has been described which distinguishes four conceptually distinct breeding population components. First, the 'base population' is chosen from which individuals are selected to be carried forward to the next generation (Cornelius *et al.*, 2006). Secondly, the 'selected population' is the subset of the initial population that is to be carried forward to the next generation. Thirdly, the 'breeding population' is composed of trees that are used to produce the next generation, and may consist of all of the selected population. Fourthly, the 'production population' is composed of the trees used to produce propagules for commercial planting, or on-farm planting in this case.

Cornelius *et al.* (2006) has articulated several ways in which these breeding programmes could lead to genetic erosion and loss of genetic diversity: (i) Because genetic variation depends on population size, intensive selection in the breeding programme will lead to loss of diversity in the base population, and possibly to cumulative genetic erosion if repeated in succeeding generations.

Ensuring that adequate numbers of trees are selected can become a challenge in small populations with desirable tree traits. (ii) Establishing a production population by intensive selection within the selected population will also lead to loss of genetic diversity, although it will not lead to long-term erosion. (iii) The use of seed sources from commercial plantations or orchards (farmers in our case) would lead in the short term to reduced diversity and inbreeding depression. Ways in which recurrent selection can achieve genetic diversity are well known, but arduous and long-term breeding investment is required. Fortunately, most of the miombo indigenous fruit trees are amenable to vegetative propagation, and efficient clonal propagation techniques that might be used for accelerated impact have been described (Akinnifesi *et al.*, 2006; Leakey and Akinnifesi, Chapter 2, this volume). Cornelius *et al.* (2006) has outlined how the propagation of many selected clones within the network of the farming community, using village cultivars and the propagation of only a few selected clones (five to ten) from each village, can be used for the rapid production of superior propagules. This participatory clonal propagation approach has been used by ICRAF in West Africa (Tchoundjeu *et al.*, 2006) and southern Africa for *Uapaca kirkiana* and *Strychnos cocculoides* (Akinnifesi *et al.*, 2006), and Latin America has applied clonal seed selection in the peach palm (Cornelius *et al.*, 2006).

Participatory identification of elite trees for clonal selection

Tree domestication is a paradigm shift, from a focus on tree improvement based on breeding and conventional forest tree selection to horticultural approaches focused on quality germplasm production for wider cultivation to serve the needs of smallholder farmers. It is an iterative process that includes a wide range of activities. The processes involved include the exploration of wild populations and the identification of superior tree species and provenances from natural variability; the evaluation and selection of suitable trees and clonal propagation to develop superior cultivars; macro- and micropropagation techniques for multiplication; scaling up to the dissemination of germplasm; and acquiring knowledge of tree management (Akinnifesi *et al.*, 2006). The domestication approaches and strategies deployed for individual species vary according to their functional use, ecology and biology, niches and biophysical limits (Simons and Leakey, 2004; Akinnifesi *et al.*, 2006; Tchoundjeu *et al.*, 2006; Leakey and Akinnifesi, Chapter 2, this volume).

Tree domestication in ICRAF Southern Africa has evolved from multipurpose tree screening in the 1980s to a more participatory domestication programme involving pomological and market-led approaches, starting from 1996. The concept of 'ideotype' is a first step towards developing an improved plant by combining characters which provide a guide to the selection of potential breeding stock in a wild population (Dickman, 1985). Its application to fruit trees, such as mango (*Mangifera indica*) (Dickman *et al.*, 1994), and the domestication of priority indigenous fruit trees, such as *Irvingia gabonensis* (bush mango) in Nigeria, has been described by Leakey and Page (2006) as an aid to the multiple-trait selection of superior trees for cultivar development, and offers an opportunity for developing a hierarchy of different ideotypes to meet different market opportunities.

In southern Africa, local knowledge of the rural communities was captured by brainstorming at village workshops about the objectives of selection with 20–30 people in each group. Tree-to-tree variation was measured, with the communities, in wild populations of *Uapaca kirkiana* and *Strychnos cocculoides*, and selection of superior trees was based on market-oriented ideotype products (Akinnifesi *et al*., 2006). Together with villagers, the authors identified the superior trees on the basis of superior traits, and these were were systematically named and tagged *in situ* according to year of collection, location and ownership. Site descriptors were documented and fruits sampled for detailed assessment of the qualitative and quantitative characteristics, including chemical and organoleptic analysis. Seeds and scions were also collected for growth and multiplication in the nursery. In some cases, duplicate materials were collected by farmers and raised in individual or group nurseries in their own communities. The superior germplasm was subsequently evaluated in clonal orchards on-station and on-farm and fruits were characterized and analysed for their chemical characteristics. This evaluation identified the trees for subsequent vegetative propagation and clonal testing, so that high-quality planting materials could be made available to farmers as soon as possible.

Through the selection and propagation of elite genotypes from the wild, new cultivars with superior or better marketable products – fruit size, sweetness and fruit load with improved uniformity – have been obtained. A total of 429 trees with superior phenotypes of priority indigenous fruits (*Uapaca kirkiana*, *Strychnos cocculoides*, *Sclerocarya birrea*, *Vitex mombassae*) was selected in the region from the wild, on the basis of criteria determined jointly with rural community dwellers (farmers, marketers, traditional chiefs, schoolchildren) in Malawi, Zambia, Zimbabwe and Tanzania. These include 107 superior *Uapaca kirkiana* and 20 *Strychnos cocculoides* phenotypes in Malawi; 108 *Uapaca kirkiana* and 34 *Strychnos cocculoides* trees in Zimbabwe; 78 *Uapaca kirkiana* and *Strychnos cocculoides* trees in Zambia; and 30 trees of *Vitex mombassae* and 20 superior *Sclerocarya birrea* trees in Tanzania.

In Malawi, the natural population of female *Uapaca kirkiana* trees was 40 trees per hectare in Dedza, 81 trees per hectare in Kasumbu and 490 trees per hectare at Phalombe. During the participatory workshops with farmers, a few outstanding trees, ideal for processing, were identified. To supplement the information received from the participatory village workshops, elite trees were identified with communities and measurements were made on fruit size, fruit sweetness (sugar content) and pulp content (Akinnifesi *et al*., 2006). The selection was based on superior fruit characteristics from different populations and land uses. For example, in Dedza, Malawi, among the elite *Uapaca kirkiana* trees identified as having high fruit yields, one tree in particular had more than 6000 fruits per tree, with large fruit size (3–4 cm diameter) and sweet taste (Akinnifesi *et al*., 2006). The big-fruit ideotype was locally described by expressions such as 'gundete okolera' in Dedza and 'mapunbu amutiye' in Phalombe, Malawi, depicting its outstanding size, appeal and taste. These same expressions were used to describe the most beautiful unmarried girl or potential village beauty queens in the communities.

In Zimbabwe, using PRA approaches, 416 participants (farmers, fruit vendors and schoolchildren, 65% of them female) identified 108 superior trees of *Uapaca kirkiana* in seven districts across the country. In addition, 200 participants (50% women) from four districts identified 34 superior phenotypes of *Strychnos cocculoides*. Total soluble sugar from *Strychnos cocculoides* trees ranged from 7 to 23% and fruit size ranged from 75 to 514 g. ICR04MapanzureZW8 had the biggest fruits (514 g) and ICR04MafungautsiZW17 had the highest total soluble sugar (23%). The variability of fruit characteristics was clearly illustrated in Zimbabwe; trees with very contrasting fruits were identified as ICR02Chimani ZW9, ICR02UrandaZW29 and ICR02MafaZW40. ICR02MafaZW40 had the highest pulp content and greatest fruit weight, but a low sugar content. In contrast, ICR02UrandaZW29 had small fruits that were very sweet and had high pulp content (Akinnifesi *et al.*, 2006). On the other hand, although ICR02ChimaniZW9 had large fruits, it was rejected in the selection process as it had low pulp content, high relative shell weight and low sugar content. It also had the highest seed weight. A consequence of domestication and cultivar development is that trees propagated clonally from mature tissues will flower and fruit earlier. They will therefore be smaller in stature and produce fewer fruits, especially in the climate of miombo woodlands. To compensate for this, trees can be grown at higher density. In our experience, marcots are bigger than grafted trees, and trees established from seedlings tend to be taller but with a smaller crown size compared with marcots. However, trees established from seedlings have a longer juvenile phase than trees established from either grafted or marcottage stocks. In high-rainfall areas of West Africa, marcots have been shown to be bigger than plants established from seedlings (Z. Tchoundjeu, personal communication).

However, it is important to recognize that, as in *Uapaca kirkiana* in Zambia, some fruit traits, including tree fruit load and pulp content per fruit, can be manipulated to a limited extent by management practices such as thinning (Mwamba, 1995b). The challenge is not only of practical improvement per se, but also of reconciling the potential genetic improvement with the practical realities of farmers' needs and perceptions and the delivery of germplasms (Simons, 1996), considering also intellectual property right (IPR) environments.

Clonal propagation

Seedling propagation is not a desired approach for commercial fruit production because of the high variability of progenies from mother trees; asexual means are therefore preferred. Both macropropagation (conventional vegetative propagation) and micropropagation (*in vitro* culture) techniques have been applied to some priority miombo indigenous fruits in southern Africa. Vegetative propagation is needed to rapidly test, select from, multiply and use the large genetic diversity in wild tree species on stations and farms. From our experience, some of the miombo fruit trees are not amenable to propagation by juvenile stem cuttings; examples are *Uapaca kirkiana*, *Parinari curatellifolia* and *Sclerocarya birrea*.

MACROPROPAGATION Grafting is the most efficient way to rapidly effect improvements in these fruit trees: *Adansonia digitata* (85–100% graft success), *Mangifera indica* (97%), *Uapaca kirkiana* (80%), *Strychnos cocculoides* (40–79%), *Sclerocarya birrea* (52–80%), *Vangueria infausta*, (100%) and *Parinari curatellifolia* (71%) (Mhango and Akinnifesi, 2001) compared favourably with exotics such as mango (90%) in the same trial. Air-layers were promising for *Uapaca kirkiana* (63%) but were not successful for *Parinari* and *Strychnos* species (Mhango and Akinnifesi, 2001). Interestingly, rooting hormone did not improve the rooting of *Uapaca* air-layers. Both grafting and air-layering set during November–December gave the best results. Top-wedge and whip methods were the most successful for grafting. The results showed that the factors determining grafting success are the skill of the person, the time of the year the scion is collected and the interval between scion collection and grafting (Akinnifesi *et al.*, 2004a, 2006).

MICROPROPAGATION As the awareness of the potential of indigenous fruit trees increases, demand will inevitably increase, and, considering the massive tree-planting initiatives in the many countries in the miombo ecoregion, explosive demands for high-value trees, including indigenous fruit trees, cannot be accommodated in the short run. All known approaches of vegetative propagation are inherently slow and are attractive for only a few thousand farmers. For instance, to produce rootstocks for grafting *Uapaca kirkiana*, at least 1 year of seedling growth is required. In order to be able to deliver high-quality propagules of superior indigenous fruits in sufficient quantities for wider adoption, it is important to explore biotechnological methods, such as tissue culture. Currently, research on tissue culture has been conducted in the ICRAF programme in southern Africa on some priority indigenous fruit trees, especially *Uapaca kirkiana*, *U. nitida* and *Pappea capensis* (plum), with the objective of developing a reproducible clonal protocol for rapid regeneration and multiplication, and to determine early graft compatibility using *in vitro* techniques (Mng'omba, 2007). Incompatibility between stock and scion in the fruit orchard could constitute a major bottleneck to production. Simons (1987) estimated that half a million grafted peach trees had died in the southeast of the USA as a result of scion/stock incompatibility. Selection of scion and stock for compatibility is important for profitable orchard establishment and management.

Phenolic compounds and *p*-cumaric acids have been implicated in early graft incompatibility in *Uapaca kirkiana*. Plants normally release phenolics to heal wounds and as a defensive mechanism against pathogen attack, lignification and protein biding. Multiplication was much easier in *P. capensis* than in *Uapaca kirkiana*. Preconditioning grafted *Uapaca kirkiana* trees and decontaminating explants in 0.1% mercuric chloride for 8 min was found to be effective in achieving a high level of *in vitro* culture asepsis in *Uapaca kirkiana*. *In vitro* propagation of *Uapaca kirkiana* is feasible with sprouts excised from the preconditioned trees (Mng'omba, 2007). Shoot multiplication was effective using ½ strength Murashige and Skoog (MS) medium plus 2.5 mg/l of IBA (indole butyric acid) medium. Results will be improved by micropropagating *Uapaca kirkiana* in such a way that stock plants are not stressed and ensuring mycorrhiza is present. Repeated exposure of the difficult-to-root microcuttings of *Pappea capensis* to 0.5 mg/l IBA improved rooting from 42 to 62%, and the number of roots per plantlet averaged 3 (Mng'omba, 2007). Somatic embryos of *P. capensis* were successfully germinated into plants (65%), and 65% of plantlets survived after hardening off in a mist chamber.

Graft compatibility increased more with homografts than with heterografts between *Uapaca kirkiana* clones, species and provenances. *Uapaca kirkiana* and *U. nitida* had weak compatibility and they may exhibit delayed incompatibility. Although *Jatropha* and *Uapaca kirkiana* belong to same family, there is outright incompatibility or early rejection. The technique seems promising for the detection of early incompatibility between close and distant related propagule sources. On the basis of a series of results, reproducible micropropagation protocols have been developed for the rapid multiplication of mature *Uapaca kirkiana* and *P. capensis* (Mng'omba *et al.*, 2007a, b).

8.3.3 Tree management systems

Tree establishment and management

Grafting and air-layering of indigenous fruit trees have addressed precocity problems, and enabled selection of superior fruit traits (Akinnifesi *et al.*, 2006). Using vegetative propagation approaches, especially air-layering and grafting, clonal fruit orchards have been established at Makoka in Malawi for *Uapaca kirkiana* from superior trees selected using participatory methods. Tree establishment protocols used for the indigenous fruit trees were those employed for mangoes. The rates of tree survival in the clonal orchards are generally high for grafted *Adansonia digitata* (100%) in Zambia, for *Uapaca kirkiana* (80%) in Malawi and for *Sclerocarya birrea* (90%) in Tanzania, but low (40%) for *Strychnos cocculoides* in Zambia (ICRAF, 2005). However, the survival of established marcots was low in Malawi, and declined over a period due to poor root development. Rejection in grafts is also attributed to stock/scion incompatibility (Mng'omba, 2007).

Tree orchards established in Makoka from grafted trees started to fruit after 2 years, but fruit load only became stable after 4 years. However, limited research has been conducted to date on management regimes of indigenous fruit trees in terms of the responses of the trees to fertilizer, manure and irrigation.

In a recent tree management trial in Makoka, Malawi the effects of fertilizer, manure and irrigation on the early growth and survival of *Uapaca kirkiana*, *Sclerocarya birrea* (subsp. *caffra*), *Vangueria infausta* and *Mangifera indica* (mango) were assessed. After 33 months, there was a striking pattern of high mortality in *Uapaca kirkiana* in plots that received all the treatments (25%) compared with those that did not receive any of the soil amendments (85%) and in plots that received only irrigation (100%) but no fertilizer amendment (Table 8.2). The trial showed that application of chemical fertilizer, compost and irrigation did not improve the growth of *Uapaca kirkiana* in Malawi, and that significant mortality may be observed with a combination of soil

Table 8.2. Mean growth increment (mm/month) and survival (%) of four fruit tree species with fertilizer, compost manure amendment and dry-season irrigation at Makoka, Malawi during the 33 months after establishment.

Species	Fertilizer	Manure	Irrigation	Height	Survival
Uapaca kirkiana	No	No	No	55.8	85
	No	No	Yes	59.3	100
	No	Yes	No	46.4	75
	No	Yes	Yes	57.8	63
	Yes	No	No	54.9	83
	Yes	No	Yes	48.6	67
	Yes	Yes	No	49.6	67
	Yes	Yes	Yes	43.7	25
Sclerocarya birrea	No	No	No	91.9	100
	No	No	Yes	102.8	83
	No	Yes	No	93.4	100
	No	Yes	Yes	94.4	69
	Yes	No	No	92.9	83
	Yes	No	Yes	98.3	75
	Yes	Yes	No	98.3	75
	Yes	Yes	Yes	90.2	83
Vangueria infausta	No	No	No	118.7	94
	No	No	Yes	116.5	100
	No	Yes	No	118.7	83
	No	Yes	Yes	113.4	100
	Yes	No	No	117.1	100
	Yes	No	Yes	115.6	100
	Yes	Yes	No	115.1	100
	Yes	Yes	Yes	118.4	96
Mangifera indica	No	No	No	92.7	100
	No	No	Yes	96.7	100
	No	Yes	No	97.9	100
	No	Yes	Yes	92.5	100
	Yes	No	No	98.8	100
	Yes	No	Yes	103.2	100
	Yes	Yes	No	107.4	100
	Yes	Yes	Yes	104.1	100

Source: F.K. Akinnifesi, J. Mhango, G. Sileshi and T. Chilanga, unpublished results.

amendment practices, except for the strategic application of irrigation of green manure alone (F.K. Akinnifesi, unpublished results). The results of this study did not support the commonly held assumption that the miombo indigenous fruit trees, especially *Uapaca kirkiana* and *Sclerocarya birrea*, require fertilization, manure and dry season irrigation to increase growth and survival. On the other hand, *V. infausta* performed better when managed with a combination of light dry-season irrigation and manuring. Similar results were reported in Botswana for *Strychnos cocculoides*, *Sclerocarya birrea* and *V. infausta* (Mateke, 2003).

However, combining the use of manure, fertilizer and irrigation is not recommended for any of the miombo fruit trees. Spacing of trees in the wild improved fruit yield (Mwamba, 1995a). In addition to tree management, an understanding of edaphoclimatic boundary requirements can help to guide management towards desired fruit orchard production in miombo indigenous fruit trees.

Vangueria infausta fruited early compared with mangoes in the fruit orchard at Makoka (Table 8.3). Although very few *Uapaca kirkiana* flowered in the second and third years, fruits were not retained. *Sclerocarya birrea* grew relatively quickly but had not started to flower or fruit after 36 months. Experience of fruit retention in another trial (Table 8.4) showed that grafted *Uapaca kirkiana* flowered and fruited but fruit started to be retained from year 4 onwards.

On-farm planting and scaling up

Many communities in southern Africa retain or nurture semi-domesticates of wild fruits in their homesteads and fields. Therefore, introducing superior cultivars to farmers is an important aspect of accelerated domestication. Farmers in the Bushbuckridge region of South Africa, for instance, planted

Table 8.3. Percentage of *Vangueria infausta* and *Mangifera indica* trees fruiting 24, 31 and 36 months after planting (MAP) at Makoka, Malawi.

Species	Date	Percentage fruiting[a]
M. indica	December 2005 (24 MAP)	33.3 (9.83)
	July 2006 (31 MAP)	99.0 (1.04)
	December 2006 (36 MAP)	70.8 (5.36)
V. infausta	December 2005 (24 MAP)	94.8 (3.00)
	July 2006 (31 MAP)	66.7 (5.14)
	December 2006 (36 MAP)	93.8 (3.10)

Source: Akinnifesi *et al.* (2006).
[a]Figures in parentheses are standard error of the mean.

Table 8.4. Relative tree growth and indicative fruiting of *Uapaca kirkiana* trees at Makoka orchard from grafted, marcotted and seedling stocks, at 4 years after establishment.

Parameter	Marcotts	Grafts	Seedling stock
Tree height (m)	2.4 ± 0.11	2.0 ± 0.13	2.7 ± 0.14
Bole height (m)	0.39 ± 0.04	0.35 ± 0.04	0.46 ± 0.64
Root collar diameter (cm)	8.50 ± 0.32	9.14 ± 0.35	10.3 ± 0.36
Crown depth (m)	2.0 ± 0.13	0.35 ± 0.04	2.4 ± 0.76
Crown spread (m)	2.7 ± 0.14	2.3 ± 0.13	2.4 ± 0.160
Number of primary branches	17.2 ± 1.33	15.8 ± 0.95	15.3 ± 2.00
Number of secondary branches	25.0 ± 2.60	19.9 ± 2.60	15.3 ± 2.00
Number of tertiary branches	15.0 ± 2.97	10.3 ± 2.91	5.6 ± 1.32
Minimum number of fruits	2	3	0
Maximum number of fruits	414	127	0
Mean number of fruits	78	52	0

Source: F.K. Akinnifesi (unpublished results).

Sclerocarya birrea after the successful liqueur made from marula (Amarula Cream) had become an incentive for rural enterprises. Half of the farmers sampled in four villages were managing self-seeded recruits of *S. birrea*, while 30% planted from seeds, truncheons or transplants of wildings. However, there has been no planting of trees with desired traits in any of the projects (Shackleton, 2004). About 79% of the households had at least one tree of *S. birrea*. This was a spin-off effect of massive project support to the communities in the Bushbuckridge region. Tree domestication efforts that build on existing farmers' knowledge and practices in the country are likely to succeed.

Several thousands of farmers have been trained in the five countries in nursery establishment, propagation and tree management. Monitoring and evaluation results confirm that an estimated total of 12,702 farmers were trained in nursery establishment and management, propagation techniques between 2001 and 2005. In 2005 alone, 758 individual and group nurseries were established and managed by farmers during the period (315 group and 443 individual nurseries): 458 in Malawi, 165 in Tanzania, 36 in Zambia and 99 in Mozambique (ICRAF, 2005). Over 6000 farmers have been testing indigenous fruits and other high-value trees in the five countries.

On-farm management showed lower survival for *Uapaca kirkiana* and *Strychnos cocculoides* in Malawi and Zambia (F.K. Akinnifesi *et al.*, unpublished results; Mhango and Akinnifesi, 2001). In a survey of farmer-managed fruit orchards of *Uapaca kirkiana*, *Ziziphus mauritiana* and *Sclerocarya birrea*, tree survival declined to 51% after 8 months and 12% after 20 months. *Strychnos cocculoides* also declined from 55% at 8 months to 18% after 20 months. At 29 months, *Sclerocarya birrea* maintained survival at 48% on farmer's fields, while *Ziziphus mauritiana* maintained survival at 50% at 29 months (Mhango and Akinnifesi, 2001). Farmers attributed poor tree survival and growth to inefficient management, lack of water, grazing and pests. About 87% of farmers did not apply any management measure. Farmers indicated that most of the trees that died did so within 6 months after establishment. The gap between on-station researcher-managed orchards and farmer-managed orchards warrants further tree management investigations and the development of appropriate on-farm management protocols. Although our work has shown that miombo indigenous fruits do not require soil fertility replenishment or intensive irrigation, weeding and protection from browsing animals are important. In addition, mycorrhizal inoculation has been stated to be a necessity, especially at the early stage.

8.4 Postharvest Utilization and Commercialization

8.4.1 Pre-harvest and postharvest handling of indigenous fruits

One aspect of indigenous tree fruit and products that has been neglected is the pre-harvest and postharvest handling (Ham *et al.*, Chapter 14, this volume). The loss of fresh fruit produce is one of the major constraints on smallholder horticultural crops in sub-Saharan Africa, and direct and indirect postharvest losses are estimated at 25–50% in developing countries (Kader, 1992). Until recently, little has been known about the effect of postharvest physiology of fruits and shelf life. There is a need to develop appropriate techniques for fruit handling, preservation and processing to maximize returns from fresh and processed fruit products. The development of such postharvest systems requires understanding of current practices and how they influence postharvest losses. In a study undertaken in Zimbabwe (Kadzere *et al.,* 2004), the authors interviewed 180 producers (collectors), 120 processors and 210 marketers. The results indicated that community dwellers had knowledge of the ripening period. The time of the year, colour changes, skin softness and abscission were the indicators used by the respondents to determine ripening, depending on species. Fruit ripening overlapped for many species, opening an avenue for having fresh fruits throughout the year (Akinnifesi *et al.*, 2004a). For *Adansonia digitata*, about 85% of users or harvesters climbed the tree to harvest because the fruit did not readily abscise when ripe (Kadzere *et al.*, 2001). Most of the harvesters collected fruits that dropped naturally, e.g. for *Parinari curatellifolia* (99%), *Sclerocarya birrea* (95%), *Uapaca kirkiana* (78%), *Ziziphus mauritiana* (61%) and *Strychnos cocculoides* (69%). Other means of harvesting included shaking the tree or throwing objects at the tree crown, hitting the stem to dislodge the fruits. Most *Uapaca kirkiana* trees had big scars and wounds on the stem because of these inappropriate harvesting methods. Fruits were broken and branches severely damaged in many cases. Fruits were harvested when unripe, just ripe or well-ripened, and were harvested several times during the season (Kadzere *et al.*, 2006a, b, c). However, allowing the fruits to drop often results in contamination and attacks from pests and diseases.

More than half of the fruits harvested were often retained for home consumption for *Adansonia digitata* (59%), *Azanza garckeana* (73%) and *Strychnos cocculoides* (76%), but less frequently for *Uapaca kirkiana* and *Ziziphus mauritiana* (28–33%), indicating that these were the most frequently sold fruits. *Uapaca kirkiana* and *Z. mauritiana* were sold by 84% of fruit marketers. Before marketing, several methods were used to add value – cleaning, grading, packaging, protecting from the sun – but these varied with species. Grading was done for most fruits before marketing (80% for *Uapaca kirkiana*), mainly based on fruit size, as bigger fruits fetch better prices. For instance, no preservation method was observed in Zimbabwe for *Uapaca kirkiana*, *Strychnos cocculoides* and *Sclerocarya birrea*, whereas 54–86% of respondents preserved *Ziziphus mauritiana*. Thirty-one per cent of respondents shaded *Uapaca kirkiana* from direct sunlight. *Ziziphus mauritiana* and *Parinari curatellifolia* were sun-dried as whole fruits before storage, or fermented as solid to extend shelf life and make fruit available during the off-season (Kadzere *et al.*, 2001). *Strychnos cocculoides* was rehydrated before consumption. Controlled dehydration has been reported to enhance palatability and boiling, which may sometimes lead to loss of vitamins (Kordylas, 1991). Marketers often performed grading, cleaning, packaging for most fruits as well.

Postharvest deterioration often results from cracks during harvesting, mechanical damage during transportation and storage, and insect pest damage (Kadzere *et al.*, 2001, 2006b). Most marketers (78–84%) indicated that postharvest losses were a major constraint for *Uapaca kirkiana*. While damage due to fruit harvesting was noted by a few people as a problem (16%), insect pests (61%), mechanical damage during transportation and storage (61–88%) were mentioned by a high proportion of respondents. Similarly, Ramadhani (2002) found that 50–75% of marketers associated market constraints with the perishability of fruits. At the market there was no agreed number of fruits a consumer could taste freely before buying. Consequently, large numbers of fruits were tasted, and 75% of wholesalers reported this as a major problem, followed by retailers (63%); the problem was less for collectors (producers) (Ramadhani, 2002).

Harvesting time has been shown to affect ripening and darkening in *Uapaca kirkiana* (Kadzere *et al.*, 2006c, d). Fruits that were harvested in December lost less weight (14%) in storage than those that were harvested 2 weeks earlier (34%). The soluble sugar content measured 6 days after harvest was also lower for fruits harvested in December (18%) compared with those harvested in November (9%). This indicates that there are benefits in delaying harvesting to improve fruit skin colour at harvest and during storage, such as reduced weight loss and maintaining a higher soluble sugar content.

8.4.2 Small-scale fruit processing enterprises

In terms of fruit processing, there have been reports of small cottage industries in different countries in the region, such as wine production from *Syzygium owariense*, *Uapaca kirkiana*, tamarinds, mangoes and other fruits, such as those commercialized in the past as Mulunguzi wine in Malawi (Ngwira, 1996). *Ziziphus mauritiana* and *Sclerocarya birrea* are produced at export quality in Lusaka, Zambia. Amarula liqueur is commercialized in South Africa and sold in more than 63 countries world-wide (Ham, 2005). In Tanzania, about 198 rural women were trained in 1998 in making wine, jam and juice from indigenous fruits, and 2 years later they had trained another 2045 processors (Saka *et al.*, 2004, Chapter 16, this volume; Akinnifesi *et al.*, 2006). Details on the nutritional characteristics and processing of the fruits have been given elsewhere (Saka *et al.*, Chapter 16, this volume). The feasibilities of four different enterprises developed by ICRAF-CP Wild in Malawi, Zambia, Tanzania and Zimbabwe were assessed and the results have been detailed by Jordaan *et al.* (Jordaan *et al.*, Chapter 15, this volume).

8.4.3 Commercialization of wild-harvested fruits

A global meta-analysis of the marketing and cultivation of wild forest products showed that farmers engaged in the cultivation of indigenous fruits had higher returns to labour, used more intensive production technologies, produced more per hectare and benefited from a more stable resource base than those that relied on wild collection (Ruiz-Perez *et al.*, 2004). The analysis suggests that cultivation of wild fruit trees will become more important as rural households move from subsistence to a cash-oriented economy. However, it is important to recognize that the best scenario occurs when the domestication and commercialization of agroforestry tree products occur in parallel (Wynberg *et al.*, 2003; Akinnifesi *et al.*, 2004a), in order that problems of the seasonality and reliability of supply, diversity and inconsistency of fruit quality are overcome throughout the supply chain (Akinnifesi *et al.*, 2006).

Market assessments for fresh fruit in the region indicate that a substantial amount of trade occurs, but that it is generally informal. Ramadhani (2002) also showed that there was equitable sharing of market margins from indigenous fruit marketing in Zimbabwe. The price of fruits charged by retailers (Z$11.2/kg) was observed to be double that charged by producers (Z$5.1/kg) and wholesalers (Z$5.24/kg). This was attributed to much higher costs for the retailer (Z$7.14/kg) than for wholesalers (Z$4.8/kg) and producers (Z$1.73/kg). Retailers made more profit than wholesalers, and producers' profits were intermediate. The relative market margin was 45% for producers, 2% for wholesalers and 53% for retailers. The wholesalers made up for their lower margin by purchasing large amounts of fruits. Women dominated the retail market, men predominated in the wholesale component, and children predominated as producers.

In another report (Ham, 2005), communities collectively harvested about 2000 tons of *Sclerocarya birrea* in South Africa and collectively earned US$180,000 annually, representing more than 10% of average household income in the communities. Collection was done when agricultural activities were minimal. The Southern African Natural Products Trade Association (Phytotrade) reported a gross revenue of US$629,500 from the sale of natural tree products by members. The key fruit tree products among them fetched US$126,420 for *Sclerocarya birrea*, US$44,120 for *Ximenia caffra*, US$22,250 for *Adansonia digitata* (baobab) and US$20,000 for *Kigelia* spp. (Phytotrade M&E Report, 2005).

Several market studies were conducted in southern Africa by CPWild and ICRAF (Ramadhani, 2002; Schomburg *et al.*, 2002; ICRAF, 2005; Ham *et al.*, Chapter 14, this volume). The marketing of indigenous fruits is characterized by the following factors:

- *Wild harvesting*. Wild harvesting of indigenous fruits predisposes it to seasonality of fruit yield, variation, and limited supply. Drought may also lead to reduction in fruit production. Consumption is greatly affected by the nature of supply in terms of quantity, quality, time, location, price, scarcity value, etc. There is a general tendency to move to year-round supply of fruits and other food products in conventional markets. A key implication is that the supply of fruit has to be adapted to demand, which calls for greater consistency in the quantity and nature of supply.
- *High transaction costs*. Marketing and postharvest handling generally have high transaction costs. Transaction costs are high because of product losses, delivery delays, the costs of monitoring transactions and agents in markets, negotiation or haggling costs, transport and logistic inefficiencies, and less informed market behaviour. Transaction costs are better addressed through improved coordination and information flows in supply chains. This is a key challenge for economic development efforts in emerging markets.

- *Generic products*. All the products that were identified and studied are generic and undifferentiated. Generic marketing certainly has a place even in modern markets, but is more associated with mass-produced products that are consumed in large quantities as opposed to niche-oriented seasonal products. Thus, having indigenous fruits marketed as generic products, even in low-income markets, is an indication of the inefficiency of the market. The main implication here is therefore the need for greater attention to product differentiation on the basis of product attributes and the nature of supply and demand.
- *Small amount of added value*. The fruits studied were all marketed with small marketing margins and the addition of little value in the form of packaging, processing, certification, quality control, and other utilities. This probably stems from the focus on low-income domestic/local markets, where consumer spending power is limited. This implies that any attempt to add value or increase the marketing margin must be associated with targeting higher-income consumers, who are able to compensate for this. The feasibility of fruit processing has been reported by Jordaan *et al.* (Chapter 15, this volume) to be positive, with high net profits of 14–28%.
- *Orientation to local markets*. Supply and consumption are entirely directed at local markets, given the traditional status of the products and consumption habits. The local markets are assumed to be low-income markets by and large, with limited ability to pay for greater sophistication and value additions that add significantly to the marketing margins. Hence, the current tendency to favour generic marketing of produce. The implication is arguably that limited benefit can be derived from much value addition in local markets. At best the improvements can be incremental. Alternatively, the improvements could radically change the product offer through processing (e.g. fruit juices, jams and preserves) while retaining product appeal. Another alternative may be to consider acceptability in higher-income markets and adapt the product accordingly.
- *Indigenous demand*. The product is undoubtedly associated with indigenous demand, which establishes its current appeal. This is a positive factor and a key element in a marketing strategy, especially in a world that is showing greater appreciation of organic and natural products. This implies that the attributes and motivations that underlie this indigenous demand could be useful selling attractions in new higher-income markets where novelty and indigenous innovation are valued.
- *Limited promotion*. The product is offered in its generic form as and when available, little is being done to promote consumption. Encouraging consumption in local and especially new markets would require much greater effort in this regard. Promotion could focus on selling product attributes beyond simply taste. These attributes may be medicinal value, uniqueness, scarcity value, novelty, etc. Promotion would largely be associated with new and especially higher-income markets dominated by corporate entities and chain stores.
- *Land tenures, policy and institutional arrangements*. Land tenure issues were investigated in Malawi and Zambia. Most of the lands belonged to members

of the households, except for small parcels of land borrowed from friends or village chiefs (German, 2004). The duration of tenure was unlimited (98%) in both countries, and the tenure was transferable. The transferability of land with indigenous fruits was less than that of land with non-indigenous fruits. The tenure type was individual ownership. Land tenures did not affect the willingness of farmers to cultivate trees as 96% of farmers in Malawi and 100% in Zambia held customary lands, and only 4% in Malawi had leasehold tenure. Patrilineal households had an average of six indigenous fruit tree species retained or planted in their gardens per household compared with households with non-indigenous fruit. This is because men felt insecure about investing in trees in matrilineal communities (e.g. Thondwe in Southern Malawi and Katete in Eastern Zambia).

- Policy research was undertaken to assess policies that affect the domestication and adoption of priority indigenous fruits, and to evaluate the effects of policies on indigenous fruit production, management, utilization and marketing in Malawi. Reviews were conducted of legislation and policies related to indigenous fruits, and incorporating the needs and priorities of farmers. The reviews were mainly done in sectors with relevant legislation and policies, such as forestry, agriculture and integrated trade and industry. A review was also carried out on the relevance of such legislation and policies to indigenous fruit production and supply systems when addressing issues of poverty alleviation as well as the quality of propagation material and genetic diversity. There are policies regarding the production of ornamental and protection of catchment areas around Zomba Mountain, but not on indigenous fruit production.

Cutting down mature indigenous fruit trees for firewood is a common problem in the region, especially in communal lands in Malawi. In South Africa Shackleton (2004) also observed that even in communities where as much as 94% were reported to be benefiting from *Sclerocarya birrea* fruit collection, trees were being felled for firewood because of fuelwood shortage. There is better scope for domesticating trees in homesteads. Shackleton (2004) noted that with much care and attention in the homesteads trees produced more fruits than in the forest reserves. Average fruit weight produced in two villages from 64 trees was more than 1 ton (1016 kg) compared with less than 100 kg in the protected forests. This confirms the earlier anecdotal assertions that cultivation and management can increase fruit production and size. This agrees with other observations for *Uapaca kirkiana* in Malawi and Zimbabwe (Maghembe, 1995; Mithöfer, 2005; J. Mhango, personal communication) and *Sclerocarya birrea* (Leakey *et al.*, 2005). Maghembe (1995) reported that the fruits from planted *Vangueria infausta* were ten times bigger than in the wild. It is often difficult to quantify the amount of fruit harvested by passers-by in the forest and the amounts scavenged by monkeys and other wild animals. The rate of abortion of fruits in unmanaged situations will be greater because there are more pest attacks and other stresses.

8.4.4 Towards improved market institutions and enterprise development

There is lack of coordination along the supply chain; in particular, quality control is absent. Some of the constraints on indigenous fruit marketing can be overcome through combined cultivation and community partnerships along the supply chain. To improve the marketing systems of indigenous fruits, the following comments and recommendations have been made regarding the future development of indigenous fruit market strategies (Ramadhani, 2002; Mithöfer, 2005; Akinnifesi *et al.*, 2006):

- There is a need to understand consumer behaviour in present and future markets.
- Target the product offer to alternative higher-income markets.
- Collaborate with retailers in high-income markets to find acceptable and appropriate marketing strategies.
- Traders may reorganize themselves into cooperatives to coordinate fruit flow to different markets, improve services and add value before sale, in the form of, for example, storage, sorting, grading, packaging and other practices that make fruits more attractive to buyers; this will in turn increase the returns.
- Formulate suitable product promotion strategies when introducing products into new markets or when revising the product offer in present markets.
- Formulate logistical arrangements to improve chain efficiency.
- Conduct more detailed studies on indigenous crops to ascertain the nature of demand, consumer characteristics, preferences and product attributes.
- Explore options for greater value addition in existing and new markets. This is already in progress but more collaboration with industry could provide valuable further impetus.
- Establish appropriate marketing- and producer-related institutions in order to entrench more efficient production, communication, logistics and technology transfer and market relations. This further entails the development of a credible marketing system.
- A key challenge in marketing and producing less known commodities is communication and information exchange with markets. Systems and strategies must be formulated to improve information flows across all interfaces.
- Each product or group of associated products requires a formulated marketing strategy.
- Opportunities for farming with selected suitable commodities must be assessed to enable standardization and consistency in the supply chain and improve marketing transparency.

8.5 Conclusions and Lessons Learned

The development of rural economic activities is pivotal to the well-being of communities in southern Africa. Indigenous fruit trees are one of the few comparative advantages that rural communities have, as custodians of the resources and knowledge about their use. The fact that indigenous fruits are widely consumed and traded and are rich in macronutrients supports the notion that they contribute to the nutritional security and livelihood of rural people. Almost two decades of research by ICRAF and partners in southern Africa on the domestication of miombo indigenous fruit trees has started to yield dividends by generating new knowledge and skills for quality germplasm production and improvement, tree management and postharvest techniques available to farmers and partners.

The research on indigenous fruit domestication by ICRAF in southern Africa has shifted away from multipurpose species screening and provenance/progeny collection and testing, and now embraces more innovative participatory domestication and clonal propagation for key priority species based on the genetic variation that exists in the wild, the science of tree biology and ecology and the local knowledge of the communities. The number of species on which intensive domestication research data are available is still small – *Uapaca kirkiana*, *Sclerocarya birrea* and *Strychnos cocculoides*. More species need to be tested using the technologies developed for these species. Research is still needed to close the gap in tree performance and adaptability in researcher-managed trials and farmer-managed trials.

Despite the fact that indigenous fruit tree species are well traded in informal markets in countries across the Southern Africa Development Community (SADC) countries, very little success has been achieved in the commercialization of fresh and processed fruits. Even though markets could be developed through tree improvement and value addition, the immediate niche for most indigenous fruits is probably in local markets and consumption. Domestication and market research and development are all essential if tangible commercial interest in indigenous fruits is to emerge beyond the current opportunistic levels at roadside and local markets. The most robust income-earning oppportunities for agroforestry tree products lie in intensified systems that mark the transition from gathering to cultivation and efforts to overcome resource depletion. Cultivation and management are ways of maintaining a sustainable supply of high-value indigenous fruits and of creating new or expanded markets. Knowledge of postharvest handling and utilization should be deployed to improve the benefits to collectors, marketers and small-scale fruit processors (and private entrepreneurs). Farmers and communities are excited about domesticating indigenous fruits and are being trained and supplied with knowledge about nurseries, propagation techniques and tree management guidelines. However, the challenge remains for scaling up the technologies to new areas without project efforts. This will also help to satisfy the quest for new high-quality products by consumers in local and global markets.

Acknowledgements

We would like to thank the BMZ (Bundesministerium für wirtschaftliche Zusammenarbeit und Entwicklung) and GTZ (Deutsche Gesellschaft für Technische Zusammenarbeit) for providing the funds required to implement the project 'Domestication and Marketing of Indigenous Fruit Trees for Improved Nutrition and Income in Southern Africa' (Project No. 2001.7860.8-001.00), and Canadian CIDA for funding the Zambezi Bazin Agroforestry Project (Project No. 050/21576). We would like to thank our partners – national agricultural and forestry research and extension organizations, private entrepreneurs and non-governmental organizations – for collaboration in the project in the four countries.

References

Agufa, C.A.C. (2002) Genetic variation in *Sclerocarya birrea* and *Uapaca kirkiana* indigenous fruit trees of the miombo woodlands. MSc Thesis, Jomo Kenyatta University of Agriculture and Technology, Nairobi, Kenya, 123 pp.

Akinnifesi, F.K., Kwesiga, F.R., Mhango, J., Mkonda, A., Chilanga, T. and Swai, R. (2004a) Domesticating priority miombo indigenous fruit trees as a promising livelihood option for smallholder farmers in southern Africa. *Acta Horticulturae* 632, 15–30.

Akinnifesi, F.K., Chilanga, T., Mkonda, A. and Kwesiga, F. (2004b) Domestication of *Uapaca kirkiana* in southern Africa: a preliminary evaluation of provenances in Malawi and Zambia. In: Rao, M.R. and Kwesiga, F.R. (eds) *Proceedings of Regional Agroforestry Conference on Agroforestry Impacts on Livelihooods in Southern Africa: Putting Research into Practice*. World Agroforestry Centre, Nairobi, pp. 85–92.

Akinnifesi, F.K., Kwesiga, F., Mhango, J., Chilanga, T., Mkonda, A., Kadu, C.A.C., Kadzere, I., Mithöfer, D., Saka, J.D.K., Sileshi, G., Ramadhani, T. and Dhliwayo, P. (2006) Towards the development of miombo fruit trees as commercial tree crops in southern Africa. *Forests, Trees and Livelihoods* 16, 103–121.

Campbell, B.M. (1987) The use of wild fruits in Zimbabwe. *Economic Botany* 41, 375–385.

Campbell, B., Luckert, M. and Scoones, I. (1997) Local level valuation of Savannah resources: a case study from Zimbabwe. *Economic Botany* 51, 57–77.

Cornelius, J.P., Clement, C.R., Weber, J.C., Sotelo-Montes, C., van Leeuwen, J., Ugarte-Guerra, L.J., Rices-Tembladera, A. and Arevalo-Lopez, L. (2006) The trade-off between genetic gain and conservation in a participatory improvement programme: the case of peach palm (*Bactris gasipaes* Kunth). *Forests, Trees and Livelihoods* 16, 17–34.

de Ferranti, D., Perry, G.E., Lederman, W.F.D. and Valdes, A. (2005) Beyond the city: the rural contribution to development. World Bank, Washington DC, 245 pp.

Dickman, D.I. (1985) The ideotype concept applied to forest trees. In: Cannell, M.R.G. and Jackson, J.E. (eds) *Attributes of Trees as Crop Plants*. Institute of Terrestial Ecology, Huntington, UK, pp. 89–101.

Dickman, D.I., Gold, M.A. and Flore, J.A. (1994) The ideotype concept and genetic improvement of tree crops. *Plant Breeding Reviews* 12, 163–193.

FAO (2001) Country Report, Malawi. FAO/WFP Crop and food assessment mission to Malawi. FAO, Rome.

Fiedler, L. (2005) Adoption of indigenous fruit tree planting in Malawi. MSc Thesis, University of Hannover, Germany, 79 pp.

Fox, F.W. and Young, M.E.N. (1982) *Food from the Veld. Edible Wild Plants of*

Southern Africa Botanically Identified and Described. Delta Books, Cape Town, 422 pp.

Franzel, S., Jaenicke, H. and Janssen, W. (1996) *Choosing the right trees: setting priorities for multipurpose tree improvement.* International Service for National Agricultural Research (ISNAR) Research Report No. 8. ISNAR, The Hague, 87 pp.

German, G.M. (2004) Land tenure, land-related factors and tree tenure systems: an investigation of farmer willingness to cultivate miombo fruit trees in Malawi and Eastern Zambia. MSc Thesis, Bunda College, University of Malawi, Malawi, 115 pp.

Ham, C. (2005) Plant for food and drink. In: Mander, M. and Mckensie, M. (eds) *Southern African Trade Directory of Indigenous Natural Products. Commercial Products from the Wild Group.* Department of Forest and Wood Science, Stellenbosch University, Stellenbosch, South Africa, pp. 17–22.

Hamisy, W.C. (2004) Promotion of effective conservation and sustainable utilization of *Uapaca kirkiana* Kuel. Arg southern African region. Final report for the Abdou Salaam Ouedraogo Fellowship (IPGRI). Tropical Pesticides Research Institute (TPRI), National Plant Genetic Resources Centre, Arusha, Tanzania.

ICRAF (2005) Annual report. Domestication and commercialization of indigenous fruit trees in southern Africa. Project Report, ICRAF/BMZ, Lilongwe, 6 pp.

Kaaria, S.W. (1998) The economic potential of wild fruit trees in Malawi. Thesis, Faculty of the Graduate School of the University of Minnesota, Minnesota, 173 pp.

Kader, A.A. (1992) Post-harvest biology and technology: an overview. In: Kader, A.A. (ed.) *Post-harvest Technology of Horticultural Crops.* University of California, California, pp. 15–20.

Kadzere, I., Chilanga, T.G., Ramadhani, T., Lungu, S., Malembo, L., Rukuni, D., Simwaza, P.P., Rarieya, M. and Maghembe, J.A. (1998) Choice of priority indigenous fruits for domestication in southern Africa: summary of case studies in Malawi, Tanzania, Zambia and Zimbabwe. In: Maghembe, J.A., Simons, A.J., Kwesiga, F. and Rarieya, M.M. (eds) *Selecting Indigenous Fruit Trees for Domestication in Southern Africa.* ICRAF, Nairobi, pp 1–39.

Kadzere, I., Hove, L., Gatsi, T., Masarirambi, M.T., Tapfumaneyi, L., Maforimbo, E., Magumise, I., Sadi, J. and Makaya, P.R. (2001) Current practices on post-harvest handling and traditional processing of indigenous fruits in Zimbabwe. Final Technical Report to Department of Agricultural Research and Technical Services. Department of Agricultural Research and Technical Services Zimbabwe, 66 pp.

Kadzere, I., Hove, L., Gatsi, T., Masarirambi, M.T., Tapfumaneyi, L., Maforimbo, E. and Magumise, I. (2004) Domestication of indigenous fruits: post harvest fruit handling practices and traditional processing of indigenous fruits in Zimbabwe. In: Rao, M.R. and Kwesiga, F.R. (eds) *Proceedings of the Regional Agroforestry Conference on Agroforestry Impacts on Livelihood in Southern Africa: Putting Research into Practice.* World Agroforestry Centre (ICRAF), Nairobi, pp. 353–363.

Kadzere, I., Watkins, C.B., Merwin, I.A., Akinnifesi, F.K., Hikwa, D., Hove, L., Mhango, J., Saka, J.D.K. (2006a) Harvesting and postharvest handling practices of *Uapaca kirkiana* (Muell. Arg.) fruits: a survey of roadside markets in Malawi. *Agroforestry Systems* 68, 133–142.

Kadzere, I., Watkins, C.B., Merwin, I.A., Akinnifesi, F.K. and Saka, J.D.K. (2006b) Harvest date affects color and soluble solids concentrations (SSC) of *Uapaca kirkiana* (Muell. Arg) fruits from natural woodlands. *Agroforestry Systems* 69, 167–173.

Kadzere, I., Watkins, C.B., Merwin, I.A., Akinnifesi, F.K. and Saka, J.D.K. (2006c) Postharvest damage and darkening in fresh fruit of *Uapaca kirkiana* (Muell. Arg.) *Postharvest Biology and Technology* 39, 199–203.

Kadzere, I., Watkins, C.B., Merwin, I.A., Akinnifesi, F.K., Saka, J.D.K. and Mhango, J. (2006d) Fruit variability and relationship between color at harvest and subsequent soluble Solids concentrations and color development during storage of *Uapaca kirkiana* (Muell. Arg.) from natural woodlands. *Horticultural Science* 41, 352–356.

Kang, B.T. and Akinnifesi, F.K. (2000) Agroforestry as alternative land-use production systems for the tropics. *Natural Resources Forum* 24, 137–151.

Karachi, M., Ruffo, N., Lemma, N. and Minae, S. (1991) *Use of multipurpose trees in western Tanzania: result of an Ethnobotanical survey of multipurpose trees in Tabora region.* ASFRENA Report, No. 40. ICRAF, Nairobi.

Kordylas, J.M. (1991) *Processing and Preservation of Tropical and Subtropical Foods.* Macmillan, London, 414 pp.

Kruise, T. (2006) Economics of in-situ conservation of indigenous trees: the case of smallholder farmers in central Malawi. MSc Thesis, University of Hannover, Germany, 103 pp.

Kwesiga, F. and Chisumpa, S.M. (1992) *Multipurpose trees of the Eastern province of Zambia: an ethnobotanical survey of their use in the farming systems.* AFRENA Report No. 49. ICRAF, Nairobi.

Kwesiga, F., Akinnifesi, F.K., Ramadhani, T., Kadzere, I. and Saka, J. (2000) Domestication of indigenous fruit trees of the miombo in southern Africa. In: Shumba, E.M., Lusepani, E. and Hangula, R. (eds) *Proceedings of a SADC Tree Seed Centre Network Technical Meeting, Windhoek, Namibia, 13–14 March 2000. SADC Tree Seed Centre Network/SADC/FAO/CIDA Strategy Workshop, Windhoek, Namibia,* pp. 8–24.

Leakey, R.R.B. and Page, T. (2006) The ‘ideotype’ concept and its application to the selection of cultivars of trees providing agroforestry products. *Forests, Trees and Livelihoods* 16, 5–16.

Leakey, R.R.B., Tchoundjeu, Z., Schreckenberg, K., Shackleton, S.E. and Shackleton, C.M. (2005) Agroforestry tree products (AFTPs): targeting poverty reduction and enhanced livelihoods. *International Journal for Agricultural Sustainability* 3, 1–23.

Maghembe, J.A. (1995) Achievement in the establishment of indigenous fruit trees of the miombo woodlands of southern Africa. In: Maghembe, J.A., Ntupanyama, Y. and Chirwa, P.W. (eds) *Improvement of Indigenous Fruit Trees of the Miombo Woodlands of Southern Africa.* ICRAF, Nairobi, Kenya, pp. 39–49.

Maghembe, J.A. and Seyani, J.H. (1991) *Multipurpose trees used by small-scale farmers in Malawi: results of an ethnobotanical survey.* AFRENA Report No. 42. ICRAF, Nairobi.

Maghembe, J.A., Simons, A.J., Kwesiga, F. and Rarieya, M. (1998) *Selecting indigenous trees for domestication in southern Africa: priority setting with farmers in Malawi, Tanzania, Zambia and Zimbabwe.* ICRAF, Nairobi, Kenya, 94 pp.

Malembo, L.N., Chilanga, T.G. and Maliwichi, C.P. (1998) Indigenous fruit selection for domestication by farmers in Malawi. In: Maghembe, J.A., Simons, A.J., Kwesiga, F. and Rarieya, M. (eds) *Selecting Indigenous Trees for Domestication in Southern Africa: Priority Setting with Farmers in Malawi, Tanzania, Zambia and Zimbabwe.* International Centre for Research in Agroforestry, Nairobi, Kenya, pp. 16 – 39.

Mateke, S. (2003) *Cultivation of Native Fruit Trees of Kalahari Sandveld: Studies on the Commercial Potential, Interactions between Soil and Biota in Kalahari Sands of Southern Africa.* Veld Products, Gaborone, Botswana, 49 pp.

Mhango, J. and Akinnifesi, F.K. (2001) On-farm assessment, farmer management and perception of priority indigenous fruit trees in southern Malawi. In: Kwesiga, F., Ayuk, E. and Agumya, A. (eds) *Proceedings of 14th Southern African Regional Review and Planning Workshop, 3–7 September 2001, Harare, Zimbabwe.* ICRAF, Nairobi, pp. 157–164

Mithöfer, D. (2005) Economics of Indigenous Fruit Tree Crops in Zimbabwe. PhD Thesis, Department of Economics and Business Administration, University of Hannover, Germany.

Mithöfer, D. and Waibel, H. (2003) Income and labour productivity of collection and use of indigenous fruit tree products in Zimbabwe. *Agroforestry Systems* 59, 295–305.

Mithöfer, D., Waibel, H. and Akinnifesi, F.K. (2006) *The role of food from natural resources in reducing vulnerability to poverty: a case study from Zimbabwe.* Paper accepted for the 26th Conference of the International Association of Agricultural Economists (IAAE), 12–18 August, 2006, Queensland, Australia.

Mmangisa, M.L. (2006) Consumer preferences for indigenous fruits in Malawi: a case study of *Uapaca kirkiana.* MSc Thesis, University of Hannover, Germany, 116 pp.

Mng'omba, S.A. (2007) Development of clonal propagation protocols for *Uapaca kirkiana* and *Pappea capensis*, two southern African trees with economic potential. PhD thesis (submitted), University of Pretoria, Pretoria, South Africa, 199 pp.

Mng'omba, S.A., du Toit, E.S., Akinnifesi, F.K. and Venter, H.M. (2007a) Repeated exposure of jacket plum (*Pappea capensis*) micro-cuttings to indole-3-butyric acid (IBA) improved in vitro rooting capacity. *South African Journal of Botany* 73, 230–235.

Mng'omba, S.A., du Toit, E.S., Akinnifesi, F.K. and Venter, H.M. (2007b) Histological evaluation of early graft compatibility in *Uapaca kirkiana* Muell. Arg. scion/stock combinations. *Horticultural Science* 42, 1–5.

Mwamba, C.K. (1995a) Variations in fruits of *Uapaca kirkiana* and the effects of in situ silvicultural treatments on fruit parameters. In: Maghembe, J.A., Ntupanyama, Y. and Chirwa, P.W. (eds) *Improvement of Indigenous Fruit Trees of the Miombo Woodlands of Southern Africa.* ICRAF, Nairobi, pp. 27–38.

Mwamba, C.K. (1995b) Effect of root-inhabiting fungi on root-growth potential of *Uapaca kirkiana* (Muell. Arg.) seedlings. *Applied Soil Ecology* 2, 217–226.

Ngulube, M.R., Hall, J.B. and Maghembe, J.A. (1995) Ecology of miombo fruit tree: *Uapaca kirkiana* (Euphorbiaceae). *Forest Ecology and Management* 77, 107–117.

Ngulube, M.R., Hall, J.B. and Maghembe, J.A. (1998) Fruit, seed and seedling variation in Uapaca Kirkiana from natural populations in Malawi. *Forest Ecology and Management* 98, 209–217.

Ngwira, T.N. (1996) Utilisation of local fruit in wine making in Malawi. In: Leakey, R.R.B., Temu, A.B., Melnyk, M. and Vantomme, P. (eds) *Domestication and Commercialisation of Non-timber Forest Products in Agroforestry Systems. Non-wood Forest Products No. 9.* FAO, Rome, pp. 188–191.

Nyland, R. (1996) *Silviculture: Concepts and Applications.* McGraw-Hill Companies, New York, 633 pp.

Ramadhani, T. (2002) Marketing of indigenous fruits in Zimbabwe. *Socio-economic Studies on Rural Development, Vol. 129.* Wissenschaftsverlag Vauk. Kiel, Germany.

Ruiz-Perez, M., Belcher, B., Achdiawan, R., Alexiades, M., Aubertin, C., Caballero, J., Campbell, B., Clement, C., Cunningham, T., Martinez, A., Jong, W. de, Kusters, K., Kutty, M.G., Lopez, C., Fu, M., Alfaro, M.A., Nair, T.K., Ndoye, O., Ocampo, R., Rai, N., Ricker, M., Schreckenberg, K., Shakleton, S., Shanley, P., Sun, T. and Young, Y.-C. (2004) Markets drive the specialization strategies of forest peoples. *Ecology and Society* 9, 1–9.

Saka, J.D.K., Mwendo-Phiri, E. and Akinnifesi, F.K. (2002) Community processing and nutritive value of some miombo indigenous fruits in central and southern Malawi. In: Kwesiga, F., Ayuk, E. and Agumya, A. (eds) *Proceedings of 14th Southern African Regional Review and Planning Workshop, 3–7 September 2001.* ICRAF, Harare, Zimbabwe, pp. 164–169.

Saka, J.D.K., Swai, R., Mkonda, A., Schomburg, A., Kwesiga, F. and Akinnifesi, F.K. (2004) Processing and utilisation of indigenous fruits of the miombo in southern Africa. In: Rao, M.R. and Kwesiga, F.R. (eds) *Proceedings of Regional Agroforestry Conference on Agroforestry Impacts on Livelihooods in Southern Africa: Putting Research into Practice.* World Agroforestry Centre, Nairobi, Kenya, pp. 343–352.

Saka, J.D.K., Rapp, I., Ndolo, V., Mhango, J. and Akinnifesi, F.K. (2006) A comparative study of the physicochemical and organoleptic characteristics of *Uapaca kirkiana, Strychnos cocculoides, Adansonia digitata* and *Mangifera indica* products. *International*

Journal of Food Science and Technology, doi:10.1111/j.1365–2621.01294.x.

Scherr, S.J. (2004) Building opportunities for small-farm agroforestry to supply domestic wood markets in developing countries. *Agroforestry Systems* 61, 357–370.

Schomburg, A., Mhango, J. and Akinnifesi, F.K. (2002) Marketing of masuku (*U. kirkiana*) and masawo (*Ziziphus mauritiana*) fruits and their potential for processing by rural communities in southern Malawi. In: Kwesiga, F., Ayuk, E. and Agumya, A. (eds) *Proceedings of 14th Southern African Regional Review and Planning Workshop, 3–7 September 2001*. ICRAF, Harare, Zimbabwe, pp.169–176.

Shackleton, C.M. (2004) Use and selection of *Sclerocarya birrea* (Marula) in the Bushbuckridge lowveld, South Africa. In: Rao, M.R. and Kwesiga, F.R. (eds) *Proceedings of Regional Agroforestry Conference on Agroforestry Impacts on Livelihooods in Southern Africa: Putting Research into Practice*. World Agroforestry Centre, Nairobi, Kenya, pp. 77–92.

Sidibe, M., Scheuring, J.F., Kone, M., Hofman, P. and Frigg, M. (1998) More on baobab's homegrown vitamin C: some trees have more than others – consistently. *Agrofrestry Today* 10(4), 10.

Simons, A.J. (1996) ICRAF's strategy for domestication of non-wood tree products. In: Leakey, R.R.B., Temu, A., Melnyk, M. and Vantomme, P. (eds) *Domestication and Commercialization of Non-Timber Forest Products in Agroforestry Systems*. FAO Non-wood Products No. 9, pp. 8–22.

Simons, A.J. and Leakey, R.R.B. (2004) Tree domestication in tropical agroforestry. *Agroforestry Systems* 61, 167–181.

Simons, R.K. (1987) Compatibility and stock-scion interactions as related to dwarfing. In: Rom, R. and Carson, R.F. (eds) *Rootstocks for Fruit Crops*. Wiley and Sons, Inc., New York, pp. 79–106.

Tchoundjeu, Z., Asaah, E.K., Anegbeh, P., Degrande, A., Mbile, P., Facheux, C., Tsoberg, A., Atangana, A.R., Ngo-Mpeck, M.L. and Simons, A.J. (2006) Putting participatory domestication into practice in West and Central Africa. *Forests, Trees and Livelihoods* 16, 53–70.

Tiisekwa, B.P.M., Ndabikunze, B.K., Samson, G. and Juma, M. (2004) Suitability of some indigenous tree fruits for manufacturing juices and jams in Tanzania. In: Rao, M.R. and Kwesiga, F.R. (eds) *Proceedings of Regional Agroforestry Conference on Agroforestry Impacts on Livelihooods in Southern Africa: Putting Research into Practice*. World Agroforestry Centre, Nairobi, Kenya, pp. 331–335.

Wiersum, K.F. (2004) Forest gardens as an 'intermediate' land-use system in the nature-culture continuum: characteristics and future potential. *Agroforestry Systems* 61, 123–134.

Wynberg, R.P., Laird, S.A., Shackleton, S., Mander, M., Shackleton, C., du Plessis, P., den Adel, S., Leakey, R.R.B., Botelle, A., Lombard, C., Sullivan, C., Cunningham, A.B. and O'Regan, D. (2003) Marula policy brief: marula commercialisation for sustainable and equitable livelihoods. *Forests, Trees and Livelihoods* 13, 203–215.

9 Domestication, Utilization and Marketing of Indigenous Fruit Trees in West and Central Africa

Z. TCHOUNDJEU,[1] A. ATANGANA,[1] E. ASAAH,[1] A. TSOBENG,[1] C. FACHEUX,[1] D. FOUNDJEM,[1] C. MBOSSO,[1] A. DEGRANDE,[1] T. SADO,[1] J. KANMEGNE,[1] P. MBILE,[1] H. TABUNA,[1] P. ANEGBEH[2] AND M. USENI[3]

[1] *ICRAF West and Central Africa Region, Humid Tropic Node, Yaounde Cameroon*; [2] *ICRAF West and Central Africa Region, Humid Tropic Node, IITA Station, Port Harcourt, Rivers State, Nigeria*; [3] *ICRAF West and Central Africa Region, Humid Tropic Node, ICRAF Office, Kinshasa, Democratic Republic of Congo*

9.1 Introduction

The participatory domestication of high-value fruit and nut trees and medicinal plants has been ongoing in West and Central Africa since 1998 (Tchoundjeu *et al.*, 1998, 2006). Key species for domestication were jointly identified by researchers and farmers after priorities had been set in four countries of the region (Cameroon, Gabon, Ghana and Nigeria) (Franzel *et al.*, 1996). Tree domestication is executed as a farmer-driven and market-led process (Simons and Leakey, 2004) encompassing socio-economic and biophysical processes research as defined during the first tree domestication conference held in Edinburgh in 1992 (Leakey and Newton, 1994). It aims to significantly increase and stabilize the income, food and health of rural households and their extended families through improvement of agroforestry tree products (Simons and Leakey, 2004; Tchoundjeu *et al.*, 2006). This is being done through the selection and propagation of trees with desirable traits (Atangana *et al.*, 2002; Tchoundjeu *et al.*, 2006) for the development of cultivars well adapted to different cropping systems, where their integration contributes to the diversity and sustainability of the farming system. Participatory tree domestication also seeks to ensure that women and other vulnerable groups obtain maximum benefit from agroforestry activities (Schreckenberg *et al.*, 2002, 2006), while enhancing environmental health at landscape level (Wiersum, 1996).

9.2 Rationale for Participatory Tree Domestication

The humid tropical region of West and Central Africa includes the Congo Basin, which is the world's second largest continuous rainforest after the Amazon. It is home to more than 20 million people, who mostly depend on natural resources for their livelihoods. With over 400 species of mammals, 1000 species of birds and 10,000 plant species, the forest harbours the most diverse assemblages of plants and animals in Africa (CARPE, 2001). About 3000 of these plant and animal species are endemic to the region.

Farmers in West and Central Africa depend on the forest for food and income. Shifting cultivation is widespread in the region but is increasingly becoming unsustainable as the population increases. To overcome this problem, new approaches are needed involving the cultivation of crops in small areas of land in ways which increase the variety and diversity of different cropping systems while assuring food security and income generation from sources other than major cash crops. To identify potential new crops in the region, the World Agroforestry Centre (ICRAF) initiated a coordinated species prioritization exercise based on farmer preference, product ranking and the market potentials of these species in Cameroon, Nigeria, Gabon and Ghana (Franzel *et al.*, 1996). This exercise led to the identification of *Irvingia gabonensis*/*Irvingia wombolu*, *Dacryodes edulis*, *Ricinodendron heudelotii*, *Garcinia kola*, *Chrysophyllum albidum* and *Cola* spp. as the top priority fruit and nut tree species for domestication in West and Central Africa (Jaenicke *et al.*, 1995; Franzel *et al.*, 1996). To this list were added two highly threatened medicinal plants: *Pausinystalia johimbe* and *Prunus africana* (Tchoundjeu *et al.*, 1998).

Standard horticultural techniques, namely marcotting (air-layering), grafting and rooting of juvenile cuttings, were developed to select, propagate and mass-produce these important species, as the traits farmers wanted to see expressed in these priority species can only be captured through vegetative propagation when they are genetically controlled.

9.3 Where Are We with Participatory Tree Domestication in West and Central Africa?

Participatory tree domestication involves many steps, namely:

- Selection of priority species based on farmers' preferences and market orientation.
- Collection of germplasm from elite trees chosen by farmers or other users.
- Establishment of passport data for trees from which germplasm was collected (for future use in tracing the owners of trees in respect of property right).
- Development of appropriate vegetative propagation techniques for the different species under domestication.
- Integration of improved propagules in different cropping systems.
- Marketing studies of products from domesticated species.

The participatory domestication project in the humid lowlands of West and Central Africa, which started in 1998, was initially limited to Cameroon and Nigeria with a focus on *Irvingia gabonensis* (Leakey *et al.*, 1998; Atangana *et al.*, 2001, 2002; Tchoundjeu *et al.*, 2005), *Dacryodes edulis* (Leakey and Ladipo, 1996; Leakey *et al.*, 2002; Mialoundama *et al.*, 2002; Wahuriu *et al.*, 2004; Anegbeh *et al.*, 2005), *Ricinodendron heudelotii* (Ngo Mpeck *et al.*, 2003), *Pausinystalia johimbe* (Tchoundjeu *et al.*, 2004) and *Prunus africana* (Dawson and Powell, 1999; Tchoundjeu *et al.*, 1999, 2002; Simons *et al.*, 2000). The project expanded progressively into Equatorial Guinea and Gabon (2002), the Democratic Republic of Congo and Ghana (2004), and the list of priority species subsequently expanded, so that it now includes *Gnetum africanum* (a leafy vegetable), *Annickia chlorantha* (medicinal tree) (Facheux *et al.*, 2003), *Vernonia*, *Alchornea*, *Dombeya*, *Myrianthus*, *Polyscias*, *Vitellaria* species and *Lophira lanceolata* (melliferous – or nectar-producing – species), and *Allanblackia* species (an indigenous fruit nut tree).

In order to capture genetic variation and select the best trees, targeted germplasm collection was organized in natural stands of priority species according to the method described by Dawson and Were (1997). The first batch of marcotts did not meet expectations when they started fruiting, as farmers did not fully understand the needs and processes of germplasm collection, and indicated to the domestication team the trees located near their homestead. Subsequently, more time was spent with farmers explaining the objectives of participatory tree domestication using training materials as visual aids, and arranging exchange visits to other participating villages to enhance their understanding of this concept.

Assessment of phenotypic variation in fruit and nut traits was carried out from 1998 to 2001 for *Irvingia gabonensis* (Leakey *et al.*, 2000, 2005; Atangana *et al.*, 2001, 2002; Anegbeh *et al.*, 2005), *Dacryodes edulis* (Leakey *et al.*, 2002; Kengni *et al.*, 2003; Wahuriu *et al.*, 2004; Anegbeh *et al.*, 2005) and *Ricinodendron heudelotii* (Ngo Mpeck *et al.*, 2003). Quantitative descriptors for variation in indigenous fruit and nut traits were identified (Leakey *et al.*, 2000), and significant tree-to-tree variation was observed in measured traits (Atangana *et al.*, 2001; Leakey *et al.*, 2002, 2005; Ngo Mpeck *et al.*, 2003; Wahuriu *et al.*, 2004; Anegbeh *et al.*, 2005). This allowed selection of superior trees for vegetative propagation using the 'ideotype' concept (Atangana *et al.*, 2002; Leakey *et al.*, 2005) for *Irvingia gabonensis* (Leakey *et al.*, 2002) and *Dacryodes edulis*. The ideotype included traits of economic importance, such as the sizes of different components of the fruit and/or kernel, visual traits such as colour of the skin or flesh, organoleptic traits (Kengni *et al.*, 2001) and nutritional traits (including protein, fatty acid and vitamin contents and food thickening properties (Leakey *et al.*, 2005)). Genetic selection also required the identification of a market-oriented ideotype, as it was based on seasonality of fruiting, yield or any other relevant trait that may enhance the value or utility of a product.

The next phase was the development of appropriate protocols for the vegetative propagation of identified superior trees to allow the multiplication of selected genotypes. The idea behind participatory domestication is to provide farmers with a package of techniques and help them adopt and use the technologies that are most appropriate for their conditions. Vegetative propagation takes many forms; the techniques implemented here were particularly simple and inexpensive, and ranged from rooting of leafy stem cuttings to air-layering and grafting. Air-layering (or marcotting) is often used to produce the first set of clonal plants from sexually mature trees with desirable traits, and when established as stock plants these are a valuable source of cuttings or scions for multiplication by rooting or grafting. Robust techniques for root cuttings using non-mist polypropagators (as described by Leakey *et al.*, 1990) were used for the vegetative propagation of top-priority species, and appropriate protocols were refined for *Irvingia gabonensis* and *Gnetum africanum* (Shiembo *et al.*, 1996a, b), *Ricinodendron heudelotii* (Shiembo *et al.*, 1997; Ngo Mpeck *et al.*, 2004), *Dacryodes edulis*, *Prunus africana* (Tchoundjeu *et al.*, 1999, 2002), and *Pausinystalia johimbe* (Tchoundjeu *et al.*, 2004).

Farmers enthusiastically adopted these techniques and made further refinements to the design of the non-mist polypropagator (Mbile *et al.*, 2004). Techniques for marcotting were also improved. This led to an increase in the post-severance survival of *Irvingia gabonensis* (from 10% to 50%) and *Dacryodes edulis* (up to 70%). This increase was attributed to the use of a giant humidity chamber for weaning marcotts. Techniques used for vegetative propagation also depend on species. For example, *Garcinia kola* has proved difficult to marcott, but it is being successively multiplied (50–60%) using grafting techniques. Grafting also provided a good success rate for *Irvingia gabonensis* (50–60% graft take).

9.4 Implementation of Participatory Tree Domestication

The innovative approach used to domesticate indigenous trees and bring them into wider cultivation was welcomed by resource-poor farmers. ICRAF staff and associated non-governmental organizations (NGOs), community-based organizations and national agricultural research systems offer assistance in the form of training that facilitated the domestication of indigenous trees and the establishment of pilot village nurseries for tree propagation. Pilot nurseries were established between 1998 and 1999 in the villages of Nkolfep, Ting Melen and Ngoumou within the forest zone and Belo in the humid savannah zone of Cameroon (Table 9.1). The farmer groups approached ICRAF to establish these nurseries, which played the role of school nurseries, farmers coming from a distance of 10–15 km for training. Since then, farmers have been acquiring the necessary skills to create satellite nurseries in their own villages. In order to reach out to villages that were unfamiliar with ICRAF activities, NGOs that were already working with these communities were used. These groups typically began with 20–30 farmers and as time passed the group declined to a core of 15 to 20 members who were committed to the training programme. As fruit, spices, barks, leaves and other vegetables are usually sold by women, they constituted about 25% of trained farmers.

During meetings held with farmers to agree on the concept and strategies for tree domestication, it was made clear to farmers that tree domestication was an opportunity to increase and diversify their income through self-employment, not an opportunity for employment by ICRAF. It was also agreed that once farmers'

Table 9.1. Location and year of establishment of village nurseries in the humid forest and savannah zones of southern Cameroon.

Humid forest zone		Savannah zone	
Village	Year of establishment	Village	Year of establishment
Abondo	1998	*Belo* (MIFACIG /CIPCRE)	1999
Essong Mintsang	2001	Dichami	2003
Abondo II	2001	*Njinikom*	
Nkolfep	1998	Mboini (CIPCRE)	2001
Elig-Nkouma	1999	Baichi	2003
Nkom-Efoufoum	2001	Kikfuini	2003
Nkef II	2001	Wombong	2003
Lekie Assi	2001	Bohim	2003
Mpong	2001	Mumifag	2004
Mbelelekie	2003	*Fundong* (CIPCRE)	
Mbagbang	2004	Abuh	2003
Kalnagha	2004	Twafundong	2003
Ting-Melen (CRATAD)	1999	Atoini	2004
Nlobisson	2002	*Upper Boyo* (CIPCRE)	
Ayo	2003	Upper Boyo	2003
Alomba	2003	*Santa*	
Essang	2003	Na'ah (MIFACIG)	2004
Makenene	2004	*Kumbo*	
Ngoumou (ATD)	1999	Riba	2004
Ottotomo	2001		
Nkon-Bibega	2003		
Yop (FONJAK)	2003		
Akoazole	2003		
Ekowondo	2003		
Le Vaillant	2004		
Bafia	2004		
Ondeck (SAILD)	2003		
Epkwassong (SAILD)	2003		

Names in italics are the pilot village in each district; names listed below them are the satellite villages/organizations. Names of associated NGOs are in parentheses.

groups had taken the decision to host a pilot nursery, ICRAF would only supply the materials that the farmers could not afford, such as nursery equipments and fertilizers. All other materials for the nurseries were provided by the farmers, attesting to the participatory nature of the domestication and also ensuring the sustainability of the activities over time. The main criteria for the choice of nursery sites made by farmers were a regular water supply throughout the year and the security of the plants produced. Farmers were responsible for the day-to-day running of the nursery, and they were asked to appoint someone within their group to be in charge of the execution of the day-to-day activities. To assure the sustainability of the project, no financial incentive was given to any of the farmers. Tasks related to nursery establishment, as well as dates for completion of these tasks, were planned with farmers. Then training in elementary nursery techniques and methods of vegetative propagation was gradually introduced to the farmer groups. Once the farmers were familiar with these techniques, they were encouraged to practise the skills they had learned on trees of their choice. As farmers are aware of trees in their own area with particular characteristics, the participatory approach has resulted in the development of off-season cultivars of *Dacryodes edulis*. Fruits from late-maturing trees of this species fetch higher prices, and thus have contributed significantly to the stabilization of the farmers' income throughout the year.

9.4.1 Expanding the network of village nurseries

When farmers from neighbouring villages who were participating in training in pilot nurseries had mastered tree propagation techniques, they created nurseries in their own village with the assistance of ICRAF staff. This has resulted in a growing network of satellite nurseries around pilot or school nurseries. The growth in the number of nurseries has been especially fast in the humid savannah zone (Table 9.1), where the most successful group, the Tantoh Mixed Farming Common Initiative Group (MIFACIG), developed a nursery that generated income that grew from US$2000 in 2002 to US$5000 in 2004 and US$5844 in 2005. This nursery contains demonstration plots of *Prunus africana*, and marcotts of *Dacryodes edulis*, *Cola nitida*, citrus, oranges and mangoes fruiting within 2–3 years after planting. The demonstration plots have stimulated the adoption of participatory tree domestication by farmers of the region. Another nursery in the humid forest zone (Lekie Assi) generated US$1300 in 2004 and has a business plan in 2005 indicating greater profits. In general, for the first 2 years, the nursery products were mainly used to satisfy the cultivation needs of the farmers themselves. Thereafter, farmers started to sell improved plants. Income generation generally depends on the location of the nursery, those having easy access to markets being the most profitable.

Satellite nurseries were started in 1999–2000 in the forest zone (Nkom Efoufoum, Elig-Nkouma, Abondo and Nkolfeb). Two years later, five other satellite nurseries were created in Essong Mintsang, Abondo II, Nkom Efoufoum, Nkef II, Lekie Assi, Mpong and Ottotomo, followed by one nursery at Nlobisson in 2002 and five more in 2003 at Mbelekie, Ayo, Alomba, Essang and Nkong-Bibega (Table 9.1). Through strategic partnerships with two NGOs (SAILD – Service d'Appui aux Initiatives Locales de Développement, Cameroon – and FONJAK – Fondation Fritz Jacob, Cameroon), five nurseries were created in the southern province of Cameroon in 2003 at Yop, Akoazole, Ondeck, Ekpwassong and Ekowondo. The multiplication of nurseries in the humid savannah has also been quite fast. Six nurseries were established in Njinikom between 2001 and 2004, two in Belo in 1999 and 2003, three in Fundong in 2003 and 2004, and one each in Santa and Kumbo in 2004. Technical support and follow-up management were implemented depending on the length of the group's experience and their degree of technical competence. Nurseries were then classified as follows:

- *Dependent*: nurseries under development needing careful attention and technical assistance from ICRAF staff and partners on weekly basis.
- *Semi-dependent*: the group members of these nurseries have mastered at least one propagation technique and only require technical assistance from ICRAF staff or partners twice a month.
- *Semi-autonomous*: nurseries requiring only occasional assistance as the members have mastered at least two vegetative propagation techniques and are applying them to other species. Members of this group are generating income from the sale of improved plants they have produced as well as from products from their outplanted cultivars.

9.4.2 Scaling up in the region

Scaling up of participatory domestication techniques is being implemented in several countries in West and Central Africa. In southeast Nigeria ten satellite nurseries have developed from an original nursery near Onne (Port Harcourt area). In the Democratic Republic of Congo 25 nurseries have been formed in Bandudu, Equateur and Bas Congo Provinces. In Gabon and Equatorial Guinea the spread has been slower than in other countries, perhaps because of the absence of NGOs in rural areas to supervise the farmers' groups. Only four nurseries are functional in the two countries. However, a lot of emphasis has been put on marketing studies in Gabon and Equatorial Guinea. In Ghana and Nigeria a fruitful partnership was developed with Unilever for the domestication of *Allanblackia* spp. Germplasm collection and vegetative propagation techniques were developed for this important species. In Liberia, Sierra Leone and Guinea-Conakry it is expected that new funding will soon stimulate tree domestication activities, as a survey undertaken in Liberia has clearly indicated that the resource-poor population of this country, which is just getting out of conflict, will need skills for tree domestication to improve their livelihood.

9.4.3 Integration of improved propagules in the farming systems

Studies on the integration of companion crops in cash crop fields are in progress (Leakey and Tchoundjeu, 2001). The plants produced in village nurseries are integrated into gaps in different cropping systems, starting from home gardens, and extending into food crops and coffee or cocoa farms. Propagation of mature trees by marcotting or grafting reduces the height of the resulting trees and many farmers have found the early fruiting characteristic of marcotts a great incentive for the adoption of this propagation technique. Moreover, the dwarf size of marcotts makes harvesting easier for farmers, who face the possibility of a fatal accident during fruit collection from taller trees. Cultivars developed by rooting cuttings from juvenile coppice shoots result in plants with more vigour and strong apical dominance. Though they fruit earlier, their height is similar to that of plants from seedlings, making them better able to shade cash crops such as cocoa and coffee than the marcotts. Therefore, through participatory tree domestication, different statures of trees can be produced. This offers opportunities for the development of multistrata cocoa/coffee agroforests, which have the advantage of buffering against fluctuations in world commodity prices while providing environmental services (Leakey *et al.*, 2003).

9.5 Market Development

To be successful, tree domestication has to be linked to commercialization and market expansion. Consequently, a strong marketing component has been developed in parallel with the tree domestication programme. Market surveys initiated to identify the prices and quantities of the main non-timber forest products in local, regional and international markets (Ndoye *et al.*, 1998; Awono *et al.*, 2002; Facheux *et al.*, 2003) have been undertaken. Studies to identify the distribution channels, processing, storage, transformation and packaging of the main products are also in progress (Kengni *et al.*, 2004; Tchoundjeu *et al.*, 2005). For species with hard-shelled nuts, such as *Ricinodendron heudelotii*, ICRAF has developed a nutcracker with a private partner based in Douala, Cameroon. The cracking machine is very efficient as more than 70% of seeds can easily be decorticated without crushing the extracted kernels. This will help farmers who find the extraction of *Ricinodendron heudelotii* kernels very labour-intensive, as kernels must be extracted one by one. Moreover, the characterization of this species (Leakey *et al.*, 2000; Atangana *et al.*, 2002) will help in the development of cultivars with the traits desired for consumption and to satisfy market preferences. To address issues related to the effects of the supply chain, quality assurance and pricing structure, farmers have been trained in group marketing. In addition, ways are being sought to provide them with market information in due time. From the group sales organized in Ondeck village in Cameroon, for example, the following results were obtained in less than 1 year of activity. The number of group enterprises increased from two in 2005 to four in 2006. Membership rose from 42 to 121 during the second group sales in 2006 and total revenue generated by the four groups rose from US$1600 in 2005 to US$6000 in 2006. The increase in sales could be explained by the improvement in the bargaining power and negotiation skills the farmers acquired during their training.

Scaling-up issues are also important, and a study of the relationship between market prices and fruit traits in *Dacryodes edulis* found that retailers charged consumers higher prices for fruit with desirable traits, whereas wholesalers only paid a negotiated price regardless of fruit characteristics (Leakey *et al.*, 2002). Thus, market studies need to be expanded to help farmers gain the right price from their production.

9.6 Impact of Participatory Tree Domestication on the Livelihood of Local Populations

Tree domestication for non-timber forest products is advocated as a potential way of helping farmers improve their livelihoods in an environmentally sustainable manner. Eight years after the programme was launched, a study was conducted to assess the potential of tree domestication in the improvement of rural livelihoods in Lekie Assi and Belo villages in the Centre and Northwest provinces of Cameroon. The study drew on the theoretical models of impact assessment and the livelihood frame, and combined both quantitative and qualitative research methods. The results clearly indicate that participatory tree domestication has effectively resulted in the adaptation of horticultural vegetative propagation techniques (rooting of cuttings, marcotting and grafting) to the multiplication, selection and mass production of high-value fruit trees. This approach has permitted the selection and maintenance of fruit traits (taste, size, fleshiness) over generations, shortened the gestation (waiting) period (for example, the gestation period for *Dacryodes edulis* has fallen from 7 to 3 years) and contributed to a halving of tree height compared with trees propagated by seeds.

Tree nursery activities also generated respectively US$25,000 and US$3000 for farmers in Belo and Lekie Assi over a period of 3 years. Farm families used a large proportion of this income for children's school fees and other needs for the household. Financial analysis of the vegetative propagation and tree nursery demonstrated the profitability of the technology. The net present value (NPV) of Lekie Assi nursery under current management conditions, is US$1000, with an internal rate of return (IRR) of 54% (Tables 9.2 and 9.3), while the NPV of Belo nursery is US$2600 with an IRR of 58% (Tables 9.4 and 9.5).

Some 30 and 13% of farmers in Lekie Assi and Belo, respectively, are successfully applying the technology to the production of fruit for household consumption. In 2006, more than 100 and 300 farmers have been trained in tree domestication techniques in Lekie Assi and Belo, respectively. Farmers have a good grasp of tree domestication techniques and use their skills to attract visitors, foreigners and students, and to gain honorific titles, awards, grants and popularity in their community. With the adoption of tree domestication, two medicinal plants, *Prunus africana* and *Pausinystalia johimbe*, and three fruit trees, *Irvingia gabonensis*, *Dacryodes edulis* and

Table 9.2. Net present value (NPV) and internal rate of return for Lekie Assi nursery.

	Year				
	2001	2002	2003	2004	2005
Total cash outflow	877,700	380,627	329,502	351,875	355,205
Total cash inflow	0	1,029,500	805,000	885,000	974,000
Net cash flow	−877,700	648,873	475,498	533,125	618,795
NPV (20%)	500,143				
Internal rate of return	54%				

Table 9.3. Effect of changes in key parameters on the profitability of Lekie Assi nursery.

Change in key parameters	Net present value (F CFA)	Internal rate of return (%)
Base analysis	500,143	54
Selling price: cutting (500 F CFA), marcott (1000 F CFA), graft (500 F CFA)	−739,954	
Valuation rate: 75%	390,600	20
Labour cost +20%	428,041	49
Labour cost –20%	566,459	58
Discount rate at 10%	843,952	54
Discount rate at 30%	278,546	54

Table 9.4. Net present value (NPV) and internal rate of return of Belo nursery.

	Year				
	2001	2002	2003	2004	2005
Total cash outflow	1,209,850	659,315	731,168	897,801	1,026,164
Total cash inflow	0	869,500	1,305,000	2,575,000	3,600,000
Net cash flow	−1,209,850	210,185	573,832	1,677,199	2,573,836
NPV (20%)	1,313,033				
Internal rate of return	58%				

Table 9.5. Effect of changes in key parameters on the profitability of Belo nursery.

Change in key parameter	Net present value (F CFA)	Internal rate of return (%)
Base analysis	1,313,033	58
Selling price: cutting (500 F CFA), marcott (1000 F CFA), graft (500 F CFA), Prunus africana cutting (50 F CFA)	−1,282,349	–
Production valuation rate: 75%	301,136	30
Labour cost +20%	1,179,391	54
Labour cost –20%	1,446,562	61
Discount rate at 10%	2,248,670	58
Discount rate at 30%	735,347	58

Ricinodendron heudelotii, are now being integrated in cropping systems in Lekie Assi and Belo. These species have been planted around the homestead for landscaping, as windbreaks, and for the provision of shade and other environmental services. The technology has proved to be sustainable as more farmers are being trained by farmers who have already acquired the skills.

Techniques used for participatory tree domestication have a wide range of advantages:

- Capture of the attributes of 'plus trees' within genetically diverse wild populations by vegetative propagation, so avoiding the long, slow process of tree breeding.
- The use of low-cost propagation systems appropriate for use in village rural areas.
- The circumvention of problems of poor and/or erratic seed supply by multiplying stockplants vegetatively.
- Incentives for farmers via early-fruiting materials.
- The use of vegetatively propagated cultivars to provide traits required by the markets.
- Empowering of farmers through a participatory approach to the acquisition of their rights to indigenous knowledge regarding the use of indigenous species, as defined by the Convention on Biological Diversity.

The problem of protection of the intellectual property rights of the farmers developing the cultivars is still to be resolved. Up to now, all marcotts and scions collected from farmers' field are labelled with the name of the farmer and a clear identification of their village. Records are also kept of the exact position of the mother tree and its characteristics, and all these 'passport data' are kept by ICRAF for future reference in case a need rises for a farmer to seek plant breeder's rights or some other form of protection of intellectual property rights.

Despite all these advantages, an immediate concern could be that large-scale companies may be involved in the establishment of newly domesticated crops, thus undermining the actions of smallholders. Moreover, action should be taken to prevent any reduction in the genetic diversity of the species by the widespread proliferation of a few cultivars and the consequent increase in susceptibility to pest and disease outbreaks. Taking all these precautions into consideration, it is believed that participatory tree domestication could significantly reduce poverty in rural areas while preserving the fragile ecosystems of West and Central Africa.

References

Anegbeh, P.O., Ukafor, V., Usoro, C., Tchoundjeu, Z., Leakey, R.R.B. and Schreckenberg, K. (2005) Domestication of *Dacryodes edulis*: 1. Phenotypic variation of fruit traits from 100 trees in southeast Nigeria. *New Forests* 29, 149–160.

Atangana, A.R., Tchoundjeu, Z., Fondoun, J.-M., Asaah, E., Ndoumbe, M. and Leakey, R.R.B. (2001) Domestication of *Irvingia gabonensis*: 1. Phenotypic variation in fruits and kernels in two populations from Cameroon. *Agroforestry Systems* 53, 55–64.

Atangana, A.R., Ukafor, V., Anegbeh, P., Asaah, E., Tchoundjeu, E., Fondoun, J.-M., Ndoumbe, M. and Leakey, R.R.B. (2002) Domestication of *Irvingia gabonensis*: 2. The selection of multiple traits for potential cultivars from Cameroon and Nigeria. *Agroforestry Systems* 55, 221–229.

Awono, A., Ndoye, O., Schreckenberg, K., Tabuna, H., Isseri, F. and Temple, L. (2002) Production and marketing of safou (*Dacryodes edulis*) in Cameroon and internationally: market development issues. *Forests, Trees and Livelihoods* 12, 125–147.

Dawson, I.K. and Powell, W. (1999) Genetic

variation in the afromontane tree *Prunus africana*, an endangered medicinal species. *Molecular Ecology* 8, 151–156.

Dawson, I.K. and Were, J. (1997) Collecting germplasm from trees – some guidelines. *Agroforestry Today* 9, 6–9.

Degrande, A., Schreckenberg, K., Mbosso, C., Anegbeh, P., Okafor, V. and Kanmegne, J. (2006) Farmers' fruit tree-growing strategies in the humid forest zone of Cameroon and Nigeria. *Agroforestry Systems* 67, 159–175.

Duguma, B., Gockowski, J. and Bakala, J. (2002) Smallholder cocoa cultivation in agroforestry systems of West and Central Africa [see http://nationalzoo.si.edu/ConservationAndScience/MigratoryBirds/Research/Cacao/duguma.cfm Accessed 18/05/05].

Facheux, C., Asaah, E., Ngo Mpeck, M.L. and Tchoundjeu, Z. (2003) Studying markets to identify medicinal species for domestication: the case of *Annickia chlorantha* in Cameroon. *HerbalGram* 60, 38–46.

Franzel, S., Jaenicke, H. and Janssen, W. (1996) *Choosing the Right Trees: Setting Priorities for Multipurpose Tree Improvement.* Research Report 8, ISNAR, The Hague, 87 pp.

Jaenicke, H., Franzel, S. and Boland, D.J. (1995) Towards a method to set priorities amongst species for tree improvement research: a case study from West Africa. *Journal of Tropical Forest Science* 7, 490–506.

Kengni, E., Tchoundjeu, Z., Tchouanguep, F.M. and Mbofung, C.M.F. (2001) Sensory evaluation of *Dacryodes edulis* fruit types. *Forests, Trees and Livelihoods* 11, 1–10.

Kengni, E., Tchoundjeu, Z., Mbofung, C.M.F., Tchouanguep, M.F. and Asaah, E.K. (2004) Value added processing of tree products in agroforestry systems: description and consumer preferences of *Dacryodes edulis* fruit pulp spread. In: Temu, A.B., Chakaredza, S., Mogotsi, K., Munthali, D. and Mulinge, R. (eds) *Rebuilding Africa's Capacity for Agricultural Development: The Role of Tertiary Education. ANAFE Symposium on Tertiary Education in Africa. April 2003.* ICRAF, Nairobi, Kenya, pp. 146–160

Leakey, R.R.B. (1991) Towards a strategy for clonal forestry: some guidelines based on experience with tropical trees. In: Jackson, J.E. (ed.) *Tree Breeding and Improvement.* Royal Forestry Society of England, Wales and Northern Ireland, Tring, UK, pp. 27–42.

Leakey, R.R.B. (1998) Agroforestry in the humid lowlands of West Africa: some reflections on future directions for research. *Agroforestry Systems* 40, 253–262.

Leakey, R.R.B. (1999) Potential for novel food products from agroforestry trees: a review. *Food Chemistry* 66, 1–14.

Leakey, R.R.B. (2004) Physiology of vegetative propagation in trees. In: Burley, J., Evans, J. and Youngquist, J.A. (eds) *Encyclopaedia of Forest Sciences.* Academic Press, London, pp. 1655–1688.

Leakey, R.R.B. and Newton, A.C. (1994) *Tropical Trees: the Potential for Domestication and the Rebuilding of Forest Resources.* HMSO, London, 284 pp.

Leakey, R.R.B. and Page, T. (2006) The 'ideotype concept' and its application to the selection of 'AFTP' cultivars. *Forests, Trees and Livelihoods* 16, 5–16.

Leakey, R.R.B. and Simons, A.J. (1998) The domestication and commercialisation of indigenous trees in agroforestry for the alleviation of poverty. *Agroforestry Systems* 38, 165–176.

Leakey, R.R.B. and Tchoundjeu, Z. (2001) Diversification of tree crops: domestication of companion crops for poverty reduction and environmental services. *Experimental Agriculture* 37, 279–296.

Leakey, R.R.B., Mesén, J.F., Tchoundjeu, Z., Longman, K.A., Dick, J.Mc.P., Newton, A., Matin, A., Grace, J., Munro, R.C. and Muthoka, P.N. (1990) Low-technology techniques for the vegetative propagation of tropical trees. *Commonwealth Forestry Review* 69, 247–257.

Leakey, R.R.B., Fondoun, J.-M., Atangana, A. and Tchoundjeu, Z. (2000) Quantitative descriptors of variation in the fruits and seeds of *Ilrvingia gabonensis. Agroforestry Systems* 50, 47–58.

Leakey, R.R.B., Atangana, A.R., Kengni, E., Wahuriu, A.N., Usoro, C., Anegbeh, P.O. and Tchoundjeu, Z. (2002) Domestication of *Dacryodes edulis* in West and Central Africa: characterisation of genetic variation. *Forests, Trees and Livelihoods* 12, 57–72.

Leakey, R.R.B., Schreckenberg, K. and Tchoundjeu, Z. (2003) The participatory domestication of West African indigenous fruits. *International Forestry Review* 5, 338–347.

Leakey, R.R.B., Tchoundjeu, Z., Smith, R.I., Munro, R.C., Fondoun, J.-M., Kengue, J., Anegbeh, P.O., Atangana, A.R., Wahuriu, A.N., Asaah, E., Usoro, C. and Ukafor, V. (2004) Evidence that subsistence farmers have domesticated indigenous fruits (*Dacryodes edulis* and *Ilrvingia gabonensis*) in Cameroon and Nigeria. *Agroforestry Systems* 60, 101–111.

Leakey, R.R.B., Greenwell, P., Hall, M.N., Atangana, A.R., Usoro, C., Anegbeh, P.O., Fondoun, J.-M. and Tchoundjeu, Z. (2005) Domestication of *Ilrvingia gabonensis*: 4. Tree-to-tree variation in food-thickening properties and in fat and protein contents of dika nut. *Food Chemistry* 90, 365–378.

Leakey, R.R.B., Tchoundjeu, Z., Schreckenberg, K., Shackleton, S. and Shackleton, C. (2006) Agroforestry tree products (AFTPs): targeting poverty reduction and enhanced livelihoods. *International Journal of Agricultural Sustainability* 16, 5–16.

Mbile, P., Tchoundjeu, Z., Degrande, A., Avana, M.-L. and Tsobeng, A. (2004) Non-mist propagation by resource-poor, rural farmers of the forest zone of Cameroon: some technology adaptations to enhance practice. *Forests, Trees and Livelihoods* 14, 43–52.

Mialoundama, F., Avana, M.-L., Youmbi, E., Mampouya, P.C., Tchoundjeu, Z., Mbeuyo, M., Galamo, G.R., Bell, J.M., Kopguep, F., Tsobeng, A.C. and Abega, J. (2002) Vegetative propagation of *Dacryodes edulis* (G. Don) H.J. Lam by marcotts, cuttings and micropropagation. *Forests, Trees and Livelihoods* 12, 85–96.

Ndoye, O., Ruiz-Perez, M., Eyebe, A. (1998) *The markets of non-timber forest products in the humid forest zone of Cameroon.* Network Paper Rural Development Forestry Network 22c. Rural Development Forestry Network, Overseas Development Institute (ODI), London.

Ngo Mpeck, M.L., Tchoundjeu, Z. and Asaah, E. (2003) Vegetative propagation of *Pausinystalia johimbe* (K. Schum.) by leafy stem cuttings. *Propagation of Ornamental Plants* 3, 3–10.

Ngo Mpeck, M.L., Asaah, E., Tchoundjeu, Z. and Amougou, A. (2004) Contributing to the domestication of *Ricinodendron heudelotii* (Baill) Pierre Ex Pax through multiplication of phenotypic variation. In: Temu, A.B., Chakeredza, S., Mogosi, K., Munthali, D. and Mulinge, R. (eds) *Rebuilding Africa's Capacity for Agricultural Development: The Role of Tertiary Education. ANAFE Symposium on Tertiary Agricultural Education, April 2003.* ICRAF, Nairobi, Kenya, pp. 196–206.

Schreckenberg, K., Degrande, A., Mbosso, C., Boli Baboule, Z., Boyd, C., Enyong, L., Kanmegne, J., Ngong, C. (2002) The social and economic importance of *Dacryodes edulis* (G. Don) H.J. Lam in southern Cameroon. *Forests, Trees and Livelihoods* 12, 15–40.

Schreckenberg, K., Awono, A., Degrande, A., Mbosso, C., Ndoye, O. and Tchoundjeu, Z. (2006) Domesticating indigenous fruit trees as a contribution to poverty reduction. *Forests, Trees and Livelihoods* 16, 35–51.

Shiembo, P.N., Newton, A.C. and Leakey, R.R.B. (1996a) Vegetative propagation of *Ilrvingia gabonensis* Baill., a West African fruit tree. *Forest Ecology and Management* 87, 185–192.

Shiembo, P.N., Newton, A.C. and Leakey, R.R.B. (1996b) Vegetative propagation of *Gnetum africanum* Welw., a leafy vegetable from West Africa. *Journal of Horticultural Science* 71, 149–155.

Shiembo, P.N., Newton, A.C. and Leakey, R.R.B. (1997) Vegetative propagation of *Ricinodendron heudelotii* (Baill) Pierre ex Pax, a West African fruit tree. *Journal of Tropical Forest Science* 9, 514–525.

Simons, A.J. (1996) ICRAF's strategy for domestication of non-wood tree products. In: Leakey, R.R.B., Temu, A.B., Melnyk, M. and Vantomme, P. (eds) *Domestication and Commercialization of Non-timber*

Forest Products in Agroforestry Systems. Non-wood Forest Products No. 9. FAO, Rome, pp. 8–22.

Simons, A.J. and Leakey, R.R.B. (2004) Tree domestication in tropical agroforestry. *Agroforestry Systems* 61, 167–181.

Simons, A.J., Jaenicke, H., Tchoundjeu, Z., Dawson, I., Kindt, R., Oginosako, Z., Lengkeek, A. and Degrande, A. (2000) The future of trees is on farms: tree domestication in Africa. In: *Forests and Society: The Role of Research. XXI IUFRO World Congress, Kuala Lumpur, Malaysia*, pp. 752–760.

Tchoundjeu, Z. (1996) Vegetative propagation of sahelian agroforestry tree species: *Prosopis africana* and *Bauhinia rufescens*. In: Dieters, M.J., Matheson, A.C., Nickles, D.G., Harwood, C.E. and Walker, S.M. (eds) *Tree Improvement for Sustainable Tropical Forestry: Proceedings QFRI-IUFRO Conference, Caloundra, Queensland, Australia*, 27 October–1 November 1996, pp. 416–419.

Tchoundjeu, Z. and Leakey, R.R.B. (1996) Vegetative propagation of African mahogany: effects of auxin, node position, leaf area and cutting length. *New Forests* 11, 125–136.

Tchoundjeu, Z. and Leakey, R.R.B. (2000) Vegetative propagation of *Khaya ivorensis* (African mahogany): effects of stockplant flushing cycle, auxin and leaf area on carbohydrates and nutrient dynamics of cuttings. *Journal of Tropical Forest Science* 12, 77–91.

Tchoundjeu, Z. and Leakey, R.R.B. (2001) Vegetative propagation of *Lovoa trichilioides*: effects of provenance, substrate, auxins and leaf area. *Journal of Tropical Forest Science* 13, 116–129.

Tchoundjeu, Z., Duguma, B., Fondoun, J.-M. and Kengue, J. (1998) Strategy for the domestication of indigenous fruit trees of West Africa: case of *Ilrvingia gabonensis* in southern Cameroon. *Cameroon Journal of Biology and Biochemical Sciences* 4, 21–28.

Tchoundjeu, Z., Avana, M.-L., Simons, A.J., Dawson, I., Were, J. and Jaenicke, H. (1999) Domestication of *Prunus africana*: the contribution of vegetative propagation. *Poster, American Society Agronomy-ASB Symposium, Salt Lake City, USA, November 1999.*

Tchoundjeu, Z., Avana, M.-L., Leakey, R.R.B., Simons, A.J., Asaah, E., Duguma, B. and Bell, J.M. (2002) Vegetative propagation of *Prunus africana*: effects of rooting medium, auxin concentrations and leaf area. *Agroforestry Systems* 54, 183–192.

Tchoundjeu, Z., Mbile, P., Asaah, E., Degrande, A., Facheux, C., Tsobeng, A., Sado, T., Mbosso, C., Atangana, A., Anegbeh, P., Mpeck, M.L., Avana, M.-L. and Tita, D. (2004a) Rural livelihoods, conservation, management and use of genetic resources of indigenous trees: ICRAF's experience and perspective in West and Central Africa. *Proceedings of the Conference and Workshop of Genetic Resources Network for West and central Africa (GRENEWECA), 26–30 April 2004, IITA, Ibadan, Nigeria* [see http://www.ipgri.cgiar.org/regions/ssa/News/WCA PGR EN.pdf].

Tchoundjeu, Z., Ngo Mpeck, M.L., Asaah, E. and Amougou, A. (2004b) The role of vegetative propagation in the domestication of *Pausinystalia johimbe* (K. Schum), a highly threatened medicinal species of West and Central Africa. *Forest Ecology and Management* 188, 175–183.

Tchoundjeu, Z., Asaah, E.K., Anegbeh, P., Degrande, A., Mbile, P., Facheux, C., Tsobeng, A., Atangana, A.R., Ngo-Mpeck, M.L. and Simons, A.J. (2006) Putting participatory domestication into practice in West and Central Africa. *Forests, Trees and Livelihoods* 16, 53–69.

Vivien, J. and Faure, J.J. (1985) *Arbres des Forêts Denses d'Afrique Centrale*. ACCT, Ministère des Relations Extérieures, Coopération et Développement, Paris, 565 pp.

Vivien, J. and Faure, J.J. (1996) *Fruitiers Sauvages d'Afrique: Espèces du Cameroun*. Nguilou-Kerou Edition, Saint-Cénéré, France, 416 pp.

Wahuriu, A.N., Kengue, J., Atangana, A.R., Tchoundjeu, Z. and Leakey, R.R.B. (2004) Domestication of *Dacryodes edulis*: 2. Phenotypic variation of fruit traits in 200 trees from four populations in the humid lowlands of Cameroon. *Food, Agriculture and Environment* 2, 340–346.

Weber, J.C., Sotelo Montes, C., Vidaurre, H., Dawson, I.K. and Simons, A.J. (2001) Participatory domestication of agroforestry trees: an example from the Peruvian Amazon. *Development in Practice* 11, 425–433.

Wiersum, K.F. (1996) Domestication of valuable tree species in agroforestry systems: evolutionary stages from gathering to breeding. In: Leakey, R.R.B., Temu, A.B., Melnyk, M. and Vantomme, P. (eds) *Domestication and Commercialisation of Non-timber Forest products in Agroforestry Systems. Non-wood Forest Products No. 9.* FAO, Rome, pp. 147–159.

10 Improving Rural Livelihoods through Domestication of Indigenous Fruit Trees in the Parklands of the Sahel

A. Kalinganire,[1] J.C. Weber,[1] A. Uwamariya[2] and B. Kone[1]

[1]World Agroforestry Centre, West and Central Africa Region, Bamako, Mali; [2]Consultant in Agroforestry and Environmental Management BP 570, Bamako, Mali

10.1 Introduction

The West African Sahel, a semi-arid landscape stretching from Niger to Senegal, is characterized by high temperatures throughout the year, with a low and highly unpredictable rainfall pattern (400–1000 mm/year), occurring during a 3-month period, and a 9-month dry season. The population growth rate is high, life expectancy is low, particularly for infants and children, and illiteracy is endemic, especially among women. Burkina Faso and Niger rank lowest on the United Nations Development Programme (UNDP) human development index. Improving peoples' livelihoods is, therefore, a challenge for Sahelian countries (UNDP, 2003). Rural people have developed strategies to adapt to this harsh environment and reduce their vulnerability to risks. Nevertheless, the rate of growth in food crop production – about 2% – is not commensurate with the population growth rate of about 2.5% per year (World Bank, 2004). More than 70% of the 44 million people in Niger, Burkina Faso, Mali and Senegal live in rural areas. How the rural population will continue to survive given the current food crop production is a daunting question. Part of the answer lies in the diversity of native trees and shrubs that people have used for generations in the parkland agroforestry system.

Parklands are mixtures of trees and shrubs that farmers select for certain functions and cultivate together with staple food crops, such as millet and sorghum. It is the principal agricultural system used by subsistence farmers in the Sahel (Bonkoungou *et al.*, 1997; Boffa, 1999). In Mali, for example, parklands occupy approximately 90% of the agricultural land. Parklands are managed to fit environmental conditions and to fulfil specific functions, so they vary in species composition and density within and among countries in the region.

Parkland trees and shrubs provide many functions for the rural poor (Appendix 10.1). They are sources of food, including fruits, fats, oils, leafy vegetables, nuts and condiments, which complement staple food crops in the local diet. Some of these foods are particularly important during the months when grains are in short supply and during years of intense drought. In addition, parkland trees and shrubs provide numerous traditional medicines that are essential for rural health-care (Fortin *et al.*, 2000). Severe micronutrient deficiencies can be alleviated by consuming indigenous fruits and vegetables (Ruel *et al.*, 2005).

They also supply fuelwood, construction materials, cordage, dyes and materials for household implements, handicrafts and clothing. Moreover, since the parkland is an essential source of forage, fodder and medicines for livestock, maintaining healthy parklands is essential for maintaining healthy animal herds. In addition to providing products, parkland trees and shrubs provide environmental services such as moderating the soil temperature, reducing soil erosion and improving soil fertility through nitrogen fixation and nutrient cycling of their leaf biomass. Since most annual food-crop plants are grown in the parkland, these service functions play a key role in maintaining annual food-crop productivity. The fact that the rural poor maintain diverse sets of species and functions in their parklands underscores the importance of this diversity in their livelihood strategies.

Unfortunately, the richness and abundance of indigenous trees and shrubs is being eroded in the parklands and other forested landscapes in the region (Eyog Matig and Ouedraogo, 1999; FAO, 2000). Less species richness means fewer distinct sources of products and services. Lower abundance means less genetic variation within species, which reduces both the capacity of trees and shrubs to adapt to environmental change and the potential gain that farmers can realize from selection. Decreases in both the richness and the abundance of these useful trees and shrubs leave the rural poor with fewer options to improve their health, nutrition and income. In addition, it reduces the available habitat for the other native plants and animals that figure importantly in local diets, medicines, etc. Since traditional knowledge is often transmitted from generation to generation by using plants, this knowledge is also being eroded as species richness and abundance decline. The loss of this knowledge will make it more difficult for future generations to establish and manage the useful species of native trees and shrubs in the region.

In 1995, the World Agroforestry Centre (ICRAF) worked with farming communities and other partners to identify the priority species for domestication programmes in the Sahel. These included baobab (*Adansonia digitata* L.), detar (*Detarium microcarpum*), néré (*Parkia biglobosa*), tamarind (*Tamarindus indica*), shea tree or karité (*Vitellaria paradoxa*) and ber (*Ziziphus mauritiana*). These species were preferred by farmers because of their nutritional, medicinal and income-generating values. In this chapter, we discuss the principal uses of these species and their potential value, and the current and future plans for their domestication.

10.2 Principal Uses of Major Species

10.2.1 Baobab (*Adansonia digitata* L.)

In the Sahel, the natural range of baobab extends from Chad to Senegal. Baobab is associated with other species in the savannahs, especially in the drier parts of West Africa (Sidibé and Williams, 2002). It is one of the most characteristic species in the region because of its massive size and its importance in people's lives. Baobab trees are one of the main sources of income, food and nutritional security during the dry season in the Sahel. All parts of the tree are exploited. The bark is used for making ropes. The leaves, bark, fruit pulp, seed and roots are used for medicines. Juice, rich in vitamin C, is prepared from the fruit pulp. The sun-dried fruit pulp can also be eaten either raw or added to sauces (A. Niang, unpublished results). The leaves are the staple vegetable used in sauces consumed with cereal-based meals. Seeds are also used in soups or roasted and consumed as snacks. Sidibé and Williams (2002) present a thorough review of the importance of the baobab in the Sahel.

Baobab fruits mature during the dry season, providing the products mentioned above, whereas the leaves are typically harvested during the rainy season, then dried and stored for further use during the long dry season. Most rural people consume the leaves in sauces. Leaves can also be produced and harvested all year round in irrigated baobab gardens, a technology developed by ICRAF and its partners in West Africa. It is widely recognized that baobab leaves are an extremely valuable source of protein, vitamin A and essential minerals. However, the traditional method of sun-drying can reduce the vitamin A content (Sidibé *et al*., 1996).

The fruits and leaves of baobab are important sources of income in the Sahel. Surveys carried out in Bamako and Ségou, Mali, show that prices range between 250 and 500 F CFA/kg for various dried leaf and fruit products (C.O. Traoré, unpublished results).

10.2.2 Detar (*Detarium microcarpum* G. et Perr.)

Detar occurs naturally in the Sudano-Sahelian zone of Senegal, Mali and Burkina Faso, as well as in Cameroon, Chad, Nigeria and Sudan. In Mali it is threatened because of extensive fruit collection, uncontrolled tree cutting, overgrazing and bushfires (Kouyaté, 2005).

Fruits of the detar are consumed raw or cooked, or processed into cakes, which sell at 5–25 F CFA per cake (Kouyaté, 2005). Fruits are sold to Senegalese merchants for 100 F CFA/kg at Kita (Mali) along the Mali–Senegal railway (M.M. Sidibé, unpublished data). The net revenue from the sale of 100 kg of fruit in western Mali can reach 200 F CFA in periods of abundance and 3000 F CFA during the off-season. The pulp is used to make an alcoholic beverage and in the preparation of couscous. The fruit has the following nutritional values: 3.2 mg vitamin C, 4.9 g protein and 64.5 g of sugar per 100 g (Kouyaté, 2005).

Detar has several other uses in rural communities. The leaves are used to thatch roofs and the wood is used for fuel, construction poles and tool handles (Kouyaté, 2005). Seeds are dried, ground and used as a fragrance. In addition, necklaces are made from the dried seeds, which are considered to have an aphrodisiac effect because of their pleasant fragrance (Kouyaté, 2005). Mosquito repellent is prepared from the roots while medicines are prepared from the seeds, fruits, leaves, roots and bark. In Burkina Faso, for example, fruits are used in the treatment of meningitis (Bationo *et al.*, 2001). Leaves and roots are also used to treat diseases of farm animals.

10.2.3 Néré (*Parkia biglobosa* (Jacq.) Benth.)

Néré is common in natural savannahs and is widely cultivated in the parklands (Teklehaimanot, 2004). Néré occurs naturally from Senegal to Uganda (Hall *et al.*, 1997). Farmers manage néré in the parkland because of its valuable non-wood products and its capacity to improve soil fertility. In the parklands, néré is associated with a range of crops, especially large leafy vegetables, but also with groundnuts and cereals such as maize and millet.

Néré seeds are valuable and are ground into a spice or condiment locally called 'soumbala', which is an important source of protein that is added to soups and stews throughout the Sahel. Hall *et al.* (1997) discuss néré as a commodity of local and regional trade in sub-Saharan Africa, especially in Benin, Burkina Faso, Cote d'Ivoire, Mali and Nigeria. Fresh soumbala is sold as balls of brown paste in the local markets. Seeds are also sold for commercial processing, which significantly increases their value. Surveys indicate that dried seed sells for about 600 F CFA/kg in Bamako and 500 F CFA/kg in Ouagadougou, Burkina Faso. Fruit pulp sells for about 200 F CFA/kg in Bamako (M.M. Sidibé, unpublished data). In 1997, the annual revenue earned from the sale of néré products was approximately 27,300 F CFA per household in Burkina Faso (Teklehaimanot, 2004).

Ouedraogo (1995) reported that the seeds are rich in protein, lipids, carbohydrates and phosphorous while the fruit pulp is high in carbohydrates and vitamin C. Fresh fruit pulp has a high sugar content and is fermented into a beverage (Kater *et al.*, 1992). The flowers are also consumed.

Néré, which is a leguminous tree, is a valuable source of fodder (Sabiiti and Cobbina, 1992). Its branches are usually lopped by farmers and fed to livestock, especially in the dry season, when good-quality feed is scarce. Moreover, fruit exocarp, seeds and leaves are applied as an organic fertilizer.

Medicines for human diseases are one of the principal uses of néré in the Sahel. All parts of the plant are used to cure many diseases, including malaria and stomach disorders. Moreover, different parts are used to treat diseases of farm animals, such as poultry lice, trypanosomes and mouth ulcers of ruminants. It is also used in traditional ceremonies (Teklehaimanot, 2004).

10.2.4 Tamarind (*Tamarindus indica* L.)

Tamarind originated in Asia, but it is widely cultivated in much of tropical Africa. According to Gunasena and Hughes (2000), it is now naturalized in Burkina Faso, Cameroon, Central Africa Republic, Chad, Ethiopia, Gambia, Guinea, Guinea-Bissau, Kenya, Madagascar, Mali, Niger, Nigeria, Senegal, Sudan, Tanzania and Uganda. Farmers commonly cultivate it in parklands in the arid and semi-arid zones of West Africa.

Although there are many uses for tamarind (Gunasena and Hughes, 2000), few are known or practised in the Sahel. In the Sahel, the fruit pulp is used primarily for sauces, porridge and juice. In Kenya, the fruit pulp is also used to tenderize meat (L. Betser, unpublished results; P. Nyadoi, personal communication), but this practice is unknown in the Sahel. Tamarind can be used as snacks, in sauces, confectionery, drinks, jam, ice cream, wine, and as a coffee substitute, a pectin, food stabilizer, dye, animal fodder, glue, edible oil and medicine (Maundu *et al.*, 1999).

Tamarind fruit pulp is nutritious, rich in tartaric acid, and used as a natural preservative in the pickle industry (Nagarajan *et al.*, 1998). The fruit pulp has low water content and high levels of protein, carbohydrate and minerals (potassium, phosphorus, calcium and iron), but it is not a significant source of vitamins A and C (Gunasena and Hughes, 2000). In eastern and western Africa, the fruit pulp is eaten raw, but local varieties generally have a strong acidic taste compared with sweet-tasting cultivars introduced from Thailand. The seeds are a rich source of protein, and have a favourable amino acid composition.

Tamarind is also a valuable timber species, used in making furniture, tool handles and charcoal and as fuelwood. In addition, the leaves, flowers, root, bark, fruit pulp and seeds are an important source of herbal medicines. The fruit pulp sells for about 400 F CFA per kilogram in Bamako, Mali (M.M. Sidibé 2006, unpublished results).

10.2.5 Karité (*Vitellaria paradoxa* C.F. Gaertn.)

Karité is a key economic fruit tree species that is very abundant across a 5000-km-wide belt of savannah between the equatorial rainforest and the Sahel (Hall *et al.*, 1996; Maranz *et al.*, 2004). Its natural range extends from eastern Senegal to the high plateau of Uganda. The best growth occurs on farmed land, where trees benefit from protection against bush fires and livestock (Kater *et al.*, 1992). Its economic importance has been analysed by Hall *et al.* (1996) and Teklehaimanot (2004), and its nutritional value by Maranz *et al.* (2004).

The tree's main product is a fat (shea butter) extracted from the nuts. It is one of the rare local sources of vegetable fat in the region (Kater *et al.*, 1992). Shea butter is sold at about 400 F CFA/kg in the Bamako market (M.M. Sidibé, unpublished results). Becker and Statz (2003) estimated that 650,000 t of karité nuts were collected throughout Africa in 2000. In addition to local uses, shea butter is exported for use in chocolate products and the pharmaceutical industry in European and other markets. Shea exports from Africa are more than 150,000 t of dry kernel per year (Becker and Statz, 2003), accounting for a market value of more than US$30 million. In 2004/2005 Mali exported about 10,000 t of dry kernel for about 800 million F CFA (AFE, 2006).

The fruits are important to the inhabitants of the Sahel because of their high nutritional value (Maranz *et al.*, 2004). Although there is much emphasis on shea butter for the international market, the use of the fruit pulp in the local diet needs to be taken into consideration during the domestication of the species.

10.2.6 Ber (*Ziziphus mauritiana* Lam.)

The natural distribution of ber extends from central Asia to Africa (Diallo, 2002). In West Africa, it occurs in all countries in the Sahel. Farmers rank it as one of the most preferred fruit tree species, but the fruits are very small: farmers are interested in Indian and Thai varieties that produce large and tasty fruits (Kalinganire *et al.*, 2007).

The main use for ber is for the fruit pulp, which is consumed fresh or dry, and also prepared into a juice. The highest potential for ber in the Sahel is for the sale of juice, fresh fruit pulp and dry fruit paste. In Bamako, the fruit pulp sells for about 350 F CFA/kg. In addition, the leaves are used for fodder and the leaves, roots and bark are used for medicinal purposes. The wood is used for handles, kitchen utensils, firewood and charcoal (Roussell, 1995; Diallo, 2001). In the Sahel ber is also used together with other tree and shrub species to make live fences that protect crops against browsing by animals in the dry season. It is recommended for planting along contour lines for erosion control.

10.3 Progress in Domestication and Improvement of Preferred Species

Deforestation, overgrazing and increasing population are resulting in the degradation of the Sahelian parklands and a decrease in the most important indigenous fruit trees, which in turn reduces the availability of wild fruits to local populations. Although many species with high potential for fruit production and a wide range of products have been identified, little work has been done in the Sahel to bring them under improved management and cultivation. In general, progress in the domestication and improvement of indigenous fruit tree species has been slow, mainly because of limited financial and skilled human resources.

The domestication programmes in the Sahel follow a farmer-driven and market-led process as discussed by Simons and Leakey (2004). The process matches the intraspecific diversity of locally important trees to the needs of subsistence farmers, product markets and agricultural environments. It also includes the exploration and manipulation of wild genetic resources to deliver uses and products for maximum social benefit. Farmers' expertise is a central part of the programme. For example, the programme includes vegetative propagation techniques to capture and multiply the characteristics that farmers prefer based on their expertise and market requirements.

The overall aim is to improve the productivity and sustainability of preferred fruit-tree species in the Sahel. The research aims at developing domestication strategies for identified species to enhance the conservation of biodiversity in the region and the use of improved fruit tree germplasm of high value for the general social benefit of producers. These participatory domestication strategies must consider not only the economic benefits and social consequences for rural communities (Faminow *et al.*, 2001; FAO, 2004; Dampha and Camera, 2005), but also the potential effects on genetic diversity within the tree species (Ledig, 1992; O'Neill *et al.*, 2001; Adin *et al.*, 2004; Hollingsworth *et al.*, 2005). For example, there is a clear trade-off between genetic improvement and genetic diversity, in that any strategy to produce genetic gain will reduce genetic diversity in the breeding population. This is particularly important for strategies based on the selection and multiplication of clones (Cornelius *et al.*, 2006).

ICRAF and its partners have initiated research to increase the production of indigenous fruit trees and make the improved genetic materials available to producers. Such research activities include mainly studies on the genetic variation of some fruit tree species, provenance/progeny trials, 'plus-tree' selection and vegetative propagation methods. The research components in the Sahel include the following:

- Evaluation of genetic variation in existing progeny/provenance trials of preferred species.
- Research supporting the effective mass-production and use of genetically improved planting stock of preferred species.
- Experimental interventions to improve seed and seedling production in partner countries, and the preparation and distribution of guidelines for the production of improved germplasm.

The programme supports the ongoing national efforts in the provision of improved germplasm of proven and promising Sahelian species. Decentralized systems for enhanced access to germplasm by rural farmers are also being promoted. These are supported by the provision of training courses for the sharing of knowledge and technology among farmers, researchers and development agents. The programme is building strong partnerships with the key research and development institutions as well as private industry to support specific interventions, which are likely to increase returns to the producers of primary agroforestry products while ensuring the preservation of the agroforestry resources.

Until relatively recently, people have collected indigenous fruits from the wild. Such fruits are mostly of very poor quality. However, for the last 2 years there has been progress in fruit tree domestication for the preferred fruit tree species in the region.

10.3.1 Studies on intraspecific variation

There are very few studies of genetic variation within the preferred fruit tree species in the Sahel. Some provenance/progeny trials have been established during the last decade, but results are not available, especially for fruit production. Such studies will provide information on geographical and genetic variation within species. A list of existing provenance and species trials for the indigenous fruit trees in the Sahel is given in Table 10.1 (trials have been established for all of the preferred species). Information about genetic variation within selected species is given briefly below.

*Baobab (*Adansonia digitata*)*

The pattern of genetic variation in the baobab has not been adequately studied. However different types, based on bark characteristics, have been classified in Mali (H. Sanou, personal communication) but not genetically confirmed. These may be considered to be ecotypes, but it is necessary to characterize the variation adequately.

Preliminary results from a 5-year-old species/provenance trial revealed better growth of *A. digitata* compared with other *Adansonia* species, namely *A. za*, *A. fony*, *A. perrieri* and *A. gregorii* (Sidibé, 2005). The study showed no significant differences in growth between two Malian provenances. Sidibé and Williams (2002) reported that leaf production is a major challenge due to its seasonality. Irrigation extends leaf production in Sahel, and Sidibé and Williams (2002) reported that the local black bark type responds well to this. The same may also apply to fruit production, but this has not been investigated. In Mali, some ecotypes were reported to have fruits with an exceptionally high vitamin C content.

*Tamarind (*Tamarindus indica*)*

Gunasena and Hughes (2000) reported high phenotypic variation in fruit traits (length of pod, pod weight, seed number, pod colour and sweetness of pulp). Nagarajan *et al.* (1998) reported extensive variation in foliage and flower production, flower colour, fruit size, fruit pulp yield and wood quality. There

Table 10.1. Provenance trials for indigenous fruit tree species in the Sahel

Species	Establishment date	No. of provenances	No. of sites	Site name and country
Adansonia spp.	2001	6	2	Samanko and Cinzana, Mali
Adansonia spp.	2001	6	1	Baguineda, Mali
Tamarindus indica	2001	8	1	Samanko, Mali
Tamarindus indica	1990	13	1	Dinderesso, Burkina Faso
Tamarindus indica	1990	13	1	Gonse, Burkina Faso
Tamarindus indica	2000	10	1	Bandia, Senegal
Vitellaria paradoxa	1997	5	1	Gonse, Burkina Faso
Ziziphus mauritiana	2000	8	1	Bandia, Senegal
Ziziphus mauritiana	2001	10	2	Samanko and Cinzana, Mali
Ziziphus mauritiana	2001	4	1	Baguineda, Mali
Ziziphus mauritiana	2001	7	1	Gonse, Burkina Faso
Ziziphus mauritiana	2002	3	1	Ouagadougou, Burkina Faso

appears to be more diversity in native African populations of tamarind than in those introduced from the South and South-east Asian regions. Diallo (2001) confirmed the high degree of diversity and phenotypic difference among the African populations, and attributed this to geographical isolation and gene mutation.

Preliminary results from tamarind provenance trials revealed considerable variation in growth and biomass production among Sahelian provenances (B.O. Diallo, unpublished results). Considerable variation in pod production within and among provenances was also reported from a 15-year-old provenance trial in Dinderesso, Burkina Faso.

*Néré (*Parkia biglobosa*)*

Results of an isozyme analysis based on samples collected from 11 countries in West Africa has shown very high genetic diversity in *Parkia biglobosa* at the inter- and intrapopulation levels (Teklehaimanot, 2004). This diversity should be considered during the planning of any genetic conservation or improvement programme.

*Karité (*Vitellaria paradoxa*)*

Bouvet *et al.* (2004) reported an analysis of molecular diversity from 80 populations covering most of the natural range from Senegal to Uganda. Using random amplified polymorphic DNA (RAPD) analysis, results from 118 individual trees indicated variation among individuals within populations. Studies carried out in 2002 across the species range indicated that human activities have affected genetic variation in this species (Teklehaimanot, 2004). Kelly *et al.* (2004) reported that populations in crop fields have the highest mean number of alleles and the highest expected heterozygosity when compared with populations in fallows and forests. On the basis of fruit characteristics and on ecological local knowledge in Mali, *Vitellaria paradoxa* can be divided into five classes (H. Sanou, unpublished results), but the differences may be due to environmental rather than genetic differences. However, H. Sanou (unpublished results) reports that selection of quality germplasm must consider the variation within populations in the improvement of karité.

Two subspecies have been proposed, *V. paradoxa* subsp. *paradoxa* and *V. paradoxa* subsp. *nilotica* (Hall *et al.*, 1996), but there is no clear distinction between them based on leaves, inflorescences and flowers, fruits or morphology. Hall *et al.* (1996) concluded that the difference is based on the origin of the populations, which originated in the eastern (subsp. *nilotica*) and western (subsp. *paradoxa*) parts of the natural range of the species. Using isozymes, Lovett and Haq (2000) found high genetic diversity within populations of *V. paradoxa* in Ghana. Fontaine *et al.* (2004), using molecular markers, recommended a separation between western and eastern populations. Moreover, chemical analysis by Maranz *et al.* (2004) indicates different fatty acid profiles across Africa and following an east–west trend among the natural populations.

It is clear that the genetic characterization of the shea tree needs further work, and genetic and environmental effects on the expression of different traits need to be understood. Regarding growth, preliminary results from an 8-year-old provenance trial established in Gonsé, Burkina Faso, revealed significant differences among provenances for tree height, collar diameter and transpiration rates (J. Bayala, unpublished results). Further observations are expected on fruit production.

*Ber (*Ziziphus mauritiana*)*

Pareek (2001) has reported high genetic variability in the growth and morphological characteristics of ber in India, and Diallo (2002) confirmed this in Africa. Early results from provenance trials in Mali and Senegal indicated significant differences in growth and biomass production among provenances. Fruit production was just beginning, but early results suggested that it also varies among provenances.

Improved cultivars are being produced from China, India and Thailand. Cultivars such as Gola, Kaithli, Umran and Seb (from India) and Sotubata (from Thailand) and various sweet Thais (from Thailand) are being tested in the Sahel, along with the best accessions collected in Brazil, Burkina Faso, Kenya, Mali, Niger and Senegal. In Burkina Faso, Ouedraogo *et al.* (2006) investigated the effects of irrigation, rock phosphate and cultivars (Gola, Seb, Umran and local as control) on growth and fruit production. At 18 months, the introduced cultivars performed better than the local cultivar.

10.3.2 Genetic improvement of preferred fruit trees

Genetic improvement using selected clones has been initiated in the Sahel. The selection of plus-trees and their clonal development may be faster means of improvement and have greater impact than conventional breeding. However, considering the trade-off between genetic gain and diversity, it is very important to ensure that genetic diversity is not severely reduced in the clonal breeding populations (Cornelius *et al.*, 2006). This is particularly important for on-farm breeding populations, because farmers tend to select very few trees/clones to establish fruit tree populations on the farm (Brodie *et al.*, 1997; Lengkeek, 2003).

Some plus-trees of *Tamarindus indica*, *Vitellaria paradoxa* and *Ziziphus mauritiana* were selected by farmers and researchers in 2004–2005, and clones were established in gene banks and regeneration plots. The initial collections (more than 150 plus-trees) were made in parts of Burkina Faso, Mali, Niger and Senegal for all three species, and also in the Tharaka district of Kenya for *T. indica*. More collections are planned in the future to ensure that the breeding population has a broad genetic base. The selected materials will be used for further improvement programmes and for the production of vegetative plant materials for large-scale propagation with collaborating farming communities.

Different ideotypes and selection criteria for the species were identified in a participatory manner with farmers and researchers. The following selection criteria were retained for the different fruit tree species:

- *A. digitata*: leaf production during dry season, fruit pulp with higher vitamin C content, high fibre production (Sidibé and Williams, 2002).

- *T. indica*: vigour, early fruit set, sweetness of the fruits, resistance to pests and disease, long, straight pods (pod size), a large, round canopy with many branches for greater fruit production, a large number of seeds, high pod production, exocarp and fibres easily removable from the fruit pulp.
- *V. paradoxa*: vigour, early and annual fruit set, sweet fruits, high oil content, resistance to pests and disease, freedom from *Tapinanthus* (a parasitic plant) attacks, young tree, uniform crown.
- *Z. mauritiana*: vigour, early fruit set, sweet fruits, resistance to pests and disease, small seeds, large round fruits, good fruit conservation, fewer thorns, high fruit production, large canopy with many branches for greater fruit production.

10.3.3 Vegetative propagation

The principal reason for using vegetative propagation is to capture and fix desirable traits, or combinations of traits, of individual trees (Leakey and Newton, 1994). Because higher yields and better products are desired, vegetative propagation is a useful tool for the domestication of indigenous fruit trees. The adapted varieties and cultivars are propagated vegetatively to maintain their desired characteristics, which would, if sexually propagated, be diluted over time (Nyambo *et al.*, 2005). For example, preliminary results from ICRAF and its research partners have shown that grafting can accelerate fruit precocity (shortening the period to first fruiting) compared with plants produced by seed. The fruiting period can be reduced from 20 to 6 years for *Vitellaria paradoxa*, from 6 to 3 years for *Parkia biglobosa*, from more than 10 years to 4 years for *Adansonia digitata*, and from 2 years to 6 months for *Ziziphus mauritiana*.

In the Sahel, vegetative propagation of fruit trees is mostly done by grafting or budding or using stem cuttings. A summary of successful methods for key fruit tree species is given below.

Preliminary results from research in Senegal and Mali indicate that *A. digitata* can be successfully grafted (with a survival rate of 85%) and multiplied by cuttings. Top- and side-grafting give the best results, but top-grafting is preferred as it is easier to do. A top-grafted plant in Mali started flowering 4 years after grafting.

According to Teklehaimanot (2004), cuttings of *P. biglobosa* are relatively easy to root if they are obtained from terminal nodes and auxins are applied.

Tamarindus indica can be vegetatively propagated by stem cuttings, shield and patch budding, grafting, air-layering or marcotting, and tissue culture (Gunasena and Hughes, 2000). The easiest and cheapest method is by using stem cuttings. The use of the growth regulator 3-indolebutyric acid (IBA) significantly increased rooting. For patch budding, rootstocks from 6- to 9-month-old seedlings should be grown in raised beds for large-scale multiplication. For grafting, although top-cleft grafting is successful under Sahelian conditions, approach grafting may be more successful (up to 85% success rate) (B. Kone, unpublished results).

Hall *et al.* (1996) reported that *V. paradoxa* can be propagated by root suckers, grafting, budding, cuttings and tissue culture. The top-cleft grafting technique gives a survival rate of about 70% on the farm (B. Kone, unpublished results). However, Sanou *et al.* (2004) reported side-cleft grafting to be an easy way of propagating this species. The recommended grafting period is in May in the Sahel, which is the end of the dry season/onset of the rainy season. Grafting on mature saplings in the field has produced successful results in Mali and is recommended as a way of enriching the parklands. Propagation by stem cuttings from hardwood, softwood and coppice shoots of mature trees can be used with 80% success. In addition, apical shoots from seedlings can be established and multiplied *in vitro* (Lovett and Haq, 2000).

Ziziphus mauritiana can be vegetatively propagated by budding and grafting techniques, but attempts to propagate it by cuttings have not yet been successful. Top-grafting is the most popular technique for propagating this species in the Sahel because it is quick and easy (the success rate is up to 95%; M. Doumbia, personal communication). For budding (Pareek, 2001), bud sticks with well swollen and recently matured (but not open) buds should be collected from juvenile shoots. In the Sahel, budding is carried out by the T (shield) method. The best time for successful budding is during the active growth period from June to September (the success rate is up to 85%; B. Kone, unpublished results). Budding should be done as close as possible to the ground to minimize the area for emergence of sprouts from the rootstock. Micrografting can also be used (Danthu *et al.*, 2001); this allows the rapid multiplication of clones but is very expensive and not readily available.

10.3.4 Fruit tree management

The management of indigenous fruits needs more investigation in the Sahel. Information about nutrition and water management, pest and disease management, tree training and pruning, and harvesting and postharvesting techniques is limited.

10.4 Conclusion

Indigenous fruit-tee species are undoubtedly important for the rural poor in the Sahel, but considerable work is needed to ensure that they contribute significantly and in a sustainable manner to the livelihoods of the rural poor. This effort requires working in a participatory manner with rural communities, research and development institutes, non-governmental organizations and private enterprise.

Acknowledgements

We wish to thank Dr Harold Roy-Macauley and an anonymous reviewer for their valuable comments on the manuscript. Mr Mamadou M. Sidibé and Doumbia Modibo of the World Agroforestry Centre, Bamako made their unpublished results available.

References

Adin, A., Weber, J.C., Sotelo Montes, C., Vidaurre, H., Vosman, B. and Smulders, M.J.M. (2004) Genetic differentiation and trade among populations of peach palm (*Bactris gasipaes* Kunth) in the Peruvian Amazon – implications for genetic resource management. *Theoretical and Applied Genetics* 108, 1564–1573.

AFE (2006) *Filière Amande de Karité au Mali: Importance, Compétitivité, Défis et Solutions Potentielles.* Action for Enterprise (AFE), Arlington, Virginia.

Bationo, A.B., Ouedraogo, S.J. and Guindo, S. (2001) Stratégies de régénération naturelle de *Detarium microcarpum* Guill. et Perr. dans la foret classée de Nazinon (Burkina Faso). *Fruits* 56, 271–285.

Becker, M. and Statz, J. (2003) Marketing of parkland products. In: Teklehaimanot, Z. (ed.) *Improvement and Management of Agroforestry Parkland Systems in Sub-Saharan Africa.* EU/INCO Project Contact IC18-CT98-0261, Final Report, University of Wales, Bangor, UK, pp. 142–151.

Boffa, J.-M. (1999) *Agroforestry Parklands in sub-Saharan Africa.* FAO Conservation Guide No. 34. Food and Agriculture Organization of the United Nations, Rome.

Bonkoungou, E.G., Ayuk, E.T. and Zoungrana, I. (eds) (1997) *Les Parcs Agroforestiers des Zones Semi-Arides d'Afrique de l'Ouest.* Centre International pour la Recherche en Agroforesterie, Nairobi, Kenya.

Bouvet, J.-M., Fontaine, C., Sanou, H. and Cardi, C. (2004) An analysis of the pattern of genetic variation in *Vitellaria paradoxa* using RAPD markers. *Agroforestry Systems* 60, 61–69.

Brodie, A.W., Labarta Chávarri, R.A. and Weber, J.C. (1997) *Tree germplasm management and use on-farm in the Peruvian Amazon: a case study from the Ucayali region, Peru.* Research Report, Overseas Development Institute, London and International Centre for Research in Agroforestry, Nairobi, Kenya.

Cornelius, J., Clement, C.R., Weber, J.C., Sotelo-Montes, C., van Leeuwen, J., Ugarte-Guerra, L.J. and Arévalo-López, L. (2006) The trade-off between genetic gain and conservation in a participatory improvement programme: the case of peach palm (*Bactris gasipaes* Kunth). *Forests, Trees and Livelihoods* 16, 17–34.

Dampha, A. and Camera, K. (2005) *Empowering Communities through Forestry: the MA&D Approach in the Gambia.* Food and Agriculture Organization of the United Nations, Rome, Italy.

Danthu, P., Hane, B., Touré, M., Sagna, P., Ba, S., De Troye, M.A. and Soloviev, P. (2001) Microgreffage de quatre espèces ligneuses sahéliennes (*Acacia senegal, Faidherbia albida, Tamarindus indica* et *Ziziphus mauritiana*) en vue de leur rajeunissement. *Tropicultura* 19, 43–47.

Diallo, O.B. (2001) Biologie de la reproduction et évaluation de la diversité génétique chez une légumineuse: *Tamarindus indica* L. (Caesalpinioideae). Thèse de doctorat. Université de Montpellier II: Sciences et Techniques du Languedoc, Montpellier, France.

Diallo, I. (2002) Etude de la biologie de la reproduction et de la variabilité génétique chez le jujubier (*Ziziphus mauritiana* Lam.). Thèse de doctorat de troisième cycle. Université Cheik Anta Diop, Dakar, Sénégal.

Eyog Matig, O. and Ouedraogo, A.S. (1999) *State of forest genetic resources in the Sahelian and north Sudanian zone of Africa.* Food and Agriculture Organization of the United Nations, Rome.

Faminow, M.D., Weber, J.C., Lariviere, S. and Martin, F. (2001) *Linking poverty and the environment through commodity chain analysis: strategic management of tree genetic resources for sustainable livelihood options in the Amazon.* Working Paper, International Centre for Research in Agroforestry, Nairobi, Kenya.

FAO (2000) *Global forest resource assessment – 2000.* Food and Agriculture Organization of the United Nations, Rome.

FAO (2004) *Simpler forest management plans for participatory forestry.* Forest Policy and Institutions Service, Working Paper No. 4. Food and Agriculture Organization of the United Nations, Rome.

Fontaine, C., Lovett, P.N., Sanou, H., Maley, J. and Bouvet, J.-M. (2004) Genetic diversity of the shea tree (*Vitellaria paradoxa* C.F. Gaertn.), detected by RAPD and chloroplast microsatellite markers. *Heredity* 93, 639–648.

Fortin, D., Lô, M. and Maynart, G. (2000) *Plantes médicinales du Sahel.* Série Etudes et Recherches No. 187–188–189. Edition révisée. Enda-Editions. Dakar, Sénégal.

Gunasena, H.P.M. and Hughes, A. (2000) *Fruits for the Future – Tamarind.* International Centre for Underutilised Crops, Southampton, UK.

Hall, J.B., Aebischer, D.P., Tomlinson, H.F., Osei-Amaning, E. and Hindle, J.R. (1996) *Vitellaria paradoxa: A Monograph.* School of Agricultural and Forest Sciences Publication No. 8. University of Wales, Bangor, UK.

Hall, J.B., Tomlinson, H.F., Oni, P.I., Buchy, M. and Aebischer, D.P. (1997) *Parkia biglobosa: A Monograph.* School of Agricultural and Forest Sciences, University of Wales, Bangor, UK.

Hollingsworth, P., Dawson, I., Goodall-Copestake, W., Richardson, J., Weber, J.C., Sotelo Montes, C. and Pennington, T. (2005) Do farmers reduce genetic diversity when they domesticate tropical trees? A case study from Amazonia. *Molecular Ecology* 14, 497–501.

Kalinganire, A., Dakouo, J.M. and Bayala, J. (2007) Rapport technique final du projet biodiversité -'Enrichissement de la biodiversité dans les parcs agroforestiers et amélioration du bien-être des populations rurales démunies au Sahel. World Agroforestry Centre, Région de l'Afrique de l'Ouest et du Centre, Bamako, Mali.

Kater, L.J.M., Kante, S. and Budelman, A. (1992) Karité (*Vitellaria paradoxa*) and néré (*Parkia biglobosa*). *Agroforestry Systems* 18, 89–105.

Kelly, B.A., Hardy, O.J. and Bouvet, J.-B. (2004) Temporal and spatial genetic structure in *Vitellaria paradoxa* (shea tree) in an agroforestry system in southern Mali. *Molecular Ecology* 13, 1231–1240.

Kouyaté, A.M. (2005) Enquête ethnobotanique sur *Detarium microcarpum* Guill. et Perr. au sud du Mali. In: Kalinganire, A., Niang, A. and Kone, B. (eds) *Domestication des espèces agroforestières au Sahel: Situation actuelle et perspectives.* ICRAF Working Paper No. 5. World Agroforestry Centre, Nairobi, Kenya.

Leakey, R.R.B. and Newton, A.C. (eds) (1994) Domestication of tropical trees for timber and non-timber products. *MAB Digest 17*, UNESCO, Paris.

Ledig, F.T. (1992) Human impacts on genetic diversity in forest ecosystems. *Oikos* 63, 87–108.

Lengkeek, A.G. (2003) 'Diversity makes a difference'. Farmers managing inter- and intraspecific tree species diversity in Meru Kenya. PhD Thesis, Wageningen University, Wageningen, The Netherlands.

Lovett, P.N. and Haq, N. (2000) Diversity of the sheanut tree (*Vitellaria paradoxa* C.F. Gaertn. F.) in Ghana. *Genetics and Crop Evolution* 47, 293–304.

Maranz, S., Kpikpi, W., Wiesman, Z., De Saint Sauveur, A. and Chapagain, B. (2004) Nutritional values and indigenous preferences for shea fruits (*Vitellaria paradoxa* C.F. Gaertn. F.) in African agroforestry parklands. *Economic Botany* 58, 588–600.

Maundu, P.M., Ngugi, G.W. and Kabuye, C.H.S. (1999) *Traditional Food Plants of Kenya.* National Museums of Kenya, Nairobi, Kenya.

Nagarajan, B., Nicodemus, A., Mandal, A.K., Verma, R.K., Gireesan, K. and Mahadevan, N.P. (1998) Phenology and controlled pollination studies in Tamarind. *Silvae Genetica* 47, 237–241.

Nyambo, A., Nyomora, A., Ruffo, C.K. and Tengnas, B. (2005) *Fruits and Nuts: Species with Potential for Tanzania. Technical Handbook No. 34.* Regional Land Management Unit (RELMA in ICRAF)/World Agroforestry Centre, Nairobi, Kenya.

O'Neill, G.A., Dawson, I.K., Sotelo Montes, C., Guarino, L., Current, D., Guariguata, M. and Weber, J.C. (2001) Strategies for genetic conservation of trees in the Peruvian Amazon basin. *Biodiversity and Conservation* 10, 837–850.

Ouedraogo, A.S. (1995) *Parkia biglobosa* (Leguminosae) en Afrique de l'Ouest. Biosystematique et amelioration. PhD thesis, Institute of Forestry and Natural Resources, Wageningen, The Netherlands.

Ouedraogo, S.J., Bayala, J., Dembele, C., Kabore, A., Kaya, B., Niang, A. and Some, A.N. (2006) Establishing jujube trees in sub-Saharan Africa: response of introduced and local cultivars to rock phosphate and water supply in a peri-urban garden in Burkina Faso, West Africa. *Agroforestry Systems* 68, 69–80.

Pareek, O.P. (ed.) (2001) *Fruits for the Future 2 – Ber (*Ziziphus mauritiana*).* International Centre for Underutilised Crops, Southampton, UK, 290 pp.

Roussel, J. (1995) *Pépinières et plantations forestières en Afrique tropicale sèche. Manuel a l'usage des ingénieurs et techniciens du reboisement.* Institut Sénégalais de Recherches Agricoles, Dakar, Sénégal, 35 pp.

Ruel, M.T., Minot, N. and Smith, L. (2005) *Patterns and determinants of fruit and vegetable consumption in sub-Saharan Africa: a multi-country comparison. Background paper for the joint FAO/WHO Workshop on Fruit and Vegetables for Health, 1–3 September 2004, Kobe, Japan.* World Health Organization, Geneva.

Sanou, H., Kambou, S., Teklehaimanot, Z., Dembele, M., Yossi, H., Sina, S., Djingdia, L. and Bovet, J.-M. (2004) Vegetative propagation of *Vitellaria paradoxa* by grafting. *Agroforestry Systems* 60, 93–99.

Sabiiti, E.N. and Cobbina, J. (1992) *Parkia biglobosa*: a potential multipurpose fodder tree legume in West Africa. *International Tree Crops Journal* 7, 113–139.

Sidibé, M.M. (2005) *Evaluation des performances de cinq espèces et provenances de baobab dans les conditions soudano-sahéliennes du Mali. Mémoire de DEA en Sciences Biologiques Appliquées.* Faculté des Sciences et Techniques, Université de Bamako, Bamako, Mali.

Sidibé, M. and Williams, J.T. (2002) *Fruits for the Future – Baobab (*Adansonia digitata *L.).* International Centre for Underutilised Crops, Southampton, UK, 96 pp.

Sidibé, M., Scheuring, J.F., Tembely, D., Sidibé, M.M., Hofman, P. and Frigg, M. (1996) Baobab – home grown vitamin C for Africa. *Agroforestry Today* 8, 13–15.

Simons, A.J. and Leakey, R.R.B. (2004) Tree domestication in tropical agroforestry. *Agroforestry Systems* 61, 167–181.

Teklehaimanot, Z. (2004) Exploiting the potential of indigenous agroforestry trees: *Parkia biglobosa* and *Vitellaria paradoxa* in sub-Saharan Africa. *Agroforestry Systems* 61, 207–220.

UNDP (2003) *Human Development Report 2003. Millennium Development Goals: A Compact among Nations to End Human Poverty.* Oxford University Press, London [available at http://hdr.undp.org/reports/global/2003/].

World Bank (2004) *World Development Indicators Database.* World Bank, New York.

Appendix

Table A10.1. Common indigenous trees that farmers cultivate for food and other uses in parklands of the West African Sahel.

Scientific, English and French names	Uses
Adansonia digitata Baobab Baobab	Food (milk substitute and cream from fruit pulp; condiment for sauces from leaves), medicine (fruit pulp and seed to treat anorexia), cordage (from bark), glue (from gum), pottery preparation (wood for firing pots), soil fertility (branches for mulch)
Balanites aegyptiaca Desert date Dattier du désert	Firewood, fodder (from leaves), food (cooked seeds, fruit pulp), oil (from seed), soap (from oil in seed and bark), medicine (oil from seed to treat ear inflammation and dermatitis, fruit pulp to treat constipation and joint pain), live fence/fodder bank, dead fence
Bombax costatum Red flowered silk cotton Kapokier rouge	Food (flower sepals for sauce, high-quality honey from flowers), household items (wood for floors, stools, chairs, window trim, basins for feeding animals, masks), fibre (from fruit for mattresses, cushions and cotton replacement for clothing), fodder (from leaves at end of dry season)
Borassus aethiopum Fan palm Rônier	Food (immature seeds, immature shoot, fruit juice, wine from sap), fibre (bath sponge from petiole, fans, house mats, baskets, hats, furniture from leave), medicine (juice from fruit to kill intestinal parasites), construction (poles and boards from stem), dune fixation and soil conservation (fibre mats from leaves)
Boscia senegalensis – Boscia du Sénégal	Food (fresh fruit, seeds), medicine (several medicines, including treatment of bilharzias, syphilis and intestinal parasites, and as purgative, tranquillizer)
Ceiba pentandra Silk cotton Fromager	Condiments (flower sepals for sauce, seeds), fibre (from fruit for cushions and mattresses), household items (wood for floors, stools, chairs, window trim, basins for feeding animals, masks), canoes (wood from stem), medicine (host for parasitic plant (*Tapinanthus pentagonia*) that is used for several medicines)
Cordyla pinnata – Poire du Cayor	Food (pulp from mature fruit), household items (wood for mortars and pestles, handles for tools)
Detarium microcarpum – Detar	Food (pulp eaten raw or cooked, processed into cakes, couscous preparation), drinks (juice, alcoholic beverage), firewood and charcoal (good quality wood, fuel wood), household items (leaves to thatch roofs, tool handles, seed used as necklaces), medicine (medicines extracted from seeds, fruits, leaves, roots, bark used to treat more than 20 different diseases)
Diospyros mespiliformis Ebony Ébenier	High-value wood (for furniture, masks, etc.), food (fresh fruit), medicine (broth from leaves to relieve fever, powder from fruit to treat gastric ulcers and haemorrhoids), pottery preparation (extract from fruit for metallic finish)

Continued

Table A10.1. *Continued*

Scientific, English and French names	Uses
Elaeis guineensis Oil palm Palmier à huile	Oil (from fruit), wine (from sap), medicine (unfermented sap to treat anorexia), fibre (from petiole for baskets)
Hyphaene thebaica Dum palm Palmier doum	Fibre (from leaves for mats, fans), food (fruit), wine (from sap), medicine (powder from fruit pulp to treat gastric ulcers), incense (from roots), construction wood, windbreaks
Lannea microcarpa – Raisinier	Food (fruit pulp, young leaves, roots), drinks (aromatic flavour from young leaves and dried fruit, beer from fermented fruit), oil (from seed), fodder (older leaves), feed for small animals (from residue after extracting oil from seed), medicine (from leaves to control vomiting and diarrhoea, from seed oil to treat dermatitis), insecticide (from seed oil), dye (from bark for cloth and skin decoration), house construction (impermeable solution to cover banco walls obtained from residue after extracting oil from seed)
Landolphia heudellotii – –	Food (fruit pulp)
Parkia biglobosa African locust bean Néré	Food (fresh fruit pulp, dried fruit pulp, protein-rich paste and cakes from seeds), medicine (from fruit pulp to treat protein deficiency in children, from bark to treat inflamed tonsils), animal food (dried fruit pulp), medicine (paste from seeds to relieve hypertension), house construction (fruit coats used as organic structural matrix for banco walls; impermeable solution from fermented fruit coats used to cover banco walls and fill cracks in banco), soil fertility improvement (nitrogen fixation)
Prosopis africana – –	Firewood and charcoal (especially for blacksmiths), food during drought (flour from seed), fodder (from fruit, young branches and leaves), medicine (from bark to treat plaque on teeth, from leaves to relieve joint pain), household items (wood for mortars and pestles, yokes for cattle, planks for house, doors, window trim, handles for tools), soil fertility improvement (leaf mulch)
Sclerocarya birrea Marula Prunier africain	Food (fresh seeds, biscuits and cakes from seed, condiment for couscous, honey from flowers), drinks (fresh fruit juice and fermented fruit juice), medicine (from bark to treat diabetes), household items (wood for chairs, handles for implements, masks, statues)
Strychnos spinosa – –	Food (mature fruit pulp, juice), medicine (note: roots, leaves, unripe fruits and seeds are poisonous), firewood, shade
Tamarindus indica Tamarind Tamarinier	Food (dried fruit pulp added to millet porridge, honey from flowers), drinks (fresh fruit juice), medicine (from fruit pulp to treat constipation, counteract fatty foods and facilitate digestion; juice from fruit pulp for antibiotic skin cream and beauty cream; from bark to treat gingivitis and other oral inflammations), household items (wood for tool handles)

Table A10.1. *Continued*

Scientific, English and French names	Uses
Vitellaria paradoxa Shea nut tree Karité, arbre à beurre	Food (butter from seeds), medicine (butter to treat gastric ulcer and dry skin, butter as base for bactericidal skin creams and beauty creams), soap (from low-quality butter), forage (host of parasitic plant (*Tapinanthus pentagonia*) which is used as forage)
Vitex doniana Black plum Prunier noir	Food (fresh fruit), firewood, wood products (softwood for planks, tables, paper)
Ximenia americana Wild plum Citron de mer	Food (fresh ripe fruit pulp), firewood, medicine (several medicines: seed oil, roots to treat syphilis and intestinal parasites, leaf for stomach-ache and diarrhoea, leprosy, tranquillizer), firewood (firewood and charcoal), fodder (from leaves), dye (extract from leaves to dye clothes), house construction (timber of good quality and very hard: poles and rafters)
Ziziphus mauritiana Jujube Jujubier	Food (fruit pulp to make cakes, porridge, high in vitamin C), drinks (from fruit pulp), medicine (low dose extract from roots to treat gastric ulcers, vomiting and syphilis; fruit pulp and other ingredients to treat anorexia; pulp and green beans for infant food), household items (wood for tool handles), firewood, live fence, fodder (from leaves), fishing (high-dose extract of roots to kill fish), soil fertility improvement (leaf mulch)

Source: B. Kone (ICRAF), based on farmer surveys in Mali (unpublished data) and Fortin *et al.* (2000).

11 The Role of Indigenous Fruit Trees in Sustainable Dryland Agriculture in Eastern Africa

Z. Teklehaimanot

School of the Environment and Natural Resources, University of Wales, Bangor, UK

11.1 Introduction

The drylands of Eastern Africa, which include arid, semi-arid and dry subhumid areas, are geographically complex. They are made up of dynamic ecosystems ranging from the lowlands of Sudan and the mountains of northern Ethiopia to the inland plains of Uganda, the coastal lands of Kenya and the offshore islands of Tanzania. They contain a wide variety of species and genetic diversity and support the largest assemblies of mammals in the world. Drylands cover 70% of Eastern Africa, making it one of the driest regions in the world. These drylands have been occupied intensively by humans for thousands of years and are currently the most densely populated areas in Eastern Africa. The population of Eastern Africa is well over 120 million and is increasing at the rate of 3.2% per annum, the highest rate of increase in Africa (IPED, 1994; UNEP, 1994; Darkoh, 2003).

In the drylands of Eastern Africa trees play a crucial role in tempering the effects of climate, while also providing shade to facilitate the growth of crops on farms and of pastures on rangelands, in addition to providing a wide range of products. Indigenous fruit trees, in particular, play an important nutritional role as a source of micro-nutrients and vitamins to augment the diets of local people. The fruits of many of these species are important as a source of income during the late dry season and early wet season, when stocks of cereal crops are usually low. Farmers selectively retain indigenous fruit trees when farmlands are established by clearing natural woodlands. Indigenous fruit trees are also deliberately maintained on rangelands to provide fodder for livestock and fruit for livestock herders. For these reasons, indigenous fruit trees constitute a major component of many of the Eastern Africa dryland agricultural systems and successfully ensure a food supply under adverse climatic and edaphic conditions. Despite the importance of indigenous fruit trees, their role in sustainable dryland agriculture is overlooked and receives little recognition from the development community. Unlike exotic trees, indigenous trees remain semi-domesticated or undomesticated in Eastern Africa. The reasons for this neglect centre around: (i) a lack of information and reliable methods for measuring the contribution of indigenous fruit trees to resident households, local communities and the rural economy, and the ecological services provided by these trees; (ii) the lack of world markets and production incentives related to markets and technology (low prices offered by local and international industries may discourage the collection and processing of these fruits); and (iii) the bias in favour of large-scale agriculture (Scoones *et al.*, 1992; Teklehaimanot, 2004).

Traditionally, local communities in the drylands of Eastern Africa have developed land-use practices that sustainably use natural resources and effectively safeguard biological diversity. However, as a result of increased human and livestock population pressures the productivity and sustainability of dryland agriculture is declining. Indigenous fruit trees are continually being destroyed by an ever-increasing population that needs new farmland, wood for fuel and construction, and conversion to irrigated agriculture. The critical role that trees have been playing both in terms of ecological services, including soil fertility and microclimate amelioration, and in securing the livelihoods and food security of the region's people is being lost. As a consequence, the communities of the drylands of Eastern Africa remain the poorest and the most food-insecure in the world (Darkoh, 2003; FAO, 2003; IPED, 2004).

Unless the degradation process of agriculture in the drylands (which are expected to get drier as a result of climate change) is reversed and agricultural methods are improved, the function of dryland agriculture in providing a direct life-support base for Eastern Africa's population of more than 120 millions may soon cease. One of the solutions to this problem is the improvement of dryland agriculture by domesticating indigenous fruit trees of economic value (Leakey *et al.*, 1999; Teklehaimanot, 2004). Improvement of dryland agriculture through the judicious planting and management of the native tree species that provide fruits that are used and traded locally and internationally can contribute to the diversification of income and enhancement of the livelihoods of rural communities. Increased planting of indigenous fruit trees also helps to restore the degraded dryland ecosystem and reverse the loss of biodiversity. This chapter presents information on the current benefits derived by local communities from some of the most common indigenous fruit trees of Eastern Africa, the role of these trees in providing ecological services to dryland agriculture, and how they can be used to restore and improve dryland agriculture and provide sustainable livelihoods for local communities.

11.2 Setting Species Priorities for Domestication

Historically, approaches to halting land degradation or restoring degraded dryland agriculture in the Sahel have focused primarily on externally driven tree-planting. As a result, many local community tree-planting efforts have focused so far on fast-growing exotics such as eucalypts, which are known to yield quick economic returns to farmers from the sale of wood (Wood and

Burley, 1991). There is, however, growing international concern about the impact of introducing species into areas outside their native range. Reasons for this concern include the reduction of local biodiversity and the introduction of pests and diseases. In addition, income from the sale of wood to farmers is insignificant because wood has a high weight-to-value ratio and transport is expensive. Indigenous fruit trees provide better returns to farmers because fruits have a lower weight-to-value ratio and are easier to transport than wood (Teklehaimanot, 2004). Therefore, the use of native fruit tree species to improve the sustainability and productivity of dryland agriculture is more attractive to local communities than the use of exotic trees.

Since domestication is concerned with the selection and management of the most highly valued indigenous fruit trees, prioritization is the first step in domestication. Historically, tree species have been prioritized by researchers with varying degrees of research self-interest. Setting priorities, however, requires understanding of user needs and preferences. It is now widely accepted that domestication is likely to be most effective when local people are involved in a participatory process of priority-setting for tree species (Franzel *et al.*, 1996). The World Agroforestry Centre (ICRAF), in collaboration with national agricultural research systems, has developed procedures and guidelines for species priority-setting in which the participation of local communities is given great emphasis (Franzel *et al.*, 1996). These methods were used in a priority-setting exercise for the domestication indigenous fruit trees carried out in Eastern Africa by the Association of Forest Research Institutions in Eastern Africa (AFREA) in collaboration with IPGRI-SSA (International Plant Genetic Resources Institute – sub-Saharan Africa) and national partner research institutions in Eastern Africa. Among the indigenous fruit tree species of the drylands of Eastern Africa, the following were identified as national priority species for domestication: *Adansonia digitata* L., *Carissa edulis* (Forssk.) Vahl, *Parinari curatellifolia* Planch. ex Benth., *Sclerocarya birrea* (A. Rich.) Hochst. subsp. *caffra* (Sond.) Kokwaro, *Tamarindus indica* L. and *Ziziphus mauritiana* Lam. (Chikamai *et al.*, 2005). These tree species are known to have potential for wider use and could contribute significantly to the improvement of dryland agriculture.

A more recent priority-setting exercise was carried out by Teklehaimanot (2005) in five countries of Eastern Africa: Ethiopia, Kenya, Sudan, Tanzania and Uganda. These countries represent a diverse range of dryland and socio-economic conditions in Eastern Africa. Two stages were involved in the exercise. The first consisted of national priority-setting through meetings and discussions with researchers and extension agents in the five countries. The second stage was a field survey conducted in the five counties. The aims of the field survey were to set priorities among indigenous fruit trees at the local level, to gather baseline information on the potential contributions of the priority species to farm households and the rural economy, and to identify problems and constraints in relation to their utilization, management and productivity, with the participation of local stakeholder networks (Teklehaimanot, 2005).

After discussions involving 31 senior scientists and forestry extension agents in the five countries, it was agreed that the field survey should focus on indigenous fruit tree species that: (i) are currently underutilized either because of suboptimal management or because their full genetic potential is not exploited; (ii) remain semi- or undomesticated; (iii) have not been the object of scientific study until fairly recently; and (iv) as a result of increasing human and animal populations, which are causing environmental pressures on the land, are declining in population size. This resulted in a total of 11 trees: *Balanites aegyptiaca* Del., *Berchemia discolor* (Klotzsch) Hemsley, *Borassus aethiopum* Mart., *Carissa edulis*, *Cordeauxia edulis* Hemsl., *Parinari curatellifolia*, *Sclerocarya birrea*, *Strychnos cocculoides* Baker, *Vangueria madagascariensis* J.F. Gmel., *Vitellaria paradoxa* and *Vitex payos* (Lour.) Merr. (Teklehaimanot, 2005).

The field survey was conducted in 26 villages located in 16 regions or districts in the five countries. A total of 167 local community members (90 farmers, 20 pastoralists, 19 traders and 38 other local community members) were interviewed during the course of the survey. The results of the survey showed that species ranking by the local people varied between the five countries. The top-ranking species were *Sclerocarya birrea* in Tanzania, *Balanites aegyptiaca* in Sudan, *Cordeauxia edulis* in Ethiopia, *Vitellaria payos* in Uganda and *Vitex payos* in Kenya (Table 11.1) (Teklehaimanot, 2005). The major reasons for such disparity may be species distribution, indigenous knowledge and economic pursuits of the local communities visited. The results of the field survey also showed that, among the national priority species, *Carissa edulis* and *Parinari curatellifolia* did not rank top according to local people's preferences. Therefore, combining the results of the two priority-setting exercises (Chikami *et al.*, 2005; Teklehaimanot, 2005) resulted in eight priority indigenous tree species for domestication with potential to improve dryland agriculture in Eastern Africa: *Adansonia digitata*, *Balanites aegyptiaca*, *Cordeauxia edulis*, *Sclerocarya birrea*, *Tamarindus indica*, *Vitellaria paradoxa*, *Vitex payos* and *Ziziphus mauritiana*.

These results are in agreement with findings of species priorities set in the drylands of southern and West Africa. Some of the above tree species are common to all three regions, so there is good scope for germplasm exchange and other forms of collaboration. For example, in West Africa *Adansonia digitata*,

Table 11.1. Ranking (top five) of indigenous tree species for domestication by local communities in Eastern Africa.

Species	Ethiopia	Kenya	Sudan	Tanzania	Uganda
Balanites aegyptiaca	3	3	1	5	3
Berchemia discolour	–	2	–	–	–
Borassus aethiopum	4	–	2	–	2
Carissa edulis	–	4	–	–	4
Cordeauxia edulis	1	–	–	–	–
Parinari curatellifolia	–	–	–	3	–
Sclerocarya birrea	5	5	3	1	5
Strychnos cocculoides	–	–	–	2	–
Vangueria madagascariensis	–	–	–	4	–
Vitellaria paradoxa	2	–	4	–	1
Vitex payos	–	1	5	–	–

Source: Teklehaimanot (2005).

Tamarindus indica, *Vitellaria paradoxa* and *Ziziphus mauritiana* were among the top five priority fruit tree species identified through farmers' input and the priority-setting exercise carried out by ICRAF (Bonkoungou *et al.*, 1998). Similarly, in southern Africa, according to Maghembe *et al.* (1998), *Adansonia digitata*, *Sclerocarya birrea*, *Tamarindus indica* and *Ziziphus mauritiana* were among the top-ranking species according to farmers' preferences.

11.3 Role of Indigenous Fruit Trees in Dryland Agriculture

11.3.1 Species profiles

The eight priority tree species (Table 11.2), with the exception of two species, occur in all five countries. The two exceptions are *Cordeauxia edulis*, which is found only in Ethiopia, and *Vitellaria paradoxa*, which is found in Ethiopia, Sudan and Uganda.

11.3.2 Contribution to household and rural economies

Both fruit pulp and seed of all the above eight species are edible and very nutritious, with high levels of energy, protein, fat, vitamins and minerals (Table 11.3), and they make important contributions to the household and local economies (Hall and Walker, 1992; Beentje, 1994; Saka and Msonthi, 1994; Hall *et al.*, 1996, 2002; Glew *et al.*, 1997; Ajiwe *et al.*, 1998; ICUC, 1999; Leakey, 1999; Maundu *et al.*, 1999; Pareek, 2001; Sidibe and Williams, 2002; Teklehaimanot, 2003). The fruits of all these species are highly valued by local communities. Farmers retain all the above trees on their farms primarily for their fruits to meet their household needs for food and income. These trees also occur on rangelands and are an important source of fodder for livestock and micronutrients and vitamins for livestock herders. All eight species are, however, underutilized in Eastern Africa compared with their utilization elsewhere. For example, *Adansonia digitata* is being domesticated in the Sahel by ICRAF and partners (ICRAF, 1997), as well as exploited commercially and industrially in India (Pareek, 2001), *Sclerocarya birrea* is being domesticated in Southern Africa (Akinnifesi *et al.*, 2006), *Tamarindus indica* in Thailand and India (ICUC, 1999), *Vitellaria paradoxa* in West Africa (Lovett and Haq, 2000; Sanou *et al.*, 2004) and *Ziziphus mauritiana* in India (Pareek, 2001). Populations of all eight species are also in decline and there are few attempts to domesticate them in Eastern Africa.

Adansonia digitata (baobab)

This is a multipurpose tree best known for its swollen trunk. The most important benefit of baobab is derived from its fruits. Seeds are eaten fresh, dried or roasted. The pod contains a very nutritious pulp, which is rich in vitamin C, calcium, potassium, phosphorus, iron and protein (Table 11.3). It is mixed with water to

Table 11.2. Description of priority indigenous fruit tree species of Eastern Africa.

Species and family	Description
Adansonia digitata Bombacaceae	Deciduous tree up to 25 m height; swollen trunk more than 10 m diameter; greyish brown and smooth bark; leaves digitate compound with 5–7 finger-like leaflets, flowers solitary, scented and white; fruits woody, yellowish brown and hairy, seeds kidney-shaped, hard and black.
Balanites aegyptiaca Balanitaceae	Evergreen tree up to 10 m height; stem diameter up to 30 cm with recurved spiny branches; leaves fleshy, succulent, distinctive grey-green pairs and ovate; flowers fragrant, yellow-green clusters, fruits oblong with both ends round, yellow when ripe with a hard pointed seed inside.
Cordeauxia edulis Leguminoceae	Evergreen nitrogen-fixing shrub or small tree up to 4 m height; both adaxial and abaxial leaf surfaces covered by thick cuticle; flowers yellow; nuts brown.
Sclerocarya birrea Anacardiaceae	Deciduous tree up to 18 m height; bark is grey and thick with irregular cracks and raised scales; leaves compound pinnate, crowded at tips of branches; flowers dioecious, male flowers pale green, female green-pink; fruits rounded and fleshy, skin cream spotted, 2–3 large oily seeds inside.
Tamarindus indica Leguminosae	Evergreen tree up to 30 m height, extensive dense crown; short bole up to 1 m diameter; bark grey; leaves compound pinnate, dull-green, oblong; flowers small, in few-flowered heads, petals gold with red veins; fruits pale brown, sausage-like, hairy pods, sticky pulp, and seeds inside.
Vitellaria paradoxa Sapotaceae	Deciduous tree up to 20 m height; leaves born at the end of branches, 20–30 leaves the terminal 5 cm of the shoot; bark white or dark, rough and deeply fissured into square scales; flowers hermaphrodite, dense fascicle containing 30–40 flowers, white, fruit oblong with 2–3 oily seeds inside.
Vitex payos Verbenaceae	Deciduous tree up to 8 m height; bark grey brown, deeply fissured leaves with 5 leaflets, occasionally 3 leaflets roughly hairy above, softly hairy beneath, flowers white, blue or mauve; fruits shortly cylindrical, black when ripe, seeds hard, grooved.
Ziziphus mauritiana Rhamnaceae	Deciduous tree up to 12 m height; diameter up to 30 cm; branches bearing paired, brown spines, one straight and the other slightly hooked; wide spreading crown; leaves alternate, shiny above, white below; flowers small and yellowish; fruits mostly rounded, reddish or dark brown.

Sources: FAO, 1983; von Maydell, 1989; Hall and Walker, 1992; Bekele-Tesemma, 1993; Beentje, 1994; Wickens, 1995; ICUC, 1999; Pareek, 2001; Hall *et al.*, 1996, 2002; Sidibe and Williams, 2002; Teklehaimanot, 2003; Mbabu and Wekesa, 2004.

Table 11.3. Nutritional composition of fruits and seeds of priority indigenous fruit trees of Eastern Africa

	Plant part	Energy (kJ/100 g)	Total protein (g/100 g)	Total fat (g/100 g)	CHO (g/100 g)	Vitamin C (mg/100 g)	Fe (mg/100 g)	Ca (mg/100 g)	P (mg/100 g)	K (mg/100 g)
Adansonia digitata	Fruit pulp	1480	3.1	4.3	79.4	270	1.7	341	73.3	2836
	Seeds	1803	19.6	30.6	4.8	–	1.8	395	614	1275
Balanites aegyptiaca	Fruit pulp	1151	5	0.1	70	35	3.1	141	62	–
	Seeds	2140	27	34	–	–	–	–	720	–
Cordeauxia edulis	Fruits	446	13.3	11.6	36.5	–	–	32	226.5	625
Sclerocarya birrea	Fruit pulp	225	3.6	0.4	12	194	2.49	481	264	317
	Seeds	2703	5.6	57.3	3.7	–	2.78	156	212	601
Tamarindus indica	Fruit pulp	1490	4.1	1.6	85	9	1.39	183	100	1226.9
	Seeds	–	17.3	5.5	–	–	2.67	185	228	–
Vitellaria paradoxa	Fruit pulp	393	4.13	–	–	–	24	184	74	1320
	Seeds	2422	8.5	39.7	–	–	1.5	82	58.7	–
Vitex payos	Fruits	1809	6.72	0.61	65.2	–	–	–	–	–
Ziziphus mauritiana	Fruits	1588	4.1	9.5	73	24	3.0	13.5	216.2	1731.8

Sources: Hall and Walker, 1992; Saka and Msonthi, 1994; Hall *et al.*, 1996, 2002; Glew *et al.*, 1997; Ajiwe *et al.*, 1998; ICUC, 1999; Leakey, 1999; Maundu *et al.*, 1999; Atkinson *et al.*, 2000; Pareek, 2001; Sidibe and Williams, 2002; Teklehaimanot, 2003.

produce beverages. Dried pulp contributes much to the economy of the localities where the tree occurs as it is processed as a powder to improve the nutritional value of food and is marketed locally and internationally (Sidibe *et al.*, 1996; Sidibe and Williams, 2002; Mbabu and Wekesa, 2004). Coloured pulp is sold as a sweet in many kiosks and supermarkets in Eastern Africa (Teklehaimanot, 2005). An oil extracted from the seeds is often used in traditional ceremonies and for cooking. Bath oil, lotions and creams have also been developed from the oil for the cosmetic industries. Fresh and dried leaves, which are rich in vitamin A, are cooked and eaten as a type of spinach and are also used as forage (Sidibe and Williams, 2002; Mbabu and Wekesa, 2004).

Balanites aegyptiaca (desert date)

This is an African dryland fruit with edible pulp and a hard woody endocarp enclosing an edible oil-rich seed (Table 11.3). *Balanites aegyptiaca* produces fruit even in unusually dry years; local communities in the drylands of Eastern Africa heavily rely on this resource as an emergency food and it is used as a regular rural market commodity. The seeds are rich in protein and energy. Oil extracted from the kernels is used for cooking and medicine. Commercially, *B. aegyptiaca* is a potential source of diosgenin and yamogenin for the manufacture of cortisone and corticosteroid (Hardman and Sofowora, 1972; von Maydell, 1989; Hall and Walker, 1992; Dirar, 1993; Mbabu and Wekesa, 2004). According to informants in the Katakwi district of Uganda, the edible leaves of *B. aegyptiaca* are the only vegetable available in the area during the dry season (Teklehaimanot, 2005). Young leaves are pounded, mixed with groundnut, cooked and eaten by the local people. The informants also report that dried leaves are the most popular trading commodity in local markets during the dry season and are sold at about UShs 170 (US$0.10) per 250 ml cup.

Cordeauxia edulis (yehib)

This shrub produces a tasty edible nut of high nutritional (Table 11.3) and economic value. The nuts are eaten fresh, dried, roasted or cooked; they have a smooth consistency and an agreeable taste like cashew nut or chestnut. The nuts are sold at local markets in Ethiopia and Somalia. The species is such a hardy species that during drought it is sometimes the only food left for the nomads. The seeds are also rich in fatty acids. During the dry season, because it is an evergreen, it is one of the few palatable fodder species available and provides abundant fodder. Leaves are infused to make a tea. Leaves also contain a brilliant red dye, cordeauxiaquinone, that stains the hands and is used in the dyeing of fabrics. Cordeauxiaquinone is also used medicinally to stimulate haemopoeisis. The plant has attracted considerable interest as a potential food crop for arid areas. It is listed as one of the most threatened tree species by IUCN (Nerd *et al.*, 1990; Bekele-Tesemma, 1993; Wickens, 1995).

Sclerocarya birrea (marula)

This species plays a very significant role in the diet and culture of people in many countries where it occurs. Both the pulp and kernel are nutritious (Table 11.3). The kernels yield highly stable oil that is useful in cooking and for manufacturing cosmetics. The oil is currently being traded internationally from the countries of Southern Africa. Local people consume the fruits and some are sold at local and regional markets (Hall *et al.*, 2002; Mbabu and Wekesa, 2004; Leakey, 2005). According to informants in the Dodoma district of Tanzania, fruits are sold at local markets at Tshs 250 (US$0.20)/kg (Teklehaimanot, 2005). In South Africa the fruits are used to produce the very popular Amarula liqueur, which is traded internationally, while a local beer/wine is culturally important (Wynberg *et al.*, 2003). A type of butter is also extracted from the kernels, which is used for the production of cosmetics by cosmetic industries in Europe and the USA (Shackleton *et al.*, 2000; Hall *et al.*, 2002; Thiong'o *et al.*, 2002; Wynberg *et al.*, 2003).

Tamarindus indica (tamarind)

This tree is thought to have originated in Eastern Africa, from where it spread to Asia and Central and South America. Thus, the tamarind has become an economically important species not only in Africa but also in Asia and the American continent. A tree produces in the range of 150–500 kg of fruit per tree (ICUC, 1999) and the edible pulp is consumed fresh and used to make syrup, juice concentrates and exotic food specialities such as chutneys, curries, pickles and meat sauces. The pulp is sticky and light brownish–red in colour and is both sweetish and acidic. The fruit pulp contains large amounts of vitamin C and sugar (Table 11.3) and is a much-valued food ingredient in many Asian and Latin American recipes. The seed is also a good source of protein and oil. The two most common varieties are a sweet and a sour variety. The sweet tamarind is produced mainly in Thailand, where it is grown on a commercial scale and is exported in fresh and processed forms. Around 140,000 t of tamarind is produced annually in Thailand. The pharmaceutical industry uses the pulp as an ingredient in cardiac and blood sugar-reducing medicines (Beentje, 1994; ICUC, 1999; Mbabu and Wekesa, 2004).

Vitellaria paradoxa (shea butter tree)

Both fruits and seeds of *Vitellaria paradoxa* are edible. The pulp is very tasty and highly nutritious (Table 11.3). The fruit is consumed fresh and sold at local markets. The primary benefit to the local people is derived from the oil present in the kernels. There are two subspecies of *V. paradoxa*. subsp. *paradoxa*, which occurs in West Africa, produces a solid fat or butter while the Eastern Africa subsp. *nilotica* produces a liquid oil that it is higher in olein and contains most of the therapeutic substances found in shea butter. Nilotica shea butter is softer and more fragrant than West African shea butter. The butter and oil are primarily used as sources of cooking fat. Shea butter is used by chocolate and cosmetic manufacturing industries worldwide. Shea kernel is a very important export commodity and contributes significantly to the generation of revenues for many countries in West Africa. It was the third most important export product of Burkina Faso in the 1980s (World Bank, 1989). Annual exports to Europe are about 40,000–75,000 t, with another 10,000–15,000 t sold to Japan. In Eastern Africa, on the other hand, shea has become an export commodity to a very small extent only in Uganda since 1990, through The Shea Project. However, the shea butter in Sudan and Ethiopia has never had access to the export market and still remains a source of household cooking fat for local communities (Hall *et al.*, 1996; Boffa, 1999; Lovett and Haq, 2000; Teklehaimanot, 2003, 2005; Schreckenberg *et al.*, 2006).

Vitex pyros (chocolate berry)

The ripe fruit of African olive is black, edible and sweet, somewhat resembling a prune in taste. Jam is made from the vitamin-rich fruits. Fruits are tasty and extensively traded locally. Fruits are sold on local markets in the Kebwizi and Embu districts of Kenya at about Kshs 150 (US$2.1)/kg. According to informants in Kebwizi, annual fruit production per tree is about 50 kg. The fruit is used in traditional medicines (FAO, 1983; Beentje, 1994; Ajiwe *et al.*, 1998; Maundu *et al.*, 1999; Mbabu and Wekesa, 2004).

Ziziphus mauritiana (ber or Indian jujube)

The ber is an economically important tree that produces fruit that has a high sugar content and high levels of vitamin C, phosphorus and calcium (Table 11.3). Fresh fruits are eaten raw. Dried fruits are sold at local markets in Kenya and Sudan (Teklehaimanot, 2005). The fruits are also boiled with rice and millet and stewed or baked or made into jellies, jams, chutneys and pickles. Oil is also extracted from the seeds. The leaves are used as forage for cattle, sheep and goats and are also used by local people as a vegetable in couscous. Fruit is produced even in unusually dry periods as the tree is drought-resistant. Thus, local communities in the drylands of Eastern Africa rely heavily on this resource as an emergency food for themselves and as fodder for their livestock. It is grown on a commercial scale in India. In 1994–1995, India produced about 1,000,000 t of fruit from an area of 88,000 ha (Pareek, 2001; Mbabu and Wekesa, 2004).

11.3.3 Contribution to the ecosystem functions of dryland agriculture

All the above indigenous fruit trees are an integral part of the dryland agricultural systems of Eastern Africa, not only because they provide food and income to the local people but also because they contribute to the ecosystem functions of dryland agriculture through the maintenance of soil fertility, water conservation and environmental protection. Traditionally, farmers in the drylands of Eastern

Africa retain indigenous trees on their farm fields after converting original woodland to cropland because they know that trees modify microclimates and thus improve agricultural crop production, and that they protect fields that are susceptible to water and wind erosion. This type of land-use practice is referred to as an agroforestry parkland system (Boffa, 1999). Thus, maintaining trees on the farm is traditionally the rule rather than the exception in the drylands of Eastern Africa. In the absence of trees, mineral fertilization alone is not sufficient to sustain crop production in dryland agriculture because of the lack of organic matter in the soil. Commercial fertilizers are beyond the economic means of many farmers.

Trees in agroforestry parkland systems have been reported to influence the fertility of dryland soils significantly by maintaining soil organic matter (Young, 1997). Evidence of the influence of trees in agroforestry parkland systems on soil fertility comes from studies that compare soil fertility status with the productivity of crops under tree crowns and in the open. Studies based on chemical analysis of soils beneath some of the priority indigenous fruit trees discussed above have shown a common pattern of superior soil fertility under isolated canopies of trees than in areas distant from the trees. For example, Kater *et al.* (1992) and Bayala *et al.* (2002) found higher levels of carbon, total phosphorus and potassium under *Vitellaria paradoxa* crowns than in the open. Higher levels of nitrogen, phosphorus, potassium and calcium under crowns of *Adansonia digitata* than in the open were also reported by Belsky *et al.* (1989). The generally higher soil nutrient status under tree canopies is also reflected in the mineral content of understorey herbaceous species. For example, Bernhard-Reversat (1982) found that the nitrogen content of aerial herb parts was higher under *Balanites aegyptiaca* than in the open.

Several factors contribute to the higher fertility of soils under tree crowns than in the open. These include the greater soil microbial activity under trees than in the open as a result of the high level of nutrient accumulation under trees through litterfall, root decay and exudation; washout and leaching of nutrients stored in tree canopies; nitrogen fixation by trees; dung deposition by livestock that use the trees as shade; and faeces dropped by birds that nest in the trees (Belsky *et al.*, 1989; Tomlinson *et al.*, 1995; Boffa, 1999).

Increases in organic matter and improved microclimatic conditions under trees enhance soil microbial activity, organic matter decomposition and soil physical characteristics. For example, Belsky *et al.* (1989) reported 35–60% higher soil microbial biomass carbon, a lower bulk density of topsoil and higher water infiltration rates under *Adansonia digitata* crowns than in the open.

Fine soil particles lost through wind erosion are intercepted by trees and deposited by throughfall and stemflow. For example, Roose *et al.* (1974, cited in Boffa, 1999) reported that rainwater collected under *Vitellaria paradoxa* canopies had higher concentrations of nitrogen, phosphorus, potassium, carbon, calcium and magnesium than in the open. Evergreen trees such as *Balanites aegyptiaca*, *Cordeauxia edulis*, *Tamarindus indica* and *Ziziphus mauritiana* may also play an important role in dust deposition as they retain their leaves during the dry season, when strong, moist, soil-laden winds prevail in drylands. However, no studies have been reported on these trees in this regard.

Trees also increase soil nitrogen availability through nitrogen fixation. *Vitex payos* and *Cordeauxia edulis* have been reported to nodulate naturally (Bekele-Tesemma, 1993; Mbabu and Wekesa, 2004). However, there are no studies that report the amount of nitrogen fixed by these tree species.

Despite the high soil fertility under trees, reports in the literature on the effects of some of the priority indigenous trees in agroforestry parkland systems on productivity of associated crops are contradictory. For example, Kater *et al.* (1992), Kessler (1992) and Boffa (1999) reported that *Vitellaria paradoxa* reduced sorghum and millet yields but the yield of cotton was unaffected. Belsky *et al.* (1989) and Grouzis and Akpo (1997), on the other hand, found higher herbaceous biomass under the crowns of *Balanites aegyptiaca* and *Adansonia digitata* than in the open. The negative effects of trees on sorghum as reported by Kater *et al.* (1992), Kessler (1992) and Boffa (1999) are generally associated with the higher tree density in agroforestry parkland systems in dry subhumid zones, where irradiance seems to be the main factor limiting growth. The positive effects of trees on herbaceous layer production reported by Belsky *et al.* (1989) and Grouzis and Akpo (1997) are, however, generally associated with low tree density in agroforestry parklands in arid and semi-arid zones, where water is the main limiting factor in primary production. By reducing air temperature, solar radiation and wind velocity, trees decrease potential evapotranspiration (PET) under their crowns, resulting in higher levels of soil moisture under trees than in the open. Lower air temperature under tree crowns because of shading has been reported by Grouzis and Akpo (1997), who observed a difference of 6°C between direct sunlight and the shade of *Balanites aegyptiaca*. Soil temperature was also substantially reduced under crowns of *Adansonia digitata* and *Vitellaria paradoxa* by at least 5°C, as reported by Belsky *et al.* (1989) and Jonsson *et al.* (1999), respectively. Agroforestry parklands have also been reported to reduce wind velocity by Jonsson *et al.* (1999), who measured a significant difference in wind speed under *Vitellaria paradoxa* trees compared with the open. As a result of reduced temperature and wind speed and consequently reduced PET, significantly greater soil moisture under tree crowns than in the open has been reported by Belsky *et al.* (1989) and Boffa (1999) under *Balanites aegyptiaca* and *Vitellaria paradoxa*, respectively.

The influence of trees in agroforestry parklands on the microclimate and soil fertility is therefore extremely favourable to the growth and development of associated crops in Eastern Africa's dryland agriculture, which is normally characterized by aridity and the depletion of nitrogen and phosphorus from the soil. This improvement of soil fertility and the microclimate by trees is exploited by farmers to sustain crop production in dryland agricultural systems with spaced indigenous trees; these are referred to as agroforestry parkland systems (Boffa, 1999; Teklehaimanot, 2003). However, agroforestry parkland systems in the drylands of Eastern Africa are undergoing losses of indigenous trees as a result of habitat alteration through monocrop agriculture and overcutting of trees for construction, fuelwood and charcoal as a result of increased population density and pressure (IPED, 1994; FAO, 2003). This loss of indigenous trees has affected the ecosystem functions of dryland agriculture in many areas, posing a threat to the livelihood and well-being of people in the region (IPED, 1994; Akhtar-Schuster *et al.*, 2000; Darkoh, 2003). Unless the degradation process is reversed and a more environmentally sound agricultural system is adopted, the economic and social factors that are driving the degradation of the ecosystem are likely to exert further pressure on dryland agriculture. Since the functioning and sustainability of dryland agriculture depend greatly on trees, as described above, the domestication and management of indigenous fruit trees in the form of the traditional agroforestry parkland system may be a solution.

11.4 Challenges and Strategies for Improvement

Traditional dryland agriculture used to be sustainable when population density was low. But the rapidly increasing population is now leading to land degradation and desertification in many areas of Eastern Africa. Dryland agriculture in Eastern Africa is losing soil fertility and biodiversity largely as a result of rising population pressure and an uncertain water supply because of frequent droughts. Traditional methods of agriculture and animal husbandry are therefore becoming difficult to sustain and new methods or approaches are needed to address the increasing pressure on natural resources. Making dryland agriculture sustainable is the greatest challenge facing the world today. There are very few answers to this as the problems of drylands are many and complex. The two major constraints on dryland agriculture are water availability, which limits crop yield almost every year, and loss of soil fertility in excess of the soil's natural rate of formation, as a result of either soil erosion or the shortening of fallow periods (Grouzis and Akpo, 1997). The priority indigenous tree species discussed above have the potential to restore the ecosystem functions of drylands by overcoming at least these two major constraints. The judicious planting of these trees in the form of agroforestry parkland systems can improve the sustainability and productivity of dryland agriculture because trees are capable of increasing the efficiency of water and nutrient use in the systems, thereby increasing associated crop production, as described in Section 11.3.3 of this chapter. These trees are well suited to farmers because they are the most preferred species locally. The trees are also capable of improving the nutritional status of societies as their fruits contain high levels of energy, protein, fat, vitamins and minerals, as discussed in Section 11.3.2. Therefore, by domesticating and cultivating the priority indigenous trees on farms, it will be possible to achieve food security, higher export earnings and sustainable and productive dryland agriculture. This approach is also attractive financially because of the low cash inputs needed and generally higher returns to land and labour. In order to achieve successful domestication and cultivation of indigenous fruit trees and to ensure that they play their role in sustaining dryland agriculture, the following strategies are deemed necessary.

11.4.1 Selection of superior quality varieties

The priority indigenous fruit trees identified in this chapter remain in a wild or semi-domesticated form. One notable characteristic of wild tree species is their enormous genetic variability (Simons, 1996; Leakey and Simons, 1998; Leakey *et al.*, 2002). It is also very well known that fruits supplied from wild sources can never be uniform and reliable because production can vary considerably from year to year and from location to location (Teklehaimanot, 1997). Therefore, there is a need to improve these trees through the selection and domestication of varieties of superior quality (Simons, 1996; Leakey and Simons, 1998; Leakey *et al.*, 2003; Leakey, 2005). Selection and domestication can be a means of ensuring quality and supply of raw materials. Selection is one of the steps of tree improvement. There is some evidence in other fruit trees that qualitative traits, such as fruit taste, are usually much more strongly inherited from selected individuals. The genetic gain we would expect in one cycle of selection for a desirable qualitative trait such as fruit taste would be much higher than that expected for a quantitative trait such as tree height, such that 60% of the progeny might be similar to their parents (Simons, 1996). Therefore, there is a good case for phenotypic selection of desirable fruits. Opportunities are even greater using vegetative propagation to develop cultivars. Although some success has been achieved in the improvement of some fruit- and nut-producing wild trees in some villages in West and Central Africa (Tchoundjeu *et al.*, 2006), scaling up these activities at country, regional and continental levels is a challenge.

Historically, several tree domestication activities have not been successful because they did not consider the participation of local people in the tree selection process. Therefore, plus-tree selection criteria, which should incorporate local people's values, knowledge and priorities, must be developed in partnership with local communities.

11.4.2 Domestication of selected superior varieties

Domestication is a set of actions aiming at improving the quality or production of a species (Harwood *et al.*, 1999). Essential paths in domestication include the development of propagation techniques, testing the genetic material, and utilizing the improved planting stock. There are two alternative approaches to domestication: the horticultural and forest-tree breeding approaches. The horticultural approach tends to emphasize the use of clonal varieties; outstanding individuals from existing populations are identified and propagated by grafting or other vegetative means (John and Brouard, 1995). Traditionally, the improvement of many tropical fruit trees, such as avocado, mango and orange, has used this approach (Harris *et al.*, 2002). Forest tree breeding, on the other hand, relies largely on sexual reproduction for propagation. Procedures are developed for crossing among selected trees to generate new populations for domestication. Therefore, considerably larger gains can be made by combining the traditional horticultural and forestry tree breeding approaches (John and

Brouard, 1995). A new model combining both approaches is actively being developed for indigenous fruit trees of Africa (Leakey *et al.*, 2005a, b; Tchoundjeu *et al.*, 2006).

11.4.3 Multiplication and distribution of improved varieties

In order to enhance farmers' participation in the planting of improved planting stock of priority indigenous fruit trees, the farmers need good access to these improved tree germplasms. The current systems of forest seed and seedling distribution to farmers remain a major constraint (Simons, 1996). Many farmers have planted trees, but the genetic quality of the material is often poor (there is a risk of inbreeding because seeds are collected from very few trees), and many farmers say that planting material of desired varieties of some species is unavailable (Tripp, 2001; A. Raebald, personal communication). This is despite the presence of large, well functioning tree seed centres in many African countries, and it means that seeds and seedlings of improved high-quality and high-yielding indigenous fruit trees cannot reach the farmers. If the domestication of indigenous fruit tree species is to contribute to the improvement of rural livelihoods, it requires that both knowledge and improved tree germplasm be transferred to farmers. Research is therefore needed to develop and test alternative (decentralized) ways in which seeds, seedlings and know-how can be distributed efficiently.

11.4.4 Incorporation of improved varieties on farms

Domestication of trees in monoculture plantations has been known to result in a loss of biological diversity, increased susceptibility to pests and diseases, and rapid depletion of soil nutrient reserves (Pedersen and Balslev, 1990). However, planting them in mixtures in the form of agroforestry parkland systems (widely spaced trees) provides continuous tree cover and contributes to both the productivity and the sustainability of farming systems by restoring the degraded ecosystem, enhancing local biodiversity and creating a more favourable microclimate for associated crops and livestock, in addition to supplying fruit and fruit products that provide income and employment in rural areas, thereby improving rural livelihoods. Research is therefore needed to determine the appropriate spacing or density of trees in order to establish each species on farms as widely spaced trees.

11.4.5 Management of trees on farms

Once a certain height is reached, some tree species tend to suppress associated crop production, particularly in the dry subhumid zone. Jonsson *et al.* (1999), during a year of exceptionally high rainfall, found no significant difference in millet yield between crops in the shade and in the open. However, under normal low-rainfall conditions in the savannah regions of West Africa, Kater *et al.* (1992), Kessler (1992) and Boffa (1999) reported that common cereal crops, such as millet and sorghum, were substantially reduced (30–60%) under mature trees of *Vitellaria paradoxa* when compared with the open field. The reduction in crop yield has been attributed to competition for light. One method of reducing light interception by the trees and improving associated cereal crop production may be crown pruning of trees (Bayala *et al.*, 2002). Results of a pruning study on *V. paradoxa* showed that associated crop production was significantly higher under pruned trees than under unpruned trees (Bayala *et al.*, 2002). Studies will therefore be needed to assess the effect of the other priority indigenous fruit tree species (pruned and unpruned) on yields of associated common food crops.

11.4.6 Improved methods of marketing and processing indigenous fruits

The marketing of indigenous tree fruits plays an important though neglected role in the sustainable use of dryland tree resources (Shanley, 2002). Although it has been reported that fruits of the eight indigenous tree species discussed in this chapter are traded locally, their trading patterns in Eastern Africa are not known. At least six of the eight tree species are also traded internationally from countries outside the Eastern Africa region. For local farmers to be successful market participants in both local and international markets, they need to monitor market demand and market prices. To engage in trade at all, they need information regarding the basic market structures (stakeholders, trade channels and patterns of value-adding along the marketing chain). If the situation is left to market dynamics alone, the economic importance of the indigenous fruit market (and with it the incentive to use indigenous tree resources sustainably) can be expected to remain weak (Statz, 2000). To become more effective market players, local farmers have to engage in networking activities and seek strategic alliances in producer–industry partnerships. Linking farmers to high-value and export markets is an important strategy for raising incomes and reducing poverty. Governments and NGOs can facilitate linkages between farmers and exporters and other buyers by helping to organize farmers' groups, establishing ground rules for farmer–buyer contracts, and gathering and disseminating lessons learned from successful marketing schemes. It is generally agreed that farmers have to be supported in collecting the required information, in getting organized amongst themselves, and in shaping partnerships with industries that further process fruit products (Koppell, 1995; Scherr *et al.*, 2002). Research has to play a key role in this process (mainly in the analysis of current markets and prevailing marketing practices).

In order to increase farmers' income from the fruits of indigenous trees, appropriate processing technologies for the fruits need to be developed to add value to the products. Research into the development of appropriate processing technology must have emphasis on simple technologies that are appropriate at village level to allow rural people the opportunity to generate income throughout the year, with maximum value added at the village level.

Acknowledgements

The author acknowledges the Leverhulme Trust, which provided the funding for a study visit conducted in Eastern Africa to identify priority species and assess their contribution to farm households and the rural economy. The assistance, support and interest of the staff of several organizations in Eastern Africa during the study visit are also gratefully acknowledged. These include the Tanzania Forest Research Institute (TAFORI), the Tanzania Tree Seed Agency, Sokoine University of Agriculture, Tanzania, the Kenya Forest Research Institute (KEFRI), the Association of Forestry Research Institutions in Eastern Africa (AFREA), Nairobi, the World Agroforestry Centre (ICRAF), Nairobi, the Forest Resources Research Institute (FORRI), Uganda, Makerere University, Uganda, the Forest Research Centre (FRC), Sudan and the Forest Research Centre (FRC), Ethiopia.

References

Ajiwe, V.I.E., Okeke, C.A., Ogbuagu, J.O., Ojukwu, U. and Onwukeme, V.I. (1998) Characterization and applications of oils extracted from *Canarium schweinfurttii, Vitex doniana* and *Xylopia aethiopica* fruits/seeds. *Bioresource Technology* 64, 249–252.

Akhtar-Schuster, M., Kirk, M., Gerstengarbe, F.W. and Werner, P.C. (2000) Causes and impacts of the declining resources in the eastern Sahel. *Desertification Control Bulletin* 36, 35–42.

Akinnifesi, F.K., Kwesiga, F., Mhango, J., Chilanga, T., Mkonda, A., Kadu, C.A.C., Kadzere, I., Mithofer, D., Saka, J.D.K., Silshi, G., Ramandhani, T. and Dhliwayo, P. (2006) Towards the development of miombo fruit trees as commercial tree crops in Southern Africa. *Forests, Trees and Livelihoods* 16, 103–121.

Atkinson, R.P.D., Macdonald, D.W. and Kamizola, R. (2000) Dietary opportunism in side-striped jackals, Canis adustus Sundevall. *Journal of Zoology* 257, 129–139.

Bayala, J., Teklehaimanot, Z. and Ouedraogo, S.J. (2002) Millet production under pruned tree crowns in a parkland system in Burkina Faso. *Agroforestry Systems* 54, 203–214.

Beentje, H. (1994) *Kenya Trees, Shrubs and Lianas.* National Museums of Kenya, Nairobi.

Bekele-Tesemma, A., Birne, A. and Tengnas, B. (1993) *Useful Trees and Shrubs in Ethiopia. Identification, Propagation and Management for Agricultural and Pastoral Communities.* Regional Soil Conservation Unit, Technical Handbook Series 5. Regal Press, Nairobi.

Belsky, A.J., Amundson, R.G., Duxbury, J.M., Riha, S.J., Ali, A.R. and Mwonga, S.M. (1989) The effects of trees on their physical, chemical and biological environments in a semi-arid savanna in Kenya. *Journal of Applied Ecology* 26, 1005–1024.

Bernhard-Reversat, F. (1982) Biogeochemical cycle of nitrogen in a semi-arid savanna. *Oikos* 38, 321–332.

Boffa, J.-M. (1999) *Agroforestry parklands in sub-Saharan Africa.* FAO Conservation Guide No. 34. Food and Agricultural Organization of the United Nations, Rome.

Bounkoungou, E.G., Djimde, M., Ayuk, E.T., Zoungrana, I. and Tchoundjeu, Z. (1998) *Taking stock of agroforestry in the Sahel: harvesting results for the future. End of phase report: 1989–96.* ICRAF, Nairobi.

Chikamai, B., Eyog-Matig, O. and Kweka, D. (2005) *Regional consultation on indigenous fruit trees in Eastern Africa.* Kenya Forestry Research Institute, Nairobi.

Darkoh, M.B.K. (2003) Regional perspectives on agriculture and biodiversity in the drylands of Africa. *Journal of Arid Environments* 54, 261–279.

Dirar, H.A. (1993) *The Indigenous Fermented Food of the Sudan: A Study in African Food and Nutrition*. CAB International, Wallingford, UK.

FAO (1983) *Fruit and fruit bearing forest trees. Examples from Eastern Africa*. FAO Forestry Paper No. 44/1. Food and Agricultural Organization of the United Nations, Rome.

FAO (2003) *Forestry outlook study for Africa – African forests: a view to 2020*. Food and Agricultural Organization of the United Nations, Rome, 92 pp.

Franzel, S., Jeanicke, J. and Jansen, W. (1996) *Choosing the right trees: setting priorities for multipurpose tree improvement*. ISNAR Research Report No. 8. International Service for National Agricultural Research (ISNAR), The Hague.

Glew, R.K., VanderJagt, D.J., Lockett, C., Grivetti, L.E., Smith, G.C., Pastuszyn, A. and Millson, M. (1997) Amino acid, fatty acid, and mineral composition of 24 indigenous plants of Burkina Faso. *Journal of Food Composition and Analysis* 10, 205–217.

Grouzis, M. and Akpo, L.E. (1997) Influence of tree cover on herbaceous above and below ground phytomass in the Sahelian zone of Senegal. *Journal of Arid Environments* 35, 285–296.

Hall, J.B. and Walker, D.H. (1992) *Balanites aegyptiaca: A Monograph*. School of Agricultural and Forest Sciences, University of Wales, Bangor, UK.

Hall, J.B., Aebischer, D.P., Tomlinson, H.F., Osei-Amaning, E. and Hindle, J.R. (1996) *Vitellaria paradoxa: A Monograph*. School of Agricultural and Forest Sciences, University of Wales, Bangor, UK.

Hall, J.B., O'Brien, E.M. and Sinclair, F.L. (2002) *Sclerocarya birrea: A Monograph*. School of Agricultural and Forest Sciences, University of Wales, Bangor, UK.

Hardman, R. and Sofowora, E.A. (1972) A reinvestigation of *Balanites aegyptiaca* as a source of steroidal sapogens. *Economic Botany* 26, 169–173.

Harris, S.A., Robinson, J.P. and Juniper, B.E. (2002) Genetic clues to the origin of the apple. *Trends in Genetics* 18, 426–430.

Harwood, C., Roshetko, J., Cadiz, R.T., Christie, B., Crompton, H., Danarto, S., Djogo, T., Garrity, D., Palmer, J., Pedersen, A., Pottinger, A., Pushpakumara, D.K.N.G., Utama, R. and van Cooten, D. (1999) Domestication strategies and process. In: Roshetko, J.M. and Evans., D.O. (eds) *Domestication of Agroforestry Trees in Southeast Asia*. Winrock International, Morrilton, Arkansas, pp. 217–225.

ICUC (1999) *Fruits for the future: Tamarind.* International Centre for Underutilised Crops Fact Sheet No 1. University of Southampton, Southampton, UK.

ICRAF (1997) *Annual report 1996*. ICRAF, Nairobi.

IPED (1994) *Effects of desertification and drought on the biodiversity of drylands.* International Panel of Experts on Desertification, Preliminary Executive Summary Report. United Nations, Geneva.

John, S.E.T. and Brouard, J.S. (1995) Genetic improvement of indigenous fruit trees in the SADC region. In: Maghembe, J.A., Ntupanyama, Y. and Chirwa, P.W. (eds) *Improvement of Indigenous Fruit Trees of the Miombo Woodlands of Southern Africa*. Proceedings of a Conference held on 23–27 January 1994 at Club Makokola, Mangochi, Malawi. ICRAF, Nairobi, pp. 12–26.

Jonsson, K., Ong, C.K. and Odongos, J.C.W. (1999) Influence of scattered néré and karité on microclimate, soil fertility and millet yield in Burkina Faso. *Experimental Agriculture* 35, 39–53.

Kater, L.J.M., Kante, S. and Budelman, A. (1992) Karité (*Vitellaria paradoxa*) and néré (*Parkia biglobosa*) associated with crops in south Mali. *Agroforestry Systems* 18, 89–105.

Kessler, J.J. (1992) The influence of karité (*Vitellaria paradoxa*) and néré (*Parkia biglobosa*) trees on sorghum production in Burkina Faso. *Agroforestry Systems* 17, 97–118.

Koppell, C. (1995) *Marketing Information Systems for Non-timber Forest Products. Community Forestry Field Manual, Forest, Trees and People Programme.* Food and Agriculture Organization of the United Nations, Rome.

Leakey, R.R.B. (1999) Potential for novel food products from agroforestry trees: a review. *Food Chemistry* 66, 1–14.

Leakey, R.R.B. (2001) Win:win landuse strategies for Africa: 1. Building on experience with agroforests in Asia and Latin America. *International Forestry Review* 3, 1–10.

Leakey, R.R.B. (2005) Domestication potential of marula (*Sclerocarya birrea* subspecies *caffra*) in South Africa and Namibia: 3. Multipurpose trait selection. *Agroforestry Sysytems* 64, 51–59.

Leakey, R.R.B. and Simons, A.J. (1998) The domestication and commercialisation of indigenous trees in agroforestry for the alleviation of poverty. *Agroforestry Systems* 38, 165–176.

Leakey, R.R.B., Wilson, J. and Deans, J.D. (1999) Domestication of trees for agroforestry in drylands. *Annals of Arid Zone* 38, 195–220.

Leakey, R.R.B., Atangana, A.R., Kengni, E., Waruhiu, A.N., Usuro, C., Anegbeh, P.O. and Tchoundjeu, Z. (2002) Domestication of *Dacryodes edulis* in West and Central Africa: characterisation of genetic variation. *Forests, Trees and Livelihoods* 12, 57–71.

Leakey, R.R.B., Schreckenberg, K. and Tchoundjeu, Z. (2003) The participatory domestication of West African indigenous fruits. *International Forestry Review* 5, 338–347.

Leakey, R.R.B., Pate, K. and Lombard, C. (2005a) Domestication potential of marula (*Sclerocarya birrea* subsp. *caffra*) in South Africa and Namibia: 2. Phenotypic variation in nut and kernel traits. *Agroforestry Systems* 64, 37–49.

Leakey, R.R.B., Shackleton, S. and Du Plessis, P. (2005b) Domestication potential of marula (*Sclerocarya birrea* subsp *caffra*) in South Africa and Namibia: 1. Phenotypic variation in fruit traits. *Agroforestry Systems* 64, 25–35

Lovett, P. and Haq, N. (2000) Evidence for anthropic selection of the sheanut tree (*Vitellaria paradoxa*). *Biomedical and Life Sciences and Earth and Environmental Science* 48, 273–288.

Maghembe, J.A., Simons, A.J., Kwesiga, F. and Rarieya, M. (1998) *Selecting indigenous trees for domestication in Southern Africa: priority setting with farmers in Malawi, Tanzania, Zambia and Zimbabwe.* ICRAF, Nairobi.

Maundu, P.M., Ngugi, G.W. and Kabuye, C.H.S. (1999) *Traditional food plants of Kenya.* National Museums of Kenya, Nairobi.

Mbabu, P. and Wekesa, L. (2004) Status of indigenous fruits in Kenya. In: Chikamai, B., Eyog-Matig, O. and Mbogga, M. (eds) *Review and appraisal on the status of indigenous fruits in eastern Africa.* Kenya Forestry Research Institute, Nairobi, pp. 35–58.

Nerd, A., Aronson, J.A. and Mizrahi, Y. (1990) Introduction and domestication of rare fruits and nuts trees for desert areas. In: Janick, J. and Simon, J.E. (eds.) *Advances in New Crops.* Timber Press, Portland, Oregon, pp. 355–363.

Pareek, O.P. (2001) *Fruits for the Future 2 – Ber* (Ziziphus mauritiana). International Centre for Underutilised Crops, Southampton, UK, 290 pp.

Pedersen, H.B. and Balslev, H. (1990) Ecuadorian palms for agroforestry. Botanical Institute, Aarhus University, Denmark.

Saka, J.D.K. and Msonthi, J.D. (1994) Nutritional value of edible fruits of indigenous wild trees in Malawi. *Forest Ecology and Management* 64, 245–248.

Sanou, H., Kambou, S., Teklehaimanot, Z., Dembélé, D., Yossi, H., Sina, S., Djingdia, L. and Bouvet, J.M. (2004) Vegetative propagation of *Vitellaria paradoxa* by grafting. *Agroforestry Systems* 60, 93–99.

Scherr, S., White, A. and Kaimowitz, D. (2002) *Making markets work for forest communities.* Policy Paper. Forest Trends, Washington and the Center for International Forestry Research (CIFOR), Bogor, Indonesia.

Schreckenberg, K., Awono, A., Degrande, A., Mbosso, C., Ndoye, O. and Tchoundjeu, Z. (2006) Domesticating indigenous fruit trees as a contribution to poverty reduction. *Forests, Trees and Livelihoods* 16, 35–51.

Scoones, I., Melnyk, M. and Pretty, J. (1992) *The hidden harvest: wild foods and agricul-*

tural systems: a literature review and annotated bibliography. International Institute for Environment and Development, London.

Sidibe, M. and Williams, J.T. (2002) *Fruits for the Future – Baobab (*Adansonia digitata*).* International Centre for Underutilised Crops, Southampton, UK

Sidibe, M., Scheuring, J.F., Tembely, D., Sidibe, M.M., Hofman, P. and Frigg, M. (1996) Baobab – homegrown vitamin C for Africa. *Agroforestry Today* 8, 13–15.

Shanley, P. (2002) *Tapping the Green Market: Certification and Management of Non-timber Forest Products.* Earthscan, London.

Shackleton, C.M., Dzerefos, C.M., Shackleton, S.E. and Mathabela, F.R. (2000) The use of and trade in indigenous edible fruits in the Bushbuckridge savanna region, South Africa. *Ecology of Food and Nutrition* 39, 225–245.

Simons, A.J. (1996) ICRAF's strategy for domestication of non-wood tree products. In: Leakey, R.R.B., Temu, A.B., Melnyk, M. and Vantomme, P. (eds) *Domestication and Commercialization of Non-timber Forest Products in Agroforestry Systems. Non-Wood Forest Products No. 9.* FAO, Rome, pp. 8–22.

Statz, J. (2000) *Development potentials for the use of non-timber forest products.* TOEB programme of the Deutsche Gesellschaft für Technische Zusammenarbeit (GTZ). TWF-22e, Germany.

Tchoundjeu, Z., Asaah, E.K., Anegbeh, P., Degrands, A., Mbile, P., Facheux, C., Tsobeng, A., Atangana, A.R., Ngo-Mpeck, M.L. and Simons, A.J. (2006) Putting participatory domestication into practice in West and Central Africa. *Forests, Trees and Livelihoods* 16, 53–69.

Teklehaimanot, Z. (1997) *Germplasm conservation and improvement of Parkia biglobosa (Jacq.) Benth. for multipurpose use.* EU/INCO Project Contract TS3-CT92–0072, Final Report. University of Wales, Bangor, UK.

Teklehaimanot, Z. (2003) *Improved management of agroforestry parkland systems in sub-Saharan Africa.* EU/INCO-DC Project Contract IC18-CT98-0261. Final Report. University of Wales, Bangor, UK.

Teklehaimanot, Z. (2004). Exploiting the potential of indigenous agroforestry trees: *Parkia biglobosa* and *Vitellaria paradoxa* in sub-Saharan Africa. *Agroforestry Systems* 61, 207–220.

Teklehaimanot, Z. (2005) *Indigenous fruit trees of Eastern Africa.* The Leverhulme Trust: a Study Abroad Fellowship report, University of Wales, Bangor, UK.

Thiong'o, M.K., Kingori, S. and Jaenicke, H. (2002) The taste of the wild: variation in the nutritional quality of marula fruits and opportunities for domestication. *Acta Horticulturae* 575, 237–244.

Tomlinson, H., Teklehaimanot, Z., Traore, A. and Olapade, E. (1995) Soil amelioration and root symbiosis of *Parkia biglobosa* (Jacq.) Benth. In West Africa. *Agroforestry Systems* 30, 145–159.

Tripp, R. (2001) *Seed provision and agricultural development.* Overseas Development Institute, London.

UNEP (1994) *World Atlas of Desertification.* Arnold, London.

von Maydell, H.J. (1989) *Trees and Shrubs of the Sahel – Their Characteristics and Uses.* GTZ, Eschborn, Germany.

Wickens, G.E. (1995) *Edible nuts.* Non-wood Forest Products No. 5. FAO, Rome.

Wood, P.J. and Burley, J. (1991) *A tree for all reasons: the introduction and evaluation of multipurpose trees for agroforestry.* ICRAF, Nairobi.

World Bank (1989) Burkina Faso. In: *World Bank Trends in Developing Economies.* World Bank, Washington, DC, pp. 45–50.

Wynberg, R.P., Laird, S.A., Shackleton, S., Mander, M., Shackleton, C., du Plessis, P., den Adel, S., Leakey, R.R.B., Botelle, A., Lombard, C., Sullivan, C., Cunningham, A.B. and O'Regan, D. (2003) Marula policy brief: marula commercialisation for sustainable and equitable livelihoods. *Forests, Trees and Livelihoods* 13, 203–215.

Young, A. (1997) *Agroforestry for Soil Management.* CAB International, Wallingford, UK.

12 Marketing of Indigenous Fruits in Southern Africa

T. RAMADHANI AND E. SCHMIDT

University of Hanover, Hanover, Germany

12.1 Introduction

An increase in population, a relative decline in agricultural productivity and the accelerated food insecurity in the southern African region have raised interest in the role of indigenous fruits in household incomes. Therefore, national governments as well as non-governmental and international organizations have been conducting research on ways to increase the production of fruits in the region. Production and marketing studies were initiated in order to better understand the prevailing production and marketing activities (Falconer, 1990; Agyemang, 1994; Cavendish, 1997; Kadzere *et al.*, 1998). These studies indicated that, traditionally, indigenous fruits were collected freely and consumed by the rural population. As time went by, local peasants – mainly women and children who lived on marginal lands and the landless population – started businesses to subsidize their household incomes (Falconer, 1990; Agyemang, 1994; Cavendish, 1997; Kadzere *et al.*, 1998). Varieties of indigenous fruits such as *Uapaca kirkiana*, *Strychnos cocculoides*, *Lannea edulis*, *Ziziphus mauritiana*, *Azanza garckeana*, *Adansonia digitata*, *Vitex mombassae* and *Berchemia discolor* were increasingly sold in villages, growth points and districts markets as well as city markets.

However, despite an increase in the number of people trading in indigenous fruits, the marketing system has remained informal, with many economic inefficiencies. For example, traditional taboos at many sites limit the production and collection of the fruits for the market, and hence hinder efforts towards commercialization. These traditional taboos were derived from a belief that fruits are a gift of God and are therefore meant to benefit the whole community. The taboos also protect the trees from unsustainable harvesting techniques. Thus, taboos, the production/collection of indigenous fruits and marketing are linked in a complex relationship.

Unsuitable means of transport are also an important obstacle faced by producers, collectors and retailers. Using buses designed to carry passengers rather than fruits increases the risk of damaging the fruits and incurs high losses. Such conditions require specialized traders who can deal efficiently with the transport of fruits. Undoubtedly, the need to use intermediate dealers significantly increases marketing costs and lowers price efficiency.

During the 1999/2000 indigenous fruit ripening season, a detailed marketing research study was conducted in Zimbabwe to generate baseline information on the prevailing marketing system for *Uapaca kirkiana* and *Strychnos cocculoides*, including consumer preferences and willingness to pay for the fruits. The study also explored the possibility of expanding marketing opportunities for these fruits. This chapter discusses the interesting findings of this study. It describes the marketing chain, characterization of the price formation system, price inefficiencies, lack of marketing transparency and the resulting imperfections in the system.

12.2 Marketing of Indigenous Fruits during the 1999/2000 Ripening Season in Zimbabwe

12.2.1 Production and the marketing chain

During the 1999/2000 ripening season for indigenous fruits there was almost no on-farm production of indigenous fruits. Producers/collectors, retailers and vendors collected the fruits for selling from communal forests and very few collected from naturally grown trees in their own fields. However, collection was limited by the lack of transparency from the government about property rights regarding the ownership and use of indigenous fruits (see Section 12.3). Furthermore, the marketing system was not characterized by a clear division of marketing activities between the actors involved. Sometimes fruit producers/collectors (those collecting the fruits for sale), as well as fruit retailers and vendors (who buy fruits from wholesalers and sell to consumers), were involved in gathering the fruits from communal forests, taking them to the markets and selling them.

However, the findings reported in this study are based on producers/ collectors who collected the fruits from communal forests and sold them to wholesalers from their homes or transported them to the urban markets. In most cases wholesalers bought the fruits from producers/collectors' homes and transported them to urban markets and sold them to retailers, vendors, and sometimes to consumers. Retailers were defined as formal traders with permanent cubicles at the central or growth point/district markets, for which they paid a monthly fee. Vendors were informal traders who sold their fruits along the highways, at roadsides, along the streets and on the market peripheries, where they did not pay taxes.

12.2.2 Product differentiation

Among the characteristics of *U. kirkiana* and *S. cocculoides* fruit marketing was lack of product differentiation at the production/collection and wholesale levels.

Unlike the marketing system for exotic fruits, the local fruits were not graded, packed or even washed. Wholesalers and producers explained that this was a result of their belief that the fruits were picked clean from the trees and hence did not need any additional preparation before sale.

In contrast, retailers sold fruits based on 'natural differentiation'. This involved different sizes (small, medium, large), different colours (brown, yellow) and different levels of freshness. Also, there was a tendency towards simple active product differentiation by means of washing, sorting and packing. Almost 90% of *U. kirkiana* fruit retailers sorted out the rotten fruits and washed the fruits. In some cases they sorted the fruits according to size and filled them into selling utensils such as plates, bowls and cups. They also packed the fruits in plastic bags weighing 300–500 g. The remaining 10% of retailers, however, did not undertake any presale activities, contending that the practice of washing and packing in bags increased perishability.

Consumers, however, increasingly preferred to buy clean, graded and packed fruits. For example they revealed preferences for even simple presale activities of sellers. Astonishingly, these changing consumer attitudes seemed not to be directly included in the presale activities of profit planning and budgeting. According to the results of a retailers' rapid appraisal, presale activities were not targeted towards price increments or profit increase, but mainly towards improving hygiene and attracting customers. In terms of marketing theory, this is at least a starting point to establish some sort of preference strategy. In order to develop measures to translate consumer preferences of indigenous fruits into price and profit increases; there is a need for further research. This seems to be particularly important because the price formation process for indigenous fruits differs significantly from the general price formation system prevailing in the market for food and other consumer goods in Zimbabwe.

12.2.3 Prices, price information and variation

During the period of this study, there was no established formal price information system for indigenous fruits. Consequently, day trading determined the market exchange of the fruits. The prevalence of short-term and more or less random contacts between sellers and buyers accounted for the use of cash as the common method of payment for both *U. kirkiana* and *S. cocculoides* fruits. In rare cases, *U. kirkiana* fruits were sold on credit, without interest or higher prices. Also, there was no evidence of contractual arrangements between buyers and sellers. According to the rapid market appraisal conducted in the 1999/2000 ripening season, the major source of price information for producers/collectors was information from neighbours, friends and colleagues who had visited the markets during the selling season. Lack of information caused pronounced price differences between regions and even within a given market place. In Gokwe, for example, both retailers and producers were unaware of the prices of similar fruits sold in other areas of the country. This might have resulted from the remoteness of Gokwe and poor communication.

On the other hand, the proximity of the markets to other towns was an advantage for Murehwa fruit traders because they were aware of prices at nearby markets. The unequal distribution of information between market participants seemed to be a widespread problem, causing severe market exchange inefficiencies.

Prices of *U. kirkiana* and *S. cocculoides* fruits varied according to the regional locations of the markets. In urban areas, for example, fruit prices were higher than in semi-urban areas/growth points and rural markets. In the case of *U. kirkiana* fruits, producer and retailer prices were lower at the Murehwa growth point markets than at the Mbare urban market. Higher prices at Mbare and Gokwe town markets created an incentive for producers and retailers in Murehwa and Gokwe rural areas to transport their fruits and sell them at these markets. Price differences at a given point in time were at least partly caused by differences in the marketing costs incurred by the different traders at the markets.

The lack of marketing transparency created leeway for price manipulation by the better-informed actors (mostly the sellers) at all stages of marketing. The absence of standard quantity/volume units of measurement has caused additional problems. First, the unit of quantity/volume traded changed with the market level. At the production level the common quantity unit was a 20 l bucket. In contrast, wholesalers measured *U. kirkiana* fruits in units of 5, 10 and 20 l tins, equivalent to 3, 6 and 11 kg, respectively. Finally, retailers packed the fruits in plastic bags, plates, bowls or cups weighing on average 150–500 g, and sometimes they sold the fruits in small tins measuring 1.5, 2.5 or 5 l. In Gokwe, retailers sold individual fruits of *U. kirkiana*. Since *S. cocculoides* fruits are large, they were sold individually at all market levels.

Second, the quantity sold at the wholesale level in a 20 l tin varied according to the supply situation. Wholesalers pressed the tin bases inwards or filled the tin with grass material/papers in order to hold fewer fruits. In any case, prices were affected by the use of different units of quantity at different sites and different market stages, as well as changes in quantities of fruits available as the supply situation changed. Hence, the commonly used procedure of price comparison was impeded and the evaluation of relative prices was very difficult. Price comparison was impeded not only for fruits between different sites, but between different market stages and different suppliers at the same market stage. Moreover, comparison of prices between different indigenous fruits and between indigenous fruits and exotic fruits (the main substitutes for indigenous fruits) traded in the same markets, but sold in standardized quantities, made business difficult.

12.2.4 Marketing costs

Marketing costs were incurred in the process of transferring the fruits from the producer to the consumer. The costs differed depending on the market location (urban or rural) and level in the marketing chain (production/collection, wholesaling or retailing).

Transport costs

Collectors usually walked from the forest to their homes. Those who sold on-farm waited for wholesalers to come to their houses, but collectors who sold at district/growth point, roadside and Mbare markets transported the fruits using Scotch carts, bicycles, buses and sometimes hired pick-up trucks. The type and number of vehicles used by the traders depended on the distance to the market and the quantity of fruits handled. The cost of transport also differed depending on the means of transport and the market location. For example, Gokwe traders who used buses to transport fruits to the Gokwe growth point market paid 50% more than their counterparts in Murehwa. This might have been a consequence of the remoteness and poor road infrastructure of Gokwe, which was served by few buses compared with Murehwa. In addition to transport costs, traders who transported the fruits to distant markets had to organize transport and to bargain for transport fares, which are sometimes unreliable and expensive. Furthermore, they had to ensure proper handling of the fruits during loading and unloading from the vehicles.

Fruit losses

Because of the unreliability of transport, perishable fruits and poor handling techniques, there was significant loss of fruit. For the 1999/2000 ripening season, *U. kirkiana* fruit losses accounted for 32% of all marketing costs incurred by producers located in Mbare market compared with 13% in Gokwe. The distance from the production sites and the large volume of fruits transported caused high losses of fruits at Mbare market. Gokwe production sites are near Gokwe market, and relatively small volumes of fruit are transported there compared with Mbare. Furthermore, buses are not designed to transport fresh produce and bus owners cannot pack the fruit bags separately from the hard luggage. This lack of flexibility of the bus workers led to high fruit losses experienced by retailers and vendors transporting fruits to Mbare market.

For fruits that did reach the market, lack of storage facilities and processing opportunities shortened the fruits' shelf life. As a result, traders were not able to recover their costs. Lack of storage facilities caused massive losses due to perishability. Almost 50% of producers, 63% of retailers and 75% of wholesalers reported that perishability was a problem. To solve this problem, Harare district council designated Tsiga open market, within the Mbare suburbs, as a seasonal market for *U. kirkiana* fruits. Also, traders incurred considerable losses in the processes of loading and packing the fruit bags on the lorries, buses and pick-ups, handling during travelling and unloading from the vehicles.

Other marketing costs and problems

Traders also paid market fees (except vendors) of varying amounts. In the 1999/2000 ripening season, producers paid from Z$3.30 to Z$5.00 per 33 kg bag (three 20 l buckets), retailers paid between Z$0.17 and Z$4.17 per 33 kg bag and wholesalers Z$2.29 per 33 kg bag. Other costs included bus fares (for the traders who transported fruits by bus), fruit acquisition/buying, time spent in presale activities by the retailers, and traders' marketing time. Also, retailers bought plastic bags for packing.

Intensive consumer tasting of *U. kirkiana* fruits also revealed serious problems at all trading levels. Tasting was the most serious problem reported by the wholesale section (75% of respondents), followed by the retail (63% of respondents) and production (50% of respondents) sections. This is because wholesalers spread their fruits on the ground and hence made them more likely to be tasted. In the retail section, fruits not packed in plastic bags were more susceptible to tasting than packed ones. In the producer section, fruits piled in buckets were tasted but not those in bags. Consumers often tasted the fruits to make sure that they were buying sweet ones. The problem was that some consumers tasted too much, to the extent of causing financial losses to the traders.

A problem similar in scale to fruit tasting was low sales in peak periods. During the ripening period the traders competed for fruits and hence harvested large quantities at the same time, a situation that caused supply to exceed demand. Low willingness to pay might also have occurred because some consumers acquired fruits from relatives from rural areas as gifts, or harvested them straight from the forest during weekend visits. Other problems reported by the traders included lack of marketing infrastructure, such as marketing sheds, storage facilities, and traditional laws limiting the collection of the fruits for sale.

12.2.5 Marketing margins

Despite the costs described above, fruit traders generated cash and subsidized their household incomes. In the 1999/2000 ripening season, producers of *U. kirkiana* made between Z$1.73/kg in Murehwa, Z$1.78/kg in Gokwe and Z$3.34/kg in Mbare (Table 12.1). Since Mbare market charged high prices, producers who sold *U. kirkiana* fruits at Mbare received 45% of the price paid by the consumer compared with 32% for their counterparts in Murehwa and Gokwe. At the production and retail levels, the large number of participants facilitated adequate competition, unlike the situation at the wholesale level. This led to a high market concentration in the wholesaling section, a sign of market power. Surprisingly, however, wholesalers received relatively low margins compared with other participants in the chain, possibly as a result of the higher costs of transport, their inability to negotiate a lower buying price and a high selling price. Therefore, to recover some of their costs, most wholesalers sold fruits in small quantities directly to consumers, in 1.5, 2.5 and 5 l tins.

12.3 Use of Indigenous Fruits, Management Policies and Property Rights

Despite the financial and social benefits enjoyed by indigenous fruits traders, rigid by-laws limiting the marketing of fruits affect the market supply. Issues of natural resource management policies and property rights are important

Table 12.1. Marketing margins (Z$/kg) for the sale of *Uapaca kirkiana* fruits.

	Murehwa market		Gokwe market		Mbare market		
Variable	Producer	Retailer	Producer	Retailer	Producer	Wholesaler	Retailer
Price or revenue/kg	3.17	9.90	3.23	10.24	5.07	5.24	11.17
Total cost/kg	1.44	3.44	1.45	3.52	1.73	4.77	7.14
Profit	1.73	6.46	1.78	6.72	3.34	0.47	4.03
Absolute marketing margin	6.73		7.01		6.10	0.17	5.93
Relative marketing margin (%)	32.02		31.54		45.39	1.52	53.09

US$1 = Z$50 in 1999/2000.
Producer absolute marketing margin = retailer price − producer price.
Retailer absolute marketing margin = retailer price − wholesaler price.
Wholesaler absolute marketing margin = wholesaler price − producer price.
Wholesaler relative marketing margin = wholesaler price − producer price/retailer price × 100.
Retailer relative marketing margin = retailer price − wholesaler price/retailer price × 100.
Source: Ramadhani (2002).

because the major source of *U. kirkiana* and *S. cocculoides* fruits for the markets is the communal forests. Access to indigenous fruits and trees located in the communal forests near villages, fields and homesteads is a fundamental aspect and needs a closer look. This is because lack of access might negatively influence support for the development of the fruit markets.

Because guidelines on property and user rights regarding indigenous fruits are not clear, traditional chiefs have imposed informal rules which allow only home consumption of indigenous fruits. Selling indigenous fruits is strictly prohibited. The rules are backed up by the traditional attitude that the fruits are a gift from God and are therefore meant for consumption by the whole community. The lack of rules that clearly specify the ownership and user rights for the trees and fruits imposes difficulties in managing and controlling the use of the fruits. Since the fruits are a common resource they are being overused. Traditional village leaders help to manage and conserve the fruit trees by practising the old taboos and beliefs, but this leads to much uncertainty for traders. These findings suggest that efforts to commercialize indigenous fruits in Zimbabwe need to take account of the ownership and use regulations relating to indigenous fruit trees and fruits.

12.4 Consumers and Buyers of Indigenous Fruits in Zimbabwe

For the purpose of this study, consumers were individuals who collected fruits for consumption from their farms or communal forests, but buyers were those who buy the fruits from the markets. The former were individuals living in the rural areas, while the latter mostly lived in urban areas. In order to be able to formulate an appropriate marketing strategy for indigenous fruits, there was a need to know the socio-demographic characteristics influencing consumption and buying behaviour. For both *U. kirkiana* and *S. cocculoides*, purchasing behaviour was associated with marriage status, education, and the market and production sites for the respective fruit ($P < 0.05$). In addition, purchase behaviour for *S. cocculoides* fruits was related to gender ($P < 0.05$).

Tables 12.2 and 12.3 show that educated people were more likely to buy *U. kirkiana* and *S. cocculoides* fruits than their counterparts. Possibly most of the people interviewed were educated. Married people were more likely than unmarried people to buy *U. kirkiana* fruits rather than *S. cocculoides*. These findings suggest that there might be a family influence in the purchasing behaviour for indigenous fruits. Further research on this aspect is needed.

While there was more purchasing activity for *S. cocculoides* fruits in Gokwe, there was less for *U. kirkiana* fruits in Murehwa and more at the Harare site. Low purchasing activity at Murehwa might be due to easy access to fruits from the nearby communal forests, farmers' fields and along the road. The inhabitants of Harare include quite a number of individuals who come from the miombo ecosystem, who are used to consuming these fruits. In addition, Harare hosts Mbare market, the central market for all types of indigenous fruits, and hence there is a high probability of finding buyers for *U. kirkiana* fruits in this market.

Gokwe town also had more purchase activities for *S. cocculoides* fruits (Table 12.3). This might be because there are few *S. cocculoides* fruit trees in

Table 12.2. Socio-demographic characteristics of buyers of *Uapaca kirkiana* indigenous fruits ($n = 255$).

		95% confidence interval	
Variable	Adjusted odds ratio*	Lower limit	Upper limit
Married	1.646	1.117	2.426
Educated	1.153	1.079	1.231
Harare	2.609	1.621	4.200
Murehwa	0.382	0.202	0.725

*Adjusted for variables in the model.
Source: Ramadhani (2002).

Table 12.3. Characteristics of buyers of *Strychnos cocculoides* fruits ($n = 91$).

		95% confidence interval	
Variable	Adjusted odds ratio*	Lower limit	Upper limit
Married	1.726	1.064	2.800
Male	1.712	1.016	2.885
Educated	1.095	1.014	1.183
Gokwe	21.484	4.870	94.770
Harare	12.734	3.019	53.717

*Adjusted for variables in the model.
Source: Ramadhani (2002, p. 121).

Gokwe town, so the inhabitants had the option of buying the fruits from the markets rather than travelling to the rural areas or waiting for fruits to be brought to them by friends and relatives. Likewise, males were almost twice as likely to buy *S. cocculoides* fruits as females. It may be that males did not spend much time collecting *S. cocculoides* fruits and their major source fruits was the market.

12.4.1 Consumer preferences and willingness to buy

Buying behaviour for the two fruits is influenced by size, colour, freshness and quantity. The importance of these characteristics to buyers of *U. kirkiana* fruits differed (Table 12.4). The most important attribute was size, which scored an average weighting of 36%. Among the different size options presented to buyers, small size had a higher positive effect on the preference structure than large size. These findings were contradictory to the findings of informal discussions on buyer attitudes and willingness to buy *U. kirkiana* fruits. In informal discussions, most buyers reported that they would buy big *U. kirkiana* fruits with a low price. However, it is possible that most people were more concerned with the notion of low price than size. Also, buyers were instructed to disregard taste, which might be incorporated in the decision-making process. However, the study results supported the view expressed in the informal discussions that small fruits are tastier than big ones; hence, according to consumers, tasty fruits are equated with good-quality fruits. Further research on this aspect is needed.

Table 12.4. Important *Uapaca kirkiana* fruit attributes identified by buyers in Zimbabwe, 1999/2000 ($n = 250$).

Attribute	Level	Utility	Average importance (%)
Colour			16
	Brown	0.630	
	Yellow	−0.630	
Quantity			30
	300 g	−0.735	
	400 g	0.072	
	500 g	0.657	
Size			36
	Small	0.844	
	Mixed	0.105	
	Big	−0.949	
Appearance			18
	Round fresh	0.317	
	Rough non-fresh	−0.317	
Kendall's tau		0.778	
Significance		0.0018	

Source: Ramadhani (2002).

The second important attribute was quantity, which had an importance value of 30%. Among the quantities presented to customers, 500 g packets were preferred and the tendency was to dislike 300 g packets. Although price was included in the research design and was expected to be perceived as fixed by the buyers, the fact that the quantities varied implied that prices were also changing, such that the larger the quantity the cheaper. Therefore, the result suggested that consumers preferred lower prices.

The appearance of the fruits was the third important attribute for buyers of *U. kirkiana*, with an average importance of 18%. Consumers preferred to buy round, fresh fruits (part worth value = 0.317) rather than rough, non-fresh fruits. The preference for fresh fruits might be derived from the experience of consumers with exotic fruits sold in the same markets. Lastly, buyers preferred brown to yellow fruits. These findings suggested that the ideal product identified by the buyers as a whole was a packet weighing 500 g containing small, brown, round, fresh *U. kirkiana* fruits. Thus, in order to increase sales, traders must provide fruits with the characteristics identified by the consumer.

The needs and desires of the buyers as a group differed from those of individual buyers. Not all consumers liked the combination of attributes identified by the group. This is because buyers differed in their background and hence their needs. In order to satisfy the needs of consumers from different backgrounds, perceptions and needs, the market was segmented to simplify the development of marketing strategies that focus on fewer groups. The segmentation process considered the utility values of the four traits identified by the aggregate consumers, and the market was divided into six segments or clusters. Findings showed that buyers in clusters 1 and 3 (for descriptions of the clusters see footnote to Table 12.5) complied with the identified ideal product (Table 12.5). Buyers in cluster 3 showed a greater preference for brown fruits than buyers in other clusters. Buyers in clusters 1 and 2 also liked brown fruits.

Table 12.5. Traits of *Uapaca kirkiana* fruits preferred by the six clusters* (*n* = 250).

	Cluster					
Preferred trait	1	2	3	4	5	6
Brown	0.34	0.63	1.24	−0.90	−0.31	−0.36
500 g packet	0.24	−1.10	0.91	−0.63	1.12	1.86
Small size	2.04	−0.97	0.60	0.92	0.06	1.31
Round, fresh	0.34	0.64	0.57	0.47	−0.57	0.89
Number of respondents	55	21	37	29	57	51

*Clusters 1 and 3 comprised individuals with four to six people in the family; about 40% of family members were younger than 6 years; medium income; 1–13 years of formal schooling; most were met in Harare.
Cluster 2 comprised individuals with family size less than or equal to 3; low income; 66% were met in Murehwa.
Cluster 5 comprised individuals with four to six people in the family; most individuals were met in Harare City Botanical Garden.
Source: Ramadhani (2002).

These findings agree with the economic theory that buyers are highly varied in their needs and hence mass marketing might not be a desirable option.

As regards quantity, clusters 1, 3, 5 and 6 preferred the largest quantity of 500 g but the remaining clusters did not. All clusters except cluster 2 preferred small fruits. While all buyers liked fresh fruits, those in cluster 5 did not appear to favour them. This indicates that a short-term plan to increase market sales could be to target clusters 1 and 3 by sorting and grading the fruits to supply the preferred combination of characteristics, which would capture the 37% of individuals in the market who preferred this type of product. Clusters 2 and 4 might not be worth targeting in the near future because they contained fewer individuals than the other groups.

Most of the people in clusters 1 and 3 were found in Harare, especially Mbare, followed by Westgate and the City Botanical Gardens. Harare had the potential for increased marketing of *U. kirkiana* fruits. In addition, most buyers bought at least 20 *U. kirkiana* fruits per week, belonged to the working class, were aged between 18 and 65, had attended between 1 and 13 years of formal schooling, and were mainly males. These are the groups who preferred all the identified fruit attributes. Cluster 2 was the opposite of clusters 1 and 3, in that 57% of them had a family size of three or fewer persons and 66% were found in Murehwa and Westgate.

12.5 Conclusions

Although there is a substantial amount of trading of indigenous fruits in both rural and urban areas of Zimbabwe, the system is still underdeveloped. Property rights guiding the use and management of the fruit trees and fruits are not transparent. Informal by-laws imposed by village and traditional leaders limit the market supply. But consumers like the fruits and are willing to pay for them. The ideal fruit for a buyer of *U. kirkiana* fruits is brown, small in size, and fresh. The information given in this chapter supports the idea that it would be beneficial to extend the production season to provide a supply of fruits outside the peak period. Planting more indigenous fruits on farms and promoting fruit storage will improve the supply during off-peak periods.

The next step is to intensify efforts towards improving the market supply of fruits and increasing demand by encouraging and facilitating a government initiative to set up a reliable institutional framework that will reduce risk and uncertainty for traders in indigenous fruits. Rules that favour private action may be introduced. Producers may either buy or be issued with free collection licences. Once the producers are free to collect and sell the fruits without any fear of punishment, they might introduce the trees on their farms. Improving marketing transparency by standardizing the units of measurement will allow prices to change at constant quantities.

Since the level of demand seems promising, traders should supply high-quality fruits to urban markets. Also, research to facilitate the introduction of the fruits to formal markets should be conducted. Traders may reorganize themselves to undertake the intensive sorting, grading and packing of the fruits.

Improving the hygienic situation by practices such as washing, grading and packing will make the fruits more attractive to consumers and hence increase sales and benefit the rural poor.

The natural resource management policies and socio-cultural settings of the southern African countries involved in the domestication project are unlike those of Zimbabwe. There is a need to compile information on the existing marketing systems and the factors that influence the system in Malawi, Tanzania and Zambia, as these countries have different natural resource management policies and socio-cultural characteristics. The pooled findings of the study described in this chapter will help in drawing appropriate conclusions to facilitate the commercialization of regional indigenous fruits.

The shelf life of *U. kirkiana* fruits could be improved by further research on ways of introducing small-scale enterprises for traders. These enterprises could cooperate with each other, apply for loans and buy storage facilities. With these storage facilities it should be possible to keep the fruits for a longer time and hence foster a steady market supply and even out the short-term inequalities between supply and demand. This might facilitate the flow of fruits to different and distant markets with minimum losses, help to increase the producers' bargaining power and improve returns to traders. Lastly, follow-up research to monitor changes in patterns of consumption and marketing activities is necessary. It would be worthwhile to document the advantages of increased production of fruits, such as increased consumption and better nutrition, and the disadvantages, such as reduced labour availability for the production of food crops, and to take appropriate measures to remedy the disadvantages identified.

Acknowledgements

The study described in this chapter was made possible by financial support from the World Agroforestry Centre (ICRAF)/Bundesministerium Fur Wirtschaftliche Zusammenarbeit (BMZ) and the Institute of Horticultural Economics, University of Hanover. The author would like to thank Professor Dr Erich Schimdt for his time and professional advice, and Drs Elias Ayuk and Freddie Kwesiga for field guidance. The author also expresses her appreciation of the help received from all the traders and consumers who provided information, and of the logistic support given by staff members of ICRAF-Zimbabwe and the Institute of Horticultural Economics, University of Hanover.

References

Agyemang, M. (1994) The Leaf Gatherers of Kwapanin, Ghana. Cited by Neumann, R. and Hirsh, E. (2000) *Commercialization of Non-Timber Forest Products: Review and Analysis of Research*. CIFOR, Indonesia, 176 pp.

Cavendish, W. (1997) The economics of natural resource utilization by communal area farmers of Zimbabwe. PhD thesis, University of Oxford, Oxford, UK.

Falconer, J. (1990) Agroforestry and household food security. In: Prinsley, R.T. (ed.) *Agroforestry for Sustainable Production: Economic Implications*. Commonwealth Science Council, London, 417 pp.

Funkhouser, S. and Lynam, T. (1999) *Masau in the Eastern Zambezi valley: final consultant report*. Harare.

Kadzere, I., Chillanga, T., Ramadhani, T., Lungu, S., Malembo, L., Rukuni, D., Simwanza, P., Rarieya, M. and Maghembe, J. (1998) *Setting priorities of indigenous fruits for domestication by farmers in Southern Africa*. ICRAF, Nairobi, 115 pp.

Ramadhani, T. (2002) *Marketing of indigenous fruits in Zimbabwe*. Socioeconomic Studies on Rural Development No. 129. Wissenschaftsverlag Vauk Kiel KG, Germany.

13 Economics of On-farm Production of Indigenous Fruits

D. Mithöfer[1] and H. Waibel[2]

[1]*International Centre of Insect Physiology and Ecology, Nairobi, Kenya;*
[2]*Leibniz University of Hanover, Hanover, Germany*

13.1 Introduction

The majority of the African population lives in rural areas, where poverty is a major factor hampering development (World Bank, 2001). Rural people frequently suffer from food shortages caused by adverse weather conditions, political instability, poor market infrastructure and other constraints. To mitigate the effects of food shortages, people in rural areas can use a wide range of products from their natural environment. For example, a variety of edible wild fruits are a popular natural resource (Maghembe *et al.*, 1998), being used for home consumption and for sale (Cavendish, 2000; Mithöfer and Waibel, 2003). Indigenous fruits are mostly collected from communal areas, from roadsides and from trees preserved in farmers' fields (Campbell, 1996). As the entry barriers to fruit collection are low, poor households are more likely to get involved in the collection of indigenous fruits (Dercon, 2000). Due to increasing population pressure and other factors, such as agricultural policies (Chipika and Kowero, 2000), the area of forest in Zimbabwe shrank by 14% from 1990 to 2000 and by 8% from 2000 to 2005 (FAO, 2005). Physio-geographical factors may also contribute towards deforestation, as shown by Deininger and Minten (2002) for Mexico. Other factors that may have contributed are increased prices for alternative sources of energy used for cooking, e.g. instead of using electricity, paraffin or gas, more households may rely on firewood or charcoal. According to traditional rules, indigenous fruit trees (IFTs) have to be preserved when clearing woodland in favour of agricultural production; nowadays, IFTs are sometimes also being felled (Rukuni *et al.*, 1998).

The declining availability of indigenous fruits in the communal areas as a result of increasing commercialization (see Ramadhani and Schmidt, Chapter 12, this volume) and non-sustainable harvesting techniques pose a threat to this indigenous natural resource. The need to ease pressure on natural resources and to maintain the availability of IFTs gave rise to the World Agroforestry Centre's

(ICRAF) domestication programme (see Akinnifesi *et al.*, 2006). Initially, Maghembe *et al.* (1998) conducted a priority-setting exercise in which farmers identified the most popular IFT species, which include *Uapaca kirkiana*, *Strychnos cocculoides* and *Parinari curatellifolia* as the top three species in Zimbabwe. Criteria for ranking those species were: (i) their role in food security; (ii) potential for commercialization; (iii) suitability for processing and conservation; and (iv) taste and abundance. The priority-setting study also included questions regarding characteristics that should be improved. According to the farmers' viewpoint, traits that need to be improved include fruit quality, fruiting precocity (early maturity, i.e. first fruit production), fruit size, and morphological characteristics such as thorniness, fruit yield, tree size and resistance to pests (Kadzere, 1998).

Research on IFT domestication has been concentrated on selecting seeds from different locations, in order to find superior genotypes, and on developing successful vegetative propagation methods. Ultimately, the domestication programme aims at encouraging the on-farm planting of IFTs with improved fruit quality and higher yields, thus enhancing farmers' income as well as contributing to the conservation of biodiversity (ICRAF, 1996). From the point of view of rural development, the ICRAF programme can be seen as an approach that promotes a strategy of crop diversification based on local resources that is contrary to those favouring green revolution-type technologies, including modern biotechnology (Leakey *et al.*, 2004). The untapped potential of wild plants is seen as a means to benefit the economies of tropical countries, to motivate improved conservation of the wild areas that supply these crops (Evans and Sengdala, 2002), and to enhance the productivity and sustainability of agroforestry systems (Simons, 1996).

Despite a widespread perception that indigenous fruits are an important source of rural incomes, their actual contribution to incomes, and the reduction of rural poverty and vulnerability, has only recently been quantified. Previous quantitative analysis showed that more information on the economics of on-farm planting of IFTs as an alternative to the collection of fruits from communal areas is required (Mithöfer, 2005). Thus, in this chapter the economics of on-farm planting and management of IFTs is assessed.

The chapter is arranged as follows. First, factors that influence the management of IFTs are described and results of other studies are reviewed. Second, the methodology for assessing the profitability of planting of IFTs is described. Third, study area and data are presented and survey results on the status of IFT management and planting, as well as farmers' reasons for planting of indigenous and exotic fruit trees (EFTs), are described. Fourth, costs and benefits of planting *U. kirkiana* trees are calculated, as well as points that show under which data constellation on-farm planting can be expected. Finally, the chapter concludes with an outline for further research and related development activities.

13.2 Factors Influencing Indigenous Fruit Tree Management

Farm households use indigenous fruits and other natural resources in order to pursue a strategy of livelihood security. Such a strategy has multiple objectives, including food self-sufficiency, social security, risk management and income. As resource availability, objectives and socio-economic household characteristics change over time, costs and benefits that accrue to farmers will change, and the household can be expected to adjust its livelihood strategy accordingly. As pointed out by Arnold and Dewees (1999), improved management of wild trees and tree planting is a function of a change in the frame conditions, such as:

- Decline of production from off-farm tree stocks due to deforestation, or more restricted access to indigenous trees.
- Growing demand for tree products due to population growth, advances in processing, and market development.
- Declining soil productivity due to increasing damage from its exposure to sun, wind or water runoff.
- Increasing need to secure rights of land tenure and use.
- To even out peaks and troughs in the seasonal flow of produce, income and seasonal labour demand, or to provide a reserve of biomass products and capital available for use as a buffer in times of stress or emergency.

A decline in the availability of indigenous fruits may influence the decision to plant trees. However, a decline in the abundance of trees that constitute a common property off-farm resource does not necessarily lead to on-farm planting. Other factors, e.g. reduction of market constraints or agricultural policies, besides resource scarcity, also play a role.

In Zimbabwe, households use IFTs primarily in their natural habitat (Campbell, 1996). This means that the practice of planting these species is uncommon; trees mostly regenerate by themselves (Minae *et al.*, 1994). Brigham (1994), Campbell *et al.* (1993, cited in Campbell, 1996, p. 133) and Price and Campbell (1998) found within rural communities that only between 1% and 10% of households had planted IFTs. More commonly, farmers nurture young saplings found on their land (Price and Campbell, 1998).

Some studies show that when land areas are cleared in favour of agricultural activities, IFTs are conserved in the crop fields (Clarke *et al.*, 1996). Indigenous fruit trees account for the majority of trees that remain standing in agricultural areas. It has been observed that tree planting and tree conservation activities around the homestead are responsible for replacing indigenous non-fruit trees with exotic and indigenous fruit trees (Price and Campbell, 1998). Trees that remain in the fields are managed by pruning, lopping and pollarding, thus providing increased compatibility with crops and providing firewood and building materials to the household (Minae *et al.*, 1994). However, the study of Rukuni *et al.* (1998) shows that the continuing deforestation process also affects IFTs. This process is limited to a varying extent by the traditional leaders, who put in place rules on IFT cutting and also enforce those rules.

Two driving forces push towards more and more active management of trees. The first is growing scarcity (i.e. on-farm resources have grown more valuable) and the second is commercialization and the potential income that can be derived from planting (McGregor, 1991). This has also clearly been demonstrated for the case of rattan cultivation in Laos (Evans and Sengdala, 2002). As the case of wild coffee in Ethiopia shows, niche markets can be an incentive to collect coffee from protected areas (Abebaw and Virchow, 2003).

Deforestation status has no influence on the planting of EFTs (McGregor, 1991). On the other hand, planting of fruit trees depends on the number of trees preserved in farmers' fields, which depends on the initial composition of woodlands before the woodlands were cleared (Wilson, 1990).

According to Musvoto and Campbell (1995), the wealth status of a household has no influence on the planting of trees. Households headed by men are likely to grow a higher number of EFTs on their farms than those headed by divorcees or widows. Similarly, households with a longer residence period have more EFTs on their farms (Price and Campbell, 1998). Among exotic fruits, mangoes are the major one. They only contribute a small share of cash income relative to crops such as maize, but they are valuable for household consumption (Musvoto and Campbell, 1995).

Exotic fruit trees are planted because they have properties that households prefer, e.g. their commercialization potential (Brigham, 1994, cited in Campbell, 1996, p. 133). Many people believe that the IFT species are inferior to exotic ones. Enhanced management of IFTs is hampered by substitution among species, which also includes planting of EFTs (Campbell, 1996, p. 133). Exotic fruit trees are increasingly being planted and utilized for construction purposes, while IFTs increasingly serve as a source of fuelwood (McGregor, 1991). In South Africa, Mander *et al.* (1996) identified lack of information as the main obstacle to commercial cultivation of traditionally used plants.

13.3 Methodology for Assessing Profitability of Tree Planting

Planting of IFTs from the small-scale farmer's point of view is an investment. Current benefits, e.g. the use labour for other purposes than tree planting, are forgone in favour of future benefits. Benefits of planting include direct benefits from the tree products (fruits, branches, leaves) and services, e.g. shade, provided by the tree including, for example, securing tenure of land, as shown for Mozambique (Unruh, 2001). In Zimbabwe, the latter type of benefits were not observed. Instead, farmers supported the idea that traders, and also farmers who want to trade in the fruits, should plant their own trees. Planted trees were considered private property by the person who planted them (Ramadhani, 2002).

Traditionally, economists have treated investments as a 'now or never' decision; that is, if the net present value is positive, investment is recommended.[1] More recently, methods have been developed that allow taking the timing of investment into account. In this approach, the investment decision constitutes a right, but not the obligation to realize the investment, which is similar to a financial call option that has not yet been exercised (Lund, 1991; Dixit and Pindyck, 1994; Trigeorgis, 1998). New investment theory, which deals with investment decisions under irreversibility, flexibility and uncertainty, is also known as the 'real options approach'. It thus becomes possible to analyse whether waiting to invest has any value. In the case of IFT planting, it then can be investigated whether farmers are better off if they wait to plant, although the net present value of the investment may be positive. This is relevant because the decision to invest in planting domesticated IFTs is characterized by irreversibility, uncertainty and flexibility. Irreversibility implies that once trees are planted, initial investment costs are sunk; uncertainty takes into account that future benefits may vary; and flexibility means that farmers can choose the time when to plant trees. Other real option values include the option to abandon, of sequential investment, to expand or contract, to temporarily shut down, and to switch outputs or inputs. Overall, flexibility has a value in itself (Trigeorgis, 1998).

Applications of the real options approach in agriculture have shown that the value of the option to postpone the investment grows with enhanced variability of input and output prices (Purvis *et al.*, 1995; Winter-Nelson and Amegbeto, 1998). The value of the option to wait is also large if expected improvements of the new technology are large, as Bessen (1999) demonstrated for a new spinning technology for the cotton industry in England in the beginning of the 20th century. This is particularly relevant for the case of IFTs where research for productivity enhancement is ongoing.

For application of this approach to the problem at hand, a positive net present value of investment is no longer sufficient for recommendation to invest in tree planting.[2] The real options approach for investments recommends investment if the present value of investment exceeds initial investment cost by a factor greater than one. In new investment theory this factor is called the hurdle rate.[3] The next best alternative to planting IFTs is to collect their products from naturally grown trees in the communal areas. Thus, the incremental benefit of planting over collection determines the decision to plant.

A major variable in investment analysis is the discount rate, which is normally derived from the opportunity costs of capital for an investment with similar risk. However, the discount rate should reflect the systematic risk associated with the investment in trees, which may differ from the average risk in the farmer's portfolio. Using the capital asset pricing model (CAPM),[4] the discount rate can be derived from the existing portfolio and its covariance with the investment at hand.[5] The result then is the risk-adjusted rate of return.

The hurdle rate increases with uncertainty and this may cause farmers to postpone the investment until they have better information on the future development of improved species and their marketing potential. Also, expected improvements of IFT via the domestication programme increases the value of waiting to invest, and therefore the hurdle rate.

In the next part of the chapter, we describe the current status of indigenous fruit tree planting in Zimbabwe, the costs and benefits related to the investment decision and, using the real options approach, demonstrate the thresholds for investment.[6]

13.4 Study Area, Data Collection and Analyses

13.4.1 Study area and data collection

Data were collected in two surveys. First, an in-depth monitoring of 19 households of Ward 16, Murehwa district and 20 households of Takawira resettlement area in Zimbabwe provided data on labour, cash and in-kind income, and expenditure of all income-generating activities. Ward 16, Murehwa District, is a communal area with a high abundance of IFTs throughout the area, whereas Takawira is a resettlement area, where households settled in 1982/1983. Households of the communal areas settled much earlier, i.e. between 1940 and 1960. In both areas, land is owned by the State and rights of use are granted to the people living there. In the communal areas, land can be subdivided and bequeathed to children, but this is not the case in the resettlement areas. Traditional leaders in the communal areas deal with conflicts in resource ownership and use, whereas elected committees take care of such issues in the resettlement areas (Rukuni *et al.*, 1994).

Indigenous fruit tree abundance in Takawira is more patchy, and this area is not as well integrated with the market as Murehwa, which is shown by less frequent buses to Harare. The households to be monitored were selected from a baseline survey sample. Fifty per cent of the households were located in two villages in the vicinity of the market and the remainder were located in two villages further from the market (five households per village). The households were selected based on their interest in IFT issues, use of the trees, and willingness to participate in the monitoring (purposive sampling). Additionally, a socio-economic survey of 303 households sampled randomly from both areas provided information on factors related to indigenous fruit use and planting. Mithöfer and Waibel (2003) described the surveys and survey findings in more detail. Supplementary information was collected in farmer workshops on the age-yield function of *U. kirkiana* and by an e-mail survey amongst IFT experts of the region.

13.4.2 Data analyses

Exploratory data analyses

Indigenous fruits contribute about 6.4% and 5.5% of cash and in-kind income of households living in Murehwa district and Takawira resettlement area, respectively (Mithöfer and Waibel, 2003). However, income from the different fruit tree species varies, as production is seasonal, covering the period from August to January. The role which indigenous fruits play in sustaining food security differs between the two areas. In Murehwa, rural households rely less on the fruits than in Takawira after a failure of the maize harvest. Differences exist between indigenous fruit tree species in terms of value; for example, *U. kirkiana* is valued for the commercialization potential of its fruits (Rukuni *et al.*, 1998), whereas *P. curatellifolia* is valued as a food source for survival (Nyoka and Rukuni, 2000) and is more frequently used to substitute for the staple maize than *U. kirkiana* and *S. cocculoides* after a failure of the maize harvest (Mithöfer and Waibel, 2003).

Although households show strong preferences for indigenous fruits, and these fruits are widely consumed (see Ramadhani and Schmidt, Chapter 12, this volume), few of the rural households have planted the trees. The socio-economic survey showed, for both locations, that a high share of households had planted EFTs, whereas they only rarely planted IFTs (Table 13.1), which supports the finding of Price and Campbell (1998). In Murehwa, avocado trees were the first choice among several species to plant; on average households would choose five seedlings from 12, whereas in Takawira households would choose four mango trees from 12. Overall, households chose on average many fewer IFTs, amongst which they would plant the highest number of *U. kirkiana*.

The reasons for choosing to plant exotic or indigenous trees varied and also differed between the two locations. In Murehwa, farmers planted EFTs to produce fruits for home consumption, or for sale as well as home consumption. Looking towards the future, 91.9% of households intended to plant more EFTs, their main reason being to increase in the variety of fruits available at their home. Farmers had not planted more IFTs because they lacked knowledge on how to plant them; the trees grow naturally, the trees are sufficiently available in the communal areas, and the seeds fail to germinate. Fewer than half of households (43.9%) intended to plant more IFTs in the future, mostly to have the fruits directly at home; 39.4% of the farmers would start to plant the IFTs if the sale of their fruits from trees in the communal areas was not permitted but was only allowed from planted trees. In order to enforce this, fruits from plantations could be certified. However, in the current, rather informal, marketing system, this would most probably not be feasible. According to Ramadhani (2002), the majority of local leaders support the idea of legalizing the sale of fruits from planted trees.

In Takawira, most farmers planted EFTs for home consumption and sale, or purely for sale. Some had not planted more of them due to unavailability of seedlings. The majority of households (86.6%) intended to plant more EFTs in the future to generate cash income. On the other hand, farmers had not

Table 13.1. Share of households who planted exotic and indigenous fruit trees and number of trees planted if households received twelve seedlings free of charge.

	Households who planted (%)[a]		Species planted if given 12 seedlings[b]	
	Murehwa	Takawira	Murehwa	Takawira
Avocado	81.9	–	5.1 (2.3)	–
Peach	–	68.3	–	2.8 (1.3)
Mango	92.8	64.6	3.2 (1.6)	3.8 (1.6)
Guava	84.2	67.1	1.7 (1.2)	2.8 (1.4)
Lemon	15.4	4.9	–	–
Uapaca kirkiana	1.8	1.2	1.3 (1.5)	1.4 (2.1)
Strychnos cocculoides	0.9	0	0.4 (0.6)	0.9 (1.1)
Parinari curatellifolia	0.5	1.2	0.4 (0.7)	0.3 (0.7)
Azanza garckeana	2.3	9.8	–	–

[a] Murehwa $n = 221$, Takawira $n = 82$.
[b] Question: If you were given twelve seedlings free of charge, how many would you choose for each species? Households who referred to a different number of seedlings from which to choose were dropped from this analysis. Thus, Murehwa $n = 190$, Takawira $n = 62$; figures are means (standard deviation in parentheses).
Source: Mithöfer (2007).

planted more IFTs because they lacked knowledge of how to do so. Still, 61% of households anticipated planting more IFTs in the future, mostly to have the fruits at their homestead. If the sale of *U. kirkiana* fruits collected from the communal areas were to be made illegal, but if the sale of fruit from planted trees were legal, 86.6% of the farmers would start to plant the trees. The higher share of Takawira farmers who are willing to plant if only the sale of fruits produced on plantations is legalized may be due to the fact that they perceive the fruits to be less available from the commons.

Reasons for planting or not planting IFTs and EFTs are similar, but the relative importance differs between the indigenous and exotic species. Overall, in Murehwa, natural regeneration and relative abundance of the IFTs weighs heavier as a reason for not planting more trees in the future than the lack of knowledge about IFT planting. In Takawira, lack of knowledge on IFT cultivation and resource constraints (e.g. no seedlings, no fencing material available) seem to be more important (Table 13.2).

Analysis of costs of establishing an indigenous fruit tree orchard

In Murehwa, households have between zero and 154 *U. kirkiana* trees of varying age and size, with an average of 24 trees, growing in their fields and homesteads. These are naturally grown trees that have been preserved in farmers' fields. If farmers start deliberate planting of the trees, investment costs, in terms of either purchase of seedlings or own production of seedlings and labour for planting, will accrue. In this study, costs of seedlings are calculated at average production costs, which include labour valued at the local wage rate in Murehwa of ZW$58 per man-day, and material inputs, i.e. tubes. Labour requirements for seedling production are available from the ICRAF Research Station in Makoka, Malawi (J.A. Maghembe, World Agroforestry Centre, Malawi, 1999, personal communication) (see Table 13.3).

Due to a germination rate of 80% (Chidumayo, 1997) and a low rate of survival (20%; Chidumayo, 1997) during the first year after planting, and a

Table 13.2. Reasons for not planting indigenous fruit trees (IFTs) in the future (%).

	Murehwa	Takawira
Number of farmers who do not intend to plant indigenous fruit trees in the future:	123	32
Lack of knowledge	20	16
IFTs sufficiently abundant in the communal areas	50	9
Resource constraints[a]	5	34
Time to maturity too long	0	22
Other[b]	6	19
No answer	20	0

[a] Lack of land, labour, fencing material, seedlings.
[b] Lack of interest, fruits difficult to sell.
Figures do not add up to 100 due to differences in rounding.
Source: Mithöfer (2005).

Table 13.3. Costs of *Uapaca kirkiana* seedling production.

Type of input	Costs (ZW$/seedling)[a]
Seedlings	
Labour	
Collecting fruits[b]	0.03
Extracting seeds	0.12
Treatment of seeds	0.12
Soil collection and transport	0.13
Filling tubes and seeding[c]	0.36
Transport	0.04
Watering	8.01
Weeding	0.22
Other labour (e.g. standing pots upright, etc.)	1.90
Material inputs	
Fruit	0.00
Soil	0.00
Water	0.00
Tubes	0.25
Non-grafted seedling costs	11.17
Costs per orchard of 35 seedlings (ZW$/orchard)	391.0
Grafting	
Labour	
Collection of scion material	8.29
Grafting	1.81
Costs of grafting per seedling	10.10
Costs per orchard of seven trees	70.7
Seedling plus grafting costs per orchard (ZW$/orchard)	461.7
Seedling plus grafting costs per tree survived[d]	92.3

[a] Valued at 1999 ZW$ (Dec. 1999 US$1 = ZW$38). These figures take the germination rate of 80% into account.
[b] Three seeds per fruit.
[c] Space requirements 1 m^2/100 tubes.
[d] Orchard of five trees.
Source: Mithöfer (2005): Labour requirements according to J.A. Maghembe (World Agroforestry Centre, Malawi, 1999, personal communication) and own information, valued at the average wage rate of Murehwa, derived from the socio-economic survey.

grafting success rate of 70% for a skilled grafter (Mhango *et al.*, 2002), a farmer initially has to produce 35 *U. kirkiana* seedlings in order to have an orchard of five grafted trees. If costs of planting the seedlings and labour for protecting them with a fence are included in initial investment costs, initial investments amount to ZW$234 (labour ZW$142, materials ZW$92) per surviving tree.[7] High costs of seedling production are mostly due to high labour costs and even more so due to the low rate of seedling survival.

Farmers perform few management tasks on naturally grown IFTs. The trees are pruned at the end of the dry season to allow better access to agricultural crops cultivated underneath. Dead or damaged branches are cut for firewood.

The area around the trees is weeded when weeding the field, which is generally done twice during the rainy season. Since most trees grow near or in agricultural fields, they benefit from manure and fertilizer application to the crops, which is carried out once at the beginning of the wet season. Only two of the 19 farmers of the households monitored in Murehwa had applied manure purposely to *U. kirkiana* trees. All farmers report that trees in their fields bear more fruits over a longer period than trees from the commons, which they attribute to the application of manure.

However, EFTs, such as avocado and mango, are managed more intensively. This includes watering, weeding, fertilizing, pruning and mulching. The trees are generally protected from livestock damage by a fence that has to be repaired regularly. Also, dead or damaged branches are regularly removed, and micro-catchments (small trenches that preserve moisture and water around the tree) are built and maintained.

It can be expected that an orchard of IFTs would require similar management to that of EFTs. Table 13.4 presents the management of EFTs as practised by households in Murehwa and Takawira. For example, watering ensures survival and growth, especially for younger seedlings. Weeding and fertilization improve the nutritional condition of the tree, whereas pruning promotes growth and facilitates harvesting. Mulching maintains moisture around the tree. Fencing is necessary to protect small seedlings from livestock damage. Pesticide spraying, however, is not necessary, as IFTs are locally adapted and are less susceptible to pests and diseases than EFTs are.

The last column of Table 13.4 describes the management measures for planted IFTs as assumed in the investment analysis. Most of the management

Table 13.4. Labour inputs for management of exotic fruit trees in Murehwa and Takawira and specification for indigenous fruit tree orchards.

	Exotic fruit tree orchards			
Task	Labour (h/tree)[a]	Frequency (/year)[b]	Season[c]	Specification for indigenous fruit tree orchards
Watering	0.8	1–18	DS	Trees <4 years: once per week, DS Trees >4 years: once per year, DS
Weeding	0.6	2	WS	As exotic fruit trees
Fertilizing	0.6	1	Before WS	As exotic fruit trees
Pruning	0.7	1	DS	As exotic fruit trees
Cutting dead and damaged branches	0.4	52	DS	Included in miscellaneous costs
Mulching	0.4	1	DS	As exotic fruit trees
Building of fences	1.1	1	DS	Once after planting
Maintenance of fences				Included in miscellaneous costs
Micro-catchments	0.4	1	WS	As exotic fruit trees

[a] Mean over all households, both locations.
[b] Mode over all households, both locations.
[c] DS: dry season; WS: wet season.
Source: Mithöfer (2005).

tasks are carried out during the dry season, weeding takes place during the wet season and dead branches are removed throughout the year. Fruits ripen during the wet season. It is assumed that an orchard of IFTs is planted close to the homestead. Harvesting labour estimates are based on data for harvesting time of indigenous fruits from trees that farmers have preserved in their fields and around the homestead.

Modelling benefits of indigenous fruit tree planting

The fruit yield of IFTs is age-dependent and varies among individual trees (Chidumayo, 1997). For *U. kirkiana*, yield sets in around years 11–16, increases up to years 25–30 (Household Monitoring and Farmer Workshop), and then declines after 45–55 years (C.K. Mwamba, National Institute for Scientific and Industrial Research, Zambia, 2000, personal communication). It was assumed that indigenous fruit yield reaches zero at an age of 90–100 years. To model the age–yield relationship, the Hoerl function (Haworth and Vincent, 1977), $u = \upsilon g^{\zeta} e^{\kappa g}$, has been specified. It describes yield, u, as a function of age, g; the coefficients υ, ζ, and κ are estimated via linear regression after transforming the equation to $\ln u = \ln \upsilon + \zeta \ln g + \kappa g$. For IFTs the age–yield function was specified as the relationship between fruit yield and the productive period of the tree (as a proxy for age) because farmers usually do not know the age of trees and tend to notice the time when a tree starts bearing fruits instead.

Three yield functions were estimated using data of farmers' estimates of the minimum, maximum and modal yield of trees that they had preserved and managed in their fields. Farmers' data were then complemented by information from a survey of IFT experts. The database comprised estimates of 38 naturally grown and farmer-managed *U. kirkiana* trees. Fruit yield functions for minimum, modal and maximum yield are given in Table 13.5.

Table 13.5. Regression results of the parameters of the age–yield function.[a]

	Parameter[b]			Criteria of estimator	
Fruit yield	$\ln \upsilon$	ζ	κ		
Maximum	2.9640***	1.1270***	−0.0861***	Adj. $R^2 = 0.751$	$F = 48$***
SE	(0.311)	(0.177)	(0.009)	SE = 0.73	$df = 29$
Mode	2.2390***	1.1550***	−0.0806***	Adj. $R^2 = 684$	$F = 42$***
SE	(0.231)	(0.150)	(0.009)	SE = 0.74	$df = 36$
Minimum	1.4840***	1.1720***	−0.0735***	Adj. $R^2 = 0.661$	$F = 31$***
SE	(0.320)	(0.009)	(0.182)	SE = 0.75	$df = 29$

[a] Age of zero years refers to the age at which fruit production begins: this is between 11 and 16 years after germination.
[b] First row: parameter values.
*: $P \leqslant 0.1$, **: $P \leqslant 0.05$, ***: $P \leqslant 0.01$.
Source: Mithöfer (2005).

In the investment analysis, multiple products of the IFTs – e.g. fruits, leaves and wood – are accounted for. The age-dependent fruit yield was treated as a random variable, with the minimum and maximum as lower and upper bounds and the modal yield function as the most likely yield for each year. Autocorrelation of fruit yield is assumed to be non-existent. For simplicity, all fruit trees in the orchard are assumed to have the same yield. Fruit prices are considered to follow a uniform distribution between ZW$0.4 and ZW$18/kg, which are the minimum and maximum farm gate price that households received in 1999/2000.

The optimal life span of the orchard has to take into account the multiple products the trees provide. It is reached in the year where returns on labour are maximized. Multiple tree products include fruit-, leaf- and wood-production functions and the prices for these products must be taken into consideration. Leaf and wood production as a function of diameter at breast height (d.b.h.) and height were applied, as estimated by Chidumayo (1997). The relationships between age and d.b.h. and between age and height were estimated by linear regression using the data from farmer-preserved and farmer-managed IFTs. Price estimates for the various products are based on our surveys. These are either market prices of the products or market prices of substitutes.

13.5 Model Outputs for Investment in IFT Planting

The returns on labour of collecting IFT products from the communal areas (ZW$506/day) are much higher than the returns on labour from planting non-domesticated trees (ZW$52/day). Thus, the incremental value of planting IFTs is negative on average. The productivity of the non-domesticated IFTs is still too low to be economic for farmers. This also explains why so few farmers have planted the trees until now. This is implicitly referred to by Murehwa farmers, who frequently state that they will not plant IFTs in the future because the perceived abundance of the trees is high; farmers of Takawira resettlement area directly cite the long time that IFTs take to produce fruits as one reason for not planting the trees in the future (Table 13.2). Without having asked farmers to rank their reasons for not planting IFTs, the frequency with which each reason was mentioned can be taken as a proxy. Since insecurity of tenure of land was never mentioned and farmers have planted other trees, this does not seem to influence the decision on whether to plant IFTs or not.

In addition, the annual rates of return on labour of indigenous (and exotic) fruit tree use show a positive covariance (0.2554) with all other activities in the income portfolio of Murehwa farm households. Thus, the theoretical benefit of enterprise diversification cannot be established through the planting of IFTs when considering annual income streams, and therefore planting does not have risk-reducing effects.[8] Overall, the risk-adjusted rate of return for planting IFTs is 15.64%, which constitutes a rate in real terms (Mithöfer, 2005; Mithöfer *et al.*, 2005). The high rate of inflation in Zimbabwe implies that not the nominal but rather the real rate has to be used for investment analysis, which was done for the present study. High inflation rates with constant or relatively slower rising nominal rates imply decreasing real rates of interest. This implies decreasing opportunity costs of capital, i.e. credits are 'cheaper', as well as decreasing profitability of investments in financial assets but increasing profitability of investments in non-financial assets. Most importantly for the present analysis, one would have to assess the effect of inflation on various activities in farmers' portfolios and adapt the risk-adjusted rate of return accordingly to the fast-changing Zimbabwean conditions.

The calculations show that to be economically attractive, domesticated IFTs would either have to be further improved or the availability of fruits from the communal areas would have to be dramatically less, resulting in higher collection costs (Mithöfer, 2005; Mithöfer *et al.*, 2005). Table 13.6 shows the improvements and changes necessary so that farmers could be expected to immediately invest. Tree improvements would have to induce early maturity and enhance yields.

Farmers can be expected to invest in planting of IFTs if trees fruit at 2 years of age in combination with a ninefold yield increase. Alternatively, increased costs of collection of indigenous fruits from the communal areas in combination with lower levels of tree improvements would induce investment, all other factors being equal. For example, if precocity could be induced at an age of 2 years, 2.8-fold higher costs of collection from the communal areas would trigger immediate investment. Such an increase in collection costs would happen if the abundance of trees decreased sharply or had already occurred in areas which have experienced high rates of deforestation in the past. If fruit quality is improved via the domestication programme so that domesticated fruits fetch prices up to three times that of non-domesticated ones, then lower levels of yield or increase in collection costs would trigger investment (Mithöfer, 2005; Mithöfer *et al.*, 2005). In order to contribute to achieving this, the domestication programme could make use of consumers' preferences and willingness to pay, as demonstrated by Ramadhani (2002), and assess fruit traits that would fetch higher prices and then concentrate on improving those.

Table 13.6. Conditions for immediate investment in on-farm indigenous fruit tree planting.

Age at first fruiting (years)	Yield level (times the current level)	Fruit quality (fruit prices 1–3 times the current level)	Collection costs (times the current level)
2	Non-improved	Non-improved	2.8–2.9
2	9–30	Non-improved	Non-changed
4	10–40	Non-improved	Non-changed
6	12–56	Non-improved	Non-changed
8	16–80	Non-improved	Non-changed
10	24–104	Non-improved	Non-changed
2	Non-improved	Improved	1.3–2
2	1.5–2	Improved	Non-changed

Source: Mithöfer (2005), Mithöfer *et al.* (2005).

13.6 Conclusions

The results presented here have implications for assessing the prospects of the domestication programme of IFTs of southern Africa. So far, Zimbabwean farmers have rarely planted IFTs and instead turn to EFT species, which are perceived to be superior due to higher economic returns. Also, the farmers of the research sites perceive *U. kirkiana* (the indigenous fruit that is most widely marketed) as sufficiently abundant in the communal areas. This is reflected in relatively high returns on labour of collecting indigenous fruits from the communal areas. Under the current level of density of IFTs in the woodlands, collection also compares well with other income-generating activities such as agriculture and horticulture, as described by Mithöfer and Waibel (2003). However, these results will change as IFTs become scarcer through deforestation.

The investment analysis suggests that further biological research on IFTs should concentrate on inducing precocity and increasing fruit quality. Increasing fruit yield is less important in improving the economics of tree planting. With respect to the increased fruit quality, further research is needed in order to assess the potential willingness of consumers to pay for enhanced fruit quality. Referring to the findings of Ramadhani (2002), domestication research work could target traits that are most preferred by consumers and imply a greater willingness to pay. With respect to inducing precocity and enhancing yield levels, the feasibility of achieving such improvements should be established.

Finally, the technology dissemination strategy would have to take into consideration low levels of farmer knowledge on IFT management, and assess and address those issues further.

Acknowledgements

We thank the ICRAF team in Zimbabwe and Malawi, in particular Dr Freddie Kwesiga, Dr Elias Ayuk and Dr Festus Akinnifesi, for advice and support during the fieldwork. The German Ministry for Economic Cooperation and Development (BMZ) provided funds through ICRAF's Domestication of Indigenous Fruit Trees Programme.

Notes

[1] This is equivalent to a benefit : cost ratio of greater than one and an internal rate of return larger than the discount rate used in the analysis. For problems concerning each approach and comparability of results, see e.g. Gittinger (1982) and Brealey and Myers (2000).

[2] According to the net present value rule, investment is profitable when the net present value exceeds zero, i.e. where accumulated discounted net benefits (the present value) exceed initial investment cost.

[3] For a more formal treatment, see Dixit and Pindyck (1994).
[4] The capital asset pricing model explains the equilibrium rate of return on capital assets given that asset's non-diversifiable risk in relation to the well-diversified market portfolio (for a formal treatment see Luenberger, 1998).
[5] For a broader analysis the risk-adjusted rate of return would have to be adjusted at regional level.
[6] More information on the technical details of this approach can be found in Mithöfer (2005) and the literature cited therein.
[7] In the investment model, costs and benefits of the two alternatives (i.e. planting versus collection) are made comparable on the basis of returns on labour, which is the major resource used in each activity.
[8] IFTs bear fruits at a time of year when income from rain-fed agriculture is zero and at that time complement income from irrigated horticultural crops. The covariance of seasonal income streams may differ from the covariance of the annual income streams; however, how this may influence the decision to invest is ambiguous, since harvesting of the fruits occurs in the same period as planting of rain-fed agricultural crops and each activity implies opportunity costs in terms of the other.

References

Abebaw, D. and Virchow, D. (2003) The micro-economics of household collection of wild coffee in Ethiopia: some policy implications for in-situ conservation of *Coffea arabica* genetic diversity. In: *Economic Analysis of Policies for Biodiversity Conservation.* BioEcon Workshop, 28– 29 August, Venice.

Akinnifesi, F.K., Kwesiga, F., Mhango, J., Chilanga, T., Mkonda, A., Kadu, C.A.C., Kadzere, I., Mithöfer, D., Saka, J.D.K., Sileshi, G., Ramadhani, T. and Dhliwayo, P. (2006) Towards the development of miombo fruit trees as commercial tree crops in Southern Africa. *Forests, Trees and Livelihoods* 16, 103–121.

Arnold, J.E.M. and Dewees, P.A. (1999) Trees in managed landscapes: factors in farmer decision making. In: Buck, L.E., Lassoie, J.P. and Fernandes, E.C.M. (eds) *Agroforestry in Sustainable Agricultural Systems.* Lewis Publishers, Boca Raton, Florida, pp. 277–294.

Bessen, J. (1999) *Real Options and the Adoption of New Technologies.* Research on Innovation, Wallingford, UK.

Brealey, R.A. and Myers, S.C. (2000) *Principles of Corporate Finance.* Irwin McGraw-Hill, Boston, Massachusetts.

Brigham, T. (1994) Trees in the rural cash economy: a case study from Zimbabwe's communal areas. MA thesis, Carleton University, Ottawa, Canada.

Campbell, B. (1996) *The Miombo in Transition: Woodlands and Welfare in Africa.* Center for International Forestry Research, Bogor, Indonesia.

Campbell, B., Grundy, I. and Matose, F. (1993) Tree and woodland resources: the technical practises of small-scale farmers. In: Bradley, P.N. and McNamara, K. (eds) *Living with Trees: Policies for Forestry Management in Zimbabwe.* The World Bank and the Stockholm Environment Institute, Washington, DC and Stockholm, pp. 29–62.

Cavendish, W. (2000) Empirical regularities in the poverty-environment relationship of rural households: evidence from Zimbabwe. *World Development* 28, 1979–2003.

Chidumayo, E. (1997) *Miombo Ecology and Management: An Introduction.* IT Publications, London.

Chipika, J.T. and Kowero, G. (2000) Deforestation of woodlands in communal areas of Zimbabwe: is it due to agricultural policies? *Agriculture, Ecosystems and Environment* 79, 175–185.

Clarke, J., Cavendish, W. and Coote, C. (1996) Rural households and miombo woodlands: use, value and management. In: Campbell, B. (ed.) *The Miombo in Transition: Woodlands and Welfare in Africa*. Center for International Forestry Research, Bogor, Indonesia, pp. 101–135.

Deininger, K. and Minten, B. (2002) Determinants of deforestation and the economics of protection: an application to Mexico. *American Journal of Agricultural Economics* 84, 943–960.

Dercon, S. (2000) *Income Risk, Coping Strategies and Safety Nets*. Centre for the Study of African Economics, Department of Economics, Oxford University, Oxford.

Dixit, A.K. and Pindyck, R.S. (1994) *Investment under Uncertainty*. Princeton University Press, Princeton, New Jersey.

Evans, T.D. and Sengdala, K. (2002) The adoption of rattan cultivation for edible shoot production in Lao PDR and Thailand: from non-timber forest product to cash crop. *Economic Botany* 56, 147–153.

FAO (2005) *Global Forest Resources Assessment 2005: Progress Towards Sustainable Forest Management*. Food and Agriculture Organization of the United Nations, Rome.

Gittinger, J.P. (1982) *Economic Analysis of Agricultural Projects*. Johns Hopkins University Press, Baltimore, Maryland.

Haworth, J.M. and Vincent, P.J. (1977) *Medium-Term Forecasting of Orchard Fruit Production in the EEC: Methods and Analysis*. European Communities Commission, Brussels.

ICRAF (1996) *Domestication of Indigenous Fruit Trees of the Miombo Woodlands of Southern Africa*. Project proposal for the Restricted Core Programme to the German Ministry for Economic Cooperation and Development (BMZ). International Centre for Research in Agroforestry, Nairobi.

Kadzere, I. (1998) *Domestication of Indigenous Fruit Trees Research Report*. Zimbabwe Agroforestry Project, International Centre for Research in Agroforestry, Harare.

Leakey, R., Tchoundjeu, Z., Smith, R.I., Munro, R.C., Fondoun, J.-M., Kengue, J., Anegbeh, P.O., Atangana, A.R., Waruhiu, A.N., Asaah, E., Usoro, C. and Ukafor, V. (2004) Evidence that subsistence farmers have domesticated indigenous fruits (*Dacryodes edulis* and *Irvingia gabonensis*) in Cameroon and Nigeria. *Agroforestry Systems* 60, 101–111.

Luenberger, D.G. (1998) *Investment Science*. Oxford University Press, New York.

Lund, D. (1991) Stochastic models and option values: an introduction. In: Lund, D. and Oksendahl, B. (eds) *Stochastic Models and Option Values*. North-Holland, Amsterdam, pp. 3–18.

Maghembe, J.A., Simons, A.J., Kwesiga, F. and Rarieya, M. (1998) *Selecting Indigenous Trees for Domestication in Southern Africa*. International Centre for Research in Agroforestry, Nairobi.

Mander, M., Mander, J. and Breen, C. (1996) Promoting the cultivation of indigenous plants for markets: experiences from KwaZulu-Natal, South Africa. In: Leakey, R.R.B., Temu, A.B., Melnyk, M. and Vantomme, P. (eds) *Domestication and Commercialization of Non-timber Forest Products in Agroforestry Systems*. Non-wood Forest Products No. 9, Food and Agriculture Organization, Rome, pp. 104–109.

McGregor, J. (1991) Woodland resources: ecology, policy and ideology. PhD thesis, Loughborough University of Technology, Loughborough, UK.

Mhango, J., Mkonda, A. and Akinnifesi, F. (2002) Vegetative propagation as a tool for the domestication of indigenous fruit trees in Southern Africa. *Proceedings of Regional Agroforestry Conference 'Agroforestry Impacts on Livelihoods in Southern Africa: Putting Research into Practise', Warmbath, South Africa, May 20–24*. International Centre for Research in Agroforestry, Nairobi.

Minae, S., Sambo, E.Y., Munthali, S.S. and Ng'ong'ola, D.H. (1994) Selecting priority indigenous fruit tree species for central Malawi using farmers' evaluation criteria. In: Maghembe, J.A., Ntupanyama, Y. and Chirwa, P.W. (eds) *Proceedings of Conference on 'Improvement of Indigenous Fruit Trees of the Miombo Woodlands of Southern Africa', Club Makokola, Mangochi,*

Malawi. International Centre for Research in Agroforestry, Nairobi, pp. 84–99.

Mithöfer, D. (2005) Economics of indigenous fruit tree crops in Zimbabwe. Doctoral thesis, Faculty of Economics and Business Administration, University of Hannover, Germany.

Mithöfer, D. and Waibel, H. (2003) Income and labour productivity of collection and use of indigenous fruit tree products in Zimbabwe. *Agroforestry Systems* 59, 295–305.

Mithöfer, D., Wesseler, J. and Waibel, H. (2005) *R&D and private investment: how to conserve indigenous fruit biodiversity of Southern Africa.* Paper presented at the German Development Economics Conference, Kiel, Germany.

Musvoto, C. and Campbell, B.M. (1995) Mango trees as components of agroforestry systems in Mangwende, Zimbabwe. *Agroforestry Systems* 32, 247–260.

Nyoka, B.I. and Rukuni, D. (2000) Progress with domestications of indigenous fruit tree species in Zimbabwe. In: Shumba, E.M., Lusepani, E. and Hangula, R. (eds) *Proceedings of the SADC Tree Seed Centre Network Technical Meeting.* SADC Tree Seed Centre Network, Food and Agriculture Organization & Canadian International Development Agency, Windhoek, Namibia, pp. 62–66.

Price, L. and Campbell, B. (1998) Household tree holdings: a case study in Mutoko Communal Area, Zimbabwe. *Agroforestry Systems* 39, 205–210.

Purvis, A., Boggess, W.G., Moss, C.B. and Holt, J. (1995) Technology adoption decisions under irreversibility and uncertainty: an *ex ante* approach. *American Journal of Agricultural Economics* 77, 541–551.

Ramadhani, T. (2002) *Marketing of Indigenous Fruits in Zimbabwe.* Wissenschaftsverlag Vauk, Kiel, Gemany.

Rukuni, M., Magadzire, G.S.T., Dengu, E., Gutu, S.C., Lupepe, G.N.H., Mafu, V.M., Mangwende, T.J.C., Mpofu, J.M., Murwira, E., Musikavanhu, E.T., Sibanda, S.K.M., Townsend, M.H.C. and Vudzijena, V. (1994) *Volume I: Main Report to His Excellency the President of the Republic of Zimbabwe of the Commission of Inquiry into Appropriate Agricultural Land Tenure Systems.* The Government Printer, Harare.

Rukuni, D., Kadzere, I., Marunda, C., Nyoka, I., Moyo, S., Mabhiza, R., Kwarambi, J. and Kuwaza, C. (1998) Identification of priority indigenous fruits for domestication by farmers in Zimbabwe. In: Maghembe, J.A., Simons, A.J., Kwesiga, F. and Rarieya, M. (eds) *Selecting Indigenous Trees for Domestication in Southern Africa.* International Centre for Research in Agroforestry, Nairobi, pp. 72–94.

Simons, A.J. (1996) ICRAF's strategy for domestication of non-wood tree products. In: Leakey, R.R.B., Temu, A.B., Melnyk, M. and Vantomme, P. (eds) *Domestication and Commercialization of Non-timber Forest Products in Agroforestry Systems.* Non-wood Forest Products No. 9, Food and Agriculture Organization, Rome.

Trigeorgis, L. (1998) *Real Options.* The MIT Press, Cambridge, Massachusetts.

Unruh, J.D. (2001) Land dispute resolution in Mozambique: institutions and evidence of agro-forestry technology adoption. CAPRI Working Paper No. 12, IFPRI, Washington, DC.

Wilson, K.B. (1990) Ecological dynamics and human welfare: a case study of population, health and nutrition in southern Zimbabwe. PhD thesis, Department of Anthropology, University of London, London.

Winter-Nelson, A. and Amegbeto, K. (1998) Option values to conservation and agricultural price policy: application to terrace construction in Kenya. *American Journal of Agricultural Economics* 80, 409–418.

World Bank (2001) *World Development Report 2000/2001. Attacking Poverty.* The World Bank, Washington, DC.

14 Opportunities for Commercialization and Enterprise Development of Indigenous Fruits in Southern Africa

C. Ham,[1] F.K. Akinnifesi,[2] S. Franzel,[3] D. du P.S. Jordaan,[4] C. Hansmann,[5] O.C. Ajayi[2] and C. de Kock[6]

[1] *University of Stellenbosch, South Africa*; [2] *World Agroforestry Centre, Lilongwe, Malawi*; [3] *World Agroforestry Centre, Kenya*; [4] *University of Pretoria, South Africa*; [5] *Agricultural Research Council, Stellenbosch, South Africa*; [6] *Speciality Foods of Africa Pvt Ltd, Harare, Zimbabwe*

14.1 Introduction

In recent times, there has been an increasing demand for natural agricultural products globally due, among other reasons, to food safety concerns and an increasing awareness of the impacts of agricultural production methods on the environment. Southern Africa is no exception, as the demand for organic products is increasing, especially among the upper and mid-income groups. During a consumer test in South Africa on kei apple (*Dovyalis caffra*) fruit juice, 86% of the consumers responded positively to the idea of a marketing strategy based on an indigenous product with no artificial flavourings or preservatives (Moelich and Muller, 2000). With more than 200 edible indigenous fruit tree species (Fox and Norwood Young, 1982), southern Africa has the potential to offer a wealth of new food products to the global market. Indigenous plant products form an integral part of the livelihoods of rural people in Africa.

The fruits of many indigenous plants are consumed mostly as snacks, except in times of famine when such fruits become the principal source of food and, as a result, fulfil a much more substantial role in the daily diet of rural people (Akinnifesi *et al.*, 2006). Snacks are especially important for children who need to eat more frequently than adults, and wild fruits and nuts are important sources of the micronutrients that are often deficient in cereal-based diets commonly consumed in the region (Ruffo *et al.*, 2002).

In addition to their use for home consumption, indigenous fruits are also traded. Families that need additional cash to pay for agricultural inputs and to buy food to supplement subsistence farm production commonly trade in indigenous fruits. Indigenous fruits sold at local and urban markets yield substantial household incomes (Schomburg *et al.*, 2001). However, despite the importance of indigenous fruits to African consumers, relatively little success has been achieved with their commercialization. Only a couple of indigenous fruit products have made it onto the international market. Amarula Cream, made from the fruits of *Sclerocarya birrea* by the Distell Corporation in South Africa, is probably one of the best-known examples. Studies that were conducted in Malawi (Kaaria, 1998) and in Zimbabwe (Ramadhani, 2002) also found that the trading of indigenous fruits is poorly developed and lags far behind the trading of exotic fruits.

14.2 Fresh Fruit Markets and Value Addition

Despite its poor state of commercial development, indigenous fruit trading and marketing are important economic activities in countries such as Malawi and Zimbabwe (Kaaria, 1998; Ramadhani, 2002). Indigenous fruits are sold in both urban and rural markets, showing that consumers like the fruits and are willing to pay for them. The above-mentioned studies also highlight some of the problems experienced in the trading of indigenous fruits.

14.2.1 Wild harvesting

A variety of methods are employed to harvest fruits from wild growing trees. Methods include the picking of fruits from the ground following abscission, climbing trees to pick fruits, throwing objects to dislodge fruits, hitting stems with heavy objects, and shaking stems or branches to dislodge fruits (Kadzere *et al.*, 2002). These crude methods of harvesting fruits not only damage the trees but also cause excessive bruising to the fruit, thus reducing the shelf life, quality and market value. The time of harvesting can be critical. Fruit collectors in some countries such as Tanzania only collect fruits that have fallen to the ground after abscission. The harvesting of such fully ripe fruits would leave only a limited consumption period and increase the chances of mechanical damage during the transport process. Producers of exotic fruits normally harvest their crops according to harvesting indices, indicating the optimum harvesting time. Fruits are harvested before they are fully ripe, and harvesting is timed so that fruits only reach full ripeness when presented to the final consumer. For indigenous fruits, there is the need for better understanding of ripening patterns and to determine whether the technique of harvesting unripe fruit for postharvest ripening can be successfully applied (Kadzere *et al.*, 2002).

One of the factors limiting the development of indigenous fruits is that they are characterized by short but intense fruiting seasons. For example, in Zambia the average collection time for *Uapaca kirkiana* fruits is only 3 months. However, some collectors did not collect throughout the season for several reasons: (i) lack of labour capacity to collect and transport fruits to the market places due to long distances and the heavy weight of the fruit loads; (ii) constraints on the type of transportation used, which is often limited to bicycles; (iii) involvement in other livelihood and income-generation activities, especially selling other merchandise and engaging in agricultural activities; and (iv) trading in *U. kirkiana* fruits was done as a part-time business mostly during the peak time of fruit supply.

14.2.2 Postharvest handling and transport

The system for postharvest handling of indigenous fruits is poorly developed, and this has inadvertently affected the commercial value of the fruits and the potential income generation to smallholder farmers. After collection of fruit, producers walk with the fruit on their heads or in two-wheeled carts to their homes, local markets or roadside collection points (Ramadhani, 2002). In Zambia only 14% of traders interviewed indicated that they incurred or paid transport costs. The majority carry the fruit to the markets, while the second most popular means of transport was by bicycle (Karaan *et al.*, 2005). Very few of the fruit collectors protect the fruit from the sun while waiting for transport. The fruits that are being marketed in urban centres went through a long handling chain before reaching the final consumer (Kadzere *et al.*, 2002). As a result, fruit collectors experience a lot of fruit losses from mechanical damage arising from cracking, compression and bruising during harvesting and transport, insect and pest damage, and overripening (Combrink, 1996; Saka *et al.*, 2002).

One of the biggest problems with transport in Malawi is its unreliability (Kaaria, 1998). Transport is infrequent and traders lose a lot of fruit while waiting for it. Postharvest decay can be a serious problem in many fruits in storage. Limiting mechanical injury should always be a primary consideration. It is important to try to minimize the load of organisms present on the fruit, using clean water for washing, and even using fungicides. These should be applied carefully to eliminate residues on consumption. The use of fungicides is usually regulated by legislation for specific products.

14.2.3 Supply chains

The supply chain normally starts with farmers or local community members who collect fruits for home consumption and for trading. They sell fruits at the roadside to other community members or to traders who transport the fruits to urban markets. These traders either function as wholesalers and sell to retailers, or sell the fruits themselves as retailers. Farmers sometimes sell their fruits directly to the public at markets, effectively eliminating all middlemen. In doing this, farmers can negotiate a better price, up to US$0.03 more per kilogram (Kaaria, 1998). The supply chains are fairly short, with the longest chain consisting of collectors, wholesalers, retailers and consumers. Figure 14.1 gives a summary of the possible supply chain combinations for indigenous fruit in a typical southern African country.

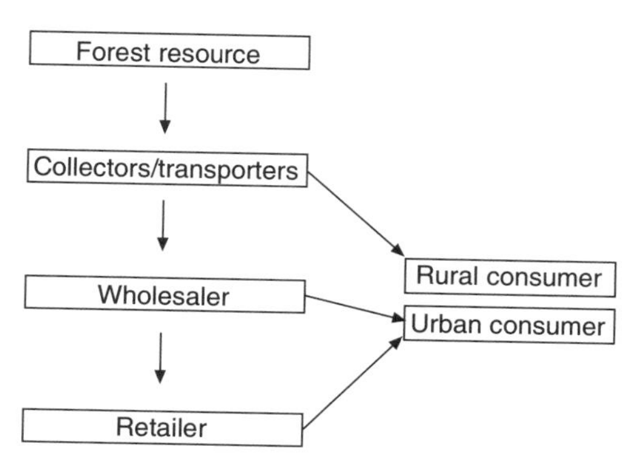

Fig. 14.1. Supply chain combinations observed in southern Africa.

Indigenous fruits are characterized by mass fruiting seasons that typically last 3 months. This seasonality can lead to excess supply over demand and hence a drastic fall in prices during the harvesting period. This is aggravated by lack of market information that results in inefficient distribution of fruit. In Malawi, a large proportion of *Ziziphus mauritiana* fruit sold in Lilongwe came from Mangochi (240 km away) instead of the much closer Salima area (Kaaria, 1998). A similar situation exists for the Limbe market, where fruits came from Mangochi (150 km away) instead of Chikwawa, which is only 70 km south of Limbe. Poor linkage of producers with consumers and the market, and the lack of market information on indigenous fruits have been identified by Schomburg *et al.* (2001) as some of the major bottlenecks to the commercial development of indigenous fruit enterprises. The implication of long-distance transportation is that mechanical damage to the fruits is increased, thereby reducing their market value. To address the fluctuations in supply, traders often diversify their merchandise by selling indigenous fruits in addition to other types of fruits, including exotic or domesticated tree and non-tree fruits such as *Mangifera indica* (mango), *Persea americana* (avocado), *Psidium guajava* (guava), *Citrus limon* (lemon), *Musa* spp. (bananas), *Vintisfera* spp. (grapes) and *Ananas* spp. (pineapple).

The most common indigenous or wild fruits that traders sell include *U. kirkiana* (masuku), *Strychnos cocculoides* (tusongole), *Strychnos pungens* or *innocua* (tugome), *Parinari curatellifolia* (mpundu), *Anisophyllea boehmii* (nfungo), *Garcinia huillensis* (nsongwa), *Diospyros mespiliformis* (chenja) and *Landolphia kirkii* (mabungo). The fruit that traders trade in the most is *U. kirkiana*, followed by *M. indica* (Karaan *et al.*, 2005).

14.2.4 Market prices

In many southern African countries, there is no defined or formal mechanism for the setting of market prices of indigenous fruits (Kaaria, 1998; Ramadhani, 2002). Prices are determined by adding a profit to all costs (purchasing costs, rental, security, tax). Seasonality, fruit taste (sweetness), labour costs (including physical hardships undergone, risks involved in collection, handling and transportation of the heavy fruits), critical need of households for immediate cash, and the quantities of fruit available, compared with demand, are the key factors determining fruit prices. Traders look at the number of suppliers bringing fruit to the market, listen to first-hand information from villagers and observe buying behaviour. It is estimated that the average profit percentage is approximately 50% (Karaan *et al.*, 2005), but this may vary depending on the need to solve prevailing and immediate domestic and other problems. Generally, prices of fruits were set jointly, with the collectors/wholesalers who sold first in the season, determining the prices. Other traders then followed by asking the same prices. Negotiation with buyers was another form of price-setting. This method of pricing was done to avoid conflict between competitors. Pricing at retail level was similar to those for the collectors/wholesalers in both method and factors considered. In order of importance, the main factors appeared to be the order price, taking the price set by other traders and total cost, and the prices set by the first sellers. Fruit prices are also affected by the lack of a standard unit of measurement for selling fruits. The quantities of fruits are measured using different utensils such as cups, bowls, plates, tins and buckets, thus making it very difficult to compare prices (Ramadhani, 2002).

14.2.5 Market conditions

One of the greatest problems associated with the trading of indigenous fruits is market infrastructure. Lack of storage facilities and protection from the weather result in fruit spoilage and lower quality of goods (Kaaria, 1998; Ramadhani, 2002). Many traders also sell their wares outside the markets because of limited space inside. These traders basically put their produce on the floor, or lay down a piece of cloth to prevent the goods from getting dirty (Kaaria, 1998). Fruits in the wholesale section of the Mbare market (Harare, Zimbabwe) are stored in large heaps on the open ground with no cover to protect them from the sun, wind and rain. This method of bulk storage has a detrimental effect on fruit quality. Fruit at the bottom of the heaps are squashed by the weight of the fruit above, while the sun bakes the fruit at the top. As a result, up to one-third of the fruit is spoiled. In Zambia, collectors and wholesalers store their fruits mostly at home and in the open. Most of them store their fruits in containers or in sacks, with only a small percentage storing fruits loose on the ground. Damage to fruits due to handling during public and private transportation and harvesting accounts for 83% of the total fruit loss and is the most economically important loss experienced by collectors/wholesalers. Other sources of loss are rotting (11%), excessive heat (3%), and the use of inappropriate containers, such as sacks instead of baskets (3%).

Retailers, like wholesalers, also stored their fruits in a variety of places, such as market buildings. Fruits are stored in containers or loose on the ground. Sales turnover was approximately 1.5 days, while fruits took about 3 days before becoming rotten. The main reasons for fruit losses among fruit retailers are rotting if fruits are not sold within 3 days (60%), breaking (33%), and theft of the fruits when stored in market areas (7%) (K. Moombe, unpublished observations). The way in which fruits are displayed, either on stands or on the ground, makes it easier for consumers to taste the fruits. Consumers taste the fruits to ensure that they are buying the sweetest types. Market traders have indicated that consumer tasting is also a big problem that leads to notable losses (Ramadhani, 2002).

14.2.6 Value adding

The indigenous fruit marketing system is characterized by a lack of sophisticated product differentiation activities. At the producer and wholesaler levels, pre-sale processes such as washing, grading, sorting and packing do not exist. Traders believe that the fruits coming from trees are clean and that there is no need for further processing (Ramadhani, 2002). In Tanzania, indigenous fruits are not displayed as prominently as other produce in rural markets. In some instances the other produce would be displayed on tables, while the indigenous fruits would be kept in containers on the ground. Indigenous fruits are sold mostly to women and children as a snack that is often eaten fresh in the market. Indigenous fruits can therefore be considered as products that consumers buy on impulse, in very small quantities, and that are consumed very soon after being purchased. People come to the market to buy other goods and see that indigenous fruits are present, and then purchase them.

Retailers of *U. kirkiana* in Zimbabwe conduct simple processing activities such as sorting rotten fruit, dividing fruits in to selling units, and sometimes washing. Other retailers, however, felt that the practice of washing and

Table 14.1. Risks and benefits of different actors along a typical indigenous fruit supply chain.

Actor	Risks	Benefits
Collector	Transport costs Danger from wild animals and snakes during collection No fixed agreements with retailers or wholesalers	Income from sale to wholesaler and retailer Access to 'free' resource
Wholesaler	Market space rental and taxes High stock losses due to spoilage Dependent on the reliability of collectors for supply Uncertainty of fruit supply	Bulk-buying allows for setting of purchase prices Freedom to buy from a large pool of collectors Relatively high profit margins
Retailer	Market rental and taxes Stock losses due to spoilage Dependent on wholesalers High transaction cost due to low sales volume	Less spoilage than wholesalers Diversification by selling not only indigenous fruits Relatively high profit margin Can buy directly from collectors

packaging in bags reduces the shelf life of the fruits (Ramadhani, 2002). In Zimbabwe, traders consider that washing of *U. kirkiana* fruits could lead to fruit discoloration (Kadzere *et al.*, 2002). In Malawi only some market traders grade their fruits based on fruit size and degree of ripeness, while in Zimbabwe collectors and traders of *Z. mauritiana* have been able to obtain better prices for their fruits by cleaning and packaging the fruit in plastic bags (Kaaria, 1998).

14.2.7 Fruit marketing strategies

Marketing and postharvest handling are subject to high transaction costs in the markets. This is not unexpected in typical developing-country markets. The transaction costs manifest in various ways including product losses, time delays, costs of monitoring transactions and agents in markets, negotiation or haggling costs, transport and logistics inefficiencies, and poor market information. Transaction costs are better contended through improved coordination and information flows in supply chains. This is a key challenge for economic development efforts in emerging markets. Supply and consumption are entirely directed at local markets, given the traditional status of the products and consumption habits. The local markets are assumed to be low-income markets, by and large, with limited ability to pay for greater sophistication and value addition that adds significantly to the marketing margins. Hence, there is a tendency to favour generic marketing of produce. The implication is, arguably, that limited benefit can be derived from much value addition in local markets; at best, the improvements would be incremental. Alternatively the improvements could radically change the product on offer through processing (e.g. juices, jams, preserves, etc.) whilst retaining the product appeal. Another alternative may be to consider acceptability in higher-income markets and adapt the product accordingly (Karaan *et al.*, 2005).

14.3 Indigenous Fruit Processing

Processing of indigenous fruits has been practised in Africa for centuries. Indigenous fruits are mainly processed to improve the fruit's taste, to preserve the fruit for later use, especially during periods of low normal fruit supply, and to obtain products which can be converted into other by-products (Kadzere *et al.*, 2002; Saka *et al.*, 2002). Indigenous fruits play an important role as food substitutes in times of drought and famine and constitute a cheap, yet rich, source of nutrients (Schomburg *et al.*, 2001). Given current knowledge that adequate nutrition can contribute positively to the quality of life for people living with HIV/AIDS, indigenous fruits can play a critical role in supplementing diets.

14.3.1 Current indigenous fruit processing practices

Four-fifths of indigenous fruit producers/collectors in the Zambezi Valley in Zimbabwe processed indigenous fruits (Kadzere *et al.*, 2002). Most of the processed products were consumed at household level while only a small percentage was traded. People employ processing techniques that were passed down to them through generations, and no major changes have been observed in terms of processing equipment and recipes. The main products from processing are alcoholic and non-alcoholic beverages, confectionary, additives in other foods, dried whole fruits, oil, butter and fruit powder (Kadzere *et al.*, 2002). Common utensils such as cooking pans, wooden mortars, bowls, sieves, cloth and plastic containers with lids, cups, wooden spoons and bottles are used to process fruit. Little consideration is given to hygienic conditions in the course of fruit processing (Mumba, *et al.*, 2002). In Tanzania, juice-making is most preferred by women processors, followed by jams and wines, because the process of juice-making is simple and affordable (Swai *et al.*, 2002). Another popular processing activity is the drying and preservation of indigenous fruits. Processed fruits need to be preserved for a maximum of 1 year until the next fruiting season (Ndabikunze *et al.*, 2000). The processing of indigenous fruits has not been limited to the household level. A number of researchers have experimented with indigenous food-processing technology. Some of the recent studies include:

- Wine making at Bunda College in Malawi (Ngwira, 1996).
- Development of recipes for juices, jams and wine in Tanzania (Swai, 2001).
- Development of recipes for juices, jams and fritters in Malawi (Saka *et al.*, 2001).
- Testing of indigenous fruit jams and juices in Tanzania (Tiisekwa *et al.*, 2002).

These studies have been complemented by research on the nutritional values of indigenous fruits. Saka (1995) and Ndabikunze *et al.* (2002) have analysed the nutritional values of a range of indigenous fruits.

14.3.2 Problems faced by processing groups

During the priority-setting workshops in Tanzania and Malawi, workshop participants were asked to list the main problems and constraints they faced in their processing activities. Table 14.2 presents a summary of these constraints in order of importance.

Workshop participants in Malawi and Tanzania indicated that lack of processing equipment and packaging materials, capital to acquire processing equipment, and absence of markets, are the main constraints faced by rural processing groups. If these constraints are seen in the context of the product priority-setting exercise, it seems that they effectively prevent rural processing groups from focusing on the production of higher-value products such as wines and oils.

Table 14.2. Summary of main constraints, in order of importance, faced by Magomero and Tabora workshop participants.

Magomero (Malawi)	Tabora (Tanzania)
Equipment/infrastructure	Processing equipment
Finance	Markets
Markets	Packaging
Packaging	Agricultural chemicals and equipment
Raw materials	Finance
Shelf life	Promotion and exposure
Training	Transport
Legislation	Training
Processing	Infrastructure
Storage space	Water
Transport	

14.3.3 Future processing strategies

Priority-setting workshops with fruit processing groups have highlighted the differences in perceptions regarding fruit trees and fruit products between commercial and community processors, as well as between processors in different areas of southern Africa. It has also shown that community processors see indigenous fruits as part of a range of fruits that can be used for processing.

An enterprise development model based on high-value preferred products such as wines and juices could form the basis for support to enterprise groups. Within such an enterprise development model, the following strategies should be considered:

1. Combine indigenous and exotic fruit trees in domestication and commercialization programmes.
2. Focus on training in the processing of high-value products such as wines and oils.
3. Support processing groups in acquiring appropriate technology.
4. Support processing groups in finding markets for their products.
5. Link processing groups with commercial enterprise partners who could assist them in producing the right products at the right quantities and quality levels.

Added to these strategies, technical research requirements to improve current processing activities should include studies to:

1. Develop and implement technology and protocols for the production of consumer-safe products (i.e. jams and dried products).
2. Develop protocols for storage of products.
3. Improve seed/nut and pulp extraction methods.
4. Develop more efficient and appropriate processing equipment.
5. Investigate physical and biochemical changes that take place during fruit maturation, processing and storage.
6. Improve the shelf life of products.

14.3.4 Processing training

The development of new and improved indigenous fruit processing technologies would only be effective if the information is disseminated to farmers and communities who can use it in their everyday lives. Traditional processing of indigenous fruits is decreasing in Zimbabwe due to changes in lifestyle and decreasing amount of fruits available for processing (Kadzere *et al.*, 2002). The World Agroforestry Centre (ICRAF) in collaboration with national partners began by training 198 farmer trainers in the Tabora region of Tanzania. The training focused on processing indigenous fruits into juices, jams and wine in four districts; as at 2002, the farmer trainers have in turn trained 2045 processors in 19 villages (Swai *et al.*, 2002).

Box 14.1. Fruits of training

The Mtendere Food Processing Club is a group of women established in Jerusalem village, approximately 30 km from Chipata, through the efforts of ICRAF. The purpose of establishing the processing group was to train the women in the vicinity of the village in the processing of fruit to improve their nutritional status, to create income-generating opportunities and to encourage the preservation and planting of trees, especially indigenous trees.

One of the Club's members was Mrs Anna Jere.* In 2001 she received 3 days of training from ICRAF in Malawi on wine-making. She started her business by producing 20 l of banana wine, using her own capital from baking bread and buns. From this first effort she proceeded to making 40 l of wine and, ever since, demand has grown to the extent that she has added jams and juices to her range of products. Mrs Jere produces products from bananas, pawpaws and mangos as well as from wild fruits such as A. boehmii, Flacourtia indica, U. kirkiana and P. curatellifolia. She collects these wild fruits in the nearby forest areas with the help of her six children. She has developed an innovative way of selling jam by baking buns and selling jam by the teaspoon to the people who buy the buns. She sells most of her products from home and found that different people prefer different products. The jam is bought by both adults and children, while juices are bought by children (who buy at any time, as opposed to adults who usually buy in the afternoon and evenings). Wine is principally bought by adults.

Additionally, wine and jam have been sold in Katete (>50 km away) and at agricultural shows in the area. The retail prices in Katete and at agricultural shows have been higher than in Jerusalem village. As the demand for the products increased, she began supplying wine, jam and juices to people within a 20 km radius around Jerusalem. Even the local traditional leader, Chief Mpezeni, is a regular client. During the harvesting season, Mrs Jere exchanges wine for maize, groundnuts (peanuts) and beans as payments in kind. One litre of groundnuts is exchanged for a 300 ml bottle of wine; a 2 l container of groundnuts is exchanged for a 750 ml bottle of wine, and a 3 l container of maize is exchanged for a 750 ml bottle of wine. Mrs Jere has managed to start a retail business, has bought three head of cattle and built a brick house with an iron-sheet roof from the sales of processed fruit products. These developments have helped to improve her standard of living and her ability to support her children, who are attending school. The purchase of cattle has had a tremendous impact, due to the fact that under Ngoni tradition women very rarely own cattle.

Mrs Jere has also started to train other women in the art of processing fruit products. She charges between US$4 and US$10 per person for training, which takes place at the women's club. So far 140 individuals have been trained. She is not worried about creating competition through the training/sharing of knowledge, since she believes that, despite an increased number of people processing fruit, standards will still differ.

*Mrs Jere passed away in 2005.

14.4 Commercial Marketing of Indigenous Fruits

14.4.1 Trends in the commercial natural product sector

Trends for commercial food products have changed drastically in the last decade, with a growing emphasis on variety and natural/organic products. Other interesting trends include a move back to traditional foods, i.e. made according to traditional recipes or customs. Of particular interest to producers in southern Africa are trends that see consumers becoming widely travelled and therefore more aware and interested in exotic foods and ethically traded products (fair trade, environmentally friendly/sustainable, etc.). Table 14.3 presents some of the more general European trends in the natural food market.

14.4.2 Certification

Certification is becoming very important in the natural products sector. Companies want to be able to trace the whole chain of natural products to be completely certain that products that claim to be natural, organic or fairly traded are indeed truly what they claim to be. Currently there are two types of certification that are most relevant for commercial companies working in the natural products sector.

Organic certification

Organic certification focuses on ensuring that crops are produced without any inorganic pesticides or fertilizers. The difficulty with organic certification is that the upfront costs are very high and few small companies or producers have the kind of money needed for organic certification. What makes it more difficult is

Table 14.3. European trends in the natural food market.

Trends	Important areas	Concerns
Organic foods and cosmetics	Herbal teas	Genetically modified foods
Natural foods and cosmetics	Ethnic foods	Misleading health claims on products
Fairly traded foods and cosmetics	Children's foods	Inaccurate ingredient and allergy labelling on packaging
Snack foods (particularly health bars)	Complementary medicines	Misleading advertising to children
Functional foods	Weight-loss products	Levels of salt in food products
Gourmet foods	Detoxification products	Over-harvesting of wild harvested natural products
Products free of genetically modified ingredients	Allergy-reducing products	
Under-utilized plant species	Organic/natural cosmetics	
Low-carbohydrate food products	Vegetarian/vegan products	
Food allergies and intolerances	Stress-relief products	
Environmentally responsible cultivation and harvesting		

Adapted from de Kock (2004).

that the costs have to be paid even though there is no guarantee that organic certification will be achieved. A first-time certification will typically incur the following costs among others: advice on the preparatory steps needed before an inspection, probably requiring a site visit by an experienced certification advisor; preparation entailing the collating of records from producers (and, indeed, further up the value chain) on the methods used and/or costs incurred in production; an inspection visit from a recognized certification agency; the establishment and implementation of a monitoring and record-keeping system to measure compliance; and repeat inspection costs. It is rare that a company or producer attains certification on the first attempt, and a second inspection is often necessary. Certification is also often an annual process requiring re-certification every year.

Fair trade certification

Fair trade is a global trading system that promotes workers' rights, protects the environment and sustains the ability of local producers to meet community needs. Essentially, it embeds social and environmental responsibility principles into economic trade chains for improved equitability and sustainability. It aims to secure a better deal for producers in the developing world through setting stable product prices that cover both their production costs and a premium that their organization can reinvest either in the business or social and environmental schemes in the wider community. The primary objective of fair trade is to develop access to niche markets in developed countries for producers in developing countries, who tend to be marginalized by conventional trading structures. The fair trade system has been established to give small-scale rural producers in developing countries the opportunity to be supported to the point where they can sustainably and equitably engage Western markets. This enables them to reach appropriate quality and quantity standards, as well as ensuring a fair price for their products.

The goals of fair trade are:

1. To improve the livelihoods and well-being of producers by improving market access, strengthening producer organizations, paying a better price and providing continuity in the trading relationship.
2. To promote development opportunities for disadvantaged producers, especially women and indigenous people, and to protect children from exploitation in the production process.
3. To raise awareness among consumers of the negative effects on producers of international trade so that they exercise their purchasing power positively.
4. To set an example of partnership through dialogue, transparency and respect.
5. To campaign for changes in the rules and practice of international trade.
6. To protect human rights by promoting social justice, sound environmental practices and economic security.

14.5 Commercial Processing of Indigenous Fruits

14.5.1 Adding value to natural products

Value addition occurs when a raw material is changed in a way that increases its selling price and profit margin, thus allowing the producer or processor to make more money from that raw material. As a general rule, most southern African countries have been exporters of raw materials or semi-processed raw materials and importers of processed materials. The reasons for this are numerous and include: lack of machinery to undertake processing; lack of skills to undertake research and development on fruit processing; lack of skills to run processing facilities; lack of funds to establish processing facilities; high maintenance costs for spare parts that have to be imported; high unit cost of running small processing facilities, making African processed products uncompetitive on world markets; and prohibitive costs of certification of processing facilities. Value addition to agroforestry products can be conducted at primary and secondary processing levels. At primary level, rural communities can become involved in processing, harvesting and grading of raw materials as well as in the extraction of pulps, herbs and oils. Processing at this level requires low inputs and skill levels, and can take place close to the source of raw materials. Generally no certification is required at this level.

Secondary processing would normally take place in a factory environment. Specialized skills and equipment are required to process pulps, herbs and oils into finished products of high quality. Factories normally need to have certification for processing of food and cosmetic products and require reliable electricity and water supplies. Secondary processing facilities are also located closer to the end product market than to the source of raw material. There are both advantages and disadvantages for producers and processors of doing primary processing at community level. Some of the advantages and disadvantages are presented in Table 14.4.

14.5.2 Current commercialization activities

The 'Commercial Products from the Wild' group conducted a regional survey of commercial indigenous fruit processors in 2003. A total of 13 organizations in Malawi, two in Botswana, 11 in Namibia and 12 in Zimbabwe were interviewed. The organizations ranged from government departments, private companies, trade organizations, private farms, nurseries and non-governmental organizations (NGOs). Some of the key results of the survey include:

1. Indigenous fruit commercialization is very poorly developed in Malawi. Only a couple of organizations are actually trading in such products, ranging from fruit juice to seedlings. Only the 'Sustainable Management of Indigenous Forests' project in Blantyre has a trade value of more than US$5000/year.

2. Zimbabwe is probably the best developed in terms of indigenous fruit commercialization in southern Africa, with high volumes of trade in indigenous products recorded in this country. Most of the organizations process their own

Table 14.4. Advantages and disadvantages of community-level primary processing.

	Advantages	Disadvantages
Producers	Reliable alternative source of income from fruits that would otherwise be wasted Reliable alternative source of income from primary processing activities Profit share and/or community projects if company is fair trade No skills or education required Opportunities for women to become more self-reliant Increased incomes result in more money being spent in the community Seasonal work so can concentrate on farming activities Recognition of contribution through acknowledgement on packaging and marketing activities of company	Quantity and quality of fruit available not reliable, therefore income not guaranteed Prices may be too low to be viable Harvesting season may interfere with other activities (e.g. farming, child-rearing) Income only during harvesting season Commercial partner may decide not to produce product every year if sales are slow Commercial partner may go out of business or discontinue product
Processors	Transportation of pulp cheaper than fruits Cheaper labour than in cities Minimal skills or training required Minimal overheads Social aspect of community level involvement (fair trade) good for marketing	Members of group may change, which could necessitate retraining each year Hygiene standards may not be good enough Group may not be efficient enough to produce required quantities of pulp Communication problems with rural areas Transportation problems: distances and state of roads Prices may be too high to be viable

products, but a couple are dependent on subcontractors and other organizations to assist with processing. The Southern Alliance for Indigenous Resources (SAFIRE) and PhytoTrade Africa play the biggest roles in supporting these institutions. Assistance is also coming from People and Plants UNESCO/WWF, the Plant Oil Producers Association (POPA), Agribusiness in Sustainable Natural African Plant Products (ASNAPP), Cultural Relations and Indigenous Awareness Associates (CRIAA), etc.

3. The Zimbabwean organizations are very well connected to international trading partners. Trading partners include fair traders: e.g. Oxfam (Australia), Commercio Alternativo (Italy), Fruxotic (Germany). Products are exported to Japan, Australia, Italy, Germany, South Africa and the UK. A large percentage of the products are sold to pharmaceutical and food-processing industries. It is interesting to note that products are targeted at higher-income, environmentally aware consumers and tourists.

4. The organizations consulted in Botswana and Namibia are all trading in fruit trees or products. Three of the organizations have an annual trade value of more than US$5000 and seven have a trade value of more than US$1000. The organizations have limited processing support, but their products are well supported by government and international organizations such as the UK Department for International Development (DFID), the German Agency for Technical Cooperation (GTZ) and the Namibia–Finland Forestry Programme. Products are mostly sold to local consumers and only a few organizations are targeting foreign tourists. Three organizations are exporting their products to Europe.

The organizations consulted focus on a large range of indigenous fruit tree species. The most preferred tree species used commercially by processors are summarized in Table 14.5.

14.5.3 Challenges related to indigenous fruit commercialization

Small-scale natural products-based businesses in the various countries of the southern African region experience different challenges. Many of the organizations consulted indicated that they need assistance, especially with aspects such as marketing, product development and the propagation of indigenous fruit trees. Some of the organizations feel isolated from the commercialization work that other organizations are doing, indicating a clear need for better networking. Some of the key areas where organizations indicated a need for assistance include:

1. Technical
   - Information on processing, packaging and marketing.
   - Improvement of processing technology.
   - Improved equipment and business management advice.
   - Product preservation.
   - Equipment shortages.
   - Appropriate processing technology.
2. Policy and institution
   - Policies or guidelines for horticultural development.
   - Lack in government technical advisory services.
   - Markets and marketing.

Table 14.5. Tree species used by commercial fruit processors.

Malawi	Zimbabwe	Namibia and Botswana
Uapaca kirkiana	*Adansonia digitata*	*Azanza garckeana*
Ziziphus mauritiana	*Ziziphus mauritiana*	*Schinziophyton rautanenii*
Adansonia digitata	*Sclerocarya birrea*	*Strychnos cocculoides*
Strychnos cocculoides	*Vignea* spp.	*Sclerocarya birrea*
Parinari curatellifolia	*Jatropha curcas*	*Vangueria infausta*
Tamarindus indica		*Ximenia caffra*
(not indigenous but naturalized)		*Ricinodendron rautanenii*

3. Raw material supply

- Propagation of indigenous fruit trees.
- Seasonality of fruits.
- Raw material supply inconsistency.
- Low quality of fruit.
- Pest and diseases related to fruit rot.

Box 14.2. Tulimara: Speciality Foods of Africa

In September 2002, Speciality Foods of Africa Pvt Ltd (SFA) was incorporated with two shareholders, SAFIRE and Caroline de Kock (the previous Marketing Director of Tulimara Pvt Ltd, now Managing Director) in Zimbabwe. SFA is the owner of the brand name Tulimara. This small indigenous plant-processing company currently has 10 products in the Tulimara range, with plans to launch several more. These current products include:

- Makoni herbal tea (natural and lemon flavours) (*Fadogia ancylantha*).
- Masau jam (*Ziziphus mauritiana*).
- Marula jelly (*Sclerocarya birrea*).
- Mazhanje jam (*Uapaca kirkiana*).
- Dried masau slices (fruit leather) (*Ziziphus mauritiana*).
- Canned nyimo beans (*Vigna subterranea*).
- Canned nyemba beans (*Vigna unguiculata*).

SFA does not have its own processing, storage or distribution facilities. It works closely with other private-sector partners who are responsible for transportation, processing and distribution. SFA's role is to ensure that all linkages in the chain function efficiently. SFA's major responsibilities are:

- Ensuring that raw materials are harvested and delivered to processors.
- New product development.
- Design and supply of packaging and labels.
- Marketing, sales and merchandising.
- Export marketing and sales.

SFA currently sells all its products in supermarkets throughout Zimbabwe. Export markets include Italy, the UK, Sweden, Australia and the USA. What makes this company unique is its community-based raw material supply networks. SFA purchases raw materials directly from rural producer groups wherever possible. SFA meets with rural producer groups in advance of harvesting to discuss quantities, prices and logistics of collection. It is the responsibility of producer groups to ensure that raw materials are of the agreed quality and quantity, and delivered at the collection points on the date and time agreed in advance.

SFA works with SAFIRE, which is linked to rural producer groups in Zimbabwe. SAFIRE is responsible for training the groups in sustainable harvesting and basic business skills. In the case of groups that are involved in primary processing (Mazhanje and Makoni) SAFIRE provides training in the processing according the SFA's specifications. SAFIRE also provides equipment where necessary for the groups through its donor-funded programmes.

SAFIRE acts as an intermediary between SFA and the producer groups in the first few years of operation until they are considered to be ready to operate on their own. When that happens, SFA and the producer group negotiate directly without the involvement of SAFIRE.

SFA has fair trade agreements with groups that are considered sustainable and able to operate on their own. The fair trade agreement includes information about prices, inflationary increases, profit share, minimum order quantities and dispute resolution.

14.6 Conclusions

Indigenous fruit commercialization activities are still in their infancy in southern Africa. Fresh fruits are being traded and many households process fruits. Commercial companies are also processing fruit products with limited success, but there are many problems along the supply and processing chains. Bad harvesting, packaging and transport activities decrease the shelf life of indigenous fruits dramatically and lead to spoilage and waste. Fruit markets lack the necessary infrastructure and support systems to function optimally. Processing technology is outdated and product quality is low. Commercial processors are experiencing problems related to government support and technology information.

Commercial fruit processing is, however, one of the few comparative advantages that rural economies have. If numerous commercialization ventures were implemented across the region, then impacts would not only improve the incomes of the participating households, but community level economies would be improved as local spending increased. Accelerated local trading (or local economic development) is beneficial in the same way that import substitution is beneficial at a national level. Increased local trading contributes to increased circulation of monies within rural communities, limiting the 'export' of capital to urban centres. The greater the number of times that monies are circulated at the rural level, the greater the benefits for local society.

Partnerships and networks between communities, farmers and commercial organizations should be encouraged. Through such partnerships and networks, the communities and farmers could benefit from expertise and already developed markets. So, for example, could the collectors and processors of fruit products benefit from linkages with organic and fair trade organizations that could assist them in selling their products into these lucrative markets. The commercial organizations could benefit through improved supply chains and access to raw materials. Successful fruit product trading could be a significant contributor to rural welfare in the region.

References

Akinnifesi, F.K., Kwesiga, F., Mhango, J., Chilanga, T., Mkonda, A., Kadu, C.A.C., Kadzere, I., Mithöfer, D., Saka, J.D.K., Sileshi, G., Ramadhani, T. and Dhliwayo, P. (2006) Towards the development of miombo frit trees as commercial tree crops in southern Africa. Forests, Trees and Livelihoods 16, 103–121.

de Kock, C. (2004) World food trends and marketing strategies. Presentation at the World Agroforestry Centre Regional Training Course on Agroforestry Marketing and Enterprise Development Support. 31 May to 4 June 2004, Commonwealth Youth Centre, Lusaka, Zambia.

Fox, F.W. and Norwood Young, M.E.N. (1982) Food from the Veld: Edible Wild Plants of Southern Africa Botanically Identified and Described. Delta Books, Craighall, South Africa, 422 pp.

Ham, C. and Akinnifesi, F. (2004) Priority Fruit Species and Products for Tree Domestication and Commercialization in Zimbabwe, Zambia, Malawi and Tanzania. Research Report of the World Agroforestry Centre and CPWild Research Alliance,

World Agroforestry Centre, Lilongwe, Malawi [see http://www.cpwild.co.za/DocsCPW.htm].

Kaaria, S.W. (1998) The economic potential of wild fruit trees in Malawi. MSc thesis, The University of Minnesota, Minneapolis, Minnesota.

Kadzere, I., Hove, L., Gatsi, T., Masarirambi, M.T., Tapfumaneyi, L., Maforimbo, E. and Magumise, I. (2002) Current status of post-harvest handling and traditional processing of indigenous fruits in Zimbabwe. In: Rao, M.R. and Kwesiga, F.R. (eds) *Proceedings of the Regional Agroforestry Conference on 'Agroforestry Impacts on Livelihoods in Southern Africa: Putting Research into Practice'.* World Agroforestry Centre (ICRAF), Nairobi, Kenya.

Karaan, M., Ham, C., Akinnifesi, F., Moombe, K., Jordaan, D., Franzel, S. and Aithal, A. (2005) *Baseline Marketing Surveys and Supply Chain Studies for Indigenous Fruit Markets in Tanzania, Zimbabwe and Zambia.* Research Report of the World Agroforestry Centre and CPWild Research Alliance [see http://www.cpwild.co.za/DocsCPW.htm].

Moelich, E. and Muller, N. (2000) *Consumer Test of Kei Apple Juices of Different Concentrations.* Report compiled by the Department of Consumer Science, University of Stellenbosch for the Department of Forest Science, University of Stellenbosch, South Africa.

Mumba, M.S., Simon, S.M., Swai, S.M. and Ramadhani, T. (2002) Utilization of indigenous fruits of miombo woodlands: a case of Tabora District, Tanzania. In: Rao, M.R. and Kwesiga, F.R. (eds) *Proceedings of the Regional Agroforestry Conference on 'Agroforestry Impacts on Livelihoods in Southern Africa: Putting Research into Practice'.* World Agroforestry Centre (ICRAF), Nairobi, Kenya.

Ndabikunze, B.K., Mugasha, A.G., Chamshama, S.A.O. and Tiisekwa, B.P.M. (2000) Nutrition and utilization of indigenous food sources in Tanzania. In: Shumba, E.M., Lusepani, E. and Hangula, S. (eds) *Proceedings of a Network Meeting on Domestication and Commercialisation of Indigenous Fruit Trees in the SADC Region.* SADC Tree Seed Centre Network, Windhoek, Namibia.

Ndabikunze, B.K., Mugasha, A.G., Chamshama, S.A.O. and Tiisekwa, B.P.M. (2002) Nutritive value of selected forest/woodland edible fruits, seed and nuts in Tanzania. In: Rao, M.R. and Kwesiga, F.R. (eds) *Proceedings of the Regional Agroforestry Conference on 'Agroforestry Impacts on Livelihoods in Southern Africa: Putting Research into Practice'.* World Agroforestry Centre (ICRAF), Nairobi, Kenya.

Ngwira, T.N. (1996) Utilization of local fruit in wine making in Malawi. In: Leakey, R.R.B., Temu, A.B., Melnyk, M. and Vantomme, P. (eds) *Proceedings of an International Conference on Domestication and Commercialization of Non-timber Forest Products in Agroforestry Systems: Proceedings of an International Conference, 19–23 February 1996, Nairobi, Kenya.* Non-wood Forest Products. FAO, Rome.

Ramadhani, T. (2002) *Marketing of Indigenous Fruits in Zimbabwe.* Socio-economic Studies on Rural Development, Vol. 129. University of Hanover, Germany.

Ramadhani, T., Chile, B. and Swai, R. (1998) Indigenous miombo fruits selected for domestication by farmers in Tanzania. In: Maghembe, J.A., Simons, A.J., Kwesiga, F. and Rarieya, M. (eds) *Proceedings of a Conference on 'Selecting Indigenous trees for Domestication in Southern Africa: Priority Setting with Farmers in Malawi, Tanzania, Zambia and Zimbabwe'.* World Agroforestry Centre (ICRAF), Nairobi, Kenya.

Ruffo, C.K., Birnie, A. and Tengas, B. (2002) *Edible Wild Plants of Tanzania.* RELMA Technical Handbook, RELMA, Nairobi, Kenya.

Saka, J.D.K. (1995) The nutritional value of edible fruit trees: present research status and future directions. In: Maghembe, J.A., Ntupanyama, Y. and Chirwa, P.W. (eds) *Improvement of Indigenous Fruit Trees of the Miombo Woodlands of Southern Africa.* International Centre for Research in Agroforestry, Nairobi, Kenya, pp. 50–57.

Saka, J.D.K., Mwendo-Phiri, E. and Akinnifesi, F.K. (2001) Community processing and nutritive value of some miombo indigenous fruits in central and southern Malawi. In: Kwesiga, F., Ayuk, E. and Agumya, A. (eds) *Proceedings of the 14th Southern African Regional Review and Planning Workshop, 3–7 September 2001.* World Agroforestry Centre (ICRAF), Harare, Zimbabwe.

Saka, J.D.K., Swai, R., Mkonda, A., Schomburg, A., Kwesiga, F. and Akinnifesi, F.K. (2002) Processing and utilisation of indigenous fruits of the miombo in southern Africa. In: Rao, M.R. and Kwesiga, F.R. (eds) *Proceedings of the Regional Agroforestry Conference on 'Agroforestry Impacts on Livelihoods in Southern Africa: Putting Research into Practice'.* World Agroforestry Centre (ICRAF), Nairobi, Kenya, pp. 343–352.

Schomburg, A., Mhango, J. and Akinnifesi, F.K. (2001) Marketing of masuku (*Uapaca kirkiana*) and masawo (*Ziziphus mauritiana*) fruits and their potential for processing by rural communities in southern Malawi. In: Kwesiga, F., Ayuk, E. and Agumya, A. (eds) *Proceedings of the 14th Southern African Regional Review and Planning Workshop, 3–7 September 2001.* World Agroforestry Centre (ICRAF), Harare, Zimbabwe.

Smock, R.M. and Neubert, A.M. (1950) *Apples and Apple Products.* Interscience Publishers, New York.

Swai, R.E.A. (2001) Efforts towards domestication and commercialisation of indigenous fruits of the miombo woodlands in Tanzania: research and development highlights. In: Kwesiga, F., Ayuk, E. and Agumya, A. (eds) *Proceedings of the 14th Southern African Regional Review and Planning Workshop, 3–7 September 2001.* World Agroforestry Centre (ICRAF), Harare, Zimbabwe.

Swai, R.E., Mbwambo, L., Maduka, S.M., Mumba, M. and Otsyina, R. (2002) Domestication of indigenous fruit and medicinal trees in Tanzania: a synthesis In: Rao, M.R. and Kwesiga, F.R. (eds) *Proceedings of the Regional Agroforestry Conference on 'Agroforestry Impacts on Livelihoods in Southern Africa: Putting Research into Practice'.* World Agroforestry Centre (ICRAF), Nairobi, Kenya.

Tiisekwa, B.P.M., Ndabikunze, B.K., Samson, G. and Juma, M. (2002) Suitability of some indigenous tree fruits for manufacturing juice and jam in Tanzania. In: Rao, M.R. and Kwesiga, F.R. (eds) *Proceedings of the Regional Agroforestry Conference on 'Agroforestry Impacts on Livelihoods in Southern Africa: Putting Research into Practice'.* World Agroforestry Centre (ICRAF), Nairobi, Kenya.

15 The Feasibility of Small-scale Indigenous Fruit Processing Enterprises in Southern Africa

D. du P.S. Jordaan,[1] F.K. Akinnifesi,[2] C. Ham[3] and O.C. Ajayi[2]

[1]*Commercial Products from the Wild Group (CP Wild), University of Pretoria, South Africa;* [2]*World Agroforestry Centre, SADC–ICRAF Agroforestry Programme, Chitedze Agricultural Research Station, Lilongwe, Malawi;* [3]*Commercial Products from the Wild Group (CP Wild), University of Stellenbosch, Matieland, South Africa*

15.1 Introduction

In the tropics, indigenous fruit trees play vital roles in the livelihoods of rural households as well as in food and nutritional security, especially during periods of famine and food scarcity (Akinnifesi *et al.* 2004, 2006, Chapter 8, this volume), although they are becoming increasingly important as a main source of food to supplement diets in better times (Saka *et al.*, 2002; Mithöfer, 2005; Saka *et al.*, Chapter 16, this volume). Indigenous fruit trees offer great potential to rural dwellers in sub-Saharan Africa who are caught in a poverty trap, living with chronic hunger and poverty, and are ravaged by ill-heath (such as HIV/AIDS, malaria and other diseases) and malnutrition; a condition that is only likely to get worse (FAO, 2002; Blair Commission Report, 2005). The per capita consumption of foods and fruits is low among rural households. Although much of the emphasis in southern Africa in the last decade has been on measures that aim at increasing staple food production for smallholder production, it is becoming increasingly clear that improving the productivity of staple crops alone will neither meet the full subsistence requirement of rural households nor provide feasible opportunities for getting out of the vicious cycle of poverty. Agricultural productivity must be linked effectively with, and be responsive to, market demand, and smallholders must see the whole farm in terms of business portfolios of options. Thus there is a need for farmers to develop business skills, acquire better access to market information, and focus greater attention on product quality and the opportunities for value adding, both for wild tree products and for tree crops grown on-farm. Recent research has suggested that rural communities can increase their nutritional well-being and incomes by utilizing and marketing fruit tree products from forests and horticultural tree

 Indigenous Fruit Trees in the Tropics: Domestication, Utilization and Commercialization (eds F.K. Akinnifesi *et al.*)

crops grown on-farm (Mithöfer, 2005; Akinnifesi *et al.*, 2006). An *ex ante* impact analysis in southern Africa (particularly in Malawi and Zimbabwe), indicates that households marketing fruits have been able to maintain income flows above the poverty line throughout the year (Mithöfer *et al.*, 2006). Tree crops are long-term assets that can help smallholder farmers diversify their household income and product portfolios, becoming capable of producing multiple streams of income and meeting domestic needs throughout the year (Akinnifesi *et al.*, 2004).

A large number of fruit trees in southern African have, throughout the ages, been used as sources of nutrition, food and medicine (Akinnifesi *et al.*, 2006). Traditionally these have been collected from the wild mainly for household consumption and, as a result, very few commercial products from indigenous plants have emerged. Many rural poor communities are located in regions of southern Africa that are rich in indigenous plants. Population growth, poverty, and a lack of alternative income-generating opportunities are, however, contributing to severe degradation of these natural resources through overgrazing, uncontrolled burning, and the gathering of fuelwood. A possible solution to the increasing trend of poverty in resource-poor communities and the accompanying destruction of natural resources is to add value to the currently underutilized natural resources available to these rural communities. Such initiatives are expected to provide income-generating opportunities for these communities and could serve as an incentive for conserving the natural resources.

In partnership with various institutions, ICRAF has developed strategies for indigenous fruit tree (IFT) domestication, product development and commercialization (Akinnifesi *et al.*, 2006). The major goal of this initiative is to improve the well-being of rural dwellers in the miombo ecosystem and develop their products. The efforts include studies along the value chain: production economics and *ex ante* analysis (Mithöfer, 2005; Mithöfer *et al.*, 2006), species and product prioritization (Ham and Akinnifesi, 2006), IFT horticulture, processing, pilot enterprise development and market chain analysis (Ramadhani, 2002; Akinnifesi *et al.*, 2006; Ham *et al.*, Chapter 14, this volume), and grassroots training and capacity-building of communities and partners (Akinnifesi *et al.*, 2004, 2006). In establishing small-scale indigenous fruit processing plants, it is recognized that natural product businesses and supply chains have unique characteristics and, consequently, problems that differentiate them from ordinary businesses and supply chains. These include the biological nature of production, seasonality, product quality, short shelf life, underdeveloped markets, geographical dispersion and distance from markets. Because of these unique characteristics and problems, a detailed feasibility study must be conducted before embarking on investment in a commercial fruit-processing venture. A feasibility study is a structured way to efficiently organize information that is needed for confident decision making regarding whether a specific proposed business venture is profitable and technically, financially and environmentally viable. The purpose of this chapter is to undertake feasibility studies for three non-wood tree product enterprises based on indigenous fruits from the miombo ecosystem in Malawi, Tanzania and Zimbabwe.

15.2 Conceptual Framework for Analysing Feasibility

The conceptual framework used to conduct these feasibility studies comprises three main components and is detailed below. Lecup and Nicholson (2000) proposed a conceptual framework to determine the feasibility of community-based tree and forest product enterprises. In their approach, four main areas of enterprise development were assessed across three levels, which include the market and economic environment, the resources environment, the social and institutional environment and the science and technology environment from community to international level. Doyer and Vermeulen (2000) proposed an extension to the framework, represented in Fig. 15.1. The Doyer and Vermeulen framework is initialized with the identification of an opportunity where the feasibility of the opportunity is assessed in the four core areas, as identified by Lecup and Nicholson (2000). If the opportunity is assessed to be viable, a business plan is compiled to define process of implementation by means of commercialization. If the opportunity is assessed to be unfeasible, the process is terminated. The following sections detail the four primary components of the feasibility assessment as proposed in the conceptual framework developed by Lecup and Nicholson (2000).

Taylor (1999) assents to these findings when he concludes that successful community-based forest product enterprises are very dependent on six central points:

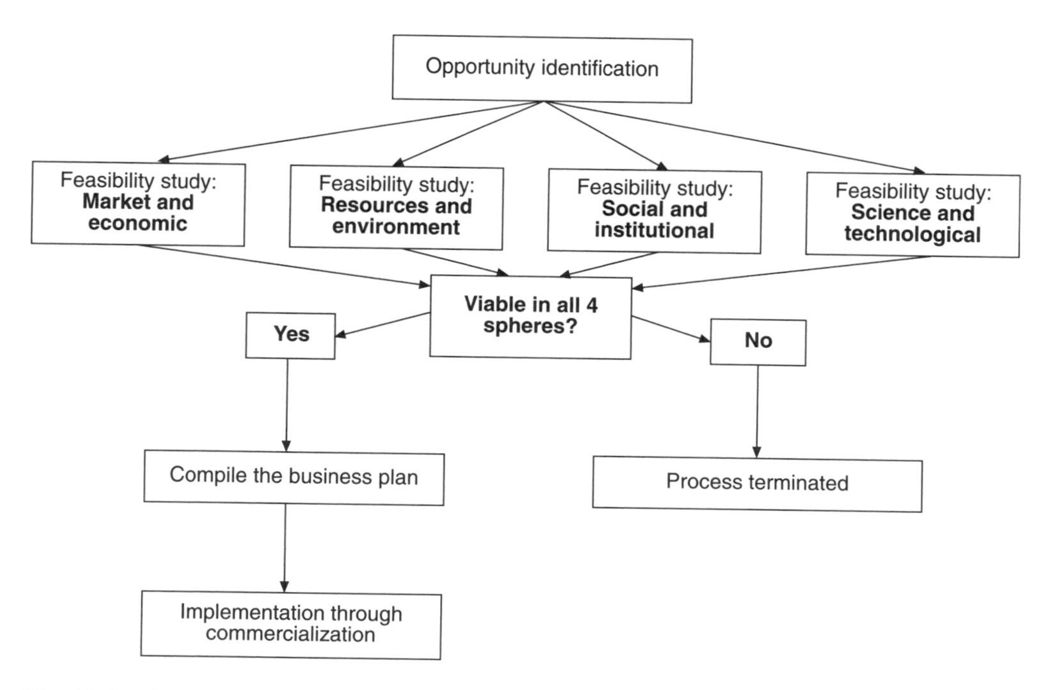

Fig. 15.1. Flow diagram for the exploration of new opportunities.

1. Support communities through clear land tenure and policy support: where local groups are well organized and can control forest access, rural enterprises tend to fare better. A clear sense of group identity, cooperative behaviour and established rights to the resource can all help.
2. Start with local markets: local markets are easier to enter and monitor than foreign markets, which often require heavy capital investment and large product volume and which tend to be vulnerable to product substitution. Enterprises may diversify to larger markets if such diversification is feasible in terms of sustainable harvests, product quality and investment requirements.
3. Focus on quality products and building management and entrepreneurial skills: these elements can be supported through coalitions involving local partners, local and national NGOs and international technical organizations.
4. Support enterprises through policies facilitating credit and trade: coherent government policies that support community-based forest product enterprises are needed, including mechanisms to make credit more available to small enterprises (such as the recognition of stands of commercial tree species as collateral) and the removal of counterproductive price controls.
5. To inspire policy makers to support rural enterprises with a coherent policy framework, FAO has proposed better accounting of the economic importance of community-based forest product enterprises, including a system for grouping community-based forest product enterprise trade statistics within existing commodity classification systems (Chandrasekharan, 1995).
6. Make the most of local knowledge and resources: maintaining cultural integrity remains an under-appreciated element of forest sustainability, particularly in remote communities and upland areas. Researchers in cultivation, marketing and processing should consider the best available knowledge from traditional as well as scientific sources, in order to optimize forest management and the contribution of community-based forest product enterprises to the lives of rural people.

15.2.1 Market and economic environment

Assessment of the market and the economic environment is an investigation into the supply of raw materials, market potential of products, competition, constraints to business entry and, margins and profitability (Lecup and Nicholson, 2000). Market and economic assessment is important to identify opportunities, strengths and constraints in the market channels and to gather information about the business environment. These opportunities, strengths and constraints need to be identified before a new marketing system can be devised or the current marketing system can be improved. Through this exercise, obstacles to the marketing of products can be identified. The goal is to gather information from key players involved in the marketing of the product. According to the opportunities identified, studies of the other areas of enterprise development are then undertaken.

15.2.2 Resources and the environment

The assessment of resources and the environment entails investigation into the availability (in time: seasonality; in space: time needed to find and harvest) of the raw materials; the regenerative potential and impact of harvesting on survival of the species; and the impact of harvesting on the environment. A fundamental concern of market analysis methodology is to identify approaches to developing products without destroying the resource base. A product will be considered for development only if its resource base will not suffer as a result of an increase in the harvesting rate, or if harvesting can be supplemented or substituted by cultivation. Therefore, it is important to get a clear picture of the status and quantity of forest resources before starting an enterprise. This can also be a tool to monitor environmental impact once an enterprise gets under way.

15.2.3 Social and institutional environment

Assessment of the social and institutional environment entails investigation into the indirect benefits of the project for the community; the contribution of the project to incomes of the community; the community's experience with the product; the potential for creating employment; and the gender impact that the project is bound to have. Social and institutional criteria must be considered equally as important as the economic, resource management and technical criteria of a potential enterprise. Potential activities should help, or at least not harm, the most economically vulnerable members of the community, especially women and children. The objectives of a social/institutional analysis are to:

1. Gather detailed information about the potential producers. This involves asking specific questions, e.g. 'To what extent is the target group able to respond to the marketing activities?' 'Are there external factors that prevent participation?'
2. Assess the direct and indirect potential impacts of the proposed products on the community in order to design socially sensitive actions and to minimize (or eliminate) potential negative impacts on women and other vulnerable groups within the community.
3. Analyse the legal aspects affecting the target group, such as access to resources and user rights.
4. Investigate sources of capital for small-scale industries and the opportunities for community members to get access to formal credit.

15.2.4 Science and technology environment

Assessment of the science and technology environment entails investigation into the suitability of a technology for members of rural target groups, the processing location and processing technology required, the status of the infrastructure and the level and availability of human resources/skills and expertise. Some of the main issues to be addressed regarding the science and technology aspects related to marketing of the products are the following:

1. What is the current level of technical skills for natural resource extraction, postharvest operations, processing and marketing?
2. How could these technical skills be improved to meet the requirements of new markets?
3. What are the costs of the technical improvements?

All the actors in the market chain should be interviewed concerning the uses and requirements of the products (such as size, colour and quality), processing techniques and related aspects. The study can be conducted with local or national professionals. For example, in the case of a small enterprise, it is usually easy to learn the price of simple equipment from local sources, but for a larger-scale enterprise it might be necessary to visit equipment distribution centres. All the components of the technical study include a cost analysis. This analysis would be indicative of whether the proposed technology, or the skills training needed to bring a competitive product to market, is too costly in relation to the value of the product or not. In this case, either the product should be eliminated or an alternative design of the product should be considered and its feasibility should be studied.

15.3 Methodology

The methodology followed in assessing the feasibility of the various enterprises is based on the conceptual framework described above. The products for which feasibility studies were conducted were chosen following general scoping studies for potential processing groups as identified by the Commercial Products from the Wild Group (CP Wild) and the World Agroforestry Centre (ICRAF–Southern Africa). The specific enterprises were selected using a broad market analysis method, taking into consideration possible market opportunities. These scoping studies took the perceived market opportunities; the capacity of the specific community enterprise, the perceived capacity of the selected community-based enterprises to reliably and consistently produce the specific product, and the resources that were readily at the disposal of the enterprise into account before deciding on a specific product. Table 15.1 details the country, the specific location of the proposed enterprise, and the product for which a feasibility analysis was conducted, to determine the viability of these enterprises.

Data pertaining to the four core areas of feasibility analysis, the market and economic environment, the resource environment, the social and institutional environment, and the science and technology environment, were collected. For each of the sites, primary data were gathered by means of formal and informal interviews and questionnaires with five–ten stakeholders in each of the selected countries. Secondary data such as country statistical data, trade statistics, demographic data and socio-economic data were also consulted to complement the primary data. In some instances it was necessary to make assumptions when data were especially hard to come by or not available. These specific assumptions were educated assumptions and are clearly documented in each of the feasibility studies.

Table 15.1. Feasibility studies conducted in selected southern African countries.

Country	Location	Product enterprise
Malawi	Magomero	Fruit juice concentrate
Tanzania	Tabora	Fruit juice concentrate
Zimbabwe	Harare	Fruit jam and cereal bars

15.4 Results

15.4.1 Malawi: fruit juice concentrate

The market and economic environment

A market opportunity was identified for the domestic production of good-quality fruit juice concentrates in Malawi to be sold to manufacturers in the growing fruit juice nectar-producing sector in Malawi. When the studies were conducted, there was no domestic competition there; all juice concentrates were imported at relatively high cost. At the time of the study, the two major difficulties in setting up enterprises in Malawi that interview respondents mentioned were lack of capital (67%) and lack of technical knowledge (13%). Analysis of the financial feasibility of the proposed enterprise revealed that (based on the assumptions that were made) the enterprise was potentially profitable, was projected to have a healthy cash flow, and would be viable over the longer term. The analysis was based on the maximum output capacity of a small-scale fruit juice concentrate processing plant and the expected output (and therefore income) and the fixed and running costs associated with producing and marketing fruit juice concentrate. A summary of the financial feasibility appears in Table 15.2.

Resource environment

There is a lack of comprehensive and detailed data on current natural resources available in Malawi. The Malawi feasibility study therefore made the assumption that the proposed enterprise could be environmentally feasible when fruit is sourced from both natural and domesticated sources.

Social and institutional environment

The social and institutional feasibility assessment of a fruit juice processing enterprise at Magomero reveals that the surrounding communities would benefit in a number of ways from its establishment. Direct benefits include the creation of employment opportunities for some of the surrounding villages as well as the creation of a 'market' for currently under-utilized resources in fruits, especially indigenous fruits. Indirect benefits include an anticipated decrease in the number of households living under the poverty and ultra-poverty line, a decrease in malnutrition, a decrease in the severity of the impact of HIV/AIDS and an increase in incomes for communities, especially for women. A lack of concise

Table 15.2. Results from indigenous fruit enterprise feasibility studies conducted in Malawi, Tanzania and Zimbabwe.

Income statement	Malawi juice concentrate	Tanzania juice concentrate	Zimbabwe Baobar	Zimbabwe jam
Gross production value ($)	107,400	137,359	61,090	20,209
Total cost ($)	61,700	108,187	33,080	11,336
Net income ($)	45,700	29,172	28,010	8,873
Tax ($)	16,000	10,210	9,803	3,105
Net profit after tax ($)	29,700	18,962	18,206	5,767
% Net profit[a]	27.7%	13.8%	29.8%	28.5%
Cash flow analysis				
Cash flow	Net positive for all months	Net positive for all months	Net positive for all months	Net positive for all months
Capital budget				
Net present value	Positive over selected period	Positive over selected period	Positive over selected period	Positive over selected period
Breakeven analysis				
Breakeven price ($)	8.50 per 20 kg can	14.64 per 20 kg can	0.34 per 50 g bar	1.22 per 410 g jar

[a] Profit as % of gross value of production.

information regarding specific communities, their levels of poverty, levels of malnutrition, the impact of HIV/AIDS and levels of income undermines the estimation of quantifiable impacts and reduces the result to a comment that merely indicates the 'direction' of the impact rather than direction and amplitude.

Science and technology environment

Processing technology that is ideally suited to the specific technical requirements of the proposed processing enterprise at Magomero, Malawi, is available from custom manufacturers in South Africa and can be readily imported into Malawi. The nature of the processing technology is also such that it is relatively easy to operate and maintain. The human resources/skills and expertise available at Magomero are also sufficient to manage and maintain the processing enterprise. The location at Magomero is further ideally suited to the establishment of a processing enterprise of this nature since it is reasonably well located with regard to resources, input supplies and access to distribution channels.

15.4.2 Tanzania: fruit juice concentrate

The market and economic environment

Like Malawi, a market opportunity was identified for the domestic production of good quality fruit juice concentrates in Tanzania to be sold to manufacturers in the growing fruit juice nectar-producing sector in the country. There is currently little domestic competition; a large proportion of juice concentrates are imported at relatively high cost. Analysis of the financial feasibility of the proposed enterprise revealed that (based on the assumptions that were made) the enterprise was potentially profitable, was projected to have a healthy cash flow and would be viable over the longer term. Once again the analysis was based on the maximum output capacity of a small-scale fruit juice concentrate processing plant and the expected output (and therefore income) and the fixed and running costs associated with producing and marketing fruit juice concentrate. A summary of the financial feasibility appears in Table 15.2.

Resource environment

As in the case of Malawi, there is a lack of comprehensive and detailed data on the currently available natural resources in Tanzania and, specifically, detailed information for Tabora. Therefore, the Tanzania feasibility study also had to make the assumption that the proposed enterprise could be environmentally feasible when fruit is imported from both natural and domesticated sources.

Social and institutional environment

Currently the fruit processing groups in Tabora are interested in processing but have a limited local market. The fruit juice enterprise would focus on exporting products out of the Tabora region. The enterprise would create employment opportunities as well as a new market for indigenous fruit products that could increase household income. Again, a lack of concise information regarding specific socio-economic indicators of the communities in the vicinity of the proposed enterprise undermines the estimation of quantifiable impacts and reduces the result to a comment that merely indicates the 'direction' of the impact rather than both the direction and amplitude.

Science and technology environment

Processing technology that is ideally suited to the specific technical requirements of the proposed processing enterprise at Tabora, Tanzania is available from custom manufacturers in South Africa and can be readily imported into Tanzania. The nature of the processing technology is also such that it is relatively easy to operate and maintain. The human resources/skills and expertise available at Tabora are questionable – but this shortcoming could be overcome through targeted training. The location at Tabora is argued to be less than ideal. Although Tabora is located close to a number of important sources of fruit, it is not well located with regard to the procurement of inputs or with regard to suitable markets for the fruit juice concentrate. Considerable costs, that reduce the viability of the processing enterprise, would need to be borne by the enterprise to transport inputs to Tabora and to transport the final products to suitable markets in the larger centres of Tanzania. To view transport costs in this instance in context, transport costs constitute 42% of the total variable costs to deliver the final product to the market, which is expected to be detrimental to the profitability and viability of the enterprise there.

15.4.3 Zimbabwe: fruit jam and fruit cereal bars

The market and economic environment

The Zimbabwean feasibility study was conducted to determine the viability of two products in collaboration with Speciality Foods of Africa (SFA), a Zimbabwean-based company specializing in the production of products from indigenous African fruit. SFA identified a market opportunity for good-quality indigenous jam made from the mazhanje (*Uapaca kirkiana*) fruit and a health bar (the 'Baobar') containing pulp from the baobab (*Adansonia digitata*) fruit to be sold in Zimbabwean, South African and European retail outlets. The analysis of the financial feasibility of the proposed products revealed that, based on the assumptions that were made, the products are potentially profitable, have reasonably healthy cash flows, and are viable over the longer term. A summary of the financial feasibility appears in Table 15.2. A major concern for the viability of these enterprises is the prevailing unfavourable economic conditions in Zimbabwe. The current operating conditions for these enterprises, amongst others, hinders the procurement of inputs and the distribution of final products and generally decreases their profitability and hence viability.

The resource environment

Again, a lack of comprehensive and detailed data on available natural resources in Zimbabwe prevents the feasibility study from making accurate assessments. It is assumed that any proposed enterprise could be environmentally feasible when fruit is obtained from both natural and domesticated sources.

The social and institutional environment

The direct benefits for communities associated with Speciality Foods of Africa products include the creation of employment opportunities, the creation of a market for currently under-utilized resources and on-the-job training of participants in growing, harvesting, processing, packaging and marketing of semi-processed fruit products. Through the efforts to improve the benefits to the communities, SFA was able to register as a Fair Trade business with the International Fair Trade Association. The assurances that such registration provides are that the harvesting of natural products used in the products under feasibility investigation adds diversity to the income of rural people, decreasing their reliance on the success and value of a single crop. It also provides work for those not engaged in farming, making use of their existing knowledge and skills and harnessing their productive potential for the benefit of themselves and the environment. Fair Trade practices ensure that communities receive the full value of their labour and command fair prices for their produce (Speciality Foods of Africa, 2005).

The science and technology environment

Speciality Foods of Africa does not process any of its products but makes use of specialist contract processors who turn fruit into final products on an outsourcing basis. Analysis of the technical feasibility of these products is therefore limited. SFA's primary role in the whole process is to coordinate the process from raw material procurement through contract processing to warehousing to distribution, marketing and export.

15. 5 Financial Analysis of Four Micro-enterprises

Results of the financial feasibility conducted for each of the respective enterprises discussed above are summarized in Table 15.2. The feasibility analysis of the two fruit juice concentrate enterprises was based on the maximum output capacity of a small-scale fruit juice concentrate processing plant and the expected output (and therefore income) and the fixed and running costs associated with producing and marketing fruit juice concentrate. The feasibility of the Baobar and jam enterprises in Zimbabwe was assessed on the basis of the perceived market available for the products. The perceived market was then used to determine the required output level (and the level of income at that output level) and the fixed and variable costs associated with that output level.

The difference in financial feasibility for the two fruit juice concentrate enterprises at Tabora and Magomero is especially noteworthy. The difference is due to the location of the processing enterprise with respect to the market. The remote location of Tabora necessitates distribution costs that are significantly higher than those of the enterprise located at Magomero, which is near potential markets. The feasibility studies of these enterprises reveal that distribution (transport) costs for Tabora amount to an estimated 42% of the total variable costs compared with 16% for Magomero. This difference makes a significant contribution to the overall feasibility of the two enterprises and is also proof of the importance of a good location to feasibility. It must, however, be taken into consideration that the robustness of the feasibility study results are heavily dependent upon the assumptions made during each of the studies and on operating conditions (political, environmental, economic, etc.) remaining relatively stable for the lifetime of the enterprise. Table 15.3 details some general risk factors for small-scale enterprises and the expected impact these factors are bound to have on feasibility. If either the assumptions or the operating environment were to differ substantially from actual circumstances, the actual feasibility of the products could differ from the current results.

15.6 Discussion

Commercial processing of indigenous fruit is one of the few competitive advantages that rural communities have. The benefits from commercializing indigenous fruits are potentially significant in terms of improved livelihoods for these communities. Results of the feasibility analyses show that the proposed ventures are potentially feasible as measured against the four criteria for feasibility.

Table 15.3 General risk factors for small-scale enterprises and expected impact on feasibility.

Risk factor	Description	Impact on feasibility
Competition	Competition from local products	Medium
	Competition from imported products	High
Quality	Poor product design	High
	Poor product quality management	High
Costs	High input costs	High
Product price	High final product price	High
Transport	Quality of transport	Medium
	Reliability of transport	Medium
	Time to market	Medium/High
Infrastructure	Poor telecommunications infrastructure	Medium
	Poor transport infrastructure	Medium
Market	Poor market acceptance	Medium
	Small market size	High
Management	Poor overall business management	High
	Poor financial management	High
	Poor marketing management	High
	Poor production management (related to quality and costs)	High
Resources	Low levels of investment into businesses	High
	Poor availability of inputs, especially natural resources	High
Operating environment	Unfavourable operating environment for small businesses	High

The enterprises are, however, vulnerable to their operating environment and the management of the enterprises and sensitivity analysis of input prices, final product prices and output levels reveal that any small changes could have a profound effect on profitability and ultimately the feasibility of these enterprises.

Based on the promising outcomes of the feasibility analyses, the establishment of numerous commercial ventures across southern Africa is anticipated to not only improve the incomes of participating households, but also the economies of communities as local spending in these communities increases. Increased local trading contributes to increased circulation of monies within rural communities, limiting the 'export' of capital to urban centres. The greater number of times that monies are circulated at the rural level, the greater the benefits for local society. There is tremendous extra growth potential through boosting rural incomes, which in turn would stimulate demand for non-tradable goods and services in rural economies, which would then bring underemployed resources into production (Ngqangweni, 1999).

Despite the potential benefits and promising future of commercial indigenous fruit processing, such enterprises and their corresponding markets are still in their infancy. Enterprises in Malawi, Tanzania and Zambia are poorly developed, while those in Zimbabwe are progressing well along the road of commercialization. Success in Zimbabwe is due to the existence of private entrepreneurs and NGOs promoting the commercialization and exporting of fruit and tree products in the country. Indigenous fruit markets are marked by a lack of product differentiation, coordination, consumer knowledge and efficiency, resulting in relatively high costs.

A number of general observations and comments emanated from the feasibility studies. Given the underdeveloped nature of the indigenous fruit processing sector in southern Africa, the key to growing viable indigenous fruit processing enterprises appears to be the empowerment of potential entrepreneurs. The training of these entrepreneurs could enable them to scale-up their operations and disseminate information to emerging processors. The feasibility studies that were conducted are indicative of their potential in southern Africa. Future research to complement these studies and aimed at building a more sustainable indigenous fruit processing sector in southern Africa should ideally consider a number of critical areas:

- The first is to study in greater detail consumer behaviour, needs and preferences that underpin existing consumption. This would make available valuable information about product benefits and characteristics that can be used in future marketing strategies, tactics and positioning.
- The second critical area is improved product innovation and marketing in the broadest sense to include market research, marketing strategies, market information systems and promotions. A key challenge in marketing and producing less-known commodities is communication and information exchange with markets. Systems and strategies must be formulated to improve information flows across all interfaces.
- The third is to ensure that viable and sustainable indigenous fruit processing enterprises is a more precisely understood resource environment. Currently there is some, if limited, understanding of the resource environment (quantity and quality) in the study sites; but there is a major shortcoming in the understanding of whether sufficient resources are available to sustainably supply the processing enterprises with raw materials at the rates that they require. More precise information would include the dispersion of species, optimal harvest rates, optimal harvest time, sustainable use practices, plant improvement and possible domestication and cultivation.

15.7 Conclusions

This chapter has investigated the feasibility of a number of non-wood forest-based community processing enterprises in three southern African countries. These feasibility studies were designed and conducted within an *ad hoc* conceptual framework. Commercial fruit processing is one of the few comparative advantages that rural communities have. Results showed that the benefits from commercializing indigenous fruits are potentially significant in terms of improved livelihoods for local communities. If numerous commercialization ventures were implemented across the region, the impacts would be an improvement in the incomes of the participating households, leading to an increase in the community-level economies as local spending increased. Increased local trading contributes to increased circulation of monies within rural communities, limiting the 'export' of capital to urban centres. The greater the frequency with which monies are circulated at the rural level, the greater the benefits for local society.

Despite the potential benefits of commercializing indigenous fruits, commercial processing enterprises and markets are still in their infancy. Enterprises in Malawi and Tanzania are poorly developed, while enterprises in Zimbabwe are progressing well along the road of commercialization. Success in Zimbabwe is due to the existence of private entrepreneurs and non-government organizations promoting the commercialization and export of fruit and tree products. Feasibility studies conducted for this project indicate that commercial indigenous fruit processing enterprises are potentially viable. These enterprises are, however, vulnerable to their operating environment and any small changes could have a profound effect on profitability.

The key to growing viable indigenous fruit processing enterprises is the empowerment of potential entrepreneurs (through training and/or fund endowment), the support of local communities through clear land tenure and policy support, starting with local markets, focusing on quality products and building management and entrepreneurial skills, the support of enterprises through policies facilitating credit and trade, making use of local knowledge and resources, clearer understanding of consumer behaviour, improving the marketing of a product in existing or new markets through innovation in product, logistics and promotion.

References

Akinnifesi, F.K., Kwesiga, F., Mhango, J., Mkonda, A., Chilanga, T. and Swai, R. (2004) Domesticating priority miombo indigenous fruit trees as a promising livelihood option for smallholder farmers in southern Africa. *Actae Horticulturae* 632, 15–31.

Akinnifesi, F.K., Kwesiga, F., Mhango, J., Chilanga, T., Mkonda, A., Kadu, C.A.C., Kadzere, I., Mithofer, D., Saka, J.D.K., Sileshi, G., Dhliwayo, P. and Swai, R. (2006). Towards developing the miombo indigenous fruit trees as commercial tree crops in southern Africa. *Forests, Trees and Livelihoods* 16, 103–121.

Blair Commission Report (2005) *Our Common Interest: Report of the Commission for Africa: Part 11*. 450 pp.

Chandrasekharan, C. (1995) Terminology, definition and classification of forest products other than wood. In: *Report of the International Expert Consultation on Non-Wood Forest Products*. Non-Wood Forest Products No. 3. FAO, Rome, pp. 345–380.

Doyer, O.T. and Vermeulen, H. (2000) Exploring new opportunities in the agribusiness industry: An approach framework for feasibility studies and business plans. Unpublished, University of Pretoria, Pretoria, Pretoria.

FAO (Food and Agricultural Organization) (2002) *The State of Food Insecurity in the World*. FAO, Rome, 36 pp.

Ham, C. and Akinnifesi, F.K. (2006) Priority fruit species and products for tree domestication and commercialization in Zimbabwe, Zambia, Malawi and Tanzania. IK Notes No. 94. The World Bank, Washington, DC, 5 pp.

Lecup, I. and Nicholson, K. (2000) *Community-based Tree and Forest Product Enterprises: Market Analysis and Development*. FAO Publication and Multimedia Service, Rome.

Mithöfer, D. (2005) Economics of indigenous fruit tree crops in Zimbabwe. PhD thesis, Department of Economics and Business Administration, University of Hannover, Germany.

Mithöfer, D., Waibel, H. and Akinnifesi, F.K. (2006) The role of food from natural resources in reducing vulnerability to poverty: a case study from Zimbabwe. Paper presented to the 26th Conference of the International Association of Agricultural Economists (IAAE), 12–18 August 2006, Queensland, Australia.

Ngqangweni, S. (1999) *Rural Growth Linkages in the Eastern Cape Province of South Africa*, MSSD Discussion Paper No. 33, Markets and Structural Studies Division, International Food Policy Research Institute.

Ramadhani, T. (2002) Marketing of indigenous fruits in Zimbabwe. Socio-economic Studies on Rural Development, V 129. PhD thesis, Hannover University, Germany, 212 pp.

Saka, J.D.K., Mwendo-Phiri, E. and Akinnifesi, F.K. (2002) Community processing and nutritive value of some miombo indigenous fruits in central and southern Malawi. In: Kwesiga, F., Ayuk, E. and Agumya, A. (eds) *Proceedings of the 14th Southern African Regional Review and Planning Workshop, 3–7 September 2001, Harare, Zimbabwe.* International Centre for Research in Agroforestry, Nairobi, pp. 164–169.

Speciality Foods of Africa (2005) *Herbal Tea* [see http://www.tulimara.co.zw/index.cfm?pg_id=2 Accessed: 22 February 2006].

Taylor, D. (1999) Requisites for thriving rural non-wood forest products enterprises. *Unasylva* 50(198), 3–8.

16 Product Development: Nutritional Value, Processing and Utilization of Indigenous Fruits from the Miombo Ecosystem

J.D.K. Saka,[1] I. Kadzere,[2] B.K. Ndabikunze,[3] F.K. Akinnifesi[4] and B.P.M. Tiisekwa[3]

[1]*Chemistry Department, Chancellor College, University of Malawi, Zomba, Malawi*; [2]*SADC-ICRAF Agroforestry Project, C/O Division of Agricultural Research and Extension, Harare, Zimbabwe*; [3]*Department of Food Science and Technology, Sokoine University of Agriculture, Morogoro, Tanzania*; [4]*SADC-ICRAF Project, Chitedze Agricultural Research Station, Lilongwe, Malawi*

16.1 Introduction

Forests and homestead farms are important sources of non-timber products. These include indigenous fruits, which are consumed by communities and also sold on rural roadsides and in urban markets to generate income. These fruits are essential for the food security, health and social and economic welfare of rural communities (FAO, 1989; Maghembe *et al.*, 1998; Dietz, 1999). For example, the shea butter nut (*Vitellaria paradoxa*), found in the dry savannah, forests and parklands of the Sudano-Sahelian regions, is used by communities as a culinary fat or oil, a soap, an ointment and in cosmetics (Boffa *et al.*, 1996). The fat is used in pastry and in confectionery as a cocoa butter substitute and as a base in cosmetic and pharmaceutical preparations. The flavoured pulp of *Theobroma grandifolia* in Brazil is used in juices, ice cream, liquor, wine-making and jellies (Velho *et al.*, 1990). The fruit of *T. grandifolia* is rich in vitamin C (28.3 mg/100 g) and iron (1.53 mg/100 g) and its seed oil contains more linoleic acid (83% of the lipids) than cocoa.

In the miombo region of southern Africa, indigenous fruits are largely a subsistence product obtained during gathering activities (Maghembe *et al.*, 1998; Kwesiga *et al.*, 2000). Most of the smallholder farmers in the Zambezi basin, particularly in Malawi, Zambia, Tanzania and Zimbabwe, are not secure in food and are chronically malnourished (Kavishe, 1993; Kwesiga *et al.*, 2000). Fruits and products made from indigenous fruits therefore constitute one of the cheapest yet richest sources of food, on which the poor (especially women and children) depend. Fruits and products from indigenous trees are particularly important during the hunger periods of the year (Kwesiga *et al.*, 2000). Thus, indigenous fruits help women in most rural households to secure food for their families, either directly or indirectly when they are sold. They generate much-needed income, which will be used for various household uses, including buying food. This chapter concerns the results of more than a decade of collaborative work between the World Agroforestry Centre (ICRAF) and its regional partners on local knowledge systems, nutritional value, product development and the processing of indigenous fruits from the miombo ecosystem of southern Africa.

16.2 Utilization of Fruits

Several studies have shown that rural people in developing countries have intimate knowledge of their natural environment and environmental processes. For example, Van Vlaenderen and Nkwinti (1993) reported that communities possess well-established systems and carefully developed techniques that over many years have allowed them to survive in harsh conditions. Van Vlaenderen (1999) has further shown that building on local knowledge and resources reduces overdependence on development interventions and promotes rapid rural economic development. Appreciation and promotion of local knowledge empowers rural people by increasing their self-reliance, confidence and capacity to utilize and manage their local resources. Gathering local knowledge from indigenous communities provides relevant information concerning opportunities for utilization and the taste, size and availability of indigenous fruits (Simons, 1996; Maghembe *et al.*, 1998). In Burkina Faso, for example, local knowledge systems indicate that indigenous fruits are used in the preparation of meals as fats, spices and soups (Ladipo *et al.*, 1996). Experience has also shown that effective utilization of indigenous knowledge and local community preferences are a key to the domestication of trees and the commercialization of their products (Kwesiga *et al.*, 2000).

In southern Africa, seed kernels or nuts of several fruit species are edible (Wehmeyer, 1966; Saka, 1994; Ndabikunze *et al.*, 2000) and are an important part of the diet of rural people. For example, *Adansonia digitata*, *Telfaria pedata*, *Terminalia catappa*, *Treculia africana*, *Parkia filicoidea* and *Parinari curatellifolia* have edible nuts which are eaten raw or after processing. These nuts are important sources of vegetable oil and are rich in protein and therefore can substitute for groundnut flour/oil in rural cooking. On the other hand, the consumption of indigenous fruits in peri-urban and urban areas of southern Africa is limited. This is evidenced by the high net imports of fruits and their products by several countries (Table 16.1). For example, Malawi imported more fruit products from South Africa, Zimbabwe and Europe than it exported. The major fresh fruits imported into Malawi were apples, oranges, grapes and pears/quinces. The fruit products, such as jams, juices and preserves, are largely

Table 16.1. Total imports and exports of tropical and temperate fruits and products (metric tonnes) during 2000–2001 from Malawi, Tanzania, Zambia and Zimbabwe.

	Exports					Imports				
Country	Fresh	Dried	Juices	Fresh, processed	Tropical fruits	Fresh	Dried	Juices	Fresh, processed	Oranges
Malawi	–	28	–	7	–	9	10	11	100	14
Tanzania	3	–	8	20	26	2	6	190	90	–
Zambia	–	–	2	15	–	20	20	870	480	–
Zimbabwe	1184	–	122	122	70	–	5	590	294	–
Total	1187	28	132	166	96	31	41	1661	964	14

–, data not available.
Source: Saka *et al.* (2004).

mixed. Wines from grapes are the second most important product imported into Malawi after fresh fruits (NSO, 1999).

Earlier efforts to promote the wider utilization of indigenous fruits in urban centres have been recorded in southern Africa. Important examples include the processing and marketing of wine from *Uapaca kirkiana* in Zambia, and the production and marketing of Mulunguzi wine from *Ziziphus mauritiana* in Malawi (Kwesiga *et al.*, 2000). The utilization of indigenous fruits was an important criterion in selecting priority tree species for domestication in southern Africa. The processing of priority key indigenous fruit products in Malawi, Tanzania and Zimbabwe is described in the following section.

16.3 Processing of Indigenous Fruits

Indigenous fruit processing is a strategy that adds value to the indigenous fruits and generates substantial cash income for rural people, thus contributing to improved household welfare. There is a potential for increasing this contribution through the development of cottage but commercially oriented industries for producing indigenous fruit products such as jams and juices. These industries would promote the improvement and efficient utilization of indigenous fruits and they could also promote the conservation of species for the sustainable supply of raw materials (Leakey and Ladipo, 1999). Furthermore, once indigenous fruit processing becomes viable for commercial purposes, the local community and the private sector might be induced to plant indigenous fruit trees as cash crops.

Indigenous fruit trees and shrubs produce large quantities of ripe fruits at different times of the year; the quantity of fruit produced generally exceeds demand, resulting in wastage (Maghembe *et al.*, 1998). This situation is exacerbated by the absence of storage or fruit-processing facilities. Fresh fruits are processed in order to: (i) provide a palatable product; (ii) preserve the product; and (iii) obtain intermediate products which can be transformed into other by-products. Transformation of fresh fruits into a dried form is advantageous because in this way they can be stored for more than 18 months and thus enhance food security in times of hunger (Akinnifesi *et al*., 2004). The ranking of products made from various fresh fruits differs among areas and countries and is influenced by different preferences among women, youth and vendors (Ham and Akinnifesi, 2006). The investment in processing methods also affects the choice of preferred products. Except for items produced by processing on the homestead, cottage and medium scales, products based on indigenous fruits are rare in southern Africa. This is largely because of the lack of a sustainable supply of indigenous fruits, inadequate technical information on technologies, and the lack of quantitative information on the market potential of the products and any assessment of the potential benefits to farmers (Saka, 1994; Dietz, 1999; Ramadhani, 2002). What is therefore needed is the development of technologies for processing fruits and a market strategy that will provide basic marketing support, including infrastructure and information on what to produce and where to sell it.

Local communities in southern Africa process different indigenous fruits into various products, such as juices, jams/jellies and alcoholic beverages, e.g. wines and beers (Fig. 16.1). Some of these products have become commercialized at local, regional and international levels. Examples of commercial alcoholic beverages from indigenous fruits include Chikoto beer (from *Uapaca*), Amarula Cream from *Sclerocarya birrea* (marula) in South Africa, *Ziziphus* and marula wines in Zambia and *Uapaca* wines in Malawi, which have reached consumers with varying levels of success (Ham and Akinnifesi, 2006). The South African

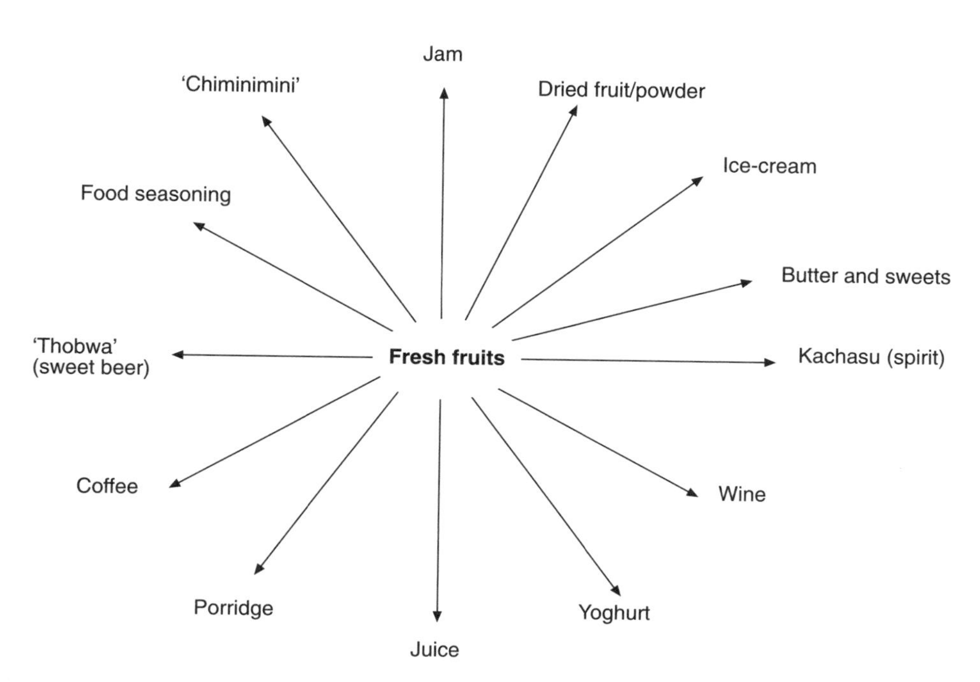

Fig. 16.1. Summary of some products from indigenous and exotic fruits in southern Africa. Source: Saka *et al.* (2004).

product Amarula Cream is marketed world-wide (Kwesiga *et al.*, 2000); it is sold in 63 countries and it is presently the second best-selling cream liqueur in the world (Akinnifesi *et al.*, 2006).

16.3.1 Processing of indigenous fruits in Malawi

Local communities in Malawi select fruits to make products based on their ability to increase and ensure food security and cash income at the household level. Such products include fresh fruits, juices, 'chiminimini' (boiled pancakes), powder, yoghurts and jams (Simons, 1996; Saka *et al.*, 2002). For example, to raise the income of rural and peri-urban households, fresh fruits, chiminimini and spirits appear to be the most important products, as they fetch good prices. In the Lower Shire of Malawi, the dominant fruit, *Ziziphus mauritiana*, is sun-dried for 3 months and stored for future use (Saka *et al.*, 2002). The dried product is brewed to give a spirit known as 'kachasu'. Its seed kernels are also roasted, pounded and added to vegetables during cooking. The seed kernel of *Trichilia emetica* is pounded and when mixed with milk and warmed is served as baby food (Williamson, 1975).

In Malawi, wine has been made from indigenous fruits such as *Uapaca kirkiana* (wild loquat), *Sclerocarya birrea*, *Ziziphus mauritiana* and *Parinari curatellifolia* (mobola plum) and the wines from indigenous fruits are of good and acceptable quality (Ngwira, 1996). *Syzygium owariense* fermented and cleared faster (within 3.5 months) than *Mangifera indica* (mango), which took 5 months (Ngwira, 1996). The mean alcohol content of wines is reported to be 12–16% for *Syzygium owariense*, 12–18% for *Psidum guajava*, 12–14% for *Tamarindus indica*, and 12–16% for *Mangifera indica* wines (Ngwira, 1996).

16.3.2 Processing of indigenous fruits in Zimbabwe

Traditional fruit processing in Zimbabwe is predominant in the drier parts of the country, where indigenous fruits supplement food requirements (Kadzere *et al.*, 2004). Rural communities in Zimbabwe have been processing baobab oil and making *Uapaca* juice and jam (Akinnifesi *et al.*, 2006). The major fruits being processed in the Zambezi valley are *Ziziphus mauritiana*, *Adansonia digitata*, *Sclerocarya birrea* and *Tamarindus indica* in the Zambezi valley, Chipinge and Gokwe. These fruits are being used directly for human consumption, while *Parinari curatellifolia* is also used as livestock feed. In a study by Kadzere *et al.* (2004), 78% of indigenous fruit processors indicated that they process in order to improve taste while 35% processed to preserve the fruits for later use, particularly during the off-season. Households were using *Strychnos cocculoides* juice and *Adansonia digitata* pulp to add a sour flavour to cereal porridge and with the intention of enriching it with vitamins. Some of the products reported from Zimbabwe are listed in Table 16.2.

The Southern Alliance for Indigenous Resources (SAFIRE) has developed several recipes for processing various products from fruits such as *Uapaca kirkiana* (known as 'mazhanje' in Shona), *Adansonia digitata*, *Strychnos cocculoides* and

Table 16.2. Indigenous fruits and their products reported in Zimbabwe.

Product type	Fruit species	Fruit part used	Vernacular name of product
Alcoholic beverage	*Parinarii curatellifolia*	Whole fruit except pip	Mahanya
	Sclerocarya birrea	Juice and pulp	Mukumbi
	Ziziphus mauritiana	Whole fruit (crushed together with the seed)	Kachasu
Non-alcoholic beverage	*Parinari curatellifolia*	Whole fruit except pip	Mahewu
	Sclerocarya birrea	Juice and pulp	Mahewu
	Ziziphus mauritiana	Whole fruit (crushed)	Mahewu
Fermented solid form	*Ziziphus mauritiana*	Whole fruit (crushed together with the seed)	Gununzvu
	Parinari curatellifolia	Pulp	
Confectionery	*Sclerocarya birrea*	Whole raw or roasted ground kernels	Nzungu, dovi
	Parinari curatellifolia	Whole raw or roasted ground kernels	Nzungu, dovi
	Adansonia digitata	Whole raw or roasted ground kernels	Nzungu, dovi
	Uapaca kirkiana	Dried pulp	Chipapata
	Ziziphus mauritiana	Dried ground pulp	
Peanut butter and oil	*Sclerocarya birrea*	Whole raw or roasted nut, ground nut	Dovi, mafuta
	Parinari curatellifolia	Whole raw or roasted nut, ground nut	Dovi, mafuta
Food additives	*Adansonia digitata*	Whole raw or roasted nut, ground nut	Dovi
	Adansonia digitata	Powder (used to flavour cereal porridge)	Bota
	Strychnos cocculoides	Juice (use the juice to prepare cereal porridge)	Bota
Dried fruit	*Azanza garckeana*	Whole fruit	–
	Ziziphus mauritiana	Whole fruit	–
	Parinari curatellifolia	Whole fruit	–
Baking soda	*Sclerocarya birrea*	Pip or stone (burn and use the ashes)	–
Seed 'cake'	*Sclerocarya birrea*	Residue after oil extraction from nuts	Chisikinya
Jam	*Sclerocarya birrea*	Rind and pulp	–
Coffee	*Adansonia digitata*	Roasted and ground seeds	Masamba
Yoghurt	*Adansonia digitata*	Powder mixed with water or milk	Bozo
Baking soda	*Adansonia digitata*	Shells (burn and use ashes)	–
Fruit powder	*Ziziphus mauritiana*	Powder	–

Source: Kadzere *et al.* (2004).

Ziziphus mauritiana (SAFIRE, 2005). For example, *U. kirkiana* makes juices, jams and fruit leathers, while *A. digitata* pulp makes juices and jams. The *A. digitata* seed kernel is processed into commercial oil which is exported to Europe through a regional network, PhytoTrade.

16.3.3 Processing of indigenous fruits in Tanzania

In the fruit-rich Uluguru mountain area of Tanzania, communities know and value the importance of indigenous fruits. Fruits are also acceptable for human consumption by urban consumers and processors in Tanzania (Tiisekwa *et al.*, 2004). The communities consume both fresh and processed forms and attach greater utility value to indigenous fruits because they are 'natural' and provide greater variety at table. Fruits are processed into juice and alcoholic drinks in Tanzania (Tiisekwa *et al.*, 2004). Consumers in the rural areas prefer processed products (jams, juices, wines, etc.) from fruits such as *Syzygium guineense* (wine), *Tamarindus indica* (juice and jam), *Adansonia digitata* (as a source of pectin for jam) and *Vitex mombassae* (juice). The urban communities were willing to consume juices and jams from *S. guineense*, *T. indica* and *A. digitata*. The wine from the cherry *Rosa roxburghii* and cashew apple (*Anacardium occidentale*) is also highly preferred (Ndabikunze *et al.*, 2000) by both rural and urban communities. A similar pattern of utilization has been reported in the Tabora region of Tanzania.

16.3.4 Increasing local capacity for fruit processing

Capacity-building is critical at all levels for the successful domestication and commercialization of indigenous fruits (Akinnifesi, 2001). Farmers need training in several areas, including processing, standards, business management and marketing information (Dietz, 1999). ICRAF and its collaborating partners in three countries have trained over 400 trainers (mostly women) in processing indigenous fruits into juices, jams and wine (Table 16.3) (Saka *et al.*, 2004). This is necessary in order to sustain greater value addition to this important regional resource. The recent discovery and subsequent utilization of ng'ongo pori (*Sclerocarya birrea*) by the Mwamko Women's Group in Kipera village in Mlali Morogoro, Tanzania, has enabled them to increase their income by processing the fruit into jam (Lyimo and Tiisekwa, 2000).

Farmers have also undergone training in basic management skills in quality assurance, entrepreneurship, marketing (including marketing research), costing and pricing, stock control, record keeping and business planning (Saka *et al.*, 2004). Training in quality assurance and management of the fresh and finished products will increase the utilization and acceptance of indigenous fruits at the household level and in urban and more affluent communities.

16.4 Nutritional Value of Fresh and Processed Products

People in drought-prone areas live on diets deficient in energy, protein and micronutrients. For a long time, these people have been depending on forest

Table 16.3. Training of indigenous fruit processors in Tanzania, Zambia and Malawi.

	First generation	Second generation	
Country	No. of farmer trainers	No. of groups	No. of farmers
Tanzania	198	43	2045
Malawi	120	150	1875
Zambia	115	5	77
Total	433	198	3997

Source: Saka *et al.* (2004).

woodland trees and shrub fruits to supplement their diet, especially during times of crop failure. Studies in areas susceptible to drought, such as Swaziland, demonstrate the importance of edible indigenous fruits as a source of food (Grivetti, 1979; Grivetti *et al.*, 1987; Humpry *et al.*, 1993; Smith *et al.*, 1995, 1996; Nordeide *et al.*, 1996).

Nutrition studies have not seriously considered the role of wild plants in local diets. Reports by several workers, such as Saka and Msonthi (1994) and Ndabikunze *et al.* (2000) have shown that wild plants contain high levels of nutrients which are important to infants and children, pregnant and lactating women, the elderly, HIV-infected people and indigenous societies. The nutritive value of indigenous fruits is therefore an important criterion for fruit tree improvement for rural community dwellers. The consumption of these fruits can help to combat malnutrition resulting from major deficiencies of vitamins A and C and essential amino acids, as well as minerals such as iron and zinc (Thiong'o *et al.*, 2002). For over 30 years, chemists and nutritionists depended on nutritive value data published by the Food and Agriculture Organization of the United Nations, most of which was largely restricted to South Africa and Eastern Africa (FAO, 1989). The southern African region has now generated data on nutritive value for various indigenous fruits in their fresh and processed forms. This section provides data on fresh and processed fruit products.

16.4.1 Nutritional value of fresh pulp and seed kernels of indigenous fruits

Concerted efforts by scientists in Malawi and Tanzania since the mid-1980s have generated nutritional data for indigenous fruits (Saka *et al.*, 1992; Saka, 1994; Ndabikunze *et al.*, 2000; E.M.T. Henry and J.D.K. Saka, unpublished results). The results for 16 fruits growing in Malawi show that *Annona senegalensis* and *Trichilia emetica* are rich in protein, while *Flacourtia indica* and *Syzygium guineense* are rich sources of iron (Saka and Msonthi, 1994). A summary of the compositions of 12 fruits growing in both Tanzania and Malawi is provided in Table 16.4. The data reveal that most fresh fruits are important sources of carbohydrates and thus have high calorific values. The highest (88.2%) and lowest (35.2%) levels were obtained for *Parinari curatellifolia* and *Azanza garckeana*, respectively. The protein content of edible portions ranged from 0.02% in *Azanza garckeana* to 17% in *Uapaca kirkiana*. The variation in fat content was also evident: three fruit species, *Uapaca kirkiana*, *Vitex mombassae* and *Syzygium guineense*, had >17% fat, while the rest had low levels. These data show the variation in nutrient composition among indigenous fruits.

The β-carotene levels varied from 21 μg/100 g in *Adansonia digitata* to 357 μg/100 g in *P. curatellifolia* (Table 16.5). *Flacourtia indica* also had very high level (303 μg/100 g). β-Carotene is of prime importance in vision, the maintenance of epithelial surfaces and also serves as an antioxidant. The total soluble solids values of indigenous fruits compare favourably with those of conventional exotic fruits (e.g. mangoes), indicating the potential of indigenous fruits for the fresh fruit market and for processing into products such as juices and jams. The vitamin C levels of the edible pulp of the fruit and the seed

Table 16.4. Chemical composition of selected edible fruits of the miombo woodlands in southern Africa.

Species	Source	Fruit part	Dry matter (%)	Ash (%)	Protein (%)	Fat (%)	Fibre (%)	Carbohydrate (%)	Reducing sugars (%)	Energy value (kJ/100 g)
Adansonia digitata	Mw	Pulp	86.8	5.0	3.1	4.3	8.3	79.4	–	1480
		Kernel	92.1	9.0	28.7	29.6	7.28	25.4	–	–
Azanza garckeana	Mw	Pulp	52.8	6.3	12.0	1.1	45.3	35.2	–	810
	Tz	Pulp	–	–	0.02	14.9	5.05		–	–
Flacourtia indica	Mw	Pulp	19.2	5.7	4.2	3.6	5.7	80.7	–	1290
Parinari curatellifolia	Mw	Pulp	27.1	1.8	3.0	1.5	5.5	88.2	–	1517
		Kernel	97.1	–	–	20.4	–	–	–	–
	Tz	Pulp	–	–	0.71	0.07	23.5	–	3.13	–
		Kernel	–	–	47.0	–	–	–	–	–
Syzygium guineense	Mw	Pulp	22.1	4.1	5.4	31.2	17.6	42.1	–	1923
	Tz	Pulp	–	–	5.43		4.75		6.98	
Sclerocarya birrea	Mw	Pulp	15.0	1.0	–		–	–	–	–
		Kernel	95.7	–	–	74.8	–	–	–	–
	Kenya	Pulp	13.7		4.7	–	–	–	–	–
		Kernel	77.8		33.1	–	–	–	–	–
Strychnos cocculoides	Mw	Pulp	27.8	3.7	11.5	6.0	17.9	61.0	–	1390
	Tz	Pulp	–	–	0.47		9.86	–	1.91	–
Uapaca kirkiana	Mw	Pulp	58.1	4.5	17.0	22.9	8.1	47.5	–	1897
	Tz	Pulp			–	6.03	4.32			
Vitex doniana	Mw	Pulp	26.5	3.4	5.7	2.6	10.2	78.1	–	1445
	Tz	Pulp	–	–	0.26	0	4.48	–	3.44	–
Vitex mombassae	Tz	Pulp	–	–	0.03	18.8	3.88	–	0.01	–
Vangueria infausta	Mw	Pulp	27.2	2.2	1.8	1.1	8.4	86.5	–	1456
	Tan	Pulp	–	–	0.01	5.19	23.5	–	–	–
Ziziphus mauritiana	Mw	Pulp	14.8	10.1	4.1	9.5	3.4	73.0	–	1588

Mw, Malawi; Tz, Tanzania; –, data not available.
Sources: Saka (1994), Ndabikunze *et al.* (2000), Tiisekwa *et al.* (2004), E.M.T. Henry and J.D.K. Saka (unpublished results).

Table 16.5. Mineral composition and vitamin content of selected edible fruits of miombo woodlands in southern Africa.

Fruit species	Source	Fruit part	Minerals and vitamins (mg/100 g)							
			Phosphorus	Calcium	Magnesium	Iron	Potassium	Sodium	Vitamin C	Vitamin A[a]
Adansonia digitata	Mw	Pulp	450	1,156	2,090	58	28,364	188	179.1	21
		Kernel	5.8	456	–	0.5	1,186	75.2	–	
Azanza garckeana	Mw	Pulp	1,476	95	1,453	84	26,190	202	20.5	67
	Tz	Pulp	220	600	170		3,130	230	24.7	–
Flacourtia indica	Mw	Pulp	1,057	354	1,380	734	24,281	589	10.2	303
Parinari curatellifolia	Mw	Pulp	339	129	830	103	10,380	252	10.4	357
		Kernel	–	–	–	–	–	–	–	–
	Tz	Pulp	180	360	160		4,230	2,640	60.7	–
		Kernel	340	1,050	140		2,640	150	61.8	–
Syzygium guineense	Mw	Pulp	1,081	149	430	136	19,683	253	11.9	226
	Tz	Pulp	20	280	210		2,740	70	81.8	–
Sclerocarya birrea	Mw	Pulp	3	16	27	1.7	309	4	–	35
		Kernel	–	–	–	–	–	–	–	–
	Kenya	Pulp	200	300	100	21.8	2,400	–	160.8	–
		Kernel	900	100	400	5.23	600	–	–	–
Strychnos cocculoides	Mw	Pulp	2,106	60	1,633	60	28,670	459	22.9	22
	Tz	Pulp	20	1,040	250		1,680	1,110	46.2	–
Uapaca kirkiana	Mw	Pulp	3,164	ND	1,129	43	13,017	146	16.8	67
	Tz	Pulp	480	720	170	–	3,090	340	98.7	–
Vitex doniana	Mw	Pulp	823	132	1,811	283	18,208	245	19.6	175
	Tz	Pulp	320	1,090	250	–	3,210	270	98.7	–
Vitex mombassae	Tz	Pulp	320	870	70	–	3,210	110	111.0	–
Vangueria infausta	Mw	Pulp	555	33	1,106	451	13,682	365	16.8	–
Ziziphus mauritiana	Tan	Pulp	320	550	140	–	2,040	270	93.7	–
	Mw	Pulp	2,162	135	507	ND	17,318	226	13.6	35

[a] Value in μg/100 g.

Mw, Malawi; Tz, Tanzania; ND not detected; –, data not available.

Sources: Saka (1994), Saka and Msonthi (1994), Ndabikunze *et al.* (2000), Tiisekwa *et al.* (2004), E.M.T. Henry and J.D.K. Saka (unpublished results).

kernel of some species are variable and in some indigenous fruits they are comparable with those in exotic and domesticated fruits. For example, the vitamin C level of *Adansonia digitata* fruits is similar to that of fruits of the same species grown in Burkina Faso (169–270 mg/100 g; Lamien *et al.*, 1996) and much higher than that of fresh oranges (50–70 mg/100 g) (Williamson, 1975). Great variation in vitamin C level among locations (Malawi and Tanzania) has also been observed by Thiong'o *et al.* (2002), who found a large amount of variation in vitamin C levels (85–319 mg/100 g) for *Sclerocarya birrea*. However, no significant differences in other food nutrients between trees, within or between provenances, existed in *S. birrea*. The composition of these edible parts of plants is thus affected by several factors, such as maturity of the fruit at harvest, age of tree, soil type, phenotype and agronomic practices. The daily adult requirement of vitamin C, as reported by Lutham (1997), ranges from 45 to 80 mg, which implies that only 50 g of the edible part of indigenous fruits is sufficient to supply the body's daily requirement of the vitamin. Indigenous fruits could therefore be used as a novel food for malnourished people and those living with HIV/AIDS in Africa (Rajabiun, 2001).

The mineral composition of pulps (Table 16.5) indicates that potassium was the most abundant element in *Adansonia digitata* and *Strychnos cocculoides*. Magnesium and phosphorus were also predominant elements, while the trace element iron was highest in *Sclerocarya birrea*. Iron is important for blood formation, hence consumers of these fruits may maintain better health. The recommended daily amount of iron for children aged 7–10 years is 23 mg. Therefore, consumption of 1 kg of any of the raw fruits per day will give the child the recommended iron intake (Lutham, 1997). This is dissimilar to conventional fruits, for which more than 10 kg is required to meet the recommended daily amount. The amounts of minerals in these fruits per 100 g are equivalent to the amounts suggested by NRC/NAS (1989) for mineral mixes to be used for nutrition rehabilitation.

The seed kernel of *Adansonia digitata* is rich in protein (28.7%) and fat (29.5%); these values (Saka, 1994) are similar to those of leguminous seeds (Brand *et al.*, 1985). This is also the case for *Sclerocarya birrea*, which also has high protein (28%) and oil (57%) levels (Thiong'o *et al.*, 2002). The seed kernels of *Adansonia digitata* and *Sclerocarya birrea* are also vital sources of trace elements such as iron, zinc and copper. The quality of the oils from the seed kernels of *A. digitata* and *S. birrea* is similar to that of groundnut oil, since the major acids present are palmitic, oleic and linoleic acids (Saka, 1994; Thiong'o *et al.*, 2002). However, unlike groundnut oil, *Adansonia digitata* oil contains malvalic acid, a toxic cyclopropanoid fatty acid, which limits its use for human consumption, but which can be removed by hydrogenation. In rural areas of the Nguru, Usambara and Ulugulu mountains in Tanzania, the seeds of *Allanblackia stuhlmanii* are used to extract cooking oil and their fat content is 68.2% (Ndabikunze *et al.*, 2000). This wild seed kernel contains more fat than coconut (36%), almonds (53.5%) and groundnuts (49%) (Ndabikunze *et al.*, 2000). This clearly indicates the potential for commercial cooking oil production from indigenous fruit seed kernels.

16.4.2 Nutritive value of indigenous fruit products

Processing affects the composition of finished fruit products. The physicochemical characteristics of fruit products (juices and jams) for indigenous fruits and non-indigenous fruits are shown in Table 16.6 (Saka *et al.*, 2007). These data indicate that mango and *Uapaca* juices have higher pH values (i.e. they are less acidic) than *Strychnos cocculoides* (kabeza) and *Adansonia digitata* (baobab) juices. The low pH values of *S. cocculoides* and *A. digitata* fruits are due to their high acid level (Saka, 1994). The baobab and kabeza juices are rich in zinc; the former has a similar copper level to mango juice. All juices exhibit high phosphorus levels (>200 μg/ml). While kabeza has the highest potassium level, baobab juice has calcium and magnesium levels consistent with higher contents of these elements in the pulp (Saka, 1994). Laboratory-made baobab and commercially available juices contain no measurable iron or vitamin C.

The jams, not unexpectedly, contained much higher nutrient levels than the juices. Processing to make jams results in water removal and thus concentration of food nutrients (Dietz, 1999; Krige *et al.*, 2006). These juices and jams would make a considerable contribution to the daily dietary requirements of children. For example, an adult consuming 200 ml of *V. mombassae* juice would take in the recommended daily amount of 60 mg.

16.4.3 Effect of provenance on chemical composition of *Adansonia digitata* fruits

The compositions of five provenances of *Adansonia digitata* from Malawi are presented in Table 16.7 (Tembo, 2006). The data show that provenance significantly affected the physicochemical properties of *A. digitata* fruits ($P < 0.001$). For example, the Chikwawa provenance had significantly higher contents of vitamin C (ascorbic acid; 347.7 mg/100 g) and reducing sugars (11.26%) than the other four provenances ($P < 0.001$). Fruits from Salima had the least ($P < 0.001$) vitamin C while Mangochi fruits had significantly ($P < 0.05$) higher iron contents than other provenances.

Generally, fruits from all locations were acidic and showed significant differences within provenances. Mangochi fruits were significantly more acidic ($P < 0.05$; 2.86%) than the others. The high values are probably due to a high ascorbic acid content (Saka, 1994). Total soluble solids ranged from 21.8% for Chikwawa to 32.5% for Dedza ($P < 0.001$). The low total soluble solids (%) contents for Chikwawa provenance is probably due to differences in the environment. Results of vitamin A analysis revealed that the Mangochi (60.9 mg/kg) and Mwanza (54.7 mg/kg) provenances had the highest vitamin A levels ($P < 0.001$). Fruits from Chikwawa had significantly lower ($P < 0.001$) levels of vitamin A. This is consistent with the results of Gross (1991), who observed that areas of high light intensity were associated with higher vitamin C but lower vitamin A levels.

The mineral content of *A. digitata* fruits varied significantly among the provenances. Mwanza and Dedza provenances had significantly higher calcium

Table 16.6. pH, acidity and mineral contents (μg/ml; mean ± standard deviation, range) of some indigenous fruit products.

Fruit product and species	Sample size (*n*)	Variable									
		Calcium	Magnesium	Potassium	Sodium	Zinc	Copper	Phosphorus	pH	Acidity (%)	Total soluble solids (%)
Juices											
Uapaca kirkiana	9	12.0 ± 2.6^{b}	57.6±22.7^{a}	140.3±31.0^{c}	54.9±6.2	–	–	536.0±55.4^{e}	3.55±0.27	1.04 0±14	21.4±1.5^{c}
		(9.1–16.3)	(31.6–105.3)	(98.1–178.8)	(43.4–64.7)			(459.2–617.1)	(3.23–4.22)	(0.99–1.32)	(20.0–24.7)
Adansonia digitata	15	37.1±6.4^{a}	79.3±11.4^{a}	382.8±17.1^{b}	69.9±10.2	4.54±0.15	4.90±0.52	240.8±42.2^{d}	3.105±0.04	1.92±0.25	14.3±1.2^{d}
		(27.0–45.7)	(61.2–95.1)	(353.9–396.4)	(57.9–90.4)	(4.35–4.81)	(4.03–5.28)	(202.6–307.9)	(3.07–3.21)	(1.63–2.36)	(13.2–16.3)
Strychnos cocculoides	6	3.67±1.14^{d}	37.8±5.5b	2809.1±160.2^{a}	13.8±5.5	2.69±0.45	–	550.2±29.5^{c}	3.53±0.05	1.18±0.11	32.3±1.9^{a}
		(2.5–5.3)	(33.0–46.7)	(2596–3000)	(7.0–22.0)	(2.13–3.28)		(518.4–597.4)	(3.48–3.61)	(1.07–1.32)	(30–34)
Mango	6	10.4±2.4^{c}	40.2±7.1^{b}	61.1±12.9^{d}	26.8±4.4	2.13±0.33	5.87±0.97	618.8±45.4	3.58±0.10	1.10±0.13	26.3±1.5^{b}
		(9.4–12.7)	(4.1–50.3)	(51.0–77.9)	(22.0–30.5)	(1.74–2.43)	(4.81–6.70)	(571.1–676.3)	(3.48–3.7)	(0.95–1.25)	(25–28.3)
Jams											
Uapaca, sweet	3	81.8±0.35	132.02±0.41	310.23±0.00	224.75±0.00	3.48±0.00	–	559.26±13.09	4.64±0.01		
Uapaca, sour	3	88.5±1.35	140.53±0.00	323.18±0.00	234.40±1.52	2.57±0.14	–	475.93±0.00	4.50±0.01		
Strychnos, sour	3	49.8±0.7	164.57±0.83	5870.15±0.00	105.73±1.52	4.38±0.14	–	809.26±26.19	3.41±0.01		
Strychnos, sweet	3	53.15±1.35	152.26±0.83	5627.35±0.00	117.52±0.00	5.79±0.14	–	911.11±13.09	3.40±0.01		
Strychnos, sweetest	3	46.46±0.00	152.84±0.00	6023.01±0.00	110.02±1.52	5.19±0.14	–	855.56±13.09	3.53±0.01		

Source of recommended daily allowances: NRC/NAS (1989).
–, below the detection limit of the spectrophotometer (0.5 μg/l).
Values with the same superscript letter in the same column are not significantly different ($P > 0.05$).
Source: Saka *et al.* (2007).

Table 16.7. Some important physicochemical properties of five provenances of *Adansonia digitata* in Malawi.

Property	Source of provenance in Malawi					$LSD_{(0.05)}$	CV (%)
	Chikwawa	Dedza	Mangochi	Mwanza	Salima		
Acidity (%)	2.58	2.20	2.86	2.40	2.68	0.15	3.10
pH	3.19	3.247	3.14	3.16	3.43	0.01	0.20
Reducing sugars (%)	11.27	8.750	10.18	10.22	9.01	0.19	1.00
Vitamin A (mg/kg)	35.83	31.08	60.92	54.71	29.86	1.249	1.6
Vitamin C (mg/100 g)	347.70	259.70	239.30	317.00	233.10	12.30	2.30
Moisture (%)	67.97	67.53	67.47	67.50	66.20	1.27	1.00
Total soluble solids (%)	21.80	32.5	32.0347	30.63	31.87	5.29	5.10
Calcium (mg/kg)	1956	2786	1958	2269.00	1978	1.46	12.80
Copper (mg/kg)	2.48	5.02	1.68	4.14	3.58	7.46	22.90
Iron (mg/kg)	19.60	15.30	22.90	17.60	16.40	4.82	21.60
Potassium (mg/kg)	979.00	766.00	1172.00	530.00	1142	10.61	25.00
Magnesium (mg/kg)	20.30	10.40	26.20	21.20	33.60	58.84	25.20
Sodium (mg/kg)	166.90	172.30	252.90	282.30	248.20	1.74	13.90
Zinc (mg/kg)	3.36	2.37	5.18	2.02	3.40	*	*

$LSD_{(0.05)}$, least squares difference at 5% level; CV, coefficient of variation.
*Almost negligible.
Source: Tembo (2006).

contents than the others, while Mangochi and Salima provenances contained significantly more potassium ($P < 0.001$) than the other provenances. The differences are probably due to variations in soil type and climate. Mangochi and Salima districts are situated along the lake shore (Lake Malawi) and thus share similar rainfall patterns and soil properties, while Dedza and Mwanza are at high altitudes.

16.4.4 Effect of processing on some physicochemical properties of indigenous fruits

The composition of fresh pulp of *Strychnos cocculoides* and derived jams is shown in Table 16.8 (Tembo, 2006). The results indicate that the composition of the fresh pulp depends on several factors, such as tree age and the extent of ripening (Krige *et al.*, 2006). For example, pre-ripe fruit pulp from middle-aged fruit trees (10–15 years) had a significantly higher ($P < 0.001$) vitamin C content (101.1 mg/100 g) than pre-ripe fruit pulp from young (5–9 years) and old (>15 years) trees. The reducing sugar levels were also significantly higher ($P < 0.001$) in pre-ripened pulp from old trees. Generally, all pre-ripe fruit pulp had a significantly higher ($P < 0.001$) vitamin C content and significantly lower acidity (%) than ripe fruit pulp. Very ripe fruit pulps had lower total soluble solids, probably because of a dilution effect.

Clearly, significant variation ($P < 0.001$) exists between the composition of fresh pulp and *Strychnos* jam. For example, ascorbic acid levels decreased

Table 16.8. Effect of processing on some physicochemical properties of *Strychnos cocculoides* pulp and jams.

	Vitamin C (mg/100 g)	Reducing sugars (%)	Total soluble solids (%)	pH	Acidity (%)	Moisture (%)
Pulp						
Young green and mature	85.5	2.90	18.24		1.39	73.5
Young ripe and mature	15.0	5.96	17.29		1.39	76.5
Mature green and mature	105.1	3.39	16.07		0.986	74.9
Mature ripe and mature	16.1	5.83	16.35		1.72	77.4
Old green and mature	34.7	3.34	19.39		1.035	74.7
Old ripe and mature	11.4	5.85	17.86		1.789	76.5
$LSD_{(0.05)}$	38.6	1.08	1.73		0.07	2.75
Jams						
Young green and mature	7.05	14.0		3.30	0.39	
Young ripe and mature	10.45	12.4		3.54	0.45	
Mature green and mature	6.51	14.4		3.43	0.42	
Mature ripe and mature	10.21	14.4		3.51	0.33	
Old green and mature	8.50	13.2		3.49	0.34	
Old ripe and mature	8.31	12.1		3.45	0.37	
$LSD_{(0.05)}$	0.00	2.04		0.01	0.02	

Source: Tembo (2006).

significantly during the production of jam ($P < 0.001$). The vitamin C contents of fresh pulp and final jam of *Strychnos cocculoides* were 82.5 and 7.05 mg/100 g, respectively. The huge loss was caused by pressing and heating during jam preparation (Krige *et al*., 2006). Klopotek *et al*. (2005) observed that major food nutrients, such as ascorbic acid, total anthocyanins and hydrophilic antioxidant capacity declined during the processing of strawberries to different products. The greater reducing sugar content in *Strychnos* jam was due to the sucrose added during production of the jam. The added sugar undergoes hydrolysis to fructose and glucose during heating, in effect increasing the level of reducing sugar in the final product. Zafrilla *et al*. (2001) also found a decrease in flavonol after processing raspberry jam.

16.5 Consumer Evaluation of Indigenous Fruit Products

The involvement of consumers is critical in determining and identifying products which have a greater chance of being accepted by rural, peri-urban and urban communities. Consumer assessment of some of the fruit products has been undertaken in southern Africa. In Malawi, eight juices sweetened with different amounts of sugar were tested organoleptically by a 16-member panel, who largely preferred juices with a sugar content of 10% (weight/volume) (Saka *et al*., 2001); individual preferences were not significantly different from the overall preference ($P > 0.05$). Consumer evaluations of two different fruit jams indicated that *Strychnos cocculoides* jam was significantly superior to *Uapaca kirkiana* jam (Saka *et al*., 2001). *S. cocculoides* jam was most preferred by the majority of consumers, who evaluated the product because the jam had better taste, flavour and texture. These quality attributes, including greater ease of spreading on bread, accounted for greater variation in consumer preference for *S. cocculoides* products. Pulp squeezed from *Uapaca kirkiana* fruits make as good a jam as *S. cocculoides* (Saka *et al*., 2007). However, when left at room temperature or under refrigeration, the *S. cocculoides* jam remained unspoiled for over 3 months while the *U. kirkiana* jam exhibited fungal growth within a week at room temperature and after 4 weeks of refrigeration. The longer shelf life of the *Strychnos* jam was attributed to its higher acid content (Saka *et al*., 1992).

Ndabikunze *et al*. (2000) reported that all fruit jams tested in a Tanzanian study were acceptable to consumers. The jams from *Adansonia digitata* and *U. kirkiana* were more strongly preferred (score 8) in all sensory attributes than that from *Tamarindus indica* (score 6–7). Mixing *A. digitata* pulp with *Hibiscus sabdariffa* produced a mixed jam with a pleasant sweet–sour taste. The banana–baobab mixed jam (mixing ratio 2:1) was most preferred ($P \leq 0.05$) for colour, while banana–baobab (3:1) jam was ranked first for spreadability. Interestingly, *A. digitata* juice was more preferred than the tamarind juice. To expand the utilization of *A. digitata* in juice and jam-making, its pulp could be used in small proportions, since it forms a very thick gel. Better methods of extracting baobab powder should be identified in order to minimize the amounts of foreign substances such as sand particles, which affect the quality of the final product (Ndabikunze *et al*., 2000).

More recently, Saka *et al.* (2007) have published results on the consumer evaluation of jams and juices from *U. kirkiana* and *S. cocculoides* fruits compared with the control products, baobab and mango juices (Table 16.9). Consumer scores were independent of the original taste of the fresh fruits. Interestingly, the non-sweet *U. kirkiana* and the sweet *S. cocculoides* fruits both gave acceptable jams. Only two variables, spreadability and appearance, were significantly different for the products from the two fruit types ($P < 0.05$). The importance of colour, which affects appearance, thus seemed to be responsible for the lower scores of *S. cocculoides* juices (Ennis *et al.*, 1979; Saka *et al.*, 2002). Compared with mango and baobab juices, *S. cocculoides* juice seemed to be the most preferred for taste, mouth feel, flavour, sweetness and overall ranking. The product was also much more preferred than *U. kirkiana* juices and yet none of the panel members were familiar with *Strychnos cocculoides* fruit, although they were familiar with *U. kirkiana*. The *U. kirkiana* and *S. cocculoides* juices were also more preferred than baobab or mango juices; the mango juice scored lowest on appearance (Ennis *et al.*, 1979).

Further processing techniques such as pasteurization and storage under different conditions affect the quality of the final products. For example, storage stability tests in Malawi and Tanzania show that there is a reduction in vitamin C with increased storage time. Tiisekwa *et al.* (2004) reported that jam prepared from *Vitex mombassae*, *U. kirkiana* and *Sclerocarya birrea* with commercial pectin and *Adansonia* powder pectin showed the recommended moisture content (<34%) after 6 months of storage.

16.6 Conclusions and Recommendations

Indigenous fruits from the miombo areas contribute to diet diversification at the household level and are an important source of food nutrients, especially vitamins and minerals. These contribute to daily nutritional needs, provide household food security and generate cash income for family welfare. The fruits can be transformed into various products, thereby adding value, reducing wastage and increasing product shelf life. Miombo indigenous fruit products are largely produced by traditional practices using locally available materials. To expand their utilization and processing, continuing efforts should be directed at the following:

- Quality assurance and standards during processing.
- Increasing the knowledge of local people about consumers' emerging priorities and appropriate technologies for the processing and utilization of fruit products.
- Developing appropriate processing techniques and diversifying the product range for priority indigenous fruits.
- Training farmers and intermediate buyers in fruit processing and business management.

Table 16.9. Average trained panel scores (n = 10) of some organoleptic characteristics of *Uapaca kirkiana* and *Strychnos cocculoides* jams and juices made from sweet and non-sweet pure or composite fruit.

	Uapaca, sweet	*Uapaca*, non-sweet	*Strychnos*, sweet	*Strychnos*, sweetest	*Strychnos*, sour
Jams					
Appearance	4.3 ± 1.1^{a}	4.5 ± 0.7^{a}	3.9 ± 0.9^{a}	4.0 ± 0.9^{a}	3.2 ± 1.0^{a}
Taste	3.3 ± 1.6^{a}	4.7 ± 0.7^{a}	4.5 ± 0.5^{a}	4.1 ± 0.7^{a}	3.3 ± 0.8^{a}
Flavour	3.5 ± 0.9^{b}	4.4 ± 0.9^{a}	3.4 ± 0.7^{b}	3.9 ± 0.6^{a}	3.3 ± 0.8^{b}
Mouth feel	3.2 ± 0.9	4.5 ± 0.5	3.3 ± 0.5	3.6 ± 0.7	3.4 ± 0.7
Texture	4.1 ± 0.8	4.7 ± 0.5	2.8 ± 0.8	3.4 ± 0.5	2.4 ± 0.5
Sweetness	4.0 ± 0.7	4.5 ± 0.5	4.0 ± 0.7	4.4 ± 0.7	3.8 ± 0.9
Spreadability	4.9 ± 0.3^{a}	4.5 ± 0.5^{b}	1.6 ± 0.7^{c}	2.5 ± 1.1^{c}	1.6 ± 0.5^{c}
Overall	4.0 ± 0.8^{b}	5.0 ± 0.1^{a}	3.5 ± 0.5^{b}	4.4 ± 0.8^{b}	3.5 ± 1.0^{b}

	Uapaca, composite	*Strychnos*, composite	Mango, composite	Baobab, composite
Juices				
Appearance	3.77 ± 0.45^{a}	3.33 ± 0.12^{a}	2.9 ± 0.5^{a}	3.45 ± 0.71^{a}
Taste	3.47 ± 0.51^{a}	4.00 ± 0.16^{a}	3.55 ± 0.15^{a}	3.41 ± 0.77^{a}
Flavour	3.36 ± 0.43^{ab}	3.80 ± 0.18^{a}	3.4 ± 0.1^{b}	3.25 ± 0.53^{b}
Mouth feel	4.33 ± 0.68^{a}	4.03 ± 0.53^{a}	3.5 ± 0.0^{b}	3.39 ± 0.24^{b}
Texture	3.53 ± 0.46^{b}	4.27 ± 0.09^{a}	3.3 ± 0.1^{b}	3.73 ± 0.40^{b}
Sweetness	3.59 ± 0.54^{b}	4.13 ± 0.09^{a}	4.1 ± 0.2^{a}	3.69 ± 0.34^{b}
Overall	3.43 ± 0.73^{ab}	4.07 ± 0.41^{a}	3.65 ± 0.15^{b}	ND

ND, not determined.
Values with different letters in each row are significantly different ($P < 0.05$).
Source: Saka *et al.* (2007).

References

Akinnifesi, F.K. (2001) Domestication: tapping the unexplored wealth of indigenous fruit trees in Malawi. *Horticulture* 3, 9–14.

Akinnifesi, F.K., Kwesiga, F., Mhango, I., Mkonda, A., Chilanga, T. and Swai, R. (2004) Domesticating priority miombo indigenous fruit trees as a promising livelihood option for smallholder farmers in southern Africa. *Acta Horticulturae* 632, 15–30.

Akinnifesi, F.K., Kwesiga, F.R., Mhango, J., Mkonda, A., Chilanga, T., Kadu, C.C.A., Kadzere, I., Mithöfer, D., Saka, J.D.K., Sileshi, G., Ramadhani, T. and Dhliwayo, P. (2006) Towards the development of miombo fruit tree as commercial tree crop in southern Africa. *Forests, Trees and Livelihoods* 16, 103–121.

Boffa, J.M., Yameogo, G., Nikiem, P. and Knudson, D.M. (1996) Shea nut (*Vitellaria paradoxa*) production and collection in agroforestry parklands of Burkina Faso. In: Leakey, R.R.B., Temu, A.B., Melnyk, M. and Vantomme, P. (eds) *Domestication and Commercialization of Non-timber Products of Agroforestry Systems*. Proceedings of an International Conference held in Nairobi, Kenya, 19–23 February 1996. Food and Agriculture Organization, Rome, pp. 110–122.

Brand, J.C., Cherikoff, V. and Truswell, A.S. (1985) The composition of aboriginal bushfoods. 3. Seeds and nuts. *Food Technology Journal of Australia* 37, 276–279.

Dietz, H.M. (1999) Opportunities and potential for processing of horticultural products by small entrepreneurs. Project 95.2254.1–001.00. Promotion of Horticulture, Malawi, March/April, 1999, 34 pp.

Ennis, D.M., Keeping, L., Chin-ting, J. and Ross, N. (1979) Consumer evaluation of the inter-relationships between the sensory components of commercial orange juices and drinks. *Journal of Food Science* 44, 1011–1016.

FAO (1989) *Forestry and Food Security*. FAO Forestry Paper 90. Food and Agriculture Organization, Rome, 128 pp.

Grivetti, L.E. (1979) Kalahari agro-pastoral hunter getherers: the Tswana example. *Ecology of Food and Nutrition* 7, 235–256.

Grivetti, L.E., Frentezl, C.J., Ginseberg, K.E., Howell, K.L. and Ogle, B.M. (1987) Bush food and edible weeds of agriculture: perspective on dietary use of wild plant in Africa, their role in maintaining human nutritional status and implication for agriculture development. In: Akhtar, R. (ed.) *Health and Diseases in Tropical Africa: Geographical and Medical Viewpoints*. Harwood, London, pp. 51–81.

Gross, J. (1991) *Pigments in Vegetables: Chlorophylls and Carotenoids*. AVI Books, Van Nostrand Reinhold, New York.

Ham, C. and Akinnifesi, F.K. (2006) *Zimbabwe, Zambia, Malawi and Tanzania: priority fruit species and products for tree domestication and commercialization*. World Bank IK Notes No. 96. The World Bank, Washington, DC, 4 pp.

Humpry, C.M., Clegg, M.S. and Keen, C.L. (1993) Food diversity and drought survival: the Hausa example. *International Journal of Food Science and Nutrition* 44, 1–16.

Kadzere, I., Hove, L., Gatsi, T., Masarirambi, M.T., Tapfumaneyi, L., Maforimbo, E. and Magumise, I. (2004) Domestication of indigenous fruits: post harvest fruit handling practices and traditional processing of indigenous fruits in Zimbabwe. In: Rao, M.R. and Kwesiga, F.R. (eds) *Proceedings of the Regional Agroforestry Conference on Agroforestry Impacts on Livelihood in Southern Africa: Putting Research into Practice*. World Agroforestry Centre (ICRAF), Nairobi, pp. 353–363.

Kavishe, F.P. (1993) *Nutrition: Relevant Actions in Tanzania*. Monograph Series No. 1. Tanzania Food and Nutrition Centre, Dar es Salaam, pp. 1–8.

Klopotek, Y., Oho, K. and Bohm, V. (2005) Processing strawberries to different products alters contents of vitamin C, total phenolics, total anthocyanins, and antioxidant capacity. *Journal of Agricultural and Food Chemistry* 53, 5640–5646.

Krige, M., Hansmann, C. and Akinnifesi, F. (2006) *Guide to Indigenous Fruit Processing*. World Agroforestry Centre and CP Wild Research Alliance, Nairobi, 23 pp.

Kwesiga, F., Akinnifesi, F.K., Ramadhani, T., Kadzere, I. and Saka, J. (2000) Domestication of indigenous fruit trees of the miombo in southern Africa. In: Shumba, E.M., Lusepani, E. and Hangula, R. (eds) *The Domestication and Commercialization of Indigenous Fruit Trees in the SADC Region*. Proceedings of a SADC Tree Seed Centre Technical Meeting held in Windhoek, Namibia, 13–15 March, 2000. SADC Tree Seed Centre Network/SADC/FAO/CIDA Strategy Workshop, Windhoek, Namibia, pp. 8–24.

Ladipo, D.O., Fondoun, J.M. and Gaga, N. (1996) Domestication of bush mango (*Irvinglia* spp.): the exploitable intraspecific varieties in west and central Africa. In: Leakey, R.R.B., Temu, A.B., Melnyk, M. and Vantomme, P. (eds) *Domestication and Commercialization of Non-timber Products of Agroforestry Systems*. Proceedings of an International Conference held in Nairobi, Kenya, 19–23 February 1996. Food and Agriculture Organization, Rome, pp. 193–205.

Lamien, N., Sidibe, A. and Bayala, J. (1996) Use and commercialisation of non-timber forest products in western Burkina Faso. In: Leakey, R.R.B., Temu, A.B., Melnyk, M. and Vantomme, P. (eds) *Domestication and Commercialization of Non-timber Products of Agroforestry Systems*. Proceedings of an International Conference held in Nairobi, Kenya, 19–23 February 1996. Food and Agriculture Organization, Rome, pp. 51–64.

Leakey, R.R.B. and Ladipo, D.O. (1999) Potential for novel food products from agroforestry trees: a review. *Food Chemistry* 66, 1–14.

Lutham, M.C. (1997) *Human Nutrition in the Developing World*. FAO Food and Nutrition Series No. 29. Food and Agriculture Organization, Rome, 508 pp.

Lyimo, M. and Tiisekwa, B.P.M. (2000) *Income generation project for women group in Kipera village. End of project report*. Sokoine University of Agriculture, Morogoro, Tanzania, 15 pp.

Maghembe, J.A., Simons, A.J., Kwesiga, F. and Rarieya, M. (1998) *Selecting Indigenous Fruit Trees for Domestication in Southern Africa: Priority Setting with Farmers in Malawi, Tanzania, Zambia and Zimbabwe*. International Centre for Research in Agroforestry, Nairobi, 94 pp.

Ndabikunze, B.K., Chamshama, S.A.O., Tiisekwa, B.P.M. and Mugasha, A.G. (2000) Nutrition and utilization of indigenous tree food sources in Tanzania. In: Shumba, E.M., Lusepani, E. and Hangula, R. (eds) *The Domestication and Commercialization of Indigenous Fruit Trees in the SADC Region*. Proceedings of a SADC Tree Seed Centre Technical Meeting held in Windhoek, Namibia, 13–15 March 2000. SADC Tree Seed Centre Network/SADC/FAO/CIDA Strategy Workshop, Windhoek, Namibia, pp. 56–61.

Ngwira, T.N. (1996) Utilization of local fruit in wine making. In: Leakey, R.R.B., Temu, A.B., Melnyk, M. and Vantomme, P. (eds) *Domestication and Commercialization of Non-timber Products of Agroforestry Systems*. Proceedings of an International Conference held in Nairobi, Kenya, 19–23 February 1996. Food and Agriculture Organization, Rome, pp. 188–191.

Nordeide, M.B., Hatloy, A., Folling, M., Lied, E. and Oshaug, A. (1996) Nutrient composition and nutritional importance of green leaves and wild food resources in agricultural district, Koutiala, in southern Mali. *International Journal of Food Science and Nutrition* 47, 455–468.

NRC/NAS (National Research Council/National Academy of Sciences) (1989) *Recommended Dietary Allowances*, 10th edn. Reports of the sub-committee on the tenth edition of RDAs, Food and Nutrition Board, Commission on Life Sciences. National Academy Press, Washington, DC, 85 pp.

NSO (1999) *National Statistics Office Draft Annual External Trade Report 1999*. National Statistics Office, Zomba, Malawi, 543 pp.

Rajabiun, S. (2001) *HIV/AIDS: A Guide for Nutrition, Care and Support*. Technical Assistance Project. Academy for Educational Development, Washington, DC, 55 pp.

Ramadhani, T. (2002) *Marketing of Indigenous Fruits in Zimbabwe*. Socio-economic Studies on Rural Development, Vol. 129. Wissenschaftsverlag Vauk., Kiel, Germany.

SAFIRE (2005) *Uapaca kirkiana Processing Technical Guide*. Southern Alliance for Indigenous Resources, Harare, 12 pp.

Saka, J.D.K. (1994) The nutritional value of edible indigenous fruits: present research and future directions. In: Maghembe, J.A., Ntupanyama, Y. and Chirwa, P.W. (eds) *Improvement of Indigenous Fruit Trees of the Miombo*. International Centre for Research in Agroforestry (ICRAF), Nairobi, pp. 50–57.

Saka, J.D.K. and Msonthi, J.D. (1994) Nutritional value of edible fruits of indigenous wild trees in Malawi. *Forestry Ecology and Management* 64, 245–248.

Saka, J.D.K., Msonthi, J.D. and Sambo, E.Y. (1992) Dry matter, acidity and ascorbic acid contents of edible wild fruits growing in Malawi. *Tropical Science* 32, 217–221.

Saka, J.D.K., Mhango, J.M. and Chilanga, T.G. (2001) Physicochemical and organoleptic characteristics of *Uapaca kirkiana* and *Strychnos cocculoides* products. *Malawi Journal of Science and Technology* 7, 123–128.

Saka, J.D.K., Mwendo-Phiri, E. and Akinnifesi, F.K. (2002) Community processing and nutritive value of some miombo indigenous fruits in central and southern Malawi. In: Kwesiga, F., Ayuk, E. and Agumya, A. (eds) *Proceedings of the 14th Southern Africa Regional Review and Planning Workshop, 3–7 September 2001, Harare, Zimbabwe*. ICRAF Regional Office, Harare, pp. 165–169.

Saka, J.D.K., Swai, R., Mkonda, A., Schomburg, A., Kwesiga, F.K. and Akinnifesi, F.K. (2004) Processing and utilisation of indigenous fruits of the miombo in southern Africa. In: Rao, M.R. and Kwesiga, F.K. (eds) *Proceedings of Regional Agroforestry Conference on Agroforestry Impacts on Livelihoods in Southern Africa: Putting Research into Practice*: World Agroforestry Centre (ICRAF), Nairobi, pp. 343–352.

Saka, J.D.K., Rapp, I., Akinnifesi, F.K., Ndolo, V. and Mhango, J. (2007) Physicochemical and organoleptic characteristics of *Uapaca kirkiana, Strychnos cocculoides, Adansonia digitata* and *Mangifera indica* fruit products. *International Journal of Food Science and Technology* 44, 836–841.

Simons, A.J. (1996) ICRAF's strategy for domestication of non-wood tree products. In: Leakey, R.R.B., Temu, A.B., Melnyk, M. and Vantomme, P. (eds) *Domestication and Commercialization of Non-timber Products of Agroforestry Systems*. Proceedings of an International Conference held in Nairobi, Kenya, 19–23 February 1996. Food and Agriculture Organization, Rome, pp. 8–22.

Smith, G.C., Clegg, M.S., Keen, C.L. and Grivetti, L.E. (1995) Mineral value of selected plant foods common to southern Burkina Faso and to Niamey, Niger, West Africa. *International Journal of Food Science and Nutrition* 47, 41–53.

Tembo, D. (2006) Physicochemical characteristics of some priority indigenous fruits growing in Malawi. MSc Thesis, University of Malawi, Zomba, Malawi, 78 pp.

Thiong'o, M.K., Kingori, S. and Jaenicke, H. (2002) The taste of the wild: variation in the nutritional quality of the marula fruits and opportunities for domestication. *Acta Horticulturae* 575, 237–244.

Tiisekwa, B.P.M., Ndabikunze, B.K., Samson, G. and Juma, M. (2004) Suitability of some indigenous tree fruits for manufacturing juices and jams in Tanzania. In: Rao, M.R. and Kwesiga, F.K. (eds) *Proceedings of the Regional Agroforestry Conference on Agroforestry Impacts on livelihoods in Southern Africa: Putting Research into Practice*. World Agroforestry Centre (ICRAF), Nairobi, pp. 331–335.

Van Vlaenderen, H. (1999) Local knowledge: what is it and why and how do we capture it? In: *Proceedings of First National Workshop on Gender, Biodiversity and Local Knowledge Systems (LinKS) to Strengthen Agricultural and Rural Development, 22–23 June, 1999 Morogoro, Tanzania*, pp. 4–9.

Van Vlaenderen, H. and Nkwinti, G. (1993) Participatory research as a tool for community development. *Development Southern Africa* 10, 211–228.

Velho, C.C., Whipkey, A. and Janick, J. (1990) Cupuassu: a new beverage crop in Brazil. In: Janick, J. and Simon, J.E. (eds) *Advances in New Crops*. Timber Press, Portland, Oregon, pp. 372–375.

Wehmeyer, A.S. (1966) The nutrient composition of some edible wild fruits in the Transvaal. *South African Medical Journal* 40, 1102–1104.

Williamson, J. (1975) *Useful Plants of Malawi.* Revised edition. University of Malawi, Zomba, Malawi, 336 pp.

Zafrilla, P., Ferreres, F. and Barberian, F.A. (2001) Effects of processing and storage on the antioxidant, ellagic acid derivatives and flavonoids of red raspberry (*Rubus idaeus*) jams. *Journal of Agricultural and Food Chemistry* 49, 3651–3655.

17 The Role of Institutional Arrangements and Policy on the Conservation, Utilization and Commercialization of Indigenous Fruits in Southern Africa

P.A. ODUOL,[1] O.C. AJAYI,[2] P. MATAKALA[1] AND F.K. AKINNIFESI[2]

[1]*World Agroforestry Centre, SADC/ICRAF Agroforestry Programme, Maputo, Mozambique*; [2]*World Agroforestry Centre, ICRAF, Lilongwe, Malawi*

17.1 Introduction

Food insecurity, health, especially HIV/AIDS, high levels of unemployment and poverty are some of the key development challenges facing the southern Africa region. To overcome these challenges, most governments are implementing programmes that promote sustainable economic growth to reduce poverty and unemployment. As part of these programmes, rural people are coping with food insecurity and sustaining their livelihoods by using forest products, including wild foods and indigenous fruit trees. These foods supplement their diets and are traded to provide cash income (FAO, 1989). The forests and natural woodlands support millions of livelihoods for people living within and neighbouring them. They provide direct and indirect benefits that include environmental services of soil, water and biodiversity conservation, animal habitats, beauty, tourism, a variety of wood and non-wood products, medicines, herbs and fruits. They are home to several indigenous fruit trees (IFTs) that offer various products and services to rural and urban communities in Africa (FAO, 1989). These IFT products and services have sustained rural and urban livelihoods for thousands of years. The biggest challenge to forest sectors in southern Africa is to promote self-sufficiency in forest products through sustained forest management and biodiversity conservation. However, increasing human populations are exerting pressure on natural resources, resulting in a general decline of woodlands through deforestation and land degradation in the region.

The management and sustainable use of forest products in general, and indigenous fruit trees in particular, have been influenced by several institutional arrangements and policies at local and national levels. In principle, legal systems exist that can be used to regulate the utilization of forest resources but, in practice, these institutions are fragmented and sometimes inconsistent with one another, making implementation difficult. Most institutions work in isolation and lack sufficient human resources, and also communication and exchange of information between them is poor. This chapter synthesizes information on the role of institutional arrangements and policy in the conservation, utilization and commercialization of indigenous fruits in the miombo woodlands of southern Africa.

17.2 Institutional Arrangements in the Management of Forest Resources and IFTs

Institutions are regulatory systems, formal and informal regulations and agreements; norms of behaviour that are generally agreed upon by communities and organizations (Harris, 1982). Several regulatory and institutional arrangements governing natural resource management are found in southern Africa. The institutions are usually complex and have persisted over time, serving collective and valued purposes. It is therefore very important to understand the governance structures or organizations and institutional framework with regard to the management of natural resources. In earlier times, communities used to decide on who could use resources, when, where and how. There exist many institutions ranging from local to national, formal to informal, within various social, cultural and traditional contexts, operating independently or alongside each other, guided by national policy, regulations and an institutional framework.

17.2.1 Dynamics of institutions governing IFTs

The types of institutions and legislation and sectors governing the management of natural recourses in the southern Africa region include government ministries, other para-state agencies, international conventions and religious or faith-based institutions. In order to develop IFTs, there is need to understand the evolution and dynamics of institutions (Campbell *et al.*, 2001). Most of the institutions in southern Africa have been affected by the changing state and administrative frameworks from colonial to post-colonial times. In colonial days, the management of forests and woodland resources was carried out by governments to the exclusion of local communities. The colonial government caused changes in existing traditional institutional structures to facilitate their development and administrative agenda. These changes undermined the traditional institutions and organizations responsible for the management of woodlands (Matose and Wily,

1996). The colonial state agencies were created to control, regulate and manage woodland resources by establishing forest reserves as 'no go' zones for local communities. Over time, however, this has changed, and policies are increasingly being geared towards co-management by the state and communities. Several countries are devolving rights and responsibilities for management of natural resources to local people (Chambers, 1983). The devolvement is born out of an appreciation that integrating formal institutions with traditional moral and political legitimacy at the local level provides more stable and effective approaches for managing natural resources. In addition, the effectiveness with which state institutions manage natural resources without the participation of local communities has been increasingly questioned. In most countries, national policies and regulations set the institutional framework. These institutions are dynamic and are pivotal in the facilitation of local empowerment. Institutions involved in the governance of land and resource management in southern Africa range from national to local, informal and formal, operating independently or alongside each other. Most of these institutions and their supporting structures fall within the state institutional framework or operate as traditional institutions (Shackleton and Campbell, 2001). Among the oldest are traditional institutions that may fall within or outside state statutes (Mukamuri *et al.*, 1999; Shackleton and Campbell, 2001). During colonial times, individuals were asked to carry out conservation work, such as building and planting trees in village woodlots, without payment.

The structural adjustment programmes embarked upon by several countries in the 1980s have greatly affected most state institutions and forced them to scale-down their activities, reduced government expenditure, and encouraged privatization and decentralization of governance. Community conservation had been institutionalized in many local communities. The decline in respect of traditional regulations is due to the emergence of new livelihood strategies in order to cope with economic problems. People are no longer interested in doing things whose benefits they are not sure of getting; they now prefer to plant trees in their individual home plots. The sale of fruits from natural forests was traditionally unacceptable. Most of these local institutions for managing collectively owned natural resources are built on controls derived from traditions, cultures and norms (Matose and Wily, 1996). However, there is evidence that local communities can successfully manage collectively owned resources (Agarwal and Yadama, 1997). In some locations, youths have turned to wood carving for tourists as a survival strategy. They are now violating traditional rules by cutting down fruit and sacred trees. The traditional rules and regulations relating to woodlands are under pressure from the commercialization of the resources. In the 1980s, individuals started to sell indigenous fruits, e.g. *Uapaca kirkiana* fruits, cakes made from *Parinari curatellifolia* and alcoholic products from *Sclerocarya birrea* in Zimbabwe.

17.2.2 Institutional and policy challenges to sustainable development of IFTs

Traditionally the miombo woodlands of southern Africa have been supporting various wildlife and plants and providing an economic lifeline to rural communities through subsistence, collection and gathering activities. Forest plants supplement and complement agriculture with household nutrition and food security. Non-wood forest products (NWFPs) include leaves and fruits that provide essential vitamins; the seeds, nuts, roots and tubers supply fats and carbohydrates, while mushrooms, gums and saps provide protein and minerals. IFTs obtained from woodlands and forests are sources of food during natural disasters such as floods and famines, when there is a general failure of agricultural crops (Akinnifesi *et al.*, Chapter 8, Mithöfer and Waibel, Chapter 13, this volume). They provide income, employment and some cultural values. The collection and processing of IFTs is suitable for equitable participation of the most vulnerable groups such as the poor, women and children. Several IFTs provide nutritious protein and fats and their edible pulps are also rich in vitamins (Saka *et al.*, 2002, Chapter 16, this volume). Farmers traditionally retain and protect several tree species in their farmlands to provide fruits and food products during food scarcity. The biggest threats to the development of the woodlands are fire and farming. Most countries in southern Africa, including Malawi, Tanzania, Zambia, Zimbabwe, Mozambique and Botswana, are net importers of fresh fruits and products from South Africa and Europe, despite the vast reserves of IFTs that they have (FAO, 2002).

Formal policies and regulatory frameworks

Despite the potential role of IFTs as enumerated above, there are several constraints on their development. Fire and farming are the biggest threat to the development of woodlands. Other constraints include policy, infrastructure support and access to technology, lack of systematic research, scarcity of equipment, quality control and regulatory arrangements. There is a lack of clear and appropriate policies that support the development of NWFP, despite their economic potential. The policies and laws governing forest access, gathering and collection have been restrictive in nature, through control for protection (Kayambazinthu *et al.*, 2003; Kowero, 2003). Resource use such as harvesting for timber and subsistence use was through law enforcement, which involved the issuance of permits by forest authorities. The management style adopted excludes most stakeholders, whose livelihoods depended on forests, resulting in conflicts with government authorities. This denies the rural communities the opportunity to manage and live in harmony with their natural resources. There are several acts and policies governing forest resources and natural resource management in the southern African countries (Kayambazinthu *et al.*, 2003; Kowero, 2003). The management of natural resources in southern Africa is regulated by several legal systems including the Agricultural Resources Act, Forest Act, Wildlife Conservation and National Parks Acts, Water, Town and Country Planning Act, Land Act, Traditional Authorities Act, Tribal Land Act, Communal Lands Act, and the Tourism and Environment Act. In addition, oversight on natural resources is exercised by several agencies ranging from Ministries of Environment and Tourism, Ministry of Parks and Wildlife Management, Ministry of Local Government, and Ministry of Water Development to para-state agencies such as Forest Departments, Departments of National Parks and Wildlife, District Councils, and the National Traditional

Healers Association. Others include Village Forest Committees (VFC), Village Natural Resources Management Committees (VNRMC), the Communal Areas Management Programme for Indigenous Resources (CAMPFIRE), Village Development Committees (VIDCO), Ward Development Committees (WARDs) and Rural Management Committees (RMCs) (Kayambazinthu *et al*., 2003). The coordination of the implementation of these legal instruments is ineffective due to poor coordination and covert competition. While these regulatory and legal systems could in principle help to protect natural resources, the drawback is that their coordination and implementation is ineffective due to the independent nature of their operations and subtle competition between the different institutions involved (Kowero, 2003). There is a need to restructure institutions and reform policies in order to address this problem. After independence, the land tenure systems in most southern African countries have also weakened traditional forest management systems (Kayambazinthu *et al*., 2003). For example, in traditional tribal lands, woodland resources are now being exploited by individuals with little respect for conservation and sustainability. Yet before independence there was great respect for the tribal chiefs who governed the use and access to woodland resources effectively. In addition, there is general lack of integration of wood and non-wood products in traditional forest and woodland management. Recently there have been legislative reforms towards joint sustainable forest management where diverse stakeholders (government, local communities and the private sector) share responsibilities in forest management in Malawi (Kayambazinthu *et al*., 2003). Such initiatives are, however, still in their nascent stages of development.

Traditional by-laws and policies

At local levels, traditional chiefs are the formal and informal management structures for natural resources including IFTs. The chiefs use rules, sacred controls and civic controls in their area of jurisdiction. Examples of these powerful and stable systems regulating the use of natural resources include *Dagashiga* in Tanzania which is very efficient in the communication and articulation of indigenous knowledge, attitudes and practices in regulating access to natural resources within the community (Johannson and Mlenge, 1993), *Ngitili* village laws that operate at communal and household level and are still effective to date (Oduol and Nyadzi, 2005). Other social and political institutions for regulating and protecting forest products and IFTs exist among the *Ngonis* of Malawi, *Bembas* of Zambia, *Lomwes* of Mozambique, *Shonas* of Zimbabwe and the *Sukumas* of Tanzania (Kayambazinthu *et al*., 2003). An understanding of the dynamics of these institutions is very important for the development of IFTs, as they may provide opportunities for conserving natural forest resources and IFTs.

On the other hand, however, they may inadvertently hinder the development of IFTs. For example, due to unclear guidelines to property and user rights of indigenous fruits, the chiefs have imposed informal rules, which allow only home consumption of the said fruits. Selling indigenous fruits is strictly prohibited. The rules are further backed up by the traditional attitude that indigenous fruits are a gift from God and therefore should neither be sold nor bought. The non-existence of rules which clearly specify the ownership and user rights of the fruit trees and fruits imposes difficulties in managing and controlling the use of the fruits. Since the fruits are still regarded as a common resource, they are being overused. Although village traditional leaders helped to manage and conserve the fruit trees through practising the old taboos and beliefs, they brought high uncertainties to traders. Efforts to commercialize indigenous fruits need to address issues related to ownership and use regulations on indigenous fruit trees as well as fruits. In Zimbabwe, for example, although there is a substantial amount of trading of indigenous fruits in both rural and urban areas, the system is still underdeveloped. Property rights guiding use and management of the fruit trees and fruits are not transparent. While consumers like the fruits and are willing to pay for them, informal by-laws enacted by village and traditional leaders limit market supply. Efforts to improve the commercialization of fruit should include encouraging and facilitating a government initiative to set up a reliable institutional framework, which will reduce risk and uncertainty for traders. Rules that favour private action can be introduced. Producers can either buy or be issued with free collection licences.

More detailed studies of natural resources management policies and socio-cultural settings on IFTs have been carried out in Zimbabwe (Mithöfer and Waibel, Chapter 13, this volume). There is a need for more information on marketing systems and factors influencing the markets in Malawi, Tanzania and Zambia, as these countries have different natural resources management policies and socio-cultural habits from Zimbabwe. A synthesis of the findings from studies in these countries will help in drawing appropriate conclusions to facilitate regional indigenous fruit commercialization.

Land tenure and property rights

Land tenure issues were investigated in Malawi and Zambia. The results show that most of the lands belong to members of the households, except for some small parcels of land borrowed from friends or village chiefs. The use designation period was unlimited (98%) in both countries, and is transferable. The transferability of lands with IFTs was, however, lower than that with non-IFTs. The tenure type is individual ownership. Land tenures did not affect the willingness of farmers to cultivate trees, as 96% of farmers in Malawi and 100% in Zambia are utilizing customary lands in both countries, and only 4% of smallholders' land in Malawi was leasehold. Patrilineal households have an average of six IFTs retained or planted in their gardens per household compared to non-IFT households. This is because men feel too insecure to invest in trees in matrilineal communities (e.g. Thondwe in southern Malawi and Katete in eastern Zambia). Cutting down matured indigenous fruit trees for firewood is a common problem in the region, especially in communal lands in Malawi.

Issues of natural resources management policies and property rights are important because the major source of *Uapaca kirkiana* and *Strychnos cocculoides* indigenous fruits for the market is the communal forests. Access to indigenous fruits and trees located in the communal forests near villages, fields and homes is an important aspect and needs a closer look. This is because lack of access might influence the preference towards the development of markets for the fruits. The profitability of investment in IFT cultivation is strongly influenced by property rights on trees and tree products. In many of the countries in southern Africa, indigenous trees and products are regarded as a common property and so incentives for private investment in their improvement is low. Incentives and profitability of planting IFTs will increase if there is a modification of the institutional issues governing the collection of fruit trees (i.e. legalizing the sale of only those fruit trees that come from plantations). Studies from Mozambique (Unruh, 2001) indicate that in the developing world, economically valuable trees are among the most common and valuable forms of customary evidence for claiming ownership of land. These valuable trees are generally permanent or perennial trees that are there to assure long-term evidence of land ownership. Due to their enduring nature, trees can be evidence that a land is still owned even though it is left fallow. In many cases, a claim to possession of valuable trees is the single most important piece of evidence for defending or asserting rights to land, irrespective of the average number of trees planted.

Trade and markets

Where local groups are well organized and can control forest access, rural enterprises tend to fare better. A clear sense of group identity, cooperative behaviour and established rights to the resource can all help. Despite both financial and social benefits enjoyed by the fruit traders, rigid by-laws limit the marketing of indigenous fruits and affect the market supply of the fruits. For example, traditional chiefs in some communities in Zimbabwe enacted by-laws that any indigenous fruit that was collected within their communities cannot be taken outside their chiefdoms. Their reasoning is that such indigenous fruits are God's gift to provide food and nutrients from the wild to every member of the community during yearly hungry seasons between December and February (before food crops are ready for harvest) and such a natural gift should not be commercialized.

The nature, extent of variation, quality, characteristics and uses of IFT products largely affects their trade. IFT markets are ephemeral in nature and have to face stiff competition from cheaper substitutes, such as from cultivated sources (FAO, 1991). IFT products are usually based on local rural enterprises, small-scale in nature and associated with traditional uses and low technology. The policies do not usually support such enterprises, as they are regarded as less important. Most IFT enterprises hardly receive any loans from credit institutions. There is a general tendency to disregard products that do not make significant contributions to national economies.

There are wide disparities in prices, with big differences between the price paid to the collector and the final price for the commodity. It is estimated that, for most IFTs, the local producer receives a minimal percentage of the price that they are sold at in developed-country markets (FAO, 1991). The price difference supports the traders and middlemen who want to maximize profits as greatly as possible. This makes market promotion very difficult. However, there is some hope due to the new attitude towards NWFP as a result of increasing demands for organic fruits, concern for the environment and preference for natural products. Knowledge and information availability is another important component of marketing; which requires improvement in the skills and attitudes of the people involved in market development. Marketing chains are complex, from the commodity production to marketing consumer products. Primary producers only manage the first stages of marketing, but it is necessary to understand the whole marketing chain and its position in order to get a fair deal and share of the benefits and wealth created by collection, extraction and, finally, sale of the goods to the consumer. It is vital to have an appropriate accessible marketing information system for NWFP and IFTs (FAO, 1991).

Indigenous knowledge of IFTs

Indigenous knowledge (IK) or local knowledge refers to the knowledge and practices developed over time, maintained by people in rural areas, who interact with their natural environment. This complex knowledge forms part of the cultural practices used in the management and use of natural resources to sustain livelihoods. This dynamic local knowledge is unique to a given culture and society, and it forms the basis for local-level decision making in agriculture, healthcare, food preparation, education, natural resource management and other activities in rural communities (World Bank, 2006). Groups of IFT collectors who include children have useful information on taste, availability and fruit size that can help in guiding the domestication of indigenous fruit trees (Simons, 1996; Maghembe *et al.*, 1998). IK is a very important resource for rural communities in fighting poverty, thus addressing the Millennium Development Goals (MDGs). It enables the development of enterprises and for the indigenous people to participate in decision making and sustainable development. However, this knowledge is being eroded and lost. It is necessary to strengthen the capacities of local people for developing their own knowledge and methodologies that promote activities to improve and sustain their livelihoods. IK should form the basis for policy development and research on domestication for IFTs (Kwesiga *et al.*, 2000).

There is variation in the organization of IFT harvests. The common method is collection by local people who have rights for such activity, for sale in local markets or to purchasing agents. Another is the employment of casual or contract labour by those with collection rights. The local collectors are often exploited by middlemen who control access to the markets, or those who control access to the resource (e.g. government officers). This makes the millions of people who are involved in NWFP collection lack incentives for best practices control and sustainable harvesting. Postharvest care is also poor in most cases, with high rates of wastage, and qualitative and quantitative losses during collection, transport and storage. The physical infrastructure and poor roads in Africa have an effect on delicate and perishable products, compared with those that can withstand rough handling and long storage. There is an urgent need to improve harvesting systems, methods and practices, by getting improved tools and techniques, training and skills, incentive systems, institutional arrangements, promotion of local facilities for processing and value addition and linking harvesting to processing and markets (FAO, 1991).

17.3 Institutional and Policy Support for Development of IFTs

Policies that support the development of NWFP hardly exist, and where they are mentioned or implemented often lack clear objectives, targets and development strategies. There are few strategic plans, programmes and projects, legal rights arrangements, incentives for development of skills, health and safety considerations, access to information (lack of databases) and support for investments from public administration. The involvement of local communities in the development of IFTs requires appropriate legislation that provides them with legal rights to encourage long-term investments and improvement. It is necessary to organize various stakeholders, government, groups of women and men, local communities and the private sector as a means to strengthening institutions to develop NWFPs for their economic and ecological benefits (FAO, 1991; Akinnifesi *et al.*, 2006). This should address gender differentiation to improve the effectiveness and benefits of IFTs programmes at the local level. The development of various sectors is necessary in order to promote healthy relationships between producers, industry and consumers. There is a need for improvements in arrangements that cover access, control, management and ownership of the resources, involvement of local people, groups, management agencies, industry, trade and international assistance. IFT resources need to be linked with national and emerging international market niches to support sustainable development. In addition, there is a need to learn from one another, exchange of information and experiences between countries, through undertaking collaborative activities and development of a regional information network (FAO, 1991). The provision of skills through training, capacity-building, technology delivery, extension support, market information systems, credit facilities, support to establish necessary infrastructure, and streamlining of forest administration are some of the development support required. Other areas of development should include harvesting, storage, processing, standardization and marketing of products often regarded as suitable for local consumption only. The sharing of benefits provides people with incentives to conserve and sustainably manage their resources. Apart from security of tenure and autonomy, there is need for provision of economic incentives, improvement in access to credit, markets and prices. Most financial assistance is usually provided to processors and exporters, but there is little financial assistance for sustainable management of the harvesting of resources (FAO, 1991). This trend needs to be reversed and improved upon.

17.4 Conclusions

The management of NWFP from natural and cultivated sources is very important for the livelihoods of people in Africa. IFTs play a key role in food security and support rural development, as most activities in this field are people-orientated, create employment, provide cash income and help in the attainment of the MDGs. Several programmes are under way in the management of these natural resources, and thousands of farmers who are now aware of the social and economic importance of IFT are reaping benefits. Most countries in southern Africa are undergoing policy and legal reforms and devolving powers regarding the management of natural resources to communities, after recognizing the benefits that follow from this in rural development. It has been realized that better-integrated institutions with traditional, social cultural and economic incentives and backed up with moral and political legitimacy at local level are more effective in the management of natural resources. Devolvement should ensure community ownership and involvement in decisions regarding the use of their natural resources. A holistic multi-sectoral approach involving communities, non-governmental organizations, governments and private sectors is needed in order to improve the development of IFTs.

While several regulations and by-laws exist to regulate IFTs, the bane of these laws is lack of effective implementation. Coherent government policies that support community-based forest product enterprises are needed, including mechanisms to make credit more available to small enterprises (such as the recognition of stands of commercial tree species as collateral) and the removal of counterproductive price controls. To inspire policy makers to support rural enterprises with a coherent policy framework, better accounting of the economic importance of community-based forest product enterprises, such as has been proposed by the FAO, should be implemented. These will include a system for grouping community-based forest product enterprise trade statistics within existing commodity classification systems (Chandrasekharan, 1993).

There is also a need to analyse the legal aspects, such as access to resources and user rights and, how they affect different social groups within the communities. Assessment of the social and institutional environment entails investigation into the indirect benefits of the project for the community, and the tradeoffs involved in IFT domestication, fruit collection and commercialization. In the development of IFTs, social and institutional criteria must be considered as equally important as the economic, resource management and technical criteria. A practical course of action must involve assessment of the direct and indirect potential impacts of proposed projects on the community in order to design socially sensitive activities and to minimize (or eliminate) potential negative impacts on women and other vulnerable groups within the community, investigate sources of capital for small-scale industries and the opportunities for community members to obtain access to formal credit.

References

Agarwal, A. and Yadama, G.N. (1997) How do local institutions mediate market and population pressures on resources? Forest panchayats in Kumaon, India. *Development and Change* 28, 435–465.

Akinnifesi, F.K., Kwesiga, F., Mhango, J., Chilanga, T., Mkonda, A., Kadu, C.A.C., Kadzere, I., Mithöfer, D., Saka, J.D.K., Sileshi, G., Ramadhani, T. and Dhiliwayo, P. (2006) Towards the development of miombo fruit trees as commercial tree crops in southern Africa. *Forests, Trees and Livelihoods* 16, 103–121.

Campbell, B.M., Mandondo, A., Nemarundwe, N., Sithole, B., De Jong, W., Luckert, M. and Matose, F. (2001) Challenges to proponents of common property resource systems: despairing voices from the social forest of Zimbabwe. *World Development* 29, 589–600.

Chambers, R. (1983) *Rural Development: Putting the First Last.* Longman, London.

Chandrasekharan, C. (1993) *Issues Involved in the Sustainable Development of Non-wood Forest Products: Expert Consultation on Non-Wood Forest Products for English-Speaking African Countries.* Arusha, Tanzania, 17–22 October 1993.

FAO (1989) *Forestry and Food Security.* FAO Forestry Paper No. 90. Food and Agricultural Organization, Rome.

FAO (1991) *Non-Wood Forest Products: The Way Ahead.* FAO Forestry Paper No. 1997. Food and Agricultural Organization, Rome.

FAO (2002) *State of Food Insecurity in the World.* Food and Agricultural Organization, Rome.

Harris, J. (1982) *Rural development: Theories of peasant economy and agrarian change.* Hutchinson University Library for Africa, London.

Johannson, L. and Mlenge, W. (1993) Empowering customary community institutions to manage natural resources in Tanzania: a case study from Bariadi District. *People, Trees and Forest Newsletter* 22, 36–42.

Kayambazinthu, D., Matose, F., Kajembe, G. and Nemarundwe, N. (2003) Institutional arrangements governing natural resource management of the miombo woodland. In: Kowero, G., Campbell, B.M. and Sumaila, U.R. (eds) *Policies and Governance Structures in Woodlands of Southern Africa.* Center For International Forestry Research, Bogor, Indonesia, pp. 45–64.

Kowero, G. (2003) The challenge to natural forest management in sub-Saharan Africa rural development: experiences from the miombo woodlands of southern Africa. In Kowero, G., Campbell, B.M. and Sumaila, U.R. (eds) *Policies and Governance Structures in Woodlands of Southern Africa.* Center For International Forestry Research, Bogor, Indonesia, pp. 1–8.

Kwesiga, F., Akinnifesi, F.K., Ramadhani, T., Kadzere, I. and Saka, J. (2000) Domestication of indigenous fruit trees on the miombo in southern Africa. In: Shumba, E.M., Lusepani, E. and Hangula, R. (eds) *Domestication and Commercialisation of Indigenous Fruit Trees in the SADC Region: Proceedings of a SADC Tree Seed Centre Technical Meeting held in Windhoek, Namibia, 13–15 March, 2000.* SADC Tree Seed Centre Network SADC/ FAO/CIDA Strategy Workshop, Windhoek, Namibia, pp. 8–24.

Maghembe, J.A., Simons, A.J., Kwesiga, F. and Rarieya, M. (1998) *Selecting Indigenous Trees for Domestication in Southern Africa: Priority Setting with Farmers in Malawi, Tanzania, Zambia and Zimbabwe.* ICRAF, Nairobi, Kenya.

Matose, F. and Wily, L. (1996) Institutional arrangements governing the use and management of miombo woodlands. In: Campbell, B.M. (ed.) *The Miombo in Transition: Woodlands and Welfare in Africa.* Center for International Forestry Research, Bogor, Indonesia, pp. 195–219.

Mukamuri, B.B., Matose, F. and Campbell, B.M. (1999) *Rural Institutions: Contextualising Community-Based Management of Natural Resources in Zimbabwe.* Institute of Environmental Studies, University of Zimbabwe and Forestry Commission, Harare.

Oduol, P.A. and Nyadzi, G. (2005) Ngitili: old tradition practice sustaining rural livelihoods in western Tanzania. *Prunus Tribute* 4, 8–9.

Saka, J.D.K, Mhango, J.M. and Chilanga, E.Y. (2001) Physicochemical and organoleptic characteristics of *Uapaca kirkiana* and *Strychnos cocculoides* products. *Malawi Journal of Science and Technology* 7, 123–128.

Saka, J.D.K., Mwendo-Phiri, E. and Akinnifesi, F.K. (2002) Community processing and nutritive value of some miombo indigenous fruits in central and southern Malawi. In: Kwesiga, F., Ayuk, E. and Agummya, A. (eds) *Proceedings of the 14th Southern Africa Regional Review and Planning Workshop, 3–7 September 2001*. ICRAF, Harare, Zimbabwe, pp. 164–169.

Shackleton, S.E. and Campbell, B.M. (2001) *Devolution in Natural Resources Management: Institutional Arrangements and Power Shifts: A Synthesis of Case Studies from Southern Africa.* USAID SADC NRM Project No. 690-0251.12, WWF–SARPO, EU/CIFOR and Common Property STEP Project, CSIR Report.

Simons, A.J. (1996) ICRAF's strategy for domestication of non-wood tree products. In: Leakey, R.R.R., Temu, A.B., Melnyk, M. and Vantomme, P. (eds) *Domestication and Commercialisation of Non-Timber Products of Agroforestry Systems: Proceedings of an International Conference held in Nairobi, Kenya, 19–23 February 1996*. FAO, Rome, pp. 8–22.

Unruh, J.D. (2001) *Land Dispute Resolution in Mozambique: Institutions and Evidence of Agro-forestry Technology Adoption.* CAPRI Working Paper No. 12. IFPRI, Washington, DC.

World Bank (2006) *Zimbabwe, Zambia, Malawi and Tanzania: Priority Fruit Species and Products for Tree Domestication and Commercialization.* IK Notes No. 94, July 2006. The World Bank, Washington, DC.

Wynberg, R.P., Laird, S.A., Shackleton, S., Mander, M., Shackleton, C.M., du Plessis, P., den Adel, S., Leakey, R.R.B., Botelle, A., Lombard, C., Sullivan, C., Cunningham, A.B. and O'Regan, D. (2003) Marula policy brief: marula commercialisation for sustainable and equitable livelihoods. *Forests, Trees and Livelihoods* 13, 203–215.

18 Ecology and Biology of *Uapaca kirkiana*, *Strychnos cocculoides* and *Sclerocarya birrea* in Southern Africa

P.W. Chirwa[1] and F.K. Akinnifesi[2]

[1]*Department of Forest and Wood Science, Stellenbosch University, Stellenbosch, South Africa*; [2]*World Agroforestry Centre (ICRAF), Lilongwe, Malawi*

18.1 Introduction

Indigenous fruit trees provide a major part of the food and nutritional requirements of people living in sub-Saharan Africa. In Zimbabwe, wild fruit trees represent about 20% of total woodland resource use by rural households (Campbell *et al*., 1997; Akinnifesi *et al*., 2006). Several studies have reported the sale of *Uapaca kirkiana* (Muell. Arg.), *Ziziphus mauritiana* (Lam.), *Strychnos cocculoides* (DC ex. Perleb) and *Parinari curatellifolia* (Planch. ex. Benth.) to generate income and as means of livelihood (Ramadhani, 2002; Akinnifesi *et al*., 2006). In South Africa, Shackleton *et al*. (2002) and Shackleton (2004) have reported the widespread use of *Sclerocarya birrea* fruits for making beer, jam and juice. Utilization and trade of fruits are integral components of local economies and culture and play important roles in household welfare. During woodland clearing prior to cultivation or settlement, important fruit trees, such as *Parinari curatellifolia*, *Strychnos cocculoides* and *Uapaca kirkiana*, are customarily left uncut and scattered around homesteads or crop fields. Packham (1993) has reported similar cases for Tanzania, Zambia and Zimbabwe, where *P. curatellifolia* and *U. kirkiana* are left deliberately in cultivated fields. With commercialization of some of these important indigenous fruit trees, over-harvesting of the target species is likely to be common and deforestation and increasing population are directly affecting the survival of these species. While recent studies in the miombo woodlands show *U. kirkiana*, *P. curatellifolia* and *S. cocculoides* to be the most important species (Kwesiga *et al*., 2000; Akinnifesi *et al*., 2004), *S. birrea* subsp. *caffra* is the single most important indigenous species in South Africa, Botswana, Namibia and Swaziland (Shackleton and Scholes, 2002; Shackleton, 2004).

The loss of habitat of the indigenous fruit trees due to expansion of agriculture and deforestation calls for management strategies appropriate for natural populations, including those on farm land. Hall *et al.* (2002) have suggested that the effective management of such populations must incorporate the ecology of the particular species. This chapter will consider the ecology and biology of *S. birrea*, *U. kirkiana* and *S. cocculoides* in southern Africa.

18.2 Ecology and Biology of the Species

18.2.1 *Sclerocarya birrea*

Sclerocarya birrea is a member of the Anacardiaceae (cashew family), along with 650 species and 70 genera of mainly tropical or subtropical evergreen or deciduous trees, shrubs and woody vines. Three subspecies of *S. birrea* are recognized: *S. birrea* subsp. *caffra* (Sond.) Kokwaro; *S. birrea* subsp. *multifoliolata* (Engl.) Kokwaro; and *S. birrea* subsp. *birrea*. The subspecies *multifoliolata* occurs in mixed deciduous woodland and wooded grassland in Tanzania (Shackleton *et al.*, 2002) while *birrea* occurs through west, north-east and east tropical Africa across a range of vegetation types, principally mixed deciduous woodland, wooded grassland and through the open dry savannahs of northern tropical Africa and the Sahelian region (Hall *et al.*, 2002). The most important subspecies in southern Africa is *caffra*.

Ecology

Despite the ubiquitous distribution and household and commercial importance of *S. birrea* subsp. *caffra* in southern Africa, there has been very little autecological research on the species (Shackleton *et al.*, 2002). *S. birrea* is one of the two species of *Sclerocarya*, the other being *Sclerocarya gillettii* Kokwaro, which is restricted to a small area of arid eastern Kenya (Hall *et al.*, 2002). The review by Hall *et al.* (2002) seems to consolidate the fact that *Sclerocarya* has its origins in Africa. Archaeological evidence indicates that the fruits of *S. birrea* subsp. *caffra* were known and consumed by humans in 10,000 BC (AFT Database). *S. birrea* is usually dioecious, a mechanism that ensures no selfing, although male and female flowers occasionally occur on the same tree (hermaphroditism). Elephants are thought to be the major dispersal agent. Other agents are humans, goats and monkeys.

Sclerocarya birrea is widely distributed across the sub-Saharan Africa, including but not restricted to the miombo woodland ecosystem stretching from Senegal to Ethiopia in the north, southward to Natal in South Africa, and eastward to Namibia, Angola and the southern part of the Democratic Republic of Congo. Hall *et al.* (2002) provide the most detailed distribution maps with respect to environmental factors such as elevation and rainfall. The species is associated with seasonal rainfall patterns, with a mean annual rainfall of 500–1250 mm. Subspecies *caffra* has been reported in more humid areas (1200–1600 mm rainfall) in transition areas between the Guineo–Congolian and Zambezian regions and in parts of Madagascar (Hall *et al.*, 2002). Subspecies *caffra* is usually associated with lower temperatures because of the high elevations common in eastern Africa, and the inclusion of higher latitudes in the range of the species means that it is often subject to lower mean annual potential evapotranspiration values (1100–1700 mm; Hall *et al.*, 2002). Most trees of the species occur in the frost-free part of its range but the populations in south-east Zimbabwe and south into South Africa experience occasional frost. While *S. birrea* occurs through most of sub-Saharan Africa, the different subspecies occupy different parts of this range. However, all the subspecies occur together only in northern Tanzania, which is the southern limit for subsp. *birrea*. Subspecies *caffra* is the most widespread and occurs in eastern Africa (Kenya, Tanzania), southern Africa (Angola, Malawi, Mozambique, Zambia and Zimbabwe) and southern Africa (Botswana, Namibia, South Africa and Swaziland) and is also recorded from Madagascar (Fox and Norwood Young, 1982; Arnold and de Wet, 1993; SEPASAL Database, 2001; Shackeleton *et al.*, 2002). The southern end of its range is the coastal belt of southern KwaZulu-Natal in South Africa, near Port Shepstone at approximately 31° S (Hall *et al.*, 2002; Shackleton *et al.*, 2002). Within South Africa, it is common in the savannah areas of northern KwaZulu-Natal, Mpumalanga, northern and north-west provinces. Subspecies *caffra* has been introduced either for experimental purposes or as an ornamental in Israel, Mauritius, Florida (USA), India and Oman (Hall *et al.*, 2002). Using random amplified polymorphic DNA (RAPD) analysis of 16 populations, Kadu *et al.* (2006) confirmed the hypothesis that the evolution of *S. birrea* is complex, with different contributions of seed and pollen in determining gene flow, and they found that Tanzania contained most chloroplast haplotypes.

Subspecies *caffra* rarely occurs in areas of lower rainfall, though it does occur in the Petersburg plateau where rainfall is 350–400 mm. It prefers well-drained, sandy, loamy soils, appearing to be generally intolerant of frost. The species has been reported to be associated with sandy-textured soils and in some cases sandy loam and loam soils. In other parts of southern Africa, heavy clayey basaltic soils have been reported to support subsp. *caffra* (Shackleton and Scholes, 2000; Hall *et al.*, 2002). Subspecies *caffra* grows on soils of variable fertility; in Malawi it grows in the fertile Shire valley and in the mopanosols. Indeed, the subspecies has been reported as an indicator of fertile soils (Hall *et al.*, 2002).

White's framework (Hall *et al.*, 2002), used in the consideration of the vegetation in which *S. birrea* occurs, is based on 20 phytochoria. *S. birrea* extends to most of these 20 phytochoria. However, in continental Africa most of the range falls within the vast Sudanian and Zambezian regional centres of endemism and the more equatorial of the two East African coastal mosaics (Zanzibar–Inhambane) adjoining the latter. Subspecies *caffra* occurs in the southern phytochoria (Hall *et al.*, 2002). As a consequence of the concentration of the *Sclerocarya* species in the drier parts of Zambezian and Sudanian regional centres, which are respectively dominated by the *Brachystegia–Julbernardia* (miombo) and *Isoberlinia* woodlands, the species is of less importance in these areas. In general *S. birrea* grows in wooded grassland, grassland and parkland vegetation formations. It may also be a part of the denser woody vegetation in the transition areas of forest consisting of woodlands, thicket, bushland and scrubland. Subspecies *caffra* has also been reported as a component of the riparian forest in Zimbabwe and South Africa (Farrell, 1968; Acocks, 1988; Hall *et al.*, 2002).

There is no well-defined set of associates for subsp. *caffra* extending across the region because of the ruggedness of the terrain and the extensive areas at elevations above 1000 m with numerous escarpments and valleys. The Zambezi drainage area is associated with the mopane woodland communities, with species such as *Albizia harveyi*, *Colophospermum mopane* and *Lonchocarpus capassa*. In the Indian Ocean north of the Limpopo basin typical species are *Acacia nigrescens*, *L. capassa* and *Xeroderris stuhlmannii* while *Afzelia quanzensis*, *Diplorhynchus condylocarpon*, *Pseudolachnostylis maprouneifolia* and *Terminalia sericea* are the associates most common in the miombo woodlands. Hall *et al.* (2002) list *Dombeya rotundifolia*, *Ozoroa paniculosa* and *Pappea capensis* as the frequent associates in the Limpopo drainage area, but these are of drier affinity than those of the more southerly drainage into the Indian Ocean, whose associates are of the dry forests, such as *Dialium schlechteri*, *Garcinia livingstonei*, *Sideroxylon inerme* and *Xylotheca kraussiana*. With the exception of *Pachypodium lealii*, the associates in the southern Atlantic coastal drainage are species widespread in the dry conditions of southern tropical Africa, which are characteristic of the Karoo–Namib regional centre of endemism.

Sclerocarya birrea usually stands out in the farming system as it is mostly left uncut because of its importance to the communities. It has been described as the most prominent species because most of the vegetation associated with it is usually of low stature (Moll and White, 1978; White, 1983; Hall *et al.*, 2002). Thus, it has widely been used as a descriptor species for plant communities in a range of vegetation types. Hall *et al.* (2002) have attributed the lack of quantitative information on its abundance to its being widely dispersed and therefore very poorly represented in inventories involving small sample plots, which are typical of ecological studies. The importance of *S. birrea* subsp. *caffra* to the system has only been reported with respect to its contribution to stand basal area (12%) and biomass (20%) (Hall *et al.*, 2002; Shackleton *et al.*, 2002). Most other studies have not mentioned the importance of the species with respect to number of stems, height or indeed diameter classes (Hall *et al.*, 2002).

Edwards (1967) has described the *Sclerocarya–Acacia* 'tree veld' with subsp. *caffra* as prominent in climax vegetations in a southern African situation. Other studies in southern Africa have also described communities where *S. birrea* subsp. *caffra* is a dominant species in climax vegetations (Hall *et al.*, 2002). This probably confirms the importance of *S. birrea* as a descriptor species, as discussed above.

Ecosystem functions

Sclerocarya birrea is important in the ecology of other plants and animals as it grows into a large tree, often a community dominant as explained earlier.

Because of its large size, it produces a large area of improved microenvironment within the subcanopy, which has been said to be a key resource area, being characterized by higher moisture and nutrient levels than open environments (Belsky *et al.*, 1989; Shackleton *et al.*, 2002). These conditions are conducive to the development of different subcanopy woody plants, grasses and other macro- and microorganisms. Removal of a large dominant species may result in the loss of these subcanopy species. The dominance and large size of *S. birrea* also make the crown of the tree an important habitat for small vertebrates and invertebrates, as well as parasitic plants (Shackleton *et al.*, 2002). Several loranthaceous parasites have been recorded on S. birrea, the most favoured being *Erianthemum dregei*, the haustorium of which is sold as a ornamental curio, and the wood rose, which are important for the livelihoods of rural curio traders (Dzerefos, 1996; Hall *et al.*, 2002; Shackleton *et al.*, 2002).

Several vertebrate and invertebrate species make use of the fruits, the most notable of these being elephants in southern Africa. Other animals reported include rhinoceroses, warthogs, kudu, baboons, velvet monkeys, zebras, porcupines and millipedes (Shackleton *et al.*, 2002). There is no work on what proportion of the diet or nutrient intake *S. birrea* contributes to these animal species. Foliage is browsed by elephants, kudu, giraffes, nyala and domestic cattle (Palmer and Pitman, 1972), as well as by the larvae of at least eight species of butterfly and moth (Pooley, 1993; Kroon, 1999), including larvae of the emperor moth *Pavonia pavonia* (Saturnidae) (Sileshi *et al.*, Chapter 20, this volume), a popular food in southern Africa. The prevalence of malaria may be associated with this species as there have been reports that the water-filled holes in the trunks of *S. birrea* are important breeding grounds for mosquitoes – more so than any other tree species in Kruger National Park (Palmer and Pitman, 1972).

Fire, livestock, tree removal and soil disturbance affect the regeneration and/or population character of *S. birrea*. Fires remain the most regular and often destructive annual event over the greater part of the area of distribution (Lamprey *et al.*, 1967; Coetzee *et al.*, 1979). Earlier observations in Zambia and Madagascar have shown that well-established subsp. *caffra* is fire-tolerant; this has been attributed to woody constituents that are highly fire-resistant, the thick bark, and a strong capacity to replace branches burnt during severe fires (Fanshawe, 1969; White, 1983; Hall *et al.*, 2002).

Biology

With no available information on species germination in the wild, there have been references to intervals of 6–10 months between seed dispersal and germination in the field, which allows a period of dormancy (Teichman, 1982; Lewis, 1987). The growth rates in the field are variable depending on the environment. In Malawi, mean heights of 1.9 and 3.9 m have been reported after 27 months and 48 months, respectively (Hall *et al.*, 2002). Earlier estimates by Poynton (1984) gave height increments of 0.6 m per year while Shone (1979) and Nerd *et al.* (1990) report increments of 1 and 2 m of height growth over the same period in South Africa and Israel, respectively. No estimates for the longevity of *S. birrea* have been established but observations have been made at ages of up to 200 years for many large individuals of the subsp. *caffra* in southern Africa (Hall *et al.*, 2002).

The onset of flowering in planted *S. birrea* subsp. *caffra* was reported to be 4–5 years in Malawi and Israel (Nerd *et al.*, 1990; Maghembe, 1995). On the other hand, Shone (1979) indicated an age of 7 years for the onset of fruiting in wild South African plants. From observation made in Israel and Malawi, early fruit crops are small and are usually very small (Nerd and Mizrahi, 1993; Bwanali and Chirwa, 2004). Bwanali and Chirwa (2004) found that the onset of fruiting also depended on the provenance or family being tested. Older planted trees (14–15 years) of subsp. *caffra* in Botswana have been reported to produce 1000 fruits (Campbell, 1986). In a *S. birrea* provenance trial in Mangochi, Malawi, the provenance from Mozambique started to fruit after 6 years (Akinnifesi *et al.*, 2006).

The leaf flush of *S. birrea* subsp. *caffra* coincides with the transition from the wet to the dry season in southern Africa (Teichman, 1982; Hall *et al.*, 2002). Trees are in full leaf during the mid to late rainy season and leaf fall is in the dry season. Hall *et al.* (2002) give details of the flowering and fruiting phenology of *Sclerocarya* in relation to the dry months (rainfall <50 mm) as flowering is a dry-month event. Flowering has been described as precocious but it is also associated with the flush of new foliage (Shone, 1979; Teichman, 1982; Hall *et al.*, 2002). Periods given in the literature for the duration of flowering are variable, with periods of up to 4 months or much less, especially for subsp. *caffra* (Hall *et al.*, 2002). The period of fruiting is usually within the rainy season but is also said to be very variable, with a period ranging from 2 to 8 weeks in Botswana (Taylor and Kwerepe, 1995). Lewis (1987) reported that the fruit-fall period ranged from 1 to 3 months.

Although male and female flowers occasionally occur on the same tree, the species is considered dioecious. There have been a few cases in a number of studies done in South Africa where monoecy has been reported. In many cases this has primarily been due to the presence of occasional female flowers in one or two of the most proximal inflorescences of the shoots on predominantly male trees (Hall *et al.*, 2002). In some cases there has been a rare occurrence of bisexual flowers. Male flowers are borne in groups of three on racemes below new leaves. The female flowers occur below the leaves on long peduncles and consist of four curling petals. They have numerous infertile stamens and a long, shiny ovary. Hall *et al.* (2002) give details of studies in which observations have been made during anthesis and pollination, with honey bees usually cited as the major pollinators. Secondary pollinators include flies and wasps.

The development of the seeds and fruit of *S. birrea* subsp. *caffra* in South Africa has been discussed by Hall *et al.* (2002). Fruits are borne in clusters of up to three at the end of the twigs and always on the new growth. The fruit is a round or oval drupe, usually wider than it is long, with a diameter of 30–40 mm. The shape and number of nuts per stone determine the final shape of the fruit. The fruit has a thick, soft leathery exocarp with tiny, round or oval spots, enclosing a juicy, mucilaginous flesh that adheres tightly to the stone and can be removed only by sucking. The flesh tastes tart, sweet and refreshing, although the fruit has a slight turpentine-like aroma and can give off a very unpleasant smell when decaying (AFT Database). Each fruit contains an exceedingly hard seed, which is covered by fibrous matter. It is usually trilocular but sometimes bilocular. Each seed locule contains a single light nut filling the entire cavity, which is sealed by a round, hard disc that protects the embryo until germination.

A fruit load averaging 500 kg/ha and as many as 18,935 fruits per tree have been reported for superior phenotypes in Botswana (S. Mateke, unpublished data). A high vitamin C content in the Nigerian population (403 mg/100 g) has been reported, twice as high as that in the Botswana population. Thiong'o *et al.* (2001) reported a vitamin C content of 90–300 mg/100 g in Kenyan populations. Hall *et al.* (2002) contend that the exceptionally high yields reported elsewhere in the literature may be from larger and older individuals.

18.2.2 *Uapaca kirkiana* Muell. Arg.

Uapaca kirkiana is a member of the family Euphorbiaceae, subfamily Phyllanthoideae, in the tribe Antidesmeae, and is the sole representative of the subtribe Uapacinae (Ngulube, 1996). The genus is distinctive within the Euphorbiaceae on account of its wood, vegetative and floral characters. The genus *Uapaca* has 61 species (Radcliffe-Smith, 1988; Mabberley, 1997). *U. kirkiana* is easily distinguished from other *Uapaca* species by its characteristically broad, leathery leaves and rounded crown. There has been no recent revision on a continental scale and the number of distinct species is probably less than has been thought previously. The greatest diversity is in the Congo basin and further south in the miombo region. *U. kirkiana* is native to the miombo ecological zone of southern Africa, including Madagascar.

Despite the popularity of *U. kirkiana* among the local population in the region, there is little information on its biology, management and utilization. A few supporting biological studies have produced information on variation in fruits (Mwamba, 1989, 1995), germination and ecological requirements (Msanga and Maghembe, 1989; Maghembe, 1995) and vegetative propagation (Jaenicke *et al.*, 2001). The only extensive work on the ecology of *U. kirkiana* has been done by Ngulube and colleagues (Ngulube *et al.*, 1995; Ngulube, 1996). The salient findings of these studies are briefly summarized below.

Ecology

Uapaca kirkiana occurs naturally south of the equator in Angola, the Democratic Republic of Congo (DRC), Burundi, Tanzania, Malawi, Mozambique, Zambia and Zimbabwe (Ngulube *et al.*, 1995; Akinnifesi *et al.*, Chapter 8, this volume). Ngulube *et al.* (1995) also give the latitudinal and longitudinal limits of its distribution. Occurrence at low elevations has been reported on the shores of Lake Malawi, while its presence at high elevations has been reported in the highlands of Benguela in Angola, Mbeya in Tanzania and Mbala in Zambia. The whole range of *U. kirkiana* broadly experiences a main dry season of

5–7 months, most areas having a mean annual rainfall between 500 and 1400 mm. Mean temperature ranges from 18 to 29°C, with frequent frosts in the more southerly areas of the range (Ngulube *et al.*, 1995).

The tree may occur in extensive pure stands in deciduous woodlands, upland wooded grasslands and along streams, often on skeletal soils at altitudes of 500–2000 m (Ngulube *et al.*, 1995). It has been used as an indicator of poor soils as it usually occurs in soils with low exchangeable cations and low in organic matter and macronutrients such as nitrogen, phosphorus and potassium. *U. kirkiana* grows on ferruginous or ferralitic soils that are generally sandy or gravely with good drainage. The species is absent in poorly drained, heavy, clayey soils (Ngulube *et al.*, 1995).

According to White's framework, *U. kirkiana* is a species typical of the Zambezian regional centre of endemism and the adjacent transitional centres, the most notable being the Guineo-Congolian and Zambezian regions (Ngulube *et al.*, 1995). It is abundant and widespread in mixed communities of *Brachystegia–Julbernardia* woodland vegetation. It is usually a dominant or codominant species and is gregarious, forming dense groves. In high-rainfall areas (>1200 mm) it forms pure stands with either closed or open canopies, becoming semideciduous forest with very sparse ground flora (Rattray, 1961; Shorter, 1989; Ngulube *et al.*, 1995).

Bush fires, a typical feature of the miombo woodlands, will affect the young coppice shoots and seedlings, especially when the fire occurs late in the dry season. Fully grown *U. kirkiana* has been said to be moderately fire-resistant (Kikula, 1986; Ngulube *et al.*, 1995). Its fire tolerance has also played a role in the succession of miombo woodlands. *U. kirkiana* was one of the species found occupying an intermediate successional stage between fire-tolerant woodland and fire-sensitive dry evergreen forest (Ngulube *et al.*, 1995).

The main associates of *U. kirkiana* are *Julbernardia* and *Brachystegia* spp., *Parinari curatellifolia*, *Pericopsis angolensis*, *Pterocarpus angolensis* and other *Uapaca* species (Ngulube, 2000). Other plants include *Annona*, *Burkea*, *Combretum*, *Ochna*, *Ximenia*, *Vangueria*, *Lannea discolor*, *Diplorhynchus*, *Dalbergia nitidula*, *Bridelia* and *Pseudolachnostylis maprouneifolia* (Ngulube *et al.*, 1995; Ngulube, 2000).

Studies that have given quantitative information on the relative abundance of *U. kirkiana* in plant populations are few, and the information given is mostly on the dominant height, basal area or the number of stems per unit area. Ngulube *et al.* (1995) gives a range of 54–75% representation of *Uapaca* in natural stands. Low values of 10–27% as well as high values of up to 90% in pure stands have been reported (Chidumayo, 1987). Under the slash-and-burn system in Zambia, natural regeneration after 6 years revealed a contribution of 42% to the basal area by *U. kirkiana* while in well-established mature stands in the miombo the contribution is less than 5% (Stromgaard, 1985; Hogberg and Piearce, 1986; Ngulube *et al.*, 1995). In mature miombo stands in Malawi, the number of *U. kirkiana* stems per hectare ranged between 27 and 95 (Ngulube *et al.*, 1995). However, the gross population structure of individuals $\geqslant$10 cm diameter at breast height (d.b.h.) was shown to be variable, the stocking rate ranging from 97 to 1073 stems per hectare (Ngulube, 2000).

Ecosystem functions

The most notable important feature of *U. kirkiana* is its association with ectomycorrhizas, some of which have a significant impact on the livelihoods of communities living near the forest where this species grows. Ectomychorrizas are conspicuous on roots at 5–10 cm depth and associations of endo-ectomychorriza have also been reported on the species (Hogberg, 1982; Thoen and Ba, 1989). *Amanita*, *Cantharellus*, *Lactarius* and *Russula* constitute the most common genera of fungi which typically form ectomycorrhizas, especially in the *Uapaca–Brachystegia* miombo woodlands (Hogberg and Piearce, 1986; Ngulube *et al.*, 1995). Ramachela (2006) recently studied the ecological interactions involved in the establishment and growth of natural woodland of *U. kirkiana* in Zimbabwe. The study showed that soil pH and potassium had significant effects on mycorrhizal diversity.

Several insects, some of which are serious pests, feed on *U. kirkiana* (Sileshi *et al.*, Chapter 20, this volume). In winter, the tree is host to the edible stinkbug *Encosternum delegorguei* (Hemiptera), which in Malawi and Zimbabwe is sold for cash in the market and is an important source of protein and money. Ngulube *et al.* (1995) have noted associations with vertebrates arising from their role as seed dispersers as they feed on the fruits. The interaction with the natural vertebrate fauna is due to browsing animals such as elephants and eland and the feeding on the fruits by monkeys and baboons, which probably also play a major role in dispersal. As a dominant or codominant tree of the miombo vegetation in hilly sites, it is useful in watershed management and the control of erosion. The tree also improves the microclimate within the canopy because of its dense rounded crown.

Biology

Uapaca kirkiana is dioecious and the unisexual inflorescences originate from axillary positions among the leaves or more often below them on the second or third season's wood of the branchlets. The spatial distributions of male and female trees in natural populations are largely unreported. In some Zambian populations, male and female reproductive individuals are randomly distributed with a male:female ratio of 1:1, and a mean distance of 11.1 m between the male and female individuals was recorded (Hans and Mwamba, 1982; Hans *et al.*, 1982). Similarly, a sex ratio of 1:1 in mature *U. kirkiana* trees has been reported in Malawi (Ngulube, 2000). Pollination is presumed to be by insects. The most common and diverse groups of insects include bees (Apidae), flies (Syrphidae), beetles (Lagriidae, Chrysomelidae and Melolonthidae), ants and wasps. Butterflies are less common but moths are potential pollinators.

Flowering coincides with the onset of the rainy season (October/ November), and the period extends over the entire 5–6 months of the rainy season. Flowering intensity is variable throughout the period, the greatest proportion of trees flowering between January and March (Ngulube, 1996). Male trees flower earlier than female individuals. Bud development takes about 16 days for male flowers and twice as long for female flowers. During anthesis, female flowers remain open 10–14 days whereas male flowers shrivel within 3–7 days of anthesis. Male flowers have a mild but non-distinctive scent while the female flowers have no detectable odour. About 45% of female inflorescences abort (20.7 and 24.8% during the bud and the anthesis stages, respectively (Ngulube, 1996).

Uapaca kirkiana fruit is described as drupaceous and borne on a thick, short peduncle, usually less than 10 mm long. The fruits are set between January and February and mature in August and November (Mwamba, 1995). Fruit production varies between harvests. Fruit loads exceeding 2000 fruits per tree have been reported in Zaire, Zambia (Mwamba, 1995) and Malawi (Akinnifesi *et al.*, 2004). There is variation in fruit load between trees within populations depending on tree size. There is also seasonal variation in fruit production within and between populations in the miombo. The average number of fruits per tree has also been shown to be variable between different tree sizes and even in the same size class. Short trees with small diameters and crowns have fewer fruits than larger trees. The physiological basis for these differences is presumably the large amount of available carbohydrates resulting from high photosynthetic activity in large trees (Cramer and Kozlowski, 1960) and variation in crown vigour between trees in the same size class. The differences reported cannot be attributed to environmental factors because the trees were growing within the same stand, indicating a possible genetic influence.

The mature fruit is described as yellow, yellow-brown or brown in colour and is 3.3 cm in diameter; it is round with a tough skin. Additional fruit colour variations have been documented to include cream and brownish-red (rufous) (Mwamba, 1995). The pulp is yellow and sweet with a pear-like taste and contains three or four seeds. In some natural populations there are some slight colour differences between individual trees: brownish-yellow and reddish (Hans, 1981). The fruit dimensions fall within the 2–4 cm range in length and breadth.

There are variations in fruit size, skin, pulp content and seed weight and volume between and within populations and trees (Mwamba, 1995). At maturity, each fruit may weigh 5–50 g. The amount of pulp ranges from 0.2 to 30 g but this varies among sites (Kwesiga *et al.*, 2000). Fruiting is biannual, i.e. alternate fruiting seasons. Some trees produce a mixture of small and large fruits while others habitually produce large or small fruits. The fruits may be thick- or thin-skinned (Mwamba, 1989). This variation is due to local competition between fruits for assimilates (Cramer and Kozlowski, 1960). At maturity, a hard, thick exocarp encloses a thin, yellow mesocarp about 1.5 mm thick when dry (Radcliffe-Smith, 1988). The skin accounts for approximately 38% of the total fruit weight (Mwamba, 1989). The fruit contains three to five seeds (pyrenes), but three or four is the most common number; they are generally whitish, cordate, carinate and apiculate with a tough fibrous sclerotesta. The four-seeded fruits are larger, have thinner skins and contain the most pulp (Mwamba, 1995). Seeds up to 2 cm long and 1.4 cm wide have been recorded (FAO, 1983). The pulp accounts for 39–45% of the fruit weight.

Uapaca kirkiana is dispersed by animals; the sugary pulp is attractive to a wide range of mammals and birds. In the natural environment, a large number of birds, ungulates and primates feed on the pulp, dispersing the seed (after sucking the pulp) as they move from the seed source (Pardy, 1951; Storrs, 1951; FAO, 1983; Dowsett-Lemaire, 1988; Seyani, 1991). The seed is white, with a rather brittle seed coat. One side is almost flat and on the opposite side there is a longitudinal ridge terminating in a sharp edge at the base. The seed has an inner membrane, endosperm and two leafy green cotyledons. There are 2500 seeds per kilogram. The seed has a high moisture content of about 48%, based on sun-dried and fresh weights (Hans, 1981). The seed has no dormancy period and once dispersed, it germinates readily. Fruit maturation and fall coincide with the rainy season, ensuring the immediate availability of appropriate germination conditions. In Zambia, up to 2039 and 1183 new and old seedlings per hectare, respectively, were recorded in a natural population (Mwamba, 1992). Seedling growth is better in the open than in the closed canopy.

Germination is intermediate between epigeal and hypogeal. The seed coat cracks and the radicle protrudes from the scar end and develops into a tap root. The seed coat splits longitudinally into equal halves and the two cotyledons unfold and expand greatly. Germination is fairly uniform, reaching 30% after 4 weeks and 90% after 6 weeks. The seed does not require pretreatment, but soaking in cold tap water overnight hastens germination (Ngulube, 1996). Up to 100% germination has been achieved at Makoka nursery (ICRAF, unpublished results).

Propagation

Until recently, vegetative propagation techniques have been a constraint to the domestication of indigenous fruit trees. Jaenicke *et al.* (2001) reported a graft take of less than 10% for *U. kirkiana.* However, these problems have been overcome and grafting success rates are now relatively high (80%) (Mhango and Akinnifesi, 2002; Akinnifesi *et al.*, 2004). *U. kirkiana* is amenable to clonal propagation and field management. Clonal propagation was used to capture superior *U. kirkiana* clones in Malawi using a participatory approach (Akinnifesi *et al.*, Chapter 8, this volume). Tree orchards of superior trees established at Makoka, Malawi, started to bear fruits after 2 years and fruiting became stable after 4 years, with a fruit load as high as 460 in some clones. Rootstocks are propagated from seeds (seedling rootstocks). A medium made of 75% forest soil and 25% sawdust has been shown to produce better soil media for rootstocks. A graft take of 80% for *U. kirkiana* using a wedge or splice technique can be obtained (Mhango and Akinnifesi, 2002; Akinnifesi *et al.*, 2004). The time of grafting and the skill of the grafter have overriding effects on the grafting success of *Uapaca*. Scions collected and grafted between October and December are best (>80% take), whereas grafting done between January and August has resulted in less than 30% graft take (Akinnifesi *et al.*, 2004, 2006). The method used to store the scion is a major factor that can affect the lifespan. Research in Malawi showed that keeping scions at room temperature was superior to other storage methods (sand, wet paper, cooler box). Air-layering has the potential for propagating plants with high vigour, and up to 63% rooting of marcotts has been achieved in the wild (Mhango and Akinnifesi, 2002). However, it is still a challenge to achieve good survival of established marcotts. The survival of the rooted propagules in the nursery and ramet survival in the field may decline with time because of the difficulty of tap root development, fertilizer and mycorrhiza requirements (Akinnifesi *et al.*, 2004). Top-grafting has also been successful in the field (Mkonda *et al.*, 2001; J. Mhango, unpublished results).

Tissue culture research has also proved to be promising for detecting early stock/scion compatibility in *U. kirkiana* (Mng'omba *et al.*, 2007; Chapter 19, this volume. Current effort involves the use of tissue culture to determine scion–rootstock compatibility. Compatibility differs between heterografts and homografts, and also between species, provenance and clones (Mng'omba *et al.*, 2007). However, excessive accumulation of phenolic compounds remains a challenge when using tissue culture to multiply *Uapaca*.

18.2.3 *Strychnos cocculoides* (Baker)

Strychnos cocculoides belongs to the family Loganiaceae. This is a semideciduous small tree, 2–9 m high, with spreading branches and a compact, rounded crown, which grows on both deep and loamy sands (Storrs, 1995). The bark is thick, creamy-brown, deeply corky and ridged longitudinally. The branchlets are hairy, purple or blackish-purple with strong paired spines curved downwards (Ruffo *et al.*, 2002). The leaves are opposite, oval to circular in shape and up to 5 cm long, green to yellowish green in colour, and conspicuously five-veined from the base.

Ecology

The genus *Strychnos* is widespread in the tropical parts of Africa, South America and India but the species *S. cocculoides* is restricted to central and southern Africa (AFT Database; Ruffo *et al.*, 2002; ICUC, 2004). In southern Africa, *S. cocculoides* grows naturally in *Brachystegia* woodlands, mixed forests, deciduous woodlands and lowlands. The biophysical limits are an altitude of 0–2000 m, mean annual temperature of 14–25°C and mean annual rainfall of 600–1200 mm. The species grows in a range of soil types from less fertile, deep sandy soil on rocky slopes, acidic clays to red or yellow–red loams.

Strychnos cocculoides is usually associated with other species in the miombo woodlands. It has been reported as one of the dominant or codominant species together with species such as *Acacia* spp., *Combretum* spp., *Terminalia* spp., *Adansonia digitata*, *Millettia stuhlmannii*, *Uapaca* spp., *Trichilia emetica*, *Sclerocarya birrea*, *Guibourtia conjugata*, *Hymenocardia ulmoides*, *Pteleopsis myrtifolia*, *Xeroderris stuhlmannii*, *Albizia versicolor*, *Albizia adianthifolia*, *Afzelia quanzensis*, *Burkea africana*, *Pterocarpus* spp., *Pseudolachnostylis maprouneifolia*, *Dialium schlechteri*, *Lonchocarpus capassa*, *Syzygium cordatum* and *Garcinia livingstonei* in Mozambique and Zimbabwe (Duarte-Mangue and Oreste, 1999; Nyoka and Muskoyonyi, 2002). It is also an associated species in the *Terminalia–Combretum* woodlands together with *Sclerocarya birrea*, *Saccharopolyspora spinosa*, *Piliostigma thonningii*, *Terminalia sericea* and *Combretum* spp. (Nyoka and Muskoyonyi, 2002).

Information on population dynamics is almost non-existent, with one study by Frost (2000) showing *S. cocculoides* contributing less than 1% to canopy cover at a site in Kataba Forest Reserve in Zambia.

Ecosystem functions

The fruit is rich in sugars, essential vitamins, minerals, oils and proteins. The ripe fruit is eaten fresh or is used to prepare a sweet–sour non-alcoholic drink (Fox and Norwood Young, 1982). The tree produces a hard wood with a straight bole suitable for construction. The soft, white pliable wood is very tough and it is used to make tool handles and building materials (ICUC, 2004). The seeds contain strychnine, which is toxic. The fruit is used to make a dye that provides protection from insects and for colouring trays and containers. *S. cocculoides* roots are chewed to treat eczema while a decoction is drunk as a cure for gonorrhoea. The fruit is used in making eardrops, and a fruit preparation is mixed with honey or sugar to treat coughs. Ground leaves are used to treat sores, and after they are soaked in water the drained liquid is used as a spray for vegetables to repel insects such as aphids and scales (ICUC, 2004).

Strychnos cocculoides has been planted as a boundary or barrier around homesteads. Farmers in semi-arid areas normally leave the species on their farm land to use as a shade tree, which is especially valuable for fruit and timber trees. It is semi-cultivated and can be raised in the nursery and planted on a cleared site. Saplings need to be protected from fire. This species seems to be affected by mycorrhizas and may therefore play an important role in the miombo ecosystem.

Biology

Strychnos cocculoides flowers during the rainy season, and the fruit ripens in the dry season. It takes 8–9 months from flower fertilization to fruit ripening. The flowers are small, about 5 mm in diameter, greenish to creamy white, borne in dense terminal cymes, up to 3.5 cm in length. The fertile flowers are said to be hermaphrodite and homostylous (styles of the same length) and pollination is by insects (entomophilous pollination). Fruits are circular, 1.6–10 cm in diameter, with a smooth, woody shell that is dark green with paler mottling when young, turning yellow after ripening. The seeds are numerous; they have a hard coat and are compressed to 2 cm in diameter and embedded in a fleshy pulp that when ripe is juicy and yellow.

Strychnos cocculoides regenerates naturally by seed, coppicing or root suckers. Seed storage is semi-orthodox, with a shelf life of at most 2 months at room temperature. Mkonda *et al.* (2004) found that fruit size varied considerably among provenances within and between countries. Fruit weight ranged from 145 g in Tanzanian provenances to more than 360 g for Murelwa and Chihoto provenances from Zimbabwe.

Knowledge about the domestication of *S. cocculoides* has increased substantially in the last decade. *Strychnos* species are known to be difficult to establish because of erratic germination and prolonged after-ripening effects in seeds. The length of storage has been shown to affect the germinability of seeds, and this varies with provenance (Mkonda *et al.*, 2004). In earlier studies, Taylor *et al.* (1996) showed that propagation by seed was successful, giving germination rates of 80% in 3 and 9 weeks in summer and winter, respectively.

The performance of *S. cocculoides* was reported to be good under on-station trials at Msekera station in Zambia; survival was 75% across provenances but contrasting results were reported under farmer management, with survival declining from 60% at 9 months to 10% at 15 months (Mkonda *et al.*, 2004). Provenance trials established from seedlings in Zambia started to produce fruits after 5 years. Taylor *et al.* (1996) reported the response of *S. cocculoides* to fertilizer application. An average fruit load of 96 kg/ha per tree has been reported by Mateke (2004) in Botswana. Fruit number raged from 258 to 946 per tree, the highest number being obtained for the Paje provenance. While cuttings have been able to sprout under 50% shade in 3 weeks, root development was reported to be poor. In Malawi, trees raised from seeds yielded fruits after 4 or 5 years, while in Botswana the period to fruition was within 3 years (Taylor *et al.*, 1996).

18.3 Summary

In this chapter we have summarized the biology and ecology of three key priority miombo fruit trees identified for southern Africa and their potential for domestication. Constraints such as lack of knowledge of the biology, ecology and propagation of indigenous fruit trees as reasons for not cultivating them have been substantially addressed in the last 10 years. Thus, domestication can be seen as a tool for saving these species from extinction and increasing their productive value to their users, and involves selection and management. Through the propagation and cultivation of indigenous fruit trees on farms, there are opportunities to further exploit the economic potentials of the three species discussed here.

However, more research is needed to develop appropriate management protocols for *U. kirkiana*, *S. birrea* and *S. cocculoides* in the region. ICRAF Southern Africa has been doing provenance testing of the three species. For *U. kirkiana* and *S. cocculoides*, some superior clones have been identified and captured vegetatively using participatory selection in Zambia and Malawi. The current challenge is to develop ideotypes for the market that have fewer seeds or are seedless, with improved fruit taste and reduced skin thickness, but this requires more strategic research tailored to the consumers' needs and preferences.

References

Acocks, J.P.H. (1988) Veld types of South Africa, 3rd edn. *Memoirs of the Botanical Survey of South Africa* 57, 1–146.

Akinnifesi, F.K., Kwesiga, F.R., Mhango, J., Mkonda, A., Chilanga, T. and Swai, R. (2004) Domesticating priority miombo indigenous fruit trees as a promising livelihood option for small-holder farmers in Southern Africa. *Acta Horticultura* 632, 15–30.

Akinnifesi, F.K., Kwesiga, F., Mhango, J., Chilanga, T., Mkonda, A., Kadu, C.A.C., Kadzere, I., Mithofer, D., Saka, J.D.K., Sileshi, G., Ramadhani, T. and Dhliwayo, P. (2006) Towards the development of miombo fruit trees as commercial tree crops in southern Africa. *Forests, Trees and Livelihoods* 16, 103–121.

Arnold, T.H. and De Wet, B.C. (eds) (1993) *Plants of Southern Africa: Names and Distribution.* National Botanical Institute, South Africa, 825 pp.

Baillon, H. (1858) *Ètude Générale du Groupe des Euphorbiacées.* Masson, Paris.

Belsky, A.J., Amundson, R.G., Duxbury, J.M., Riha, S.J., Ali, A.R. and Mwonga, S.M. (1989) The effects of trees on their physical, chemical and biological environments in a semi-arid savanna. *Journal of Applied Ecology* 26, 1005–1024.

Bwanali, R. and Chirwa, P.W. (2004) Growth and development of a five year old *Uapaca kirkiana* Muell. Arg. provenance trial at Nauko in Malawi. FRIM Report 04001. Forestry Research Institute of Malawi, Zomba, Malawi.

Campbell, A. (1986). The use of wild food plants, and drought in Botswana. *Journal of Arid Environments* 11, 81–91.

Campbell, B., Luckert, M. and Scoones, I. (1997) Local level valuation of savannah resources: a case study from Zimbabwe. *Economic Botany* 51, 57–77.

Chidumayo, E.N. (1987) Woodland structure, destruction and conservation in the copperbelt area of Zambia. *Biological Conservation* 40, 22, 89–100.

Coetzee, B.J., Englebrecht, A.H., Jourbert, S.C.J. and Retief, P.F. (1979) Elephant impact on *Sclerocarya caffra* trees in *Acacia nigrescens* tropical plains thornveld of the Kruger National Park. *Koedoe* 22, 39–60.

Cramer, P.J. and Kozlowski, T.T. (1960) *Physiology of trees.* McGraw Hill, New York.

Dowsett-Lemaire, F. (1988) Fruit choice and seed dissemination by birds and mammals in the evergreen forests of upland Malawi. *Review of Ecology* 43, 251–285.

Duarte-Mangue, P. and Oreste, M.N. (1999) *Non wood forest products in Mozambique. Forestry Statistics and Data Collection.* EC/FAO ACP Data Collection Project Technical Report AFDCA/TN/04.

Dzerefos, C.M. (1996) Distribution, establishment, growth and utilisation of mistletoes (Loranthaceae) in the Mpumalanga lowveld. MSc thesis, University of the Witwatersrand, Johannesburg.

Edwards, D. (1967) A plant ecological survey of the Tugela basin. *Memoirs of the Botanical Survey of South Africa* 35, 1–285.

Fanshawe, D.B. (1969) The vegetation of Zambia. *Forestry Research Bulletin* 7, 1–67.

FAO (1983) *Food and fruit bearing forest species. 1: Examples from Eastern Africa.* FAO Forestry Paper No. 44/1. FAO, Rome.

Farrell, J.A.K. (1968) Preliminary notes on the vegetation of the lower Sabi-Lundi Basin, Rhodesia. *Kirkia* 6, 223–248.

Fox, F.W. and Norwood Young, M.E. (1982) *Food from the Veld.* South African Institute for Medical Research, Johannesburg.

Frost, P.G.H. (2000) *Vegetation structure of the MODIS validation site, Kataba Forest Reserve, Mongu, Zambia.* Report prepared for the NASA Southern African Validation of EOS (SAVE) project, IGBP Kalahari Transect programme, and SAFARI 2000. Institute of Environmental Studies, University of Zimbabwe, Harare, Zimbabwe.

Hall, J.B., O'Brien, E.M. and Sinclair, F.L.

(2002) Sclerocarya birrea*: a Monograph.* Publication No. 19. School of Agricultural and Forest Sciences, University of Wales, Bangor, UK, 157 pp.

Hans, A.S. (1981) *Uapaca kirkiana* Muell. Arg. (Euphorbiaceae). Technical Report, Tree Improvement Research Centre. Zambia National Council for Scientific Research, Lusaka, 48 pp.

Hans, A.S. and Mwamba, C.K. (1982) Spatial relationships between male and female trees of *Uapaca kirkiana* Muell. Arg. Research Paper, Tree Improvement Research Centre. National Council for Scientific Research, Lusaka.

Hans, A.S., Chembe, E.E. and Mwanza, L.K. (1982) *Mathematical treatment of vegetation trustland and miombo woodland of Chipata district with reference to multipurpose fruit species*. Research Paper, Tree Improvement Research Centre. National Council for Scientific Research, Lusaka, 19 pp.

Hogberg, P. (1982) Mycorrhizal associations in some woodland and forest trees and shrubs in Tanzania. *New Phytologist* 92, 407–415.

Hogberg, P. and Piearce, G.D. (1986) Mycorrhizas in Zambian trees in relation to host taxonomy, vegetation type and successional patterns. *Journal of Ecology* 74, 775–785.

ICUC (International Centre for Underutilised Crops) (2004) *Fruits for the Future: Monkey Orange* [see http://www.icuc-iwmi.org/files/Resources/Factsheets/strychnos%20factsheet.pdf Accessed 19 July 2007].

Jaenicke, H., Simons, A.J., Maghembe, J.A. and Weber, J.C. (2001) Domesticating indigenous fruit trees for agroforestry. *Acta Horticulturae* 523, 45–51.

Kadu, C.A.C., Imbuga, M., Jamnadas, R. and Dawson, I. (2006) Genetic management of indigenous fruit trees in Southern Africa: A case study of *Sclerocarya birrea* based on nuclear and chloroplast variation. *South African Journal of Botany* 72, 421–427.

Kikula, I.S. (1986) The influence of fire on the composition of miombo woodland of SW Tanzania. *Oikos* 46, 317–324.

Kroon, D.M. (1999) *Lepidoptera of Southern Africa: Host-plants and Other Associations. A Catalogue*. Lepidopterist's Society of Africa, Jukskei Park, South Africa.

Kwesiga, F., Akinnifesi, F.K., Ramadhani, T., Kadzere, I. and Saka, J. (2000) Domestication of indigenous fruit trees of the miombo in southern Africa. In: Shumba, E.M., Lusepani, E. and Hangula, R. (eds) *The Domestication and Commercialization of Indigenous Fruit Trees in the SADC Region*. SADC Tree Seed Centre Network, Harare, pp. 8–24.

Lamprey, H.F., Bell, R.H.V., Glover, P.E. and Turner, M.J. (1967) Invasion of the Serengeti national park by elephants. *East African Wildlife Journal* 5, 151–166.

Lewis, D.M. (1987) Fruiting patterns, seed germination and distribution of *Sclerocarya birrea* in an elephant-inhabited woodland. *Biotropica* 19, 50–56.

Mabberley, D. (1997) *The Plant Book: A Portable Dictionary of the Vascular Plants*, 2nd edn. Cambridge University Press, Cambridge, 858 pp.

Maghembe, J.A. (1995) Achievements in the establishment of indigenous fruit trees of the miombo woodlands of southern Africa. In: Maghembe, J.A., Ntupanyama, Y. and Chirwa, P.W. (eds) *Improvement of Indigenous Fruit Trees of the Miombo Woodlands of Southern Africa*. Proceedings of a Conference held on 23–27 January 1994, Mangochi, Malawi. ICRAF, Nairobi.

Mhango, J. and Akinnifesi, F.K. (2002) On-farm assessment, farmer management and their perceptions of priority indigenous fruit trees in southern Malawi. In: Kwesiga, F.R., Agumya, A. and Ayuk, E. (eds) *Proceedings of the 14th ICRAF Southern Africa Regional Review and Planning Workshop, 1–7 September, 2001, Harare*, pp. 157–164.

Mkonda, A., Akinnifesi, F.K., Swai, R., Kadzere, I., Kwesiga, F.R., Maghembe, J.A., Saka, J., Lungu, S. and Mhango, J. (2004) Towards domestication of 'wild orange' Strychnos cocculoides in southern Africa: a synthesis of research and development efforts. In: Rao, M.R. and Kwesiga, F.R. (eds) *Proceeding of the Regional Agroforestry Conference on Agroforestry Impacts on Livelihoods in Southern Africa: Putting research into practice*. World Agroforestry Centre (ICRAF), Nairobi, pp. 77–76.

Mng'omba, S.A., du Toit, E.S., Akinnifesi, F.K. and Venter, H.M. (2007) Early recognition of graft compatibility in *Uapaca kirkiana* fruit tree clones, provenances and species using *in vitro* callus technique. *HortScience* (in press).

Moll, E.J. and White, F. (1978) The Indian Ocean coastal belt. *Monographiae Biologicae* 31, 561–598.

Msanga, H.P. and Maghembe, J.A. (1989) Physical scarification and hydrogen peroxide treatments improve germination of *Vangueria infausta* seed. *Forest Ecology and Management* 28, 301–308.

Mwamba, C.K. (1989) Natural variation in fruits of *Uapaca kirkiana* in Zambia. *Forest ecology and Management* 26, 299–303.

Mwamba, C.K. (1992) *Influence of crown area on empirical distribution of growth parameters and regeneration density of* Uapaca kirkiana *in a natural miombo woodland forest*. Research Paper No. 14, Tree Improvement Research Centre. National Council for Scientific Research, Lusaka, 1–12.

Mwamba, C.K. (1995) Variations in fruits of *Uapaca kirkiana* and effects of in situ silvicultural treatments on fruit parameters. In: Maghembe, J.A., Ntupanyama, Y. and Chirwa, P.W. (eds) *Improvement of Indigenous Fruit Trees of the Miombo Woodlands of Southern Africa*. Primex Printers, Nairobi, pp. 27–38.

Nerd, A. and Mizrahi, Y. (1993) Domestication and introduction of marula (*Sclerocarya birrea* subsp. *caffra*) as a new crop for the Negev desert of Israel. In: Janick, J. and Simon, J.E. (eds) *New Crops*. Wiley, New York, pp. 496–499.

Nerd, A., Aronson, J.A. and Mizrahi, Y. (1990) Introduction and domestication of rare and wild fruit and nut trees for desert areas. In: Janick, J. and Simon, J.E. (eds) *Advances in New Crops*. Timber Press, Portland, Oregon, pp. 355–363.

Ngulube, M.R. (1996) Ecology and management of *Uapaca kirkiana* in southern Africa. PhD thesis, School of Agricultural and Forest Sciences, University of Wales, Bangor, UK.

Ngulube, M.R. (2000) Population structures of *Uapaca kirkiana* (Euphorbiaceae) in the miombo woodlands of Malawi: status and management prospects for fruit production. *Journal of Tropical Forest Science* 12, 459–471.

Ngulube, M.R., Hall, J.B. and Maghembe, J.A. (1995) Ecology of a miombo fruit tree: *Uapaca kirkiana* (Euphorbiaceae). *Forest Ecology and Management* 77, 107–118.

Ngulube, M.R., Hall, J.B. and Maghembe, J.A. (1996) *Uapaca kirkiana* (Euphorbiaceae): a review of silviculture and resource potential. *Journal of Tropical Forest Science* 8, 395–411.

Nyoka, B.I. and Muskoyonyi, C. (2002) *State of forest and tree genetic resources in Zimbabwe*. Forest Genetic Resources Working Paper No. 35, 23 pp.

Packham, J. (1993) *The Value of Indigenous Fruit-bearing Trees in Miombo Woodland Areas of South-Central Africa*. Rural Development Forestry Network Paper. [http://www.odi.org.uk/fpeg/publications/rdfn/15/rdfn-15c-ii.pdf Accessed 19 July 2007].

Palmer, E. and Pitman, N. (1972) *Trees of Southern Africa*, Volume 2. Struik, Cape Town.

Pardy, A.A. (1951) Notes on indigenous trees and shrubs of southern Rhodesia. *Rhodesian Agricultural Journal* 48, 261–266.

Peter, C.R. (1988) Notes on the distribution and relative abundance of *Sclerocarya birrea* (A. Rich.) Hochst. (Anacardiaceae). *Missouri Botanical Garden Monographs in Systematic Botany* 25, 403–410.

Pooley, E.S. (1993) *The Complete Guide to Trees of Natal, Zululand and Transkei*. Natal Flora Publications Trust, Durban.

Poynton, R.J. (1984) *Characteristics and uses*

of trees and shrubs cultivated in South Africa. Directorate of Forestry, Bulletin No. 39. Government Printer, Pretoria, 201 pp.
Radcliffe-Smith, A. (1988) Euphorbiaceae (part 2): Uapaca. In: Polhill, R.M. (ed.). *Flora of Tropical East Africa*. Balkema, Rotterdam, pp. 566–571.
Ramachela, K. (2006) Studies on the mycorrhizosphere and nutrient dynamics in the establishment and growth of *Uapaca kirkiana* in Zimbabwe. PhD thesis, University of Stellenbosch, South Africa, 123 pp.
Ramadhani, T. (2002) *Marketing of Indigenous Fruits in Zimbabwe. Socio-economic Studies on Rural Development, Volume 129*. Wissenschaftsverlag Vauk. Kiel, Germany.
Rattray, J.M. (1961) Vegetation types of Southern Rhodesia. *Kirkia* 2, 68–93.
Ruffo, C.K., Birnie, A. and Tengnäs, B. (2002) *Edible Wild Plants of Tanzania*. Technical Handbook No. 27. RELMA, Nairobi, 766 pp.
SEPASAL Database (2001) *Sclerocarya birrea (A. Rich.) Hochst. subsp. caffra (Sond.) Kokwaro*. Royal Botanic Gardens, Kew [see http://www.rbgkew.org.uk/ceb/ sepasal/birrea.htm Accessed 19 July 2007].
Seyani, J.H. (1991) Uapaca kirkiana*: underutilized multipurpose tree species in Malawi worth some development*. Paper presented at the International Workshop on the biodiversity of traditional and underutilized crops, 12–15 June 1991, Valletta, Malta, 10 pp.
Shackleton, C.M. (2004) Use and selection of *Sclerocarya birrea* (marula) in the Bushbuckridge lowveld, South Africa. In: Rao, M.R. and Kwesiga, F.R. (eds) *Proceedings of Regional Agroforestry Conference on Agroforestry Impacts on Livelihoods in Southern Africa: Putting Research into Practice*. World Agroforestry Centre, Nairobi, pp. 77–92.
Shackleton, C.M. and Scholes, R.J. (2000) Impact of fire frequency on woody community structure and soil nutrients in the Kruger National Park. *Koedoe* 43, 75–81.
Shackleton, S.E., Shackleton, C.M., Cunningham, A.B., Lombard, C., Sullivan, C.A. and Netshiluvhi, T.R. (2002) Knowledge on *Sclerocarya birrea* subsp. *caffra* with emphasis on its importance as a non-timber forest product in South and southern Africa: a summary. Part 1: Taxonomy, ecology and role in rural livelihoods. *Southern African Forestry Journal* 194, 27–41.
Shone, A.K. (1979) Notes on the marula. *South African Department of Forestry Bulletin* 58, 1–89.
Shorter, C. (1989) *An Introduction to the Common Trees of Malawi*. The Wild Society of Malawi, Lilongwe, Malawi, 115 pp.
Storrs, A.E.G. (1995) *Know your Trees: Some Common Trees Found in Zambia*. Regional Soil Conservation Unit.
Stromgaard, P. (1985) Biomass, growth, and burning of woodland in a shifting cultivation area of south central Africa. *Forest Ecology and Management* 33, 163–178.
Taylor, F.W. and Kwerepe, B. (1995) Towards domestication of some indigenous fruit trees in Botswana. In: Maghembe, J.A., Ntupanyama, Y. and Chirwa, P.W. (eds) *Improvement of Indigenous Fruit Trees of the Miombo Woodlands of Southern Africa. Proceedings of a conference, 23–27 January 1994, Mangochi, Malawi*. ICRAF, Nairobi.
Taylor, F.W., Matoke, S.M. and Butterworth, K.J. (1996) A holistic approach to the domestication and commercialization of non-timber forest products. In: *International Conference on Domestication and Commercialization of Non-Timber Forest Products in Agroforestry Systems, Nairobi, Kenya, 19 to 23 February 1996*. Non-Wood Forest Products Series No. 9.
Teichman, I. von (1982) Notes on the distribution, morphology, importance and uses of the indigenous Anacardiaceae: 1. The distribution and morphology of Sclerocarya birrea (the marula). *Trees in South Africa* October–December, 35–41.
Thiong'o, M.K., Kingori, S. and Jaenicke, H. (2002) The taste of the wild: variation in the nutritional quality of the marula fruits and opportunities for domestication. *Acta Horticulturae* 575, 237–244.

Thoen, D. and Ba, A.M. (1989) Ectomycorrhizas and putative ectomycorrhizal fungi of Afzelia africana Sm. and Uapaca guineensis. *New Phytologist* 113, 549–559.

White, F. (1983) *The Vegetation of Africa. A Descriptive Memoir to Accompany the UNESCO/AETFAT/UNSO Vegetation Map of Africa. UNESCO, Natural Resources Research 20.* UNESCO, Paris, 356 pp

19 Germplasm Supply, Propagation and Nursery Management of Miombo Fruit Trees

F.K. Akinnifesi,[1] G. Sileshi,[1] A. Mkonda,[2] O.C. Ajayi,[1] J. Mhango[3] and T. Chilanga[4]

[1]*World Agroforestry Centre, ICRAF, Lilongwe, Malawi*; [2]*Zambia-ICRAF Agroforestry Project, Chipata, Zambia*; [3]*Mzuzu University, Mzuzu, Malawi*; [4]*Bvumbwe Agricultural Research Station, Limbe, Malawi*

19.1 Introduction

The improvement and domestication of indigenous fruit tree germplasm constitute a component of agroforestry in southern Africa. Yet the germplasm base is currently being eroded as forests all over the region are cleared and the remaining miombo woodland is under increasing pressure (Kwesiga *et al.*, 2000). Several studies have shown that it is possible to augment income for smallholder farmers through the domestication of high-value tree crops (Leakey *et al.*, 2005; Akinnifesi *et al.*, 2006; Tchoundjeu *et al.*, 2006). Germplasm improvement of miombo fruit trees entails the application of silvicultural, tree breeding and horticultural skills to obtain the most valuable fruit trees as quickly and as inexpensively as possible. The participatory selection and propagation approach adopted by the World Agroforestry Centre (ICRAF) for the domestication of indigenous fruit trees has been described (Leakey and Akinnifesi, Chapter 2, this volume).

One of the major challenges to the widespread adoption of agroforestry is the availability of high-quality planting stock of priority trees. The production of high-quality planting stock is even more critical to the domestication of indigenous fruit trees, as farmers have neither the knowledge and skills needed for their propagation nor the appropriate technologies for nursery management. According to Carandang *et al.* (2007), there is much interest in producing planting material of indigenous fruit trees but activity is constrained by the availability of suitable planting material, the distance from seed sources, the lack of propagation technologies, and the lack of awareness among farmers of the potential and niche of indigenous fruit trees on the farm. Interest is further constrained by the lack of technical skills among potential nursery

 Indigenous Fruit Trees in the Tropics: Domestication, Utilization and Commercialization (eds F.K. Akinnifesi *et al.*)

operators and the inadequacy of extension services to facilitate the acquisition of such skills.

Research into the propagation of indigenous fruit trees through seeds and vegetative material has been investigated for the past decade in southern Africa (Maghembe, 1995; Mhango and Akinnifesi, 2001). Some advances have been made in the nursery management of selected miombo fruit trees. Significant progress has been made in the establishment of fruit tree nurseries (Akinnifesi *et al.*, 2004, 2006, Chaper 8, this volume). Since the early 1990s, ICRAF has been promoting decentralized germplasm production and distribution systems at the grass roots level, where support to farmer nurseries is emerging as one of the main pathways (Böhringer *et al.*, 2003).

In this chapter we review the state of knowledge about germplasm delivery and the propagation and nursery management of miombo fruit trees in Africa. The chapter will be limited to examples relevant to small-farm agroforestry systems in southern Africa as opposed to highly intensive monocultural orchard production systems. We discuss the factors involved in determining the most appropriate methods of propagation for particular species, and the associated constraints and limitations in scaling up the domestication of indigenous fruit trees in southern Africa.

19.2 Germplasm Delivery and Conservation

Germplasm may be defined as genetic material, which can be seed, pollen, vegetative propagules or other material, though it is normally thought of as seed (Dawson and Were, 1997). The supply of germplasm in sufficient quantities and quality at the small-scale user's level is one of the factors that limit agroforestry development (Simons, 1996). While the trend in the past has been to put more emphasis on central supply systems such as national tree seed centres and government nurseries, policies are now being adopted to encourage the decentralized supply of seeds, seedlings and vegetative propagules. Germplasm quality has a direct impact on tree survival, growth and production. Tree germplasm quality has physical, genetic and physiological components (Mulawarman *et al.*, 2003). Valuable tree genetic resources are declining around farming communities in southern Africa, especially in Malawi, where annual deforestation is very high (1%), therefore limiting options for livelihoods and economic development. The ICRAF Medium Term Plan 1998–2000 states that domestication is not about breeding *per se* and that many advances are made through studies of indigenous knowledge, the dynamics of germplasm utilization on farms, nursery techniques, genetic selection, pomology, marketing and processing (ICRAF, 1997, p. 26). Through an understanding of both supply of and demand for superior germplasm of priority species, it is possible to increase the rate of adoption of agroforestry technologies. To promote the adoption of quality germplasm, information has to be disseminated along with the materials.

Propagation by seed (sexual propagation) is often the easiest, cheapest and most common means for most agroforestry trees, and is generally considered the default approach (Mudge and Brennan, 1999). On the other hand, vegetative (asexual) propagation has shown to be most relevant to the objective of germplasm improvement and domestication of miombo indigenous fruit trees for various reasons. The first reason is the need to achieve desirable fruit and tree traits in a reasonable time, such as superior fruit size, improved taste and increased pulp content (Akinnifesi *et al.*, Chapters 8 and 21, Tchoundjeu *et al.*, Chapter 9, this volume). This involves the development of cultivars with high productivity, high harvest index, superior quality and improved food value and acceptability (Leakey *et al.*, 2005; Akinnifesi *et al.*, 2006). The second reason is the need to obtain fruits and products through improved precocity, i.e. a shorter period before first fruiting. In a priority-setting exercise in the region, farmers indicated the desire to cultivate indigenous fruit trees with early-bearing traits and superior fruit traits (Maghembe *et al.*, 1998). The third reason is the need to overcome the natural variability that could affect fruit quality and acceptability to consumers and markets. It is also necessary to devise innovative ways of meeting the increasing demand for quality planting material by farmers and the development community.

Tree germplasm is unequivocally the primordial input to agroforestry (Simons, 1996). Early efforts to domesticate trees have followed the example of the forestry approach based on seeds, in which the emphasis is on timber. This problem is particularly apparent with miombo fruit trees, which are little known in the commercial sector. Therefore, adequate seed is simply not available. Even where collections exist of some species there is general under-representation of the genetic diversity.

19.2.1 Germplasm collection

Germplasm collection is a key step in any process to domesticate trees. Germplasm collection aims at capturing the natural genetic variation in a priority species to enable selection and improvement in their ultimate domestication (Dawson and Were, 1997). In any germplasm collection strategy it is important to standardize collection practices, to sample as much variation as possible and to prevent introducing biases in the evaluation stage. Most agroforestry trees are grown from seed, while fruit trees are often propagated vegetatively. In collecting germplasm of miombo fruit tree species, different strategies have been employed or recommended (Ndugu *et al.*, 1995; Dawson and Were, 1997; Mulawarman *et al.*, 2003). It is recommended that germplasm collection should pay attention to the following: (i) trees should have abundant fruits (at least 500) to provide sufficient seed for multilocation trials; (ii) farmers should determine which trees are superior; (iii) at least 25 trees should be sampled per provenance; (iv) trees should be separated by at least 50 m to avoid coancestry; and (v) seed should be kept separate by using the open-pollinated family system (Kwesiga *et al.*, 2000). In 1995, ICRAF trained a total of 30 germplasm collectors from Botswana, Malawi, Mozambique, Namibia, Swaziland, Tanzania, Zambia and Zimbabwe using funds received through the Southern Africa Development Community (SADC), Southern Tree Seed Centre Network (STSCN), and facilitated the collection and exchange of over 10 t of germplasm of *Uapaca kirkiana* and *Sclerocarya birrea* provenances (Kwesiga *et al.*, 2000).

Protocols that complied with the FAO code of conduct for germplasm were developed following the germplasm collection of some of the major species, including *U. kirkiana*, *Sclerocarya birrea*, *Strychnos cocculoides*, *Parinari curatellifolia* and *Vitex mombassae* (Akinnifesi *et al.*, 2004; Mkonda *et al.*, 2004; Swai *et al.*, 2004). Germplasm collection may be either for tree improvement or for conservation (Ndugu *et al.*, 1995). The object of the collection mentioned at the beginning of this paragraph was improvement and domestication, but conservation was an implicit responsibility. For conservational collection, the strategy is to ensure sampling of the major variation patterns in the wild in the most economical way. The goal is to ensure that diverse genotypes are conserved for future availability and use. Ndugu *et al.* (1995) suggested the need to conserve at least 200 superior clones in field gene banks in the region, considering that most miombo fruits are recalcitrant and that long-term seed storage in seed banks would be impossible. On-farm conservation by farmers is another way of maintaining live genes at low cost.

Normally, the aim of collection is to sample germplasm that is genetically as representative as possible of a population (or provenance). This is called 'systematic sampling', and it means collecting seed from many trees in an individual population (Dawson and Were, 1997). In addition to systematic sampling within populations, several populations have been sampled for each species, the idea being to get germplasm that represented the geographical ranges of the tree species. Sometimes in the course of the collection, if researchers decide that important characteristics have high heritability and can therefore be selected for at the time of sampling, they can do phenotypic selection of trees within populations. This method, called 'targeted collection', increases the chances of capturing superior material for improvement programmes. This has been the approach used for collecting germplasm of *Sclerocarya birrea* and *Uapaca kirkiana* in southern Africa (Dawson and Were, 1997). For both species, seed from trees that villagers identified as producing fruit with superior characteristics were collected. Occasionally, collectors may also do vegetative sampling to collect scions or cuttings from superior phenotypes, or because trees do not bear seed at the time of sampling. This approach may be useful for the targeted collection of fruit trees such as *S. birrea* and *U. kirkiana*, for which mature, seed-bearing trees can be difficult to find in some locations (Dawson and Were, 1997). Details of the management and performance of these provenances are covered by Akinnifesi *et al.* in Chapter 8 of this volume and have been reported by Akinnifesi *et al.* (2004).

Since most of the miombo fruits have a wide natural range involving several countries, regional issues may arise in germplasm collection. Securing approval from the respective national genetic centres to collect germplasm may be time-consuming and may take as long as 6 months (Ndungu *et al.*, 1995). The exchange of valuable indigenous fruit germplasm across nations with potential economic gain may face resistance. It may be easier to collect germplasm across long distances in seed form rather than by vegetative means, which may require special transport and storage requirements.

19.2.2 Germplasm delivery pathways

Lack of adequate understanding of germplasm delivery pathways increases frustration among users (Chilanga *et al.*, 2002). Therefore, the delivery pathways currently available, or those in need of creation, need to be identified to ensure the availability of improved germplasm. The pathways of germplasm delivery include distribution to national agricultural research systems (NARS), non-governmental organizations (NGOs), communities, the private sector, dissemination to farmers and diffusion by farmer-to-farmer exchange (Simons, 1996). Very little information is available on the demand and supply of miombo fruit tree germplasm. It is crucial that germplasm demand is real and not just perceived. Therefore, practical methods for accurately forecasting the amount of miombo fruit tree germplasm that will be available have been developed (A.J. Simons, unpublished results). ICRAF uses a number of case studies to examine how much germplasm will be needed, by whom, when, and for how long. In Malawi, for instance, at the beginning of each planting season, germplasm demand by farmers was established through questionnaires sent to the various partners involved in promoting agroforestry. In this way seed demand for each year was established and plans to source the seeds were drawn up. Seed was sourced through local collections from seed stands, purchase from national tree seed centres and the Forestry Research Institute of Malawi (Chilanga *et al.*, 2002). More than 100 t of seeds have been distributed to farmers through partners in the region in the last decade. Key constraints in tree seed supply have been identified and ways to reduce their effect are being developed. Technical requirements for decentralized germplasm supply systems have also been elaborated elsewhere (Böhringer *et al.*, 2001).

Farmers' nurseries as a pathway for germplasm delivery

Farmers have consistently cited the lack of quality germplasm as a major constraint on the diversification and expansion of their agroforestry practices (Simons, 1996; Weber *et al.*, 2001; Akinnifesi *et al.*, 2006; Carandang *et al.*, 2007). Supporting farmer nurseries was suggested to be one pathway for promoting decentralized tree seedling production in an efficient way, while at the same time providing opportunities for building natural resource, human and social capital, all three being considered equally crucial in developing more sustainable land use systems (Böhringer *et al.*, 2003). Roshetko and Verbist (2000) recognized three pathways in which improved germplasm can be made available to farmers: (i) through distribution of seed originating from national research centres, community-based NGOs and the private sector; (ii) through the dissemination of selected seeds originating from farmers and farmer groups; and (iii) by direct diffusion through informal farmer-to-farmer exchange. In indigenous fruit tree domestication, the extent to which the establishment of nurseries by smallholder farmers can be strengthened through training will determine the success and sustainability of new tree crop development. This includes training smallholder farmers to collect quality germplasm (seed, buds, cuttings, marcotts and/or scions) from superior trees in a way that ensures genetic quality; the efficient establishment, management of nurseries; tree establishment and management on the farm; harvest and postharvest handling activities; and the marketing of germplasm and tree products.

In Malawi the majority of nurseries were of the group type (86%). This makes it easy to train farmers in groups, and farmer training in nursery management has been ongoing for the last 10 years. Nearly 13,000 farmers have been trained in tree propagation, nursery establishment and management, and farm management. Indigenous fruit tree seedlings have been disseminated to farmers in Malawi, Zambia, Zimbabwe and Tanzania since the late 1990s. Farmer-to-farmer exchange and farmer training have been important methods of training larger numbers of farmers in nursery establishment and tree management. Special emphasis has been given in each community to the training of a few people in grafting techniques, as this is central to the domestication programme. These trainees can then offer services to other farmers and farmer groups, as well as manage their own commercial nurseries.

19.2.3 Germplasm conservation

For most of the species collected, live gene banks have been established, which have conservation value as well as a utilization purpose. These are typically set up as provenance trials, progeny trials and orchards to allow the determination of genetic parameters.

For the conservation of genetic resources of indigenous fruits, seed must be collected from 25 unrelated individuals, corresponding to an effective population size of 100 (Kwesiga *et al.*, 2000; Eyog-Matig *et al.*, 2001). The effective population size is the number of individuals in an ideal population in which each individual has an equal number of expected progeny (assuming random gamete distribution). The factors that determine the effective population size include differences in fertility between individuals, non-random mating, unequal numbers of male and female individuals in dioecious species, such as *Uapaca kirkiana* and *Sclerocarya birrea*, and fluctuations in population size from generation to generation (Falconer, 1989).

The traditional approach to tree improvement involves a number of sequential steps: species selection trials; provenance trials to identify the best sources of a species; progeny tests to identify the best mother trees to establish seedlings and clonal orchards; and the production of high-quality germplasm for dissemination to users (Weber *et al.*, 2001; Akinnifesi *et al.*, 2004; Leakey and Akinnifesi, Chapter 2, this volume). This process is time-consuming and expensive to undertake for indigenous fruit trees species, and there is a danger that the genetic diversity may be compromised (Weber *et al.*, 2001). Recently, participatory collections of germplasm (seed and scion) based on multitrait selection were done for *Uapaca kirkiana* and *Strychnos cocculoides* in Malawi, Tanzania, Zambia and Zimbabwe (Akinnifesi *et al.*, 2006). In addition, information was collected on the ecology, soil, and geographical location and fruit traits of the selected trees and stored. The Malawi collections involved 102 superior *Uapaca kirkiana* trees clonally propagated and raised in the nursery and established in a clonal gene bank at Makoka (Akinnifesi *et al.*, Chapter 8, this volume). The aim of participatory domestication is not only to involve farmers in selecting the most promising cultivars and planting materials, but also to conserve, multiply and disseminate these materials.

19.3 Propagation by Sexual Techniques

Most miombo fruit trees can be propagated by seed. *Uapaca* seed does not require pretreatment but soaking in cold tap water overnight hastens germination (Ngulube, 1996). Germination is fairly uniform, reaching 30% after 4 weeks and 90% after 6 weeks. Up to 100% germination has been achieved at Makoka nursery (Akinnifesi *et al.*, 2007).

Propagation of *Strychnos cocculoides* by seed has been successful, with 80% germination for seeds sown in the propagation box during summer (Taylor *et al.*, 1996; Mkonda *et al.*, 2004). In Botswana, seeds sown in winter took more than 9 weeks to germinate, whereas those sown in summer germinated within 3 weeks. Germination of *Vangueria infausta* is reportedly sporadic until the seeds are scarified or treated with hydrogen peroxide (Msanga and Maghembe, 1989). At Veld Products Research in Botswana, no difficulty in germination of *Vangueria infausta* has been observed with seeds sown in the summer months. Over 70% germination has been recorded within 21 days of sowing in sandy beds (Taylor *et al.*, 1996). In the following sections we provide a brief review of progress made in understanding seed viability in storage and the seed treatments required to obtain uniform germination of miombo fruit trees in southern Africa.

19.3.1 Viability in storage

Seed viability under storage condition and germination vary from species to species and depend on various factors. The seeds of some miombo fruit tree species lose viability during storage, whether at room temperature (around 20°C) or in a cold room (7–10°C) (Hans, 1981; Ngulube and Kananjii, 1989; Prins and Maghembe, 1994). Characteristics such as a very hard seed coat, physiological dormancy and low viability contribute to the poor germination of many indigenous fruit species, such that it is difficult to obtain sufficient propagules for planting (Kwapata, 1995). Some species, such as *Uapaca kirkiana*, have recalcitrant seeds, i.e. seeds that do not tolerate low temperatures and lose their viability if stored for any length of time, even under conditions that are normally conducive to seed longevity (Dawson and Were, 1997). In *U. kirkiana* seed viability is limited to a maximum of 4 months after removal from the fruit (Hans, 1981). According to Hans (1981), the viability of seeds of *U. kirkiana* can drop from 80 to 20% within 4 months. This is due to the high water content of the seed (48% of sun-dried seeds on a fresh weight basis). The seed has an inner membrane, an endosperm and two leafy green cotyledons. It has no dormancy period once dispersed. In the miombo woodlands, *U. kirkiana* seeds normally germinate as soon as the fruit ripens (Prins and Maghembe, 1994).

Uapaca kirkiana has desiccation-sensitive non-orthodox seeds, with a high water content and active metabolism. Studies in Malawi (Ngulube and Kananjii, 1989) also showed that the viability of *U. kirkiana* seeds carefully stored in cold room conditions was lost within a month of collection. The embryo would deteriorate quickly during storage because of the high water content of the seed and its active metabolism, which encourages the internal production of heat and the rapid growth of microorganisms, especially storage fungi (Hans, 1981). During the short-term storage of non-orthodox seeds, the proliferation of microorganisms must be curtailed to keep the seeds viable (Sutherland *et al.*, 2002).

Many other miombo fruit tree species have orthodox seeds. Orthodox seeds are seeds that can remain viable for long periods if processed and stored in the appropriate manner; normally seed should have a low moisture content and be kept at low temperature (Dawson and Were, 1997). For example, *Azanza garckeana*, *Flacourtia indica*, *Sclerocarya birrea* and *Parinari curatellifolia* have orthodox seeds, and their seed may store well for a year or more (Teichman *et al.*, 1986; Msanga, 1998). In orthodox seeds, ageing (which results in loss of viability) is a function of time, temperature and moisture content; with a controlled storage environment seed survival can be prolonged. A dry, cool environment normally lowers the risks of fungal proliferation and infestation by seed pests that affect the viability of orthodox seeds.

Seeds of *Strychnos cocculoides* are intermediate between the orthodox and recalcitrant types (Msanga, 1998). Mkonda *et al.* (2004) monitored the viability of four *S. cocculoides* provenances from Zambia stored for 0–15 months at room temperature and −4°C and found significant differences between provenances and storage temperatures. The provenance Kalulushi had lower viability under both storage conditions. When stored at room temperature, it lost viability more quickly compared with the other provenances. Over 70% of the seeds of Kasama and Serenje provenances remained viable for up to 15 months when stored at −4°C, while their viability dropped dramatically within 9 months of storage at room temperature (Mkonda *et al.*, 2003).

19.3.2 Seed pretreatment and germination

Seeds of some miombo fruit tree species germinate immediately, while others require special seed pretreatment. For example, up to 80% germination can be achieved by simply cleaning and soaking the seeds of *Diospyros mespiliformis*, *Strychnos spinosa*, *Syzygium cordatum*, *Uapaca kirkiana*, *Ziziphus abyssinica* and *Ziziphus mauritiana*, while less than 20% germination could be achieved with different pretreatments of seeds of *Flacourtia indica* and *Parinari curatellifolia* (Prins and Maghembe, 1994). Nicking the seeds of *Azanza garckeana* to partially remove the seed coat gave 100% germination (Maghembe, 1995). For *Ziziphus* spp., *Tamarindus indica*, *Parkia filicoides* and *Bridelia cathartica*, direct sowing of seeds after removing pulp gave 93–100% germination (Maghembe, 1995). For

Uapaca kirkiana, complete removal of the seed coat gave 70–100% germination (Hans, 1981; Maghembe, 1995; Mwabumba and Sitaubi, 1995), although simply cleaning followed by soaking in cold water for 24 h yielded a similar results, i.e. 90–100% (Maghembe, 1995; J. Mhango, personal communication). This was due to the presence of water-soluble germination inhibitors in the endocarp (A.S. Hans, unpublished results). The endocarp extract has been shown to be acidic (pH 3.7–5.8), which encourages fungal attack of germinating seeds when the seeds are within the endocarp.

Germination can fail for abiotic or biotic reasons. An investigation of failure of germination may reveal that the seeds have no embryos, the seeds are dead because of ageing or treatment, or the seeds are in dormancy – all abiotic factors. For seeds of some species, the inability to germinate is due to the impenetrability of the hard seed coat by water (Werker, 1980). *Azanza garckeana*, *Canthium foetidum*, *Diospyros mespiliformis* and *Bridelia cathartica* have hard seed coats, which prevents easy seed germination. In other species, such as *Sclerocarya birrea* and *Parinari curatellifolia*, the seed is embedded in a hard stone-like structure (Mojeremane and Tshwenyane, 2004) that prevents the seed from imbibing water and eventual germination. However, treatment of *Sclerocarya birrea* seeds by manual nicking has resulted in 75% germination (T. Chilanga, personal communication). If the operculum has not been opened the seed may take up to 9 months or more to germinate depending on the moisture status of the site and the rate of stone breakdown (Mojeremane and Tshwenyane, 2004). If a seed is not exposed to sufficient moisture, the proper temperature, oxygen and, for some species, light, the seed will not germinate.

In Zambia, Mkonda *et al.* (2004) showed that *Strychnos cocculoides* needs no treatment as different scarification techniques yielded no improvement. However, germination rates seem to be optimum 1–2 months after sowing, and the rate can vary between provenances from 58 to 83% (Mkonda *et al.*, 2004). Swai *et al.* (2004) obtained 91% germination for *S. cocculoides* in Tanzania after thorough cleaning and soaking in cold water for 24 h.

Some miombo woodland fruit species still have germination problems. For instance, Mwabumba and Sitaubi (1995) obtained no germination for *Flacourtia indica* and *Parinari curatellifolia* in Malawi. *Vitex doniana* had a low germination rate regardless of different treatments (Mwabumba and Sitaubi, 1995), but *Vitex doniana* gave 69% in Tanzania (Swai *et al.*, 1995). There were low germination rates for *Flacourtia indica* (0–16%), *Parinari curatellifolia* (0–23%) and *Boehemia discolor* (0–9%) regardless of treatment (Swai *et al.*, 2004). Maghembe (1995) obtained only 20% germination for *Parinari curatellifolia* and 17% for *Flacourtia indica* when seeds were sown directly after removing the pulp but without treatment.

Viable seeds are considered dormant when they fail to germinate under normal conditions suitable for germination. The expression of dormancy is under genetic control, but it is also strongly influenced by environmental factors (Osborne, 1981; Naylor, 1983). Seed dormancy is nature's way of setting a clock that allows seeds to initiate germination when conditions are favourable for germination and to ensure survival of the seedlings. Seeds from indigenous fruits may also contain germination inhibitors that may be removed or deactivated with time or by certain seed treatments. Delayed and irregular germination is therefore a serious constraint in the large-scale propagation of miombo fruit trees. Quickly germinated seeds may have a greater chance of producing seedlings than seeds that take longer to germinate, which may be exposed to an increased likelihood of fungal, insect or rodent attack. In practical terms, delayed germination will also tie up nursery beds for prolonged periods, increasing nursery costs and reducing seedling vigour (Prins and Maghembe, 1994). It may also increase variability in growth after establishment. Hence, special seed pretreatment is required to induce quick and uniform germination.

Various seed pretreatments have been studied for some of the miombo fruit trees, and selected recommended practices are presented in Table 19.1. Various methods, including seed cleaning, soaking in cold or hot water, mechanical scarification and acid scarification, are often recommended to overcome dormancy.

In the case of *Flacourtia indica*, small seeds have either no embryo or have a very hard seed coat, which makes imbibition difficult (Prins and Maghembe, 1994). Nicking and removal of the seed coat should have a marked positive effect on germination. However, the small size of the seed and the hard seed coat often results in embryo damage during pretreatment, resulting in poor germination when seeds are nicked or the seed coat is removed, probably because of unimpeded swelling resulting in internal damage. The difficulty with *Parinari curatellifolia* is that seeds deteriorate quickly at high temperatures. The endocarp is also very hard, making it difficult to extract the seed without damage (Prins and Maghembe, 1994), and often the seed is no longer viable before sowing. To make up for this, nature has made the roots of *Parinari curatellifolia* sproutable.

Seeds of *Sclerocarya birrea* do not germinate well without seed pretreatment. Hall *et al.* (2002) summarize studies conducted by various workers on seed pretreatment and the germination of *Sclerocarya birrea* (Table 19.1). Teichman *et al.* (1986) investigated the influence of several factors on the germination of *Sclerocarya birrea*, including scarification with concentrated sulphuric acid and soaking in potassium hydroxide for 24 h, and the effects of temperature and the exposure of intact and de-operculated endocarps to white light. Scarification with sulphuric acid was ineffective. Germination was higher at temperatures ranging between 27 and 32°C. White light was concluded to have an inhibitory effect on germination. Removal of the operculum led to a pronounced increase in germination, while soaking in potassium hydroxide solution and leaching with cold water further increased germination (Teichman *et al.*, 1986). The endocarp was suspected to protect the seed from leaching, which might remove germination-inhibiting substances (Hall *et al.*, 2002). The germination rate of seeds from endocarp soaked in cold and hot water has also been shown to be higher than that of those with untreated endocarps (Msanga, 1998; Mateke and Tshikae, 2002). Many workers have also investigated prolonged seed storage as a way to increase germination success (Hall *et al.*, 2002). According to Teichman *et al.* (1986), de-operculated endocarps of *S. birrea* subsp. *caffra* stored for 14–23 months had a germination rate of over 80% in 6 days compared with <50% for endocarps stored for 2 weeks to 2 months. In Botswana, removal of the operculum has resulted in very high levels of germination, as the seeds have multiple embryos. Research has shown that removal of the operculum is not essential for successful germination, but it is very important to keep seeds dry during the cold months and to germinate them when the mean maximum temperature is at least 21°C. Keeping seeds dry in the cold season after fruit ripening is important in releasing them from dormancy as moisture in winter causes seeds to develop deeper dormancy.

Table 19.1. Seed pretreatments used to achieve best germination of miombo fruit tree seeds.

Fruit tree species	Seed pretreatment	Germination rate (%)	References
Adansonia digitata	No treatment; scarification by cold water, hot water or manual partial nicking	82–96	Mwabumba and Sitaubi, 1995
Azanza garckeana	Nicking to partially remove seed coat (i.e. making a hole in the seed coat)	100	Prins and Maghembe, 1994
Diospyros mespiliformis	Removing the fruit pulp before sowing (no treatment) Nicking to partially remove seed coat	80–87	Prins and Maghembe, 1994
Flacourtia indica	Removing the fruit pulp before sowing	20	Prins and Maghembe, 1994
Parinari curatellifolia	Nicking seeds or complete removal of seed coat	17–30	Prins and Maghembe, 1994; Swai *et al.*, 2004
Sclerocarya birrea	Removal of the operculum for 14–23 months Soaking in potassium hydroxide (1 mol) for 24 h	80	Teichman *et al.*, 1986
Strychnos cocculoides	Cleaning seeds by removing all the mesocarp and soaking for 24 h	80–90	Prins and Maghembe, 1994; Swai *et al.*, 2004; Mkonda *et al.*, 2004; Mateke and Tshikae, 2002
Syzygium cordatum	Cleaning seeds by removing all the mesocarp and soaking for 24 h	93	Prins and Maghembe, 1994
Uapaca kirkiana	Cleaning seeds by removing all the mesocarp and soaking for 24 h Nicking seeds	100	Prins and Maghembe, 1994
Vangueria infausta	Nicking seeds	40	Prins and Maghembe, 1994
Vitex mombassae	Complete removal of seed coat	69	Swai *et al.*, 2004
Ziziphus mauritiana	Removing the fruit pulp before sowing Cleaning seeds and placing in 0.5 l warm water at 65°C, then leaving to cool for 24 h; manual scarification by nicking	70	Mwabumba and Sitaubi, 1995

19.4 Vegetative Propagation

Vegetative (asexual) propagation is defined as the regeneration of a new individual from a portion (ramet) of a stock plant (ortet) by a process involving mitotic cell division, and the subsequent regeneration of complementary cells, tissues and or organs, or an entire plant, to replace those missing from the ramet (Mudge and Brennan, 1999). This can be effected by grafting, cutting, air-layering and shoot micropropagation.

Although many miombo fruit tree species can be propagated by seed, progeny obtained in this way may not be true to type. Natural variability in trees propagated through seedlings is a major bottleneck in fruit orchards. Sexual outbreeding through cross-pollination has evolved as the dominant natural reproductive strategy (breeding system) in trees because it ensures genetic recombination, heterozygosity, and concomitant seedling variation on which natural selection may act to bring about adaptation favourable to survival (Mudge and Brennan, 1999). The domestication of normally outcrossing species depends on increasing uniformity because it makes selection of improved genotypes possible. Genetic gains associated with deliberate selection of superior genotypes of normally outcrossing species can be captured immediately, via cloning of selected individual putative trees (Simons and Leakey, 2004; Leakey and Akinnifesi, Chapter 2, this volume), thereby avoiding the loss of the selected trait in subsequent seedling generations due to segregation of alleles (Mudge and Brennan, 1999). Therefore, vegetative propagation techniques are indispensable for the capture and multiplication of the phenotypic variation expressed by superior individuals of miombo fruit trees with desirable characters.

Vegetative propagation could also help in shortening the period to first fruit as most miombo indigenous fruit trees may start fruiting in 12–16 years when propagated by seed (Taylor and Kwerepe, 1995). It is also possible to select fruits for seedlessness. This aspect has not been researched for miombo fruits, but it may be worthwhile in future. Vegetative propagation of trees can be done using a variety of techniques, which include grafting, budding, air-layering, rooting juvenile stem cuttings and *in vitro* tissue culture.

19.4.1 Grafting

The history of improvement of many tropical fruit tree crops around the world, such as mangoes, citrus and avocado, indicates that grafted stock is the best way to effect improvements rapidly. Because there is evidence that phenotypic fruit traits are similar to genotypic characteristics (Ndugu *et al.*, 1995) and can be fixed by vegetative propagation (Simons, 1996; Mudge and Brennan, 1999; Simons and Leakey, 2004), there is a good case for phenotypic selection of desirable fruit traits (Akinnifesi *et al.*, 2006).

Fruit and nut trees have been grafted because: (i) it is difficult to propagate important clones by cuttings; (ii) the grafted crops are superior and have high value; (iii) it is also possible to benefit from improved rootstocks (double-working), when carefully selected, and from changing cultivars of certain indigenous fruit trees; (iv) it increases plant growth rate and reduces the times to production and reproductive maturity, leading to earlier fruit production; and (v) it enables the repair of damaged parts of trees (Hartmann *et al.*, 1997). Grafting involves placing two similar or dissimilar plant organs (stem/stem, stem/root, root/root) from genetically compatible plants in intimate contact, with sufficient pressure and cambial alignment to induce the formation of an anatomically and physiologically functional graft union between scion and stock (Mudge and Brennan, 1999). It is the art of connecting two pieces of living plant tissue together in such a manner that they will unite and subsequently grow into one composite plant (Hartmann *et al.*, 1997). It is more expensive than cutting or air-layering. However, the selection of scion and rootstock may affect the outcome of grafting. The most common grafting method for miombo fruits is the whip and tongue method. Others are the whip (splice graft), cleft graft, wedge graft, saddle graft, four-flap graft, side stub, side tongue, side veneer graft, bark grafting (detached scion), approach grafting and bridge grafting (Macdonald, 1986; Hartmann *et al.*, 1997).

Until recently, limited success with vegetative propagation has been a constraint to the domestication of miombo indigenous fruit trees in southern Africa. For example, a graft take of less than 10% was earlier reported for *Uapaca kirkiana* (Jaenicke *et al.*, 2001). However, this problem was overcome once the conditions necessary for scion collection were properly understood (Akinnifesi *et al.*, 2006). In Malawi, grafting success has been reported to be 100% in *Vangueria infausta*, 85–100% in *Adansonia digitata*, 80% in *Uapaca kirkiana*, 40–79% in *Strychnos cocculoides*, 52–80% in *Sclerocarya birrea*, and 71% in *Parinari curatellifolia* (Mhango and Akinnifesi, 2001; Swai *et al.*, 2004) and 74% for *Strychnos cocculoides* in Zambia (Mkonda *et al.*, 2004). In Botswana and South Africa the best results have been obtained when scions of *Sclerocarya birrea* are grafted in September–October (Taylor *et al.*, 1996). Mature scions of *Sclerocarya birrea* grafted onto spring-sown seedling rootstocks have developed into plants that yield first fruits in the fourth or fifth year after grafting. Plants from seedlings not grafted have been observed to take 8–10 years before the first signs of flowering (Taylor and Kwerepe, 1995).

Some of the major determinants of grafting success include the skill of the person doing the grafting, the time of the year the scion is collected and the interval between scion collection and grafting (Akinnifesi *et al.*, 2004; Nkanaunea *et al.*, 2004). The best time for grafting or collecting the scion is from August to December. In the field, growth rates of grafted trees improve greatly with increased moisture availability. Other factors include proper alignment of the vascular tissues between scion and rootstock, stock plant nutrition, scion respiration rate, the disease status of scion and rootstock, and the relative humidity of the propagation environment (Akinnifesi *et al.*, 2006).

Budding is a form of grafting, except that the scion is reduced in size to contain only one bud. As done in oranges, budding has been found to be very effective in propagating *Ziziphus mauritiana*. Budding can be done by T-budding or shield budding or using the inverted T method (Macdonald, 1986).

There are two types of stock: seedling rootstocks and clonal rootstocks. Seedling rootstocks have the advantage of being able to penetrate the soil and grow more firmly. Their disadvantage is genetic variation that may lead to variability in the performance of the scion of the grafted plant, especially when the seedling is raised from an unknown seed source (Hartmann *et al.*, 1997). Careful selection of the rootstock is necessary to reduce such variability. On the other hand, clonal rootstocks are genetically the same, and all the plants in a clone can be expected to have identical growth characteristics.

In both macro- and micropropagation, delayed, early graft incompatibility or outright stock/scion rejection may occur as a result of adverse physiological responses between grafting partners (Mng'omba *et al.*, 2007a, b), virus or phytoplasma transmission, anatomical mismatching or poor craftsmanship of the grafter, the phenomenon known as 'green finger,' and anatomical abnormalities of the vascular tissue in the callus bridge (Macdonald, 1986; Hartmann *et al.*, 1997).

19.4.2 Top-working

Top-working is a form of grafting done in the field by connecting tissues of a scion from a desired tree to a growing tree. A fruit tree may be an undesirable cultivar, such as an old, unproductive plant or cultivar, or one that is susceptible to diseases. Top-working is done to extend the production cycle or to increase resilience or the value of the cultivar. For indigenous fruit trees, top-working could help to rejuvenate old trees and make them more productive. In addition, certain trees that are dioecious may be unproductive because of a lack of male (staminate) flowers. Such trees could be made more productive by introducing male branches into an orchard with predominantly female (pistillate) stands and vice versa.

Just as in grafting, top-working between and within the same family, species and clone of a species is always feasible, but compatibility is greater for clones (Hartmann *et al.*, 1997). Top-working has been successful in *Uapaca kirkiana* at Makoka. Scions of superior trees were grafted onto growing *Uapaca kirkiana* planted a few years earlier. Similarly, A. Mkonda *et al.* (unpublished) obtained 65% graft take in *in situ* grafting of *Uapaca kirkiana* in the wild. This technique would be particularly useful for improving the production quality of seedlings of established trees on the farm. One major advantage of top-working is that it will overcome problems associated with tree establishment and poor survival.

19.4.3 Air-layering

Air-layering is a method of propagation performed directly on the branches of growing parent trees *in situ*. It involves the rooting of cuttings in which adventitious roots are initiated on a stem while it is still attached to the plant (Hartmann *et al.*, 1997). The rooted stem (layer) is then detached and transplanted, and becomes a separate plant on its own roots. The physical attachment of the stem during rooting provides a continual supply of water, nutrients, minerals and carbohydrates through the intact xylem and phloem to the rooting area.

Air-layering has been found to be feasible for many of the miombo indigenous fruit tree species, especially *Uapaca kirkiana*, in which a success rate of about 63% was obtained (Mhango and Akinnifesi, 2001; Akinnifesi *et al.*, 2006). The best time to set *Uapaca kirkiana* marcotts is between August and October. Our results with *Uapaca kirkiana* showed that no root-initiation hormone was required for rooting. However, the survival of established marcotts may pose a minor challenge, as mortality increases with time in the first 6–12 months in some cases (Akinnifesi *et al.*, 2006). Where field survival is good, trees established from air-layers were more vigorous than grafted trees, and produced more fruits starting after 24 months in Makoka, Malawi. Similar results were obtained in the humid zones of Cameroon (Tchoundjeu *et al.*, 2006).

19.4.4 Propagation by cuttings

The use of cuttings is the simplest and probably the most important form of vegetative propagation in the developing world. For species that can be propagated by cutting, it has numerous advantages (Hartmann *et al.*, 1997). The success of cuttings depends on their ability to produce adventitious roots that can be easily acclimatized to the environment after rooting. The advantages of cuttings include the following: (i) the method is inexpensive, rapid and simple compared with other vegetative propagation methods; (ii) many new plants can be established in a limited space from a few stock plants; (iii) no special skills are required, unlike in grafting, budding, air-layering and micropropagation; (iv) the problem of incompatibility due to poor union of stock and scion in grafted plants is avoided; and (v) the plants produced are more uniform because the variation imposed by the stock in grafting is avoided. Stem, leaf or root cuttings can be used. Stem cuttings are classified according to the nature of the wood, i.e. the degree of lignification: hardwood, semi-hardwood, soft wood or herbaceous.

Rooting of juvenile stem cuttings seems to be more successful than mature cuttings for most miombo indigenous fruit tree species. Rooting success of cuttings of up to 60% has been observed in *Parinari curatellifolia* (R. Swai, personal communication). Few of the miombo woodland indigenous fruit trees have been tested with leaf cuttings under macropropagation.

Azanza garckeana and *Vitex mombassae* respond favourably to propagation by root cuttings. In Botswana, attempts at propagating *Sclerocarya birrea* by cuttings have not succeeded either in the open or under 50% shade netting (Taylor *et al.*, 1996). Although cuttings in the propagation box responded favourably in terms of vegetative growth within 3 weeks, no roots were observed after 3 months in the box, and rooting hormone (Seradix No. 2) did not seem to induce rooting (Taylor *et al.*, 1996).

Likewise, cuttings of *Strychnos cocculoides* sprouted within 3 weeks under 50% shade, but no root development was observed (Taylor *et al.*, 1996). For *Sclerocarya birrea*, the use of truncheons (hardwood cuttings) has been found to be an effective method of propagation (Shackleton, 2004). In South Africa, Shackleton (2004) reported that farmers plant *Sclerocarya birrea* using truncheons (25%), transplanting wildings (31%) or seeds directly (44%). Likewise, truncheons of *Adansonia digitata* sprout readily and roots when pushed into the soil. Although propagation of *Sclerocarya birrea* by truncheons has been successful, this method has poor commercial prospects (Taylor *et al.*, 1996).

Many miombo fruit trees fail to undergo active rooting because of low genetic and physiological capacity for adventitious root formation, which limits the effectiveness of propagation with cuttings. Mng'omba (2007) and Mng'omba *et al.* (2007a, b) have alluded to the possibility of phenolic compound formation at the cut surface, limiting vegetative propagation – in grafting this might lead to a weak union or incompatibility between scion and stock; when propagation is being attempted from cuttings it might cause the development of calluses without rooting. Excessive formation of calluses has also been indicated as a reason for poor rooting of cuttings in *Uapaca kirkiana* and *Sclerocarya birrea* in southern Africa (Jaenicke *et al.*, 2001).

19.4.5 *In vitro* micrografting

Tissue culture is an aseptic laboratory procedure that requires unique facilities and special skills. *In vitro* regeneration involving the use of relatively small propagules, referred to as explants, cultured under controlled environments in small containers of a nutrient medium in which all nutrients required for growth are provided artificially within an aseptic environment (Hartmann *et al.*, 1997; Mudge and Brennan, 1999). The multiplication process may involve the stimulation of normal (non-adventitious) elongation and/or the branching of the original explant (shoot/root), and/or the *de novo* induction of adventitious organs (shoots and/or roots), including but not limited to axillary or nodal shoots (Mudge and Brennan, 1999). Shoot culture involves using growth regulators during a multiplication stage to promote repeated cycles of shoot elongation and/or branching, followed by repeated cycles of subdivision and reculture on fresh medium; the proliferated roots are then rooted as microcuttings either *in vitro* or *ex vivo*. Therefore, it is not feasible for use by small-scale farmers in decentralized nursery systems.

In the past, it was thought that tissue culture was inappropriate for small-scale farmers, and less emphasis has been placed on it in domestication work compared with other propagation methods. Recently, *in vitro* micropropagation has been pursued as one of the methods that should be exploited for the mass and rapid vegetative multiplication of superior priority indigenous fruit trees (Akinnifesi *et al.*, Chapter 8, this volume). Reasons for using tissue culture (Hartmann *et al.*, 1997) include: (i) micropropagation involves very small amounts of plant tissue; (ii) somatic embryogenesis is possible; (iii) embryo rescue is possible; (iv) anthers (microspores) can be cultured to produce haploid plants for breeding; (v) micrografting is possible; and (iv) tree crops can be improved. In addition, it makes possible the mass multiplication of specific clones, the production of pathogen-free plants, and the clonal propagation of parent stock for hybrid seed production (Hartmann *et al.*, 1997).

The potential of meeting the great demand for germplasm for scaling up in the short run is low for most macrovegetative propagation approaches. All known approaches to vegetative propagation are inherently slow and are feasible for only a few hundreds or thousands of farmers. For instance, rootstocks of *Uapaca kirkiana* need to attain pencil thickness before grafting is possible, and this requires about 1 year of growth. In order to be able to deliver high-quality propagules of superior indigenous fruit trees in sufficient quantities it is important to exploit the potential of tissue culture.

Recently, *in vitro* culture of *Uapaca kirkiana* and *Pappea capensis* (plum) was undertaken, with the objective of developing a reproducible clonal protocol for rapid regeneration and multiplication, and to determine early graft compatibility using *in vitro* techniques (Mng'omba, 2007; Mng'omba *et al.*, 2007a, b). Based on a series of results, reproducible micropropagation protocols have been developed for the rapid multiplication of mature *U. kirkiana* and *P. capensis* (Mng'omba, 2007; Mng'omba *et al.*, 2007a, b). The technique is promising for the detection of early incompatibility between close and distantly related propagule sources. Graft compatibility was greater for homografts than for heterografts among and between *U. kirkiana* clones, species and provenances (Mng'omba, 2007).

Results also showed that *in vitro* propagation of *U. kirkiana* is feasible with sprouts excised from preconditioned trees (Mng'omba, 2007). Better results were obtained when micropropagation of *U. kirkiana* was undertaken in such a way that stock plants were not stressed. Repeated exposure of the difficult-to-root microcuttings of *P. capensis* increased the success of *in vitro* culture (Mng'omba, 2007). Somatic embryos of *P. capensis* were successfully germinated into plants (65%), and there was 65% plantlet survival after hardening off in a mist chamber.

Nkanaunea *et al.* (2004) investigated the effects of age and type of rootstock and light intensity on the success of micrografting *U. kirkiana* in Malawi. Comparison of a factorial combination of three rootstock ages (2, 3 and 4 months), two types of rootstock (rooted and unrooted) and four light intensities (photosynthetic photon flux density 0, 45, 135 and 225 $\mu mol/m^2/s$) indicated a significant effect of those factors on the success of micrografts. The best micrograft success was achieved when 2-month-old rooted rootstocks were used and the cultures were incubated under illumination at 45 $\mu mol/m^2/s$ (Nkanaunea *et al.*, 2004).

19.5 Nursery Management

Until recently, techniques for seed treatment and enhancing growth in the nursery using ideal propagation media and optimum fertilizer regimes have not been available to farmers. There has also been a lack of information on the ecological requirements of many miombo fruit tree species. For example, the requirement for inoculation with rhizobia and mycorrhizal symbionts has not been studied.

Nursery management is a crucial aspect of providing good-quality planting materials in adequate quantities (Wightman, 1999). Tree seedling quality has two aspects. The first is the genetic quality or the source of the seed and the second is its physical condition. The characteristics of the parent tree can greatly influence the characteristics of the seedlings. According to Wightman (1990), a good-quality seedling has the following characteristics:

- It is healthy, growing vigorously and free of diseases.
- It has a robust and woody single stem that is free of deformities.
- The stem is sturdy and root diameter is large.
- The root system is dense, with many fine roots and fibrous hairs.
- There is a balance between shoot and root mass.
- It is accustomed to short periods of drought and full sunlight.

The physical condition of the seedling can be improved by appropriate management, which includes nursery site selection, the use of appropriate media, fertilization and pest management.

19.5.1 Nursery site selection

Selection of the nursery site is an important consideration when seedlings are raised on seedbeds or in containerized nurseries. Miombo fruit tree seedlings were exclusively raised in polyester bags by Böhringer *et al.* (2003). However, nursery site selection in terms of source of shade and water supply is crucial in raising planting material of miombo fruit trees. Studies of the effects of shading showed that seedlings of *Sclerocarya birrea* raised under 25% shade had better growth, so that within 12 months the seedlings were big enough for grafting. In the 1995 grafting session, 100% uptake was recorded for the seedlings that were raised under 25% shade. Seedlings raised under 50 and 100% shade were too thin and too short, respectively (Taylor *et al.*, 1996). Similarly, *Uapaca kirkiana* seedlings grew well under partial shade (Ngulube *et al.*, 1995).

The supply of water to plants in the nursery appears to be an important determinant of the nursery site in southern Africa. A study conducted in 1998/1999 on nursery location and productivity (Böhringer *et al.*, 2003) showed that most tree nurseries are located close to permanent water sources such as dambos and river banks. The study indicated significant differences between dambos and riverbanks. A steady supply of water for irrigation is crucial for an indigenous fruit tree nursery.

19.5.2 Potting mixture

The potting mixture used can have a significant influence on the growth of seedlings. Normally, forest soil from ordinary miombo woodland or from a natural forest stand of the species is used for raising indigenous fruit tree seedlings. Mhango (2000) compared various potting mixtures in southern Malawi, and recommended a mixture of 75% forest soil and 25% sawdust as the best for growth of *Uapaca kirkiana*. Compared with forest soil alone, this mixture is reported to be lighter and to have higher total porosity, water-holding porosity and infiltration rates (Mhango, 2000). However, in Zambia this mixture resulted in poor seedling performance (Sileshi *et al.*, 2007), probably because the sawdust had immobilized the nutrients in the soil, leading to nutrient stress. In the same study, forest soil sterilized with heat and a mixture of unsterilized forest soil was compared with sawdust and unsterilized forest soil. Height growth, root length, number of leaves and shoot dry weight of seedlings were greater in unsterilized forest soil than in sterilized forest soil or forest soil mixed with sawdust (Table 19.2). On the other hand, growth in stem diameter, number of primary roots and root dry weight were less in unsterilized forest soil. This suggests that there is a need for further research on appropriate soil media for different species.

There is evidence that early growth and survival of certain indigenous fruit trees are negatively affected when they are grown without inoculation with associated microsymbionts such as mycorrhizas (Ramachela, 2001). To achieve the inoculation of seedlings with symbiotic microorganisms to form mycorrhizas, forest soil from mature stands of the fruit tree species is often used in the potting mixture. In addition, although most priority miombo indigenous fruit trees do not belong to the family Leguminosae (Maghembe, 1995), which is highly associated with nitrogen-fixing rhizobia, they may benefit from the associated group of bacteria which are free-living and are found in the vicinity of grasses, and the free-living group of bacteria which is present and active in easily available organic

Table 19.2. Effect of potting mixture, foliar and soil applied fertilizer on the growth of *Uapaca kirkiana* seedlings in the nursery at Msekera.

Factor and level	Height (cm)	Leaf number	Stem diameter (mm)	Root length (mm)	Primary roots (number)	Root dry weight (g)	Shoot dry weight (g)	Foliar disease incidence (proportion)
Potting mixture								
UFS	12.3[a]	7.9[a]	3.4[b]	27.5[a]	18.3[b]	2.8[b]	4.1[a]	0.85
FS + SD	10.9[b]	6.7[b]	4.4[a]	20.8[b]	8.3[c]	9.7[a]	2.8[a]	0.73
SFS	10.8[b]	7.6[ab]	4.3[a]	28.5[a]	23.8[a]	10.0[a]	3.4[a]	0.87
P value	0.008	0.009	0.001	0.0001	0.0001	0.0001	0.072	0.029
Soil fertilizer (D-compound)								
Without	10.9[b]	7.5[a]	4.3[a]	27.7[a]	20.4[a]	7.6[a]	3.8[a]	0.79
With	12.1[a]	7.5[a]	3.6[b]	24.8[b]	15.1[b]	6.2[a]	3.3[a]	0.85
P value	0.025	0.635	0.002	0.001	0.0001	0.439	0.610	0.290
Foliar fertilizer								
Without	10.6[b]	7.2[a]	3.6[a]	25.5[a]	18.4[a]	5.7[a]	2.4[b]	0.83
With	12.2[a]	7.7[a]	4.2[a]	26.9[a]	17.2[b]	8.0[a]	4.5[a]	0.80
P value	0.027	0.493	0.153	0.743	0.033	0.120	0.001	0.136

UFS, unsterilized forest soil; FS + SD, mixture of forest soil and sawdust; SFS, sterilized forest soil.
Treatments followed by the same letter are not significantly different at the 5% probability level according to the *t* test (LSD).
Source: Sileshi *et al.* (2007).

matter (Muller-Samann and Kotschi, 1994). In some species there is a special symbiotic relationship between the roots and fungi and/or bacteria. The fine roots of most miombo indigenous fruit trees form symbiotic associations – mycorrhizas – with various fungi. The symbiotic relationship between plant roots and certain mycorrhizal species can increase the plant's uptake of important nutrients, as well as make the plant more tolerant of drought and disease (Bolan, 1991). It might be worthwhile to collect microsymbionts when collecting scions of indigenous fruit trees from the wild.

Trees mainly develop two types of mycorrhizas: ectomycorrhizas (ECM) and vesicular–arbuscular mycorrhizas (VAM), which are distinct in morphology (Gerdemann, 1968). In Zambia, *Uapaca kirkiana*, *Uapaca nitida*, *Uapaca sansibarica* and *Parinari curatellifolia* form ectomycorrhizal associations while *Strychnos innocua*, *Strychnos spinosa* and *Syzygium guineense* form endomycorrhizas in their natural stands in the miombo (Högberg and Piearce, 1986). *Uapaca kirkiana* has also been shown to form endomycorrhizas in Tanzania (Högberg, 1982). A genuine dual ECM/VAM association in *Uapaca staudtii* has also been reported in Cameroon (Moyersoen and Fitter, 1999), although a tendency towards greater fractional colonization with ECM and lower colonization with VAM has been observed in samples from dry areas than from waterlogged habitats (Moyersoen and Fitter, 1999). A natural mycorrhizal association was observed on roots of *Sclerocarya birrea* subsp. *birrea* in Mali (Soumaré *et al.*, 1994). In Zimbabwe, McGregor (1995) has reported an association between *S. birrea* subsp. *caffra* and a mushroom. Table 19.3 gives the types of mycorrhiza associated with selected indigenous fruit trees of the miombo.

In natural stands of *Uapaca kirkiana* in Zambia, the roots of the seedlings are associated with seven species of fungus (Mwamba, 1995). Though certain

Table 19.3. Mycorrhizas associated with selected indigenous fruit trees of the miombo and related species.

Miombo fruit tree species	Mycorrhizal status	Country of report	Reference
Diospyros consolata	VAM	Tanzania	Högberg, 1982
Sclerocarya birrea subsp. *caffra*	VAM	Botswana	Mateke and Tshikae, 2002
Sclerocarya birrea subsp. *caffra*	ECM	Zimbabwe	McGregor, 1995
Strychnos henningsii	VAM	Tanzania	Högberg, 1982
Strychnos innocua	VAM	Zambia	Högberg and Piearce, 1986
Strychnos madagascarenis	VAM	Tanzania	Högberg, 1982
Strychnos spinosa	VAM	Zambia	Högberg and Piearce, 1986
Uapaca kirkiana	Dual ECM/VAM	Tanzania	Högberg, 1982
Uapaca kirkiana	ECM	Zambia	Högberg and Piearce, 1986
Uapaca nitida	ECM	Zambia	Högberg and Piearce, 1986
Uapaca sansibarica	ECM	Zambia	Högberg and Piearce, 1986
Uapaca guineensis	ECM	Senegal	Thoen and Ba, 1989
Uapaca staudtii	Dual ECM/VAM	Cameroon	Moyersoen and Fitter, 1999

ECM, ectomycorrhiza; VAM, vesicular–arbuscular mycorrhiza.

preferences have been observed in some trials, mycorrhizal fungi do not need a specific host nor do host plants require a particular mycorrhizal species (Muller-Samann and Kotschi, 1994). The stimulating effects of different fungi can vary considerably in the same host plant (Mwamba, 1995). In *Sclerocarya birrea* subsp. *caffra*, early growth in height was enhanced by association with mycorrhizal fungi of the genera *Acaulospora* and *Glomus* (Bâ *et al.*, 2000; Mateke and Tshikae, 2002). Research on growth and survival of *Strychnos cocculoides* has indicated difficulties in raising seedlings in the nursery (Taylor *et al.*, 1996; Mkonda *et al.*, 2004). This species seems to be affected by mycorrhizal factors as well as shade.

19.5.3 Seedling fertilization

There is a dearth of information on the fertilizer requirement of indigenous fruit tree seedlings. In a study on the effect of soil-applied and foliar fertilization on the growth of *Uapaca kirkiana*, we compared application of D-compound fertilizer versus no fertilizer application, and foliar application versus no application. Compared with the untreated seedlings, those with soil fertilization showed significantly increased height growth, decreased growth in stem diameter and root length and fewer primary roots, while fertilization did not affect leaf number and root dry weight. Foliar application of fertilizer increased height growth and shoot dry weight but did not affect leaf number, growth in diameter, root length and root dry weight (Table 19.2).

19.5.4 Pest management

Several species of pests affect miombo fruit tree seeds and seedlings in the nursery (Sileshi *et al.*, Chapter 20, this volume). Seed-borne microorganisms may reduce seed viability, germination and seed longevity in storage of all types of seeds. Diseases of seeds and germinants are often overlooked in the nursery as the affected seeds are not visible and consequently the losses are often attributed to poor seed. To ensure germination in the nursery, seeds need to be examined and the problem characterized before management methods to reduce the problem can be developed. Indigenous fruit tree seedlings also suffer from attack by a wide range of fungi in the nursery. Fungi that cause foliage diseases in the nursery usually require conditions of high moisture and free water for long periods of time to establish infection. Such moisture occurs during extended periods of rain or fog, or during continued use of overhead irrigation. Fungal diseases may affect the roots, stem or foliage of seedlings, and can reduce seedling survival and growth. For instance, survival of *Uapaca kirkiana* seedlings in the nursery has been very low (<10%) at Msekera, Zambia (G. Sileshi, unpublished results) compared with greater than 80% in Makoka, Malawi (T. Chilanga, unpublished results). The low survival in Zambia was attributed to various foliar diseases, nutrient deficiencies and possibly inadequate mycorrhizal inoculants. Studies on the effects of three potting mixtures (unsterilized forest soil, unsterilized forest soil plus sawdust, and sterilized forest soil), foliar fertilization (with and without) and soil application of fertilizers (with and without) on the incidence of these diseases were conducted at Msekera. The incidence differed with soil application of fertilizer but not with foliar application (Sileshi, 2007). There were also differences among potting mixtures, where unsterilized soil mixed with sawdust had lower incidence of disease compared with sterilized forest soil (Table 19.2).

19.6 Conclusion and Way Forward

There is emerging interest in growing indigenous fruit trees in southern Africa, but this enthusiasm is limited by the lack of a supply of quality germplasm and may lead to frustration among prospective users. Therefore, appropriate germplasm delivery pathways are needed to meet the demands for germplasm by smallholder farmers.

Adequate information on requirements for the maintenance of the physiological and genetic quality of seed is basic to the development of miombo fruit trees as commercial species. Therefore, there is a need to establish seed zones and seed transfer guidelines for the collection of propagules between provenances based on genetic and environmental criteria, in order to minimize the risk of maladaptation of indigenous fruits in new niches. Selection and management of seed sources based on their phenotypic traits coupled with deployment in environments similar to those of the respective source could be considered as the easiest option for seed transfer. This can be implemented by establishing seed zone systems in locations with sufficiently uniform ecological conditions so that the phenotypic or genetic characters within a species in the zone are similar (Barner and Willan, 1983). Germplasm collection needs to focus on superior traits, and efforts must be made to credit the farmers and local communities in whose custody these materials have been preserved for centuries (Akinnifesi *et al.*, 2006, Chapter 8, this volume). Farmers' and breeders' rights must also be upheld in case large-scale commercial ventures with new tree crops are established. This can be achieved using participatory selection and domestication approaches, as detailed by Leakey and Akinnifesi, Chapter 2, Akinnifesi *et al.*, Chapter 8 and Tchoundjeu *et al.*, Chapter 9, this volume.

There is a considerable amount of undocumented indigenous knowledge on miombo fruit trees. As germplasm is collected, the indigenous knowledge associated with it needs to be systematically collected and documented. However, this must be done in compliance with the Convention on Biological Diversity and farmers must be allowed to maintain the rights to their indigenous knowledge and the genetic materials.

Conservation of forest genetic resources, of which indigenous fruits are an integral part, is regarded as constituting the actions and policies that ensure the continued existence, development and availability of these resources in the future. In addition to conserving the genes it is necessary that development must proceed in such a way as to maintain existing knowledge and the landscape, in order to buffer them against environmental changes (Eriksson *et al.*, 1993; FAO, 1993).

It is well known that long-term storage of the seeds of tree species currently plays a very minor role in the conservation of indigenous fruit trees. *Ex situ* conservation of germplasm in gene banks with duplicate collections in various centres is essential. Centres of diversity should also be identified and *in situ* conservation activities should be concentrated in these areas. The conservation of indigenous fruits could be achieved in a manner similar to that of forest genetic resources, through a diversity of approaches combining strictly forest protected areas with multiple-use areas managed by local people, with natural forest extensively managed primarily for the production of fruits and planted orchards under intensive management (Kemp, 1992; Palmberg-Lerche, 1997). Range-wide collection and orchard establishment of superior phenotypes of the various species from the SADC could achieve some of these conservation objectives. Encouragement of the use of community-owned orchards of selected superior phenotypes is an indirect way of conserving genetic resources on the farm.

Seedlings of some miombo fruit trees, such as *Uapaca kirkiana*, show high mortality in the nursery. Mortality could result from various stress factors, including diseases, inappropriate growing media and nutritional deficiencies (Mhango, 2000; Swai, 2002), that directly or indirectly affect plant growth and biomass allocation, commonly expressed as the root-to-shoot ratio. Therefore, factors leading to the mortality of seedlings should be worked out in detail and methods to reduce mortality need to be devised. Pests are responsible for perhaps the most noticeable damage to seeds and seedlings. Therefore, integrated nursery pest management needs to be developed (Sileshi *et al.*, Chapter 20, this volume).

Vegetative propagation techniques, especially grafting and to lesser extent air-layering, need to be given greater attention for priority species, especially *Vangueria infausta*, *Adansonia digitata*, *Uapaca kirkiana*, *Ziziphus mauritiana*, *Strychnos cocculoides*, *Sclerocarya birrea* and *Parinari curatellifolia*, which appear to be amenable to grafting (Akinnifesi *et al.*, 2004). Grafting and rooting of cuttings should be tried on a large scale for species such as *Uapaca kirkiana*, in which seed viability and seedling survival are poor, and for *Flacourtia indica* and *Parinari curatellifolia*, in which germination is more erratic and slow.

As farmer nurseries evolve and move from the production of low-cost tree seedlings for basic food security and ecosystem system services to higher-value ones for income generation and market, it can be expected that the quality of individual tree seedlings produced will gain much higher importance. Decentralized tree germplasm systems may perform well in terms of providing the quantities required but could fall short on the quality and diversity goals of society (Place and Dewees, 1999). Tree seedling quality needs to be given more attention in future assessments in order to keep pace with the anticipated evolution of farmer nurseries, where diversity in production is expected to become more important.

Large-scale domestication of high-value indigenous fruit trees has not yet been promoted by governments in southern Africa (Akinnifesi *et al.*, Chapter 8, this volume), unlike in Latin America, where there are many initiatives (C. Clement, personal communication; Clement *et al.*, Chapter 6, this volume). As in the cases of South-east Asia (Carandang *et al.*, 2007) and West Africa

(Tchoundjeu *et al.*, Chapter 9, this volume), where external intervention in the form of funds and other material inputs may not be crucial determinants of domestication, technical support on appropriate and low-cost tree propagation methods, awareness information and extension material is considered critical in the scaling up of indigenous fruit trees in southern Africa.

References

Akinnifesi, F.K., Kwesiga, F.R., Mhango, J., Mkonda, A., Chilanga, T. and Swai, R. (2004) Domesticating priority miombo indigenous fruit trees as a promising livelihood option for smallholder farmers in southern Africa. *Acta Horticulturae* 632, 15–30.

Akinnifesi, F.K., Kwesiga, F., Mhango, J., Chilanga, T., Mkonda, A., Kadu, C.A.C., Kadzere, I., Mithofer, D., Saka, J.D.K., Sileshi, G. and Dhliwayo, P. (2006) Towards the development of miombo fruit trees as commercial tree crops in Southern Africa. *Forests, Trees and Livelihoods* 16, 103–121.

Akinnifesi, F.K., Mhango, J., Sileshi, G. and Chilanga, T. (2007) Early growth and survival of three miombo indigenous fruit tree species under fertilizer, manure and dry-season irrigation in southern Malawi. *Forest Ecology and Management* 248.

Bâ, A.M., Plenchette, C., Danthu, P., Duponois, R. and Guissou, T. (2000) Functional compatibility of two arbuscular mycorrhizae within thirteen fruit trees in Senegal. *Agroforestry Systems* 50, 95–105.

Barner, H. and Willan, R.L. (1983) *The concept of seed zones.* Technical Note No. 16. Danida Forest Seed Centre, Halebaek, Denmark.

Böhringer, A., Ayuk, E., Katanga, R. and Ruvuga, S. (2003) Farmer nurseries as a catalyst for developing sustainable land use systems in southern Africa. Part A: Nursery establishment, productivity and early impact. *Agricultural Systems* 77, 187–201.

Bolan, N.S. (1991) A critical review on the role of mycorrhizal fungi in the uptake of phosphorus by plants. *Plant and Soil* 134, 189–207.

Carandang, W.M., Tolentino, E.L. and Roshetko, J.M. (2007) Smallholder tree nursery operations in southern Philippines: supporting mechanisms for timber tree domestication. *Forests, Trees and Livelihoods* 17.

Chilanga, T., Akinnifesi, F.K. and Brandi-Hanssen, E. (2002) Tree germplasm demand and supply for wider adoption of agroforestry by smallholders in southern Malawi. In: Kwesiga, F., Ayuk, E. and Agumya, A. (eds) *Proceedings of the 14th Southern African Regional Planning and Review Meeting.* SADC-ICRAF Zambezi Basin Agroforestry Project, 3–7 September 2001, Harare, pp. 140–147.

Dawson, I. and Were, J. (1997) Collecting germplasm from trees – some guidelines. *Agroforestry Today* 9, 6–9.

Eriksson, G., Namkoong, G. and Roberds, J. (1993) Dynamic gene conservation for uncertain futures. *Forest Ecology and Management* 62, 15–37.

Eyog-Matig, O., Kigomo, B. and Boffa, M.J. (eds) (2001) Recent research and development in genetic resources. In: *Proceedings of the Training Workshop on the Conservation and Sustainable Use of Forest Genetic Resources in Eastern and Southern Africa, 6–11 December 1999, Nairobi, Kenya.* International Plant Genetic Resources Institute, Nairobi.

Falconer, D.S. (1989) *Introduction to Quantitative Genetics*, 3rd edn. Longman, London, 340 pp.

FAO (1993) *Conservation of Genetic Resources in Tropical Forest Management.* FAO Forestry Paper No. 107. FAO, Rome.

Gerdemann, J.W. (1968) Vesicular-arbuscular mycorrhiza and plant growth. *Annual Review of Phytopathology* 6, 397–418.

Hall, J.B., O'Brien, E.M. and Sinclair, F.L. (2002) *Sclerocarya birrea: A Monograph.*

School of Agricultural and Forest Science Publication No. 19, University of Wales, Bangor, UK, 157 pp.

Hans, A.S. (1981) *Uapaca kirkiana:* distribution, uses, ecology, variation and genetics, physiology of propagation, disease, proposals and recommendations, and other Uapaca species. Technical Report, Tree Improvement Centre. National Council for Scientific and Industrial Research, Kitwe, Zambia.

Hartmann, H.T., Kester, D.E., Davies, F.T., Jr and Geneve, R.L. (1997) Propagation by specialized stems and roots. In: Hartmann, H.T., Kester, D., Davies, F. and Geneve, R. (eds) *Plant Propagation: Principles and Practices*, 6th edn. Prentice Hall, New Jersey, pp. 520–540.

Högberg, P. (1982) Mycorrhizal associations in some woodland and forest trees and shrubs in Tanzania. *New Phytologist* 92, 407–415.

Högberg, P. and Piearce, G.D. (1986) Mycorrhizas in Zambian trees in relation to host taxonomy, vegetation type and succession patterns. *Journal of Ecology* 74, 775–785.

ICRAF (1997) *Annual Report 1996.* International Centre for Research in Agroforestry (ICRAF), Nairobi.

Jaenicke, H., Simons, A.J., Maghembe, J.A. and Weber, J.C. (2001) Domesticating indigenous fruit trees for agroforestry. *Acta Horticulturae* 523, 45–51.

Kemp, R.H. (1992) The conservation of genetic resources in managed tropical forests. *Unasylva* 43(169), 34–40.

Kwapata, A.J. (1995) The potential use of tissue culture. In: Maghembe, A.J., Ntupanyama, Y. and Chirwa, P.W. (eds) Improvement of indigenous fruit trees of the miombo woodlands of southern Africa. In: *Improvement of Indigenous Fruit Trees of the Miombo Woodlands of Southern Africa.* Proceedings of a Conference held on 23–27 January 1994, Mangochi, Malawi. ICRAF, Nairobi.

Kwesiga, F. and Mwanza, S. (1995) Underexploited wild genetic resources: the case of indigenous fruits in Zambia. In: Maghembe, A.J., Ntupanyama, Y. and Chirwa, P.W. (eds) *Improvement of Indigenous Fruit Trees of the Miombo Woodlands of Southern Africa.* Proceedings of a Conference held on 23–27 January 1994, Mangochi, Malawi. ICRAF, Nairobi.

Kwesiga, F., Akinnifesi, F.K., Ramadhani, T., Kadzere, I. and Saka, J. (2000) Domestication of indigenous fruit trees of the miombo in southern Africa. In: Shumba, E.M., Luseani, E. and Hangula, R. (eds) *Proceedings, of a SADC Tree Seed Centre Network Technical Meeting, Windhoek, Namibia, 13–14 March 2000.* Co-sponsored by CIDA and FAO, pp. 8–24.

Leakey, R.R.B., Tchoundjeu, Z., Schreckenberg, K., Shackleton, S.E. and Shackleton, C.M. (2005) Agroforestry tree products (AFTPs): targeting poverty reduction and enhanced livelihoods. *International Journal for Agricultural Sustainability* 3, 1–23.

Macdonald, B. (1986) *Practical Woody Plant Propagation for Nursery Growers. Vol. 1.* Timber Press, Portland, Oregon, 669 pp.

Maghembe, A.J. (1995) Achievements in the establishment of indigenous fruit trees of the miombo woodlands of southern Africa. In: Maghembe, A.J., Ntupanyama, Y. and Chirwa, P.W. (eds) *Improvement of Indigenous Fruit Trees of the Miombo Woodlands of Southern Africa.* Proceedings of a Conference held on 23–27 January 1994, Mangochi, Malawi. ICRAF, Nairobi, pp. 39–49.

Maghembe, A.J., Simons, A.J. and Kwesiga, F. (eds) (1998) Selecting indigenous fruits trees for domestication in Southern Africa: priority setting with farmers in Malawi, Tanzania, Zambia and Zimbabwe. International Centre for Research in Agroforestry, Nairobi, 94 pp.

Maliro, M.F.A. and Kwapata, M.B. (2004) Impact of deforestation on diversity of wild and semi-wild edible fruit tree species in southern Malawi. In: Rao, M.R. and Kwesiga, F.R. (eds) *Proceedings of the Regional Agroforestry Conference on Agroforestry Impacts on livelihoods in Southern Africa: Putting Research into Practice.* World Agroforestry Centre (ICRAF), Nairobi, pp. 27–33.

Mateke, S. (2003) *Cultivation of Native Fruit Trees of Kalahari Sandveld: Studies on the Commercial Potential, Interactions between Soil and Biota in Kalahari Sands of Southern Africa*. Veld Products, Botswana, 49 pp.

Mateke, S.M. and Tshikae, P.B. (2002) Selection and cultivation of Kalahari native edible fruit producing trees: Botswana's eight years experience. Paper presented at the Southern Africa Agroforestry Conference, 20–24 May 2002. Warmbaths, South Africa.

McGregor, J. (1995) Gathered produce in Zimbabwe's communal areas: changing resource availability and use. *Ecology of Food and Nutrition* 33, 163–193.

Mng'omba, S.A. (2007) Development of clonal propagation protocols for *Uapaca kirkiana* and *Pappea capensis*, two southern African trees with economic potential. PhD thesis, University of Pretoria, South Africa, 199 pp.

Mng'omba, S.A., du Toit, E.S., Akinnifesi, F.K. and Venter, H.M. (2007a) Histological evaluation of early graft compatibility in *Uapaca kirkiana* Muell Arg. scion/stock combinations. *HortScience* 42, 1–5.

Mng'omba, S.A., du Toit, E.S., Akinnifesi, F.K. and Venter, H.M. (2007b) Repeated exposure of jacket plum (*Pappea capensis*) micro-cuttings to indole-3-butyric acid (IBA) improved in vitro rooting capacity. *South African Journal of Botany* 73, 230–235.

Mhango, J. (2000) Soil and nutrient requirements for early growth of *Uapaca kirkiana* seedlings. In: Akinnifesi, F.K. *et al.* (eds) *Achievements in agroforestry research and development in Malawi*. Annual Report 2000. SADC-ICRAF, Harare, pp. 62–67.

Mhango, J. and Akinnifesi, F.K. (2001) On-farm assessment, farmer management and perception of priority indigenous fruit trees in southern Malawi. In: Kwesiga, F., Ayuk, E. and Agumya, A. (eds) *Proceedings of 14th Southern African Regional Review and Planning Workshop, 3–7 September 2001, Harare, Zimbabwe*. International Centre for Research in Agroforestry, Nairobi, pp. 157–164.

Mkonda, A., Lungu, S., Maghembe, J.A. and Mafongoya, P.L. (2003) Fruit- and seed germination characteristics of *Strychnos cocculoides* an indigenous fruit tree from natural populations in Zambia. *Agroforestry Systems* 58, 25–31.

Mkonda, A., Akinnifesi, F.K., Maghembe, J.A., Swai, R., Kadzere, I., Kwesiga, F.R., Saka, J., Lungu, S. and Mhango, J. (2004) Towards domestication of 'wild orange' *Strychnos cocculoides* in southern Africa: a synthesis of research and development efforts. In: Rao, M.R. and Kwesiga, F.R. (eds) *Proceedings of Regional Agroforestry Conference on Agroforestry Impacts on Livelihooods in Southern Africa: Putting Research into Practice*. World Agroforestry Centre (ICRAF), Nairobi, pp. 67–76.

Mojeremane, W. and Tshwenyane, S.O. (2004) The role of morula (*Sclerocarya birrea*): a multipurpose indigenous fruit tree of Botswana. *Journal of Biological Sciences* 4, 771–775.

Moyersoen, B. and Fitter, A. (1999) Presence of arbuscular mycorrhizas in typically ectomycorrhizal host species from Cameroon and New Zealand. *Mycorrhiza* 8, 247–253.

Msanga, H.P. and Maghembe, J.A. (1989) Physical scarification and hydrogen peroxide treatments improve germination of *Vangueria infausta* seed. *Forest Ecology and Management Journal* 28, 301–308.

Msanga, H.P. (1998) *Seed Germination of Indigenous Trees in Tanzania*. Canadian Forest Service, Edmonton. UBC Press, 292 pp.

Mudge, K.W. and Brennan, E.B. (1999) Clonal propagation of multipurpose and fruit trees used in agroforestry. In: Buck, L.E., Lassoie, J.P. and Fernandes, E.C.M. (eds) *Agroforestry in Sustainable Agricultural Systems*. CRC Press, Boca Raton, Florida, pp. 157–190.

Mulawarman, A., Roshetko, J.M., Sasongko, S.M, and Irianto, D. (2003) *Tree Seed Management – Seed Sources, Seed Collection and Seed Handling: A Field Manual for Field Workers and Farmers*. International Centre for Research in Agroforestry (ICRAF) and Winrock International, Bogor, Indonesia.

Muller-Samann, K.M. and Kotschi, J. (1994) *Sustaining Growth: Soil Fertility Management in Tropical Smallholdings.* Technical Centre for Agricultural and Rural Cooperation (CTA) and Deutsche Gesellschaft fur Technische Zusammenarbeit (GTZ), Mergraf Verlag, Germany, 498 pp.

Mwabumba, L. and Sitaubi, L. (1995) Seed pretreatment, growth and phenology of some indigenous fruit trees of the miombo ecozone at Naugu Forestry Reserve, Malawi. In: Maghembe, A.J., Ntupanyama, Y. and Chirwa, P.W. (eds) *Improvement of Indigenous Fruit Trees of the Miombo Woodlands of Southern Africa.* Proceedings of a Conference held on 23–27 January 1994, Mangochi, Malawi. ICRAF, Nairobi, pp. 66–72.

Mwamba, C.K. (1995) Variation in fruits of *Uapaca kirkiana* and effects of in-situ silvicultural treatments on fruit parameters. In: Maghembe, A.J., Ntupanyama, Y. and Chirwa, P.W. (eds) *Improvement of Indigenous Fruit Trees of the Miombo Woodlands of Southern Africa.* Proceedings of a Conference held on 23–27 January 1994, Mangochi, Malawi. ICRAF, Nairobi, pp. 27–38.

Naylor, J.M. (1983) Studies on the genetic control of some physiological processes in seeds. *Canadian Journal of Botany* 61, 3561–3567.

Ndugu, J., Janaeke, H. and Boland, D. (1995) Considerations for germplasm collection of indigenous fruit trees in the miombo. In: Maghembe, A.J., Ntupanyama, Y. and Chirwa, P.W. (eds) *Improvement of Indigenous Fruit Trees of the Miombo Woodlands of Southern Africa.* Proceedings of a Conference held on 23–27 January 1994, Mangochi, Malawi. ICRAF, Nairobi, pp. 1–11.

Ngulube, M.R. (1996) Ecology and management of *Uapaca kirkiana* in southern Africa. PhD thesis. School of Agricultural and Forest Sciences, University of Wales Bangor, UK.

Ngulube, M.R. and Kananjii, B. (1989) *Seed problems of edible indigenous fruits in Malawi.* Forestry Research Institute of Malawi, Zomba, Malawi, 16 pp.

Ngulube, M.R., Hall, J.B. and Maghembe, J.A. (1995) Ecology of a miombo fruit tree: *Uapaca kirkiana* (Euphorbiaceae). *Forest Ecology and Management* 77, 107–117.

Nkanaunea, G.A., Kwapata, M.B., Bokosi, J.M. and Maliro, M.F.A. (2004) The effect of age and type of rootstock, and light intensity on the success of *Uapaca kirkiana* micrografts. In: Rao, M.R. and Kwesiga, F.R. (eds) *Proceedings of Regional Agroforestry Conference on Agroforestry Impacts on Livelihooods in Southern Africa: Putting Research into Practice.* World Agroforestry Centre, Nairobi, pp. 55–58.

Osborne, D.J. (1981) Dormancy as a survival stratagem. *Annals of Applied Biology* 98, 525–531.

Palmberg-Lerche, C. (1997) *Towards a Coherent Framework for the Conservation and Sustainable Utilization of Forest Genetic Resources.* Forest Genetic Resources Information No. 25, pp. 15–18.

Place, F. and Dewees, P. (1999) Policies and incentives for the adoption of improved fallows. *Agroforestry Systems* 47, 323–434.

Prins, H. and Maghembe, J.A. (1994) Germination studies of fruit trees indigenous to Malawi. *Forest Ecology and Management* 64, 111–125.

Ramachela, K. (2001) Ecology and patterns of distribution of mycorrhizal fungi in the *Uapaca kirkiana* woodland ecosystem. *Discovery and Innovation.* Special edition, 127–138.

Roodt, V. (1998) *Trees and Shrubs of the Akavango Delta: Medicinal Uses and Nutritional Value.* Shell Oil Botswana, Gaborone.

Roshetko, J. and Verbist, B. (2000) *Tree Domestication.* Lecture Note 6. International Centre for Research in Agroforestry (ICRAF), South Asian Regional Research Programme, Bogor, Indonesia, 17 pp.

Shackleton, C.M. (2004) Use and selection of *Sclerocarya birrea* (marula) in the Bushbuckridge lowveld, South Africa. In: Rao, M.R. and Kwesiga, F.R. (eds) *Proceedings of Regional Agroforestry Conference on Agroforestry Impacts on Livelihooods in Southern Africa: Putting*

Research into Practice. World Agroforestry Centre (ICRAF), Nairobi, pp. 77–92.

Sileshi, G. (2007) Evaluation of statistical procedures for efficient analysis of insect, disease and weed abundance and incidence data. *East African Journal of Science* 1, 1–9.

Sileshi, G., Akinnifesi, F.K., Mkonda, A. and Ajayi, O.C. (2007) Effect of growth media and fertilizer application on biomass allocation and survival of *Uapaca kirkiana* Müell Arg seedlings. *Scientific Research and Essay* 2(8), in press.

Simons, A.J. (1996) ICRAF's strategy for domestication of non-wood tree products. In: Leakey, R.R.B., Temu, A.B., Melnyk, M. and Vantomme, P. (eds) *Domestication and Commercialization of Non-timber Forest Products in Agroforestry Systems. Non-Wood Forest Products No. 9*. FAO, Rome, pp. 8–22.

Simons, A.J. and Leakey, R.R.B. (2004) Tree domestication in tropical Agroforestry. *Agroforestry Systems* 61, 167–181.

Soumaré, A., Groot, J.J.R., Koné, D. and Radersma, S. (1994) Structure spatiale du système racinaire de deux arbres du Sahel: *Acacia seyal* and *Sclerocarya birrea*. Rapports du project production Soudano-Saheliènne, Agricultural University, Wageningen 5, 1–36.

Sutherland, J.R., Diekmann, M. and Berjak, P. (2002) *Forest Tree Seed Health*. IPGRI Technical Bulletin No. 6. International Plant Genetic Resources Institute, Rome, 85 pp.

Swai, R.E.A. (2002) Efforts towards domestication and commercialization of indigenous fruits of the miombo woodland in Tanzania: research and development highlights. In: Kwesiga, F., Ayuk, E. and Agumya, A. (eds) *Proceedings of the 14th Southern Africa Regional Review and Planning Workshop, 3–7 September 2001, Harare, Zimbabwe*. ICRAF, Nairobi, pp. 149–156.

Swai, R.E.A., Maduka, S.M., Mbwambo, L., Mumba, M. and Otsyina, R. (2004) Domestication of indigenous fruit and medicinal tree species in Tanzania: a synthesis. In: Rao, M.R. and Kwesiga, F.R. (eds) *Proceedings of Regional Agroforestry Conference on Agroforestry Impacts on Livelihooods in Southern Africa: Putting Research into Practice*. World Agroforestry Centre (ICRAF), Nairobi, pp. 59–65.

Taylor, F. and Kwerepe, B. (1995) Towards the domestication of some indigenous fruit trees in Botswana. In: Maghembe, A.J., Ntupanyama, Y. and Chirwa, P.W. (eds) *Improvement of Indigenous Fruit Trees of the Miombo Woodlands of Southern Africa*. Proceedings of a Conference held on 23–27 January 1994, Mangochi, Malawi. ICRAF, Nairobi, pp. 113–134.

Taylor, F.W., Matoke, S.M. and Butterworth, K.J. (1996) A holistic approach to the domestication and commercialization of non-timber forest products. In: Leakey, R.R.B., Temu, A.B., Melnyk, M. and Vantomme, P. (eds) *Domestication and Commercialization of Non-timber Forest Products in Agroforestry Systems. Non-Wood Forest Products No. 9*. FAO, Rome.

Tchoundjeu, Z., Asaah, E.K., Anegbeh, P., Degrande, A., Mbile, P., Facheux, C., Tsobeng, A., Atangana, A.R., Ngo-Mpeck, M.L. and Simons, A.J. (2006) Putting participatory domestication into practice in West and Central Africa. *Forests, Trees and Livelihoods* 16, 53–69.

Teichman, I. von, Small, J.G.C. and Robbertse, P.J. (1986) A preliminary study on the germination of *Sclerocarya birrea* subsp. *caffra*. *South African Journal of Botany* 52, 145–148.

Thoen, D. and Ba, A.M. (1989) Ecto-mycorrhizas and putative ecto-mycorrhizal fungi of *Afzelia africana* Sm. and *Uapaca guineensis* Mull. Arg. in southern Senegal. *New Phytologist* 113, 549–559.

Weber, J.C., Sotelo-Montes, C., Vidaurre, H., Dawson, I.K. and Simons, A.J. (2001) Participatory domestication of agroforestry trees: an example from Peruvian Amazon. *Development in Practice* 11, 425–433.

Werker, E. (1980) Seed dormancy as explained by the anatomy of embryo envelopes. Impermeability to water, impermeability to oxygen, and mechanical resistance to radicle protrusion. *Israel Journal of Botany* 29, 22–44.

Wightman, K.E. (1999) Good tree nursery practices: practical guidelines for community nurseries. International Centre for Research in Agroforestry, Nairobi, 95 pp.

20 Pest Management in Miombo Fruit Trees

G. Sileshi,[1] P. Barklund,[2] G. Meke,[3] R.R. Bandeira,[4] C. Chilima,[3] A.J. Masuka,[5] R.K. Day[6] and F.K. Akinnifesi[1]

[1]*World Agroforestry Centre, ICRAF, Lilongwe, Malawi;* [2]*Swedish University of Agricultural Sciences, Uppsala, Sweden;* [3]*Forestry Research Institute of Malawi, Zomba, Malawi;* [4]*Edwardo Mondelane University, Maputo, Mozambique;* [5]*Kutsaga Research Company, Harare, Zimbabwe;* [6]*CABI Africa Centre, Nairobi, Kenya*

20.1 Introduction

According to the International Plant Protection Convention (IPPC), a pest is defined as any species, strain or biotype of plant, animal or pathogenic agent injurious to plants or plant products (IPPC, 2006). Pests may affect fruit productivity either directly through fruit infestation or indirectly by attacking the roots, shoots and foliage, thereby interfering with the normal physiological processes of the plant. Pests of fruit trees include weedy plants and parasitic higher plants that compete with fruit trees for water, light and nutrients; herbivorous mites, insects, birds and mammals that physically feed on the plant; and pathogenic organisms (viruses, bacteria, mycoplasmas, fungi, certain algae and plant-parasitic nematodes, etc.) that cause diseases. A disease can be defined as any physiological disturbance of the normal functioning of a plant as a result of a detrimental interaction between the pathogen, environment and host. Diseases affect the production and utilization of all fruits by reducing the health of the fruit-bearing plant and directly reducing yield, quality or storage life (Ploetz *et al*., 1994). Disease can be caused by living agents (biotic diseases) such as fungi, bacteria, viruses, or by non-living agents (abiotic diseases) such as fire, drought, frost and lightning. Decline diseases, the third category of diseases, are those whose causal factors are both biotic and abiotic agents.

Many of the potential pests that affect miombo fruit trees are natural components in the ecosystem of the miombo and, until the commercialization of some of the species, were regarded as inevitable but acceptable factors determining tree species mix, growth rate, and so on. However, as soon as forest species are regarded as commercially valuable, the perception of the importance of insects and diseases takes on a new dimension. For example, in Botswana the domestication of *Vangueria infausta* is seriously hampered by a mite that causes galls on the leaves. The mite spreads easily and quickly if the trees are grown at high density and severe infection will probably affect production adversely (Taylor *et al.*, 1996).

In their native habitat, indigenous fruit trees grow in mixed age classes along with a mixture of other species (Campbell, 1987). However, domestication of fruit trees changes the focus and introduces artificial, user-executed selection, promotion of the domesticated species and gradual elimination of 'useless' species from the ecosystem. Such a process tends to produce less resilient monocultures, prone to increased levels of damage by pests and pest outbreaks as a result of lost pest-buffering capacity of the system. A good example is the commercialization of *Ziziphus mauritiana*, an indigenous fruit tree of the Indian subcontinent (Pareek, 2001). The expected increase in adoption and expansion of tree planting carries the potential risk that new pests may occur (Mchowa and Ngugi, 1994; Leakey and Simons, 1998). The commercialization of indigenous fruits is also bound to lead to a demand for higher-quality produce and lower tolerance of pest damage.

Information on pests and diseases of the edible wild fruit trees of the miombo in Southern Africa is limited. The records that are available are based on *ad hoc* studies of the pests of these fruit trees. In this review we summarize the available information on pests and diseases of priority miombo fruit tree species (Maghembe *et al.*, 1994) based on published literature and current studies, identify gaps and suggest management options. The pest complex of most wild miombo fruit tree species is not known and there is little, if any, published information on pest biology and population dynamics. The bulk of the information in the following discussion is based on observations from our ongoing work in eastern Zambia, southern Malawi, southern Mozambique and Zimbabwe.

20.2 Seed and Seedling Pests

Seed damage is almost exclusively caused by fungi, among which two biological groups can be recognized. One of these includes numerous non-specific moulds, such as species of the genera *Alternaria*, *Fusarium* and *Penicillium*, which, when the air humidity is high, infest the seed coat, or penetrate the seed coat after damage by insects or other agents. The other group of seed fungi consists of specialists from the genera *Rhizoctonia*, which can attack intact seeds and cause internal rotting. *Stemphylium* sp. has been mentioned as a seed-borne fungus affecting *Parinari* (Masuka *et al.*, 1998). Diseases of seeds and germinants are often overlooked in the nursery as the affected seeds are not visible and consequently the losses are often attributed to 'poor seed'. However, to improve germination in the nursery, seeds need to be examined and the problem characterized before management methods to reduce the problem can be applied.

A number of insects and diseases (see Section 20.3) attack the seedlings of indigenous fruit trees. Disease of rotting of seeds and germinants is often characterized as damping-off, a common nursery disease that affects seeds, germinants and seedlings of most plant species. Damping-off is caused by a number of fungi, which can be either soil-borne or seed-borne. Seed-borne fungi are carried with, on or in the seed. The symptoms of seedling rot caused by damping-off fungi vary with the age and stage of development. Root rot of older seedlings is a late seedling rot that manifests itself as partial or complete destruction of the root system. Stunted growth can be a prolonged effect of a partially destroyed root system. This late-stage seedling rot is often not recognized, being taken for drought or nematode damage. At least 30 different species of fungi are known to cause damping-off in tree seedlings, mainly representatives of the genera *Fusarium*, *Pythium*, *Phytophthora* and *Rhizoctonia*. Several different species of *Fusarium* can cause root rot of tree seedlings. Among the different *Fusarium* fungi, *F. oxysporum* is the most common seedling fungus and it usually causes post-emergence damping-off. One of the main sources of *Fusarium* inoculum is the seed, although other sources are also important. Some species are more aggressive than others. Less aggressive fungi, although living on the plant, may not develop disease until the seedling becomes stressed by moisture, drought or heat. Therefore, the cultural practice of moisture-stressing seedlings to harden them to drier conditions has to be applied gradually so as not to cause fungal attack.

The water mould fungi, species of *Pythium* and *Phytophthora*, are commonly occurring pathogens that cause damping-off root rot in nurseries, and tree seedlings (like other plant seedlings) are susceptible. For instance, damping-off and root collar restriction are common in *Sclerocarya birrea* nurseries (Hall *et al.*, 2002). The most important of the factors that increase damping-off in container nurseries is dirty seed. Fungal contamination is more common in dirty seed lots, which produce weak germinants that are particularly susceptible to damping-off. Because many of the fungi responsible for damping-off are seed-borne, seeds can be sanitized prior to sowing.

20.3 Field Pests of Miombo Fruit Trees

Pests attack every part of mature fruit trees, causing damage of varying degree. Many ecologists have found it useful to group pests into feeding guilds in order to study the ecological interactions between pests, their hosts, their natural enemies, and climate (Speight *et al.*, 1999). Accordingly, pests of miombo fruit trees have been conveniently divided into three major guilds: root-damaging; stem- and leaf-damaging; and flower-, fruit- and seed-damaging. Tentative numbers of insect pest species recorded as causing damage to various parts of the major miombo fruit trees are presented in Table 20.1. These records were generated from the Agroforestry Pest Database (G. Sileshi, unpublished results).

20.3.1 Root-damaging pests

Fungal diseases

Among the root-infecting fungi, the honey fungus *Armillaria* is probably the most important in miombo fruit trees. *Parinari curatellifolia* subsp. *curatellifolia*,

Table 20.1. The number of insect pest species recorded on the priority miombo fruit tree species of Southern Africa.

Priority miombo fruit tree species	Root-damaging species	Leaf- and stem-damaging species			Fruit- and seed-damaging species		Total species
		Defoliating	Sap-sucking	Stem boring	Fruit-feeding	Seed-feeding	
Uapaca kirkiana	1	17	11	4	7	1	41
Parinari curatellifolia	1	14	4	?	3	1	23
Adansonia digitata	1	3	8	1	1	–	14
Strychnos cocculoides	–	–	1	–	–	–	1
Anisophyllea boehmii	–	1	–	–	–	–	1
Azanza garckeana	–	16	2	2	1	–	21
Flacourtia indica	–	5	4	–	–	–	9
Syzygium guineense	–	2	1	–	1	–	4
Syzygium cordatum	–	17	1	–	–	–	18
Uapaca nitida	–	2	–	–	–	–	2
Vangueria infausta	–	6	–	–	3	–	9
Annona senegalensis	–	6	1	–	5	–	12
Sclerocarya birrea	–	25	5	16	5	–	51

–, records not available.

P. curatellifolia subsp. *mobola*, *Uapaca kirkiana* and *U. nitida* are reported to be susceptible to *Armillaria* root rot in Zambia (Parker, 1978) and in Malawi (Lee, 1970). *Armillaria* consists of several species, which may differ in host range as well as in pathogenicity (Strouts and Winter 1994). The fungus kills the inner bark and cambium and later invades the wood. The tree will die as a result of girdling at the root collar or extensive root killing. *Armillaria* is a potential threat to proposed orchards of wild fruit, particularly where miombo woodlands have been cleared to provide plantation sites (Parker, 1978). The reason is that *Armillaria* species affect many species in the indigenous forest. For instance, in Zimbabwe they are endemic in miombo forests.

Other root and butt rot fungi such as *Fomes* and *Ganoderma* spp. develop their fruit-bodies, called brackets, close to the stem base or on roots (Parker, 1978). *Ganoderma lucidum* affects *Parinari curatellifolia* in Zimbabwe (Masuka *et al.*, 1998). These fungi probably infect roots by means of airborne spores washed into the soil. A different type of root rot is caused by *Phytophthora* spp., also attacking seedlings in the nursery and causing rot of succulent roots. In affected trees many small roots die and necrotic brown lesions in the inner bark are often present on the larger roots and in the lower trunk. Trees suffering from such root rots often begin showing symptoms of drought, and become weakened and susceptible to attack by other pathogens or various other causes that are then mistakenly taken as the causes of the death of the tree. These fungi grow and produce spores under moist soil conditions, and as the spores disperse with the soil water the roots of susceptible hosts will be infected. Losses caused by *Phytophthora* are great on fruit trees, especially in soils with bad water drainage. There is a dearth of quantitative data on the losses caused on miombo fruit trees. However, *Phytophthora* root rot is generally regarded as the most serious soil-borne disease of citrus and avocados (Ploetz *et al.*, 1994). In South Africa, different *Phytophthora* species attack *Eucalyptus* and *Acacia mearnsii* (Roux and Wingfield, 1997) and in this way these exotics can be a source of infection also for miombo fruit trees. Root damage caused by insects can also give access to the tree root system for fungi which are not able to enter through undamaged bark. For example, *Fusarium* spp. can under such circumstances cause substantial damage.

Insects

Termites are probably the most serious pests causing mortality of tree seedlings after outplanting in the field. Termites attack plants at the base of the stem, ring-barking or cutting them completely. Termites are not host-specific and most often attack stressed plants (Sileshi, 2000). The most important termite pest genera in Africa include *Odontotermes*, *Macrotermes*, *Pseudacanthotermes*, *Microtermes*, *Ancistrotermes*, *Allodontermes*, *Amitermes*, *Trinervitermes* and *Hodotermes* (Uys, 2000). Damage is due to feeding either under soil sheeting on the outer surface of the plants or on the roots. White grubs (larvae of scarabaeid beetles) also affect roots of tree seedlings such as *Sclerocarya birrea* (G. Sileshi, unpublished results).

Plant-parasitic nematodes

Nematodes are tiny, worm-like, animals mostly living freely in water and soil. Plant-parasitic nematodes are major pests of fruit trees, interacting with pathogenic microorganisms in disease complexes, and in some instances constituting the main cause of damage to plants (Keetch, 1989; Kleynhans *et al.*, 1996). Much of the damage associated with nematodes goes unnoticed or is attributed to other factors, mainly because growers are unaware of nematodes (Kleynhans *et al.*, 1996). Plant-parasitic nematodes can affect fruit trees in a variety of ways. Uptake of nutrients and water by a plant can be drastically affected by direct feeding or root infestations of nematodes. Nematode feeding also creates open wounds, which provide entry to a wide variety of plant-pathogenic fungi and bacteria. These microbial infections may result in greater losses than the damage from nematodes. Because nematodes damage roots, any condition which stresses the plant, such as drought, flooding, nutrient deficiencies or soil compaction, will tend to amplify the damage symptoms noted above. Failure of plants to respond normally to fertilizers and slower than normal recovery from wilting are signs of nematode infestation. Below-ground attacks are mostly recognized as wilting of above-ground parts.

Little is known about root-attacking nematodes in the miombo forest. Plant-parasitic nematodes have been found in soil samples from baobab (*Adansonia digitata*) in East Africa (Coleman *et al.*, 1991). In Senegal, the baobab has been found to be a host of root-knot nematodes (*Meloidogyne* sp.) and *Rotylenchulus reniformis*, and therefore baobab trees, although not showing damage, may be an important reservoir for these nematodes (Taylor *et al.*, 1978). Nematodes cause shortening and malformation of roots and subsequent death of *Sclerocarya birrea* trees (Hall *et al.*, 2002). Nematode problems are well known in most exotic fruit trees. For instance, apple, peach, citrus and mango are respectively affected by 40, 41, 35 and 20 different species of nematodes in southern Africa (Kleynhans *et al.*, 1996). On the other hand, the plant-parasitic nematodes affecting fruit trees of the miombo are little known.

20.3.2 Trunk- and leaf-damaging pests

Fungal diseases

Trunk-damaging fungi largely cause diseases of three kinds: dieback and canker diseases; vascular wilt diseases; and wood rot diseases. Bark-infecting fungi are seen to cause fluxing (oozing), necrosis, and canker of small branches and shoots. In Zambia such damage has been reported to contribute to dieback of *Uapaca* spp. (Parker, 1978). The bark-infecting genus *Botryosphaeria* includes many species, some of which are parasitic and cause dieback and canker diseases on woody hosts. Some are host-specific and others have a wider host range. The dominating species, *Botryosphaera dothidea*, is known to attack over 100 species of importance in the region, among them apple, eucalypts, grevillea, guava, loquat, macadamia, mango, maple, mulberry, papaya, peach and pear (Sinclair *et al.*, 1989). *Botryosphaeria ribis* attacks avocado, guava and pear in Malawi (Lee, 1970). *Botryosphaeria* species infect trees which are weakened by drought. Although there are many other dieback and canker fungi, *Botryosphaeria* spp. are obvious candidates for attacking miombo fruit trees.

Cryphonectria cubensis is the causal agent of eucalyptus canker in many countries throughout the world. The miombo species *Syzygium cordatum* is host of *Cryphonectria cubensis* in South Africa and a source of infection for eucalyptus canker there. *Cryphonectria cubensis* has been shown to be a potential pathogen of guava (*Psidium guajava*) and other exotic and indigenous members of the family Myrtaceae (Swart *et al.*, 1991).

A well-known vascular wilt disease of fruit trees such as avocado is caused by *Verticillium*. The causal organism, *Verticillium dahliae*, is soil-borne and enters the roots and invades the tree's vascular system, causing sudden wilt of branches or whole trees. *Ceratocystis* causes wilt disease but mostly infects the trunk with the help of a vector, as in the case of Dutch elm disease, transmitted by a bark beetle in Europe and North America. Also, pruning wounds can be infection points. Fungal decay of the wood is a common feature where the trunk or limbs of a tree have been damaged by other agents. A wound is the necessary entry point into the wood for these fungi. *Phellinus* spp. and *Daedalea* spp. are examples of pathogenic decay fungi, evident as brackets on trunks and branches. When the fruit bodies appear on the trunk a considerable rot has already developed in the trunk. Saprophytic wood decay fungi such as *Schizophyllum commune* frequently fruit on dead wood on trunks and can be seen throughout the year on branches of *Uapaca kirkiana* (Parker, 1978). However, the decomposition of wood starts in living trees.

On foliage, fungal infections give rise to leaf spots, shot-holes, blotch or more extensive necrosis. Tar spot of leaves caused by *Phaeochorella parinari* is common in *Parinari* species in Zambia (Parker, 1978) and Zimbabwe (Masuka *et al.*, 1998). Leaf spot of *Sclerocarya birrea* caused by *Alternaria longissima*, *Cochliobolus bicolor* and *Glomerella cingulata* is common in Malawi (G. Meke, unpublished results). In southern Malawi, *Colletotrichum gloeosporioides*, *Alternaria obtecta* and *Phoma* spp. caused leaf spots on *Uapaca kirkiana* (G. Meke, unpublished results; G. Sileshi, unpublished results). *Colletotrichum* also affects *Annona* species in Zimbabwe (Masuka *et al.*, 1998). A smaller leaf spot of *Uapaca kirkiana* caused by *Cercospora* is common (Parker, 1978). Brown leaf spot is very common on *Uapaca kirkiana* and almost all plants observed had this symptom. *Phoma* sp., *Fusarium pallidoroseum* and *Fusarium solani* have been isolated from infected *Uapaca kirkiana* plants but *Fusarium* species are unlikely to be the causal agent of the leaf spot symptoms (G. Meke, unpublished results). Widespread necrosis of older leaves of *Uapaca kirkiana* has been linked to *Pestalotiopsis versicolor. Pestalotiopsis* fungi are mostly non-pathogenic, and probably are opportunistic invaders of tissues damaged by other agents (Parker, 1978).

Rust fungi heavily infect leaves of *Flacourtia indica* (G. Sileshi, unpublished results). The powdery mildew (*Phyllactinia* sp.) that attacks seedlings of *Sclerocarya birrea* in the nursery also attacks saplings and mature trees in the field. For instance, *Phyllactinia guttata* was observed on over 80% of plants, with over 50% leaf infection in Malawi (G. Meke, unpublished results). Rusts and powdery mildews are host-specific obligate parasites, and may cause serious damage. It is necessary to study their ecology and biology to be able to find the best way to keep infections under control.

Colletotrichum gloeosporioides has been reported to cause necrotic leaf spot of *Annona cherimola* (Pennisi and Agosteo, 1994), leading to leaf fall. *C. gloeosporioides* causes anthracnose in mango, and it affects leaves, flowers and fruits. In India *C. gloeosporioides* also causes fruit rot of *Ziziphus mauritiana* (Gupta and Madaan, 1977). *C. gloeosporioides* is an example of a fungus that has spread worldwide and it is a highly variable species with many strains. This fungus can spread between different tree species and can cause great losses. Many tropical fruits are affected by the fungus, and other miombo fruit trees are not likely to be exempt.

Sooty mould fungi can sometimes be found covering the surface of the leaf and surrounding the petiole and shoot of *Uapaca*, *Sclerocarya* and *Strychnos*, especially those infested with aphids and scale insects (G. Sileshi, unpublished results). However, sooty moulds are not parasitic. The presence of sooty moulds is usually of rather minor importance to the health of the plant, but it does indicate the presence of insects and may be a warning of a severe aphid or scale problem. On the other hand, sooty mould of baobab (*Adansonia digitata*), possibly caused by *Antennulariella* (or its *Capnodendron* state), has been cited as a killing disease. Since the mid-1990s large numbers of baobab trees in Zimbabwe have been reported to be dying. Affected trees exhibit a strikingly blackened or burnt appearance, hence the colloquial term 'sooty baobab'. The phenomenon is now regarded as episodic and related to lengthy periods of below-average rainfall, aggravated lately by increasingly intensive land use in arid areas (Piearce *et al.*, 1994).

Parasitic higher plants

In natural stands and on farms, branches of *Uapaca kirkiana* are sometimes attacked by the parasitic flowering plants of the family Loranthaceae. The species commonly seen are *Agelanthus subulatus*, *Phragmanthera cornetii*, *Tapinanthus dependens* and *Viscum congdonii* (Ngulube *et al.*, 1995). *Sclerocarya birrea* is similarly attacked by *Agelanthus crassifolius*, *A. prunifolius*, *Erianthemum dregei*, *Helixanthera garciana*, *Pedistylis galpinii* and *Tapinanthus globiferus* (Hall *et al.*, 2002). Many of these parasitic plants cause the death of twigs and branches, but without any counter-measure the entire tree can also be killed in a few years. The effect of these parasites on the fruit load of miombo trees is still not quantified.

Mites and insects

Mites are ubiquitous pests of agricultural crop plants in southern Africa (Meyer, 1981). Several species of mites are known to attack crop plants, some of which are among the most serious pests. Some mites, such as *Calacarus citrifolii*, *Brevipalpus* spp., *Oligonychus coffeae*, *Tetranychus lombardinii* and *T. cinnabarinus*, are known to attack hundreds of different plant species each (Meyer, 1981), and miombo fruit trees are likely to be affected by some of these species. However, little is known about the mite pests of indigenous miombo fruit trees. The only reported cases concern unidentified mites damaging young shoots of *Sclerocarya birrea*, causing malformation of leaves and stems (Hall *et al.*, 2002), and *Tetranychus* mites on *Ziziphus mauritiana* (G. Sileshi, unpublished results).

Insects are responsible for perhaps the most noticeable and extensive damage to trees in miombo woodlands, such as malformations of shoots, stems and leaves. For instance, insect-induced shoot galls can stimulate the formation of a witch's broom of adventitious shoots or foliage, which can also be a result of fungal infection. For instance, leaf and shoot galls are frequent defects of *Parinari* spp. (Parker, 1978), although the causal agents have not been identified. Gall midges (Diptera: Cecidiomyiidae), gall wasps (Hymenoptera: Cynipidae) and eriophyid mites (Acarina) can induce gall formation.

Among the most destructive of foliage-feeding arthropods are insects of the orders Orthoptera (grasshoppers, locusts, etc.), Lepidoptera (butterflies and moths) and Coleoptera (beetles). Grasshoppers and locusts are highly mobile and their characteristic damage is seen in the much-eaten leaves that remain after the pests have gone. Caterpillars (lepidopteran larvae) are most easily located during their feeding, and large colonies on a single host tree have been observed to rapidly cause a high degree of defoliation (Parker, 1978). Among the Lepidoptera, perhaps the most notable defoliators are caterpillars of emperor moths (Saturniidae). For instance, caterpillars of *Bunaea alcinoe* completely defoliate *Uapaca kirkiana* trees in parts of Malawi (Meke, 1998) while those of *Argema mimosae* cause substantial defoliation of *Sclerocarya birrea*, which can reduce growth and fruit production (Van Den Berg, 1990). In the miombo woodlands of northern Zambia, caterpillars of the emperor moths *Gynanisa maja* and *Gonimbrasia zambesina* defoliate *Uapaca kirkiana*, *U. banguelensis*, *U. sansibarica*, *Anisophyllea boehmii*, *Parinari curatellifolia* and *Syzygium guineense* (Mbata *et al.*, 2002), though their main hosts are the *Julbernardia* species. Larvae of *Imbrasia belina* are known to defoliate *Diospyros mespiliformis* and *Sclerocarya birrea* (Parker, 1978; Mughogo and Munthali, 1995), though its preferred host is *Colophospermum mopane*. Several species of *Deudorix* defoliate *Parinari curatellifolia* (Kroon, 1999). Damage by adult beetles and their larvae is also widespread. For instance, the leaf beetle *Mycrosyagrus rosae* and scarab beetles (*Euphoresia* sp.) commonly defoliate *Uapaca kirkiana* in Zambia. A mosaic of holed leaves, often showing 10–20% defoliation and occasionally up to 50%, can be observed on *U. kirkiana* trees. However, by virtue of continued foliar production through most of the year, *U. kirkiana* trees are remarkably resilient to the effects of their defoliators (Parker, 1978).

A significant problem of crown dieback is associated with shoot-boring Lepidoptera and Coleoptera, especially Curculionidae, Scolytidae and Cerambycidae. The timber borer *Euryops burgeoni* has been recorded as a pest of *Parinari* species in forests of eastern and central Africa. In Zambia it burrows into a root of a living healthy tree, but apparently it is not yet a serious problem (Parker, 1978). As often happens with tropical trees, any basal damage to the trunk invites termite damage to the wood, which, together with the activities of decay fungi, is likely to produce a hollow core in the butt.

Among insects with sucking mouth parts, the plant bugs, scale insects and mealybugs commonly attack the leaves, shoots or branches of fruit trees. Many of the plant bugs may not cause serious damage other than limited necrosis at the sites where they feed, caused by toxic saliva. Sap-sucking by adults and nymphs of the coreid bug *Leptoglossus membranaceus* can cause wilting of branches in *Uapaca kirkiana* (Meke, 1998). Scale insects commonly infest leaves and shoots throughout the year. On *U. kirkiana*, *Ledaspis* sp. and *Ceroplastes uapacae* are abundant in Zambia and Malawi (Parker, 1978; G. Meke, unpublished results; G. Sileshi, unpublished results). *Ceroplastes uapacae* sucks sap from the lower surface of the leaf but also causes chlorotic flecks on the upper surface, and thus may be a virus transmitter.

Mammals

Wild and domestic animals cause damage by browsing and trampling. This is critical during the juvenile phase of fruit trees. In the eastern province of Zambia, mortality of *Sclerocarya birrea* seedlings planted on farmers' fields was high as a result of browsing by goats (A. Mkonda, personal communications). In the wild, elephants, eland and zebra are the main browsers of *Uapaca kirkiana* (Shorter, 1989). Though elephants have been widely reported as browsing and damaging *S. birrea*, their impact has been described as negligible (Hall *et al.*, 2002).

20.3.3 Flower- and fruit-damaging pests

Bacterial and fungal diseases

There is a dearth of information on bacteria affecting the reproductive parts of miombo fruit trees. The only example is the bacterial spot caused by *Pseudocercospora purpurea*, which affects *Sclerocarya birrea* fruits. Small green spots appear on the fruit, which enlarges, breaks open, turns black and gives off a gummy liquid (Hall *et al*., 2002). The fungi that affect flowers of miombo fruit trees are also little known. The well-known anthracnose fungus *Colletotrichum gloeosporioides* affects the fruit of mango, avocado and papaya (Ploetz *et al*., 1994), *Ziziphus* spp., *Parinari curatellifolia* and probably *S. birrea*. The fungus may disfigure the skin of 80% of *P. curatellifolia* fruits and may cause pitted, irregular-shaped or distorted fruits. Scab, corking and raised fissured areas result from infection of *Uapaca kirkiana* fruits (Parker, 1978). Thus, the quality of wild fruits may be seriously reduced by this important and widespread fungus. Other fruit-damaging fungi included rusts (*Puccinia* sp.) on fruits of *Strychnos spinosa* and *S. madagascariensis*, and *Fusarium* sp. affecting fruits of *Vangueria infausta* in Mozambique.

Mites and insects

A variety of mites and insects, such as beetles, flies and caterpillars, are involved in the destruction of blossoms, and, together with blight of the flower stalk, are responsible for flower fall and fruit failure. For instance, mites damage the flower buds of *Sclerocarya birrea*, causing flower malformation (Hall *et al*., 2002). In southern Mozambique, the pyralid caterpillar *Spatulipalpia monstrosa* and the nitidulid beetles *Carpophilus* spp. attack flowers and fruits of *Annona senegalensis* and *Vangueria infausta* (R.R. Bandeira, unpublished results). In Zambia and Zaire the main problem of *Parinari curatellifolia* is infestation of immature fruits by larvae of the weevil *Balaninus* sp., which can cause up to 50% seed destruction and fruit fall (Parker, 1978). These could considerably reduce the prospects of tree regeneration. Feeding by the thrips *Heliothrips haemorrhoidalis* and *Selenothrips rubrocinctus* (Thysanoptera: Thripidae) causes light brown blemishes on the skin of *S. birrea* fruits.

Carpophilus spp. bore into maturing fruits to feed on the sweet pulp of most miombo fruit trees. *C. hemipterus* and *C. fumatus* attack *Uapaca kirkiana*, *Sclerocarya birrea* and *Parinari curatellifolia* fruits in Zambia (Parker, 1978; G. Sileshi, unpublished results). In Malawi, *C. fumatus* attacks *Ficus sycomorus* and other wild and cultivated fruits, while *C. hemipterus* attacks *Adansonia digitata*, *S. birrea* and *Morus alba* (Lee, 1971).

The larvae of fruit flies and moths attack mainly mature or ripe fruits. Moth larvae such as *Deudorix* sp., *Cryptophlebia* and fruit flies (*Ceratitis* spp.) invade the mature fruit pulp of *Parinari curatellifolia* (Parker, 1978). *Sclerocarya birrea* is the primary natural host of the marula fruit fly *Ceratitis cosyra*, which can infest its fruits heavily. The same fruit fly species is the major pest of mango in Africa (Lux *et al*., 1998, 2003). In addition, in Kenya, Tanzania and several other countries, *S. birrea* is now heavily attacked by an additional fruit fly species, *Bactrocera invadens*, a new invasive pest recently detected in East Africa (Lux *et al*., 2004). The pest originates from Sri Lanka, where it attacks fruit of mango and *Terminalia catappa*. The latter is also attacked by this pest in Africa. *Bactrocera invadens* is rapidly spreading within Africa. Recently it has been recorded also from Sudan, Ethiopia, Democratic Republic of the Congo, Nigeria, Ghana and other countries in West Africa. Wherever it is present, *B. invades* tends to outcompete and displace the indigenous *Ceratitis cosyra*, and to become a dominant pest of both mango and morula. Like *Ceratitis cosyra*, it can cause fruit infestation reaching up to 70%. Thus, both the indigenous fruit fly, *Ceratitis cosyra*, and the new invasive one, *B. invadens*, if not controlled, can substantially reduce the profits from the utilization of morula fruits and mango cultivation alike. In Zambia, *Ceratitis cosyra* also attacks *Uapaca kirkiana* and *Parinari curatellifolia* (G. Sileshi, unpublished results) and also mangoes (Javaid, 1986). *Strychnos* species are the native hosts of *Ceratitis pedestris* and *Scleropithus* spp. in southern Africa (Hancock, 1989). *Ceratitis* spp. have been observed attacking *Annona senegalensis* and *Garcinia livingstonei* fruits in Mozambique (R.R. Bandeira, unpublished results). Other miombo fruits are likely to be infested by specific sets of other native fruit fly species common in Africa, such as *Ceratitis anonae*, *C. rosa*, *C. fasciventris*, *C. capitata* and many others, each of them highly polyphagous, with diverse but partially overlapping host ranges.

In general, moths are little known and greatly underrated pests of fruits in Africa, both cultivated and miombo fruits. In South Africa, *Sclerocarya birrea* fruits are also damaged by *Mussidia melanoneura* and *Cortyta canescens* (De Lange *et al*., 2001). Moth larvae such as *Deudorix* sp. and *Cryptophlebia* invade mature fruit pulp of *Parinari curatellifolia* (Parker, 1978). The false codling moth *Cryptophlebia leucotreta* is one of the more serious pests. It has a wide range of host plants in southern Africa. It has been reared on a number of occasions from *Uapaca kirkiana*, *Parinari curatellifolia* and *Sclerocarya birrea* in eastern Zambia (G. Sileshi, unpublished results). It has also been reported to attack miombo species such as *Ziziphus mucronata*, *Sclerocarya birrea* and *Ximenia caffra*, and exotic fruits such as citrus, mango, avocado, peach and guava in Malawi (Lee, 1971) and South Africa (Begemann and Schoeman, 1999). Infestations lead to premature fruit drop in citrus, and it can damage as much as 90% of a citrus crop.

Packing of fruit infested by either fruit fly or moth larvae causes postharvest decay and may lead to export rejection due to international phytosanitary regulations (Begemann and Schoeman, 1999). Although fruit-damaging insects are important pests of horticultural crops as well as miombo fruits in southern Africa, little work has been done on them, particularly with regard to their biology and ecology.

Birds and mammals

Among the birds that feed on *Sclerocarya birrea*, the parrot *Poicephalus meyeri* damages the fruit to feed on the nuts (Hall *et al*., 2002). Frugivorous bats also feed widely on fruits of *Parinari curatellifolia* in eastern Zambia (G. Sileshi, unpublished results). Baboons, monkeys, zebras, bushbucks, elands, elephants, bush pigs, giraffes, duikers, lemur, rats, squirrels and domestic animals such as cattle and goats are known to feed on fruits of *Sclerocarya birrea* (Hall *et al.*, 2002). A wide range of mammals have been reported to feed on *Uapaca kirkiana* fruits, including baboons, blue monkeys, velvet monkeys, galagos, bush pigs, warthogs, squirrels, elephants, elands and zebras (Ngulube *et al.*, 1995; Seyani, 1996). Of these, primates could be expected to cause the most damage because of their numbers and their ability to climb fruit trees; they often cause excessive fruit drop from the shaking that occurs as they climb and jump from one branch to another.

20.4 Postharvest Pests

Fruits may undergo significant deterioration during storage, transport to local and distant markets and in the market until their disposal. Miombo fruits normally do not undergo postharvest treatment. In most parts of sub-Saharan Africa, storage, packaging, transport and handling technologies for fruits are practically non-existent and hence considerable amounts of produce are lost or contaminated. Fruits may lose weight and shrivel, change colour and ultimately deteriorate. It was a general concern among farmers that most miombo fruits have a short shelf life (Malambo *et al.*, 1998). For example, *Uapaca kirkiana* was reported to store for less than 2 weeks. The short shelf life of many species makes it difficult for households to enjoy the fruits year-round. Microflora and insects may lead to postharvest deterioration of fruit and shortening of the shelf life. Little is known about the microflora causing postharvest deterioration in miombo fruit trees in southern Africa. According to studies in India, *Aspergillus niger*, *A. sydowii*, *Rhizopus oryzae*, *Penicillium chrysogenum*, *Alternaria tenuissima*, *Cladosporium chartarum*, *Phoma hissarensis*, *Botryodiplodia theobromae* and *Curvularia* sp. cause postharvest spoilage in *Ziziphus mauritiana* (Kainsa *et al.*, 1978; Gupta, 1983). In addition, proliferation of some of the microflora could lead to contamination of fruit products with mycotoxins.

Damage by *Carpophilus* sp., fruit fly larvae and false codling moth also continues during storage and transport (G. Sileshi, unpublished results). These insects attack ripe fruits, rapidly degrading the pulp and causing fruit rot. Fruit can be infested before harvest and may be sold before the larvae emerge. Some field diseases may also continue developing during the postharvest period. Anthracnose caused by *Colletotrichum gloeosporioides* reduces the fruit shelf life of mango, avocado and peach during storage and transport (Ploetz *et al.*, 1994). It is the most important postharvest disease of avocado in South Africa, causing losses of up to 37% of fresh fruit. Decay due to anthracnose develops as the length of fruit storage increases, and occurs during the time fruit is being transported and marketed. *Alternaria*, which causes blossom disease in the field, also leads to postharvest rot in ripening fruit.

Microflora and insects also leave behind contaminants such as mycotoxins, pathogens and allergens, which compromise the quality of fruit products. Mycotoxins are fungal metabolites which are potent toxins to humans and animals. Mycotoxins are produced mainly by the fungal genera *Aspergillus*, *Penicillium* and *Fusarium*. Under favourable conditions these fungi grow and produce mycotoxins in certain foodstuffs such as dried fruits, nuts, spices and a range of cereals. For instance, patulin, produced by a number of moulds, occurs in apple juice and other apple-based products contaminated by *Penicillium expansum*. Mycotoxins attract world-wide attention because of the significant economic losses associated with their impact on human health, animal productivity and domestic and international trade.

Contamination of juices with pathogenic microorganisms has caused numerous illnesses and fatalities. Fruit juices with pH below 4.6 were once deemed a minor health threat as a result of their high acidity. Furthermore, refrigeration temperatures below 5°C represented an additional hurdle to pathogen growth. The emergence of hitherto unsuspected food pathogens with acid resistance combined with increases in the numbers of susceptible and immune-compromised individuals, such as HIV/AIDS patients, the chronically ill, the very young and the very elderly, has dramatically changed this picture. Fruit products can also be contaminated with allergenic sensitizers of plant, microbiological and animal origin. Of all the allergenic contaminants known to occur in postharvest products, it is the storage mites which have received the most attention, though beetles, moths and moulds are also known. Thus, for safety and economic reasons commercial fruit products are subject to strict regulatory control.

20.5 Strategies for Pest Management

From the discussion above it can be seen that major pest problems could occur in the orchard from time to time. The other picture emerging from the discussion is that tree ill-health may be caused by a host of stress factors which may be mutually reinforcing. Stress factors could include drought, browsing by animals, damage by insects and pathogens, weed competition, nutrient imbalances, water stress or excess physical damage (e.g. fire, frost and heat). In order to reduce pest problems in fruit trees farmers should adopt 'Good Agricultural Practices' (GAPs). GAPs are the outcome of good crop management, including integrated pest management (IPM). IPM may be defined as an adaptable range of pest control methods which is cost-effective whilst being environmentally acceptable and sustainable (Perrins, 1997). IPM uses pest management techniques such as sound crop management, enhancing natural enemies and the planting of pest-resistant crops, and uses pesticides judiciously. IPM is generally targeted against the entire pest complex of an agro-ecosystem rather than individual pest species (Sileshi and Ajayi, 2002). Generally, the emphasis is on anticipating and preventing pest problems whenever possible.

The development of a customized IPM programme takes years of research and involves many participants including farmers, industry, researchers and extension. However, the domestication of miombo fruit trees only began in the 1980s (Wehmeyer, 1986) and the short time since then has not allowed much research on pests and pest management. Therefore, most of the IPM tactics discussed here are based on studies conducted on the management of pests of exotic fruit trees and forestry species. The recommendations given below should be selected and adapted to suit the farming system, geographical and climatic conditions of the location, and the prevailing pest complex.

20.5.1 On-farm pest management

In order to reduce pest problems and ensure a healthy crop in the orchard, first and foremost fruit tree growers should adopt GAPs. Growers should integrate suitable techniques and procedures into one harmonious and concerted strategy for the effective and efficient management of fruit trees. This means that IPM practices for all classes of pests in the orchard, including insects, mites, diseases, weeds, birds and rodents, must be woven together with all of the horticultural practices that are consistent with an overall sound approach to crop production (Prokopy, 1994). The major steps to be taken are discussed below.

Site requirements

Tree species vary in their site requirements, soil physical and chemical properties, soil microbiological associations such as mycorrhizas, and cultural procedures (Cordell *et al*., 1989). Ecological studies have shown that although many miombo fruit tree species grow naturally over a range of environments, they are sensitive to site conditions (Mwamba, 1983; Ngulube *et al*., 1995; van Wyk and van Wyk, 1997; Hall *et al*., 2002). Pest problems become most severe when the requirements of a tree species are not fully met. Therefore, the first step in managing pests is to establish the orchard on an appropriate site.

Cultural practices

Several horticultural practices may be applied to complement pest management in orchards. These include maintaining tree stability in the soil to permit the establishment of a strong root system, ensuring adequate nutrient and moisture availability for tree and fruit growth, and pruning trees in a way that optimizes fruit yield and quality. Too much nutrient flux into trees, especially uptake of nitrogen, may prevent good coloration of fruit and lead to poor storage quality. This may also reduce the efficiency of mycorrhizal fungi. This in turn may lead to more rapid development of arthropod pests, such as aphids, that require succulent foliage (Prokopy, 1994). Pruning of excessive growth of succulent foliage and woody tissue will improve fruit colour, and by lowering relative humidity within the tree canopy it can reduce fungal diseases. Roots should not be damaged in the orchard as this may become an avenue for infection by wood rot fungi. When a fruit orchard is established on a former forest area, all tree stumps down to 1 cm should be dug out and taken away to reduce the risk of infection by *Armillaria*.

Orchard ground cover competes with trees for nutrients and moisture. Therefore, it is advisable to keep understorey vegetation to a minimum.

Damage by rodents can be reduced by managing understorey plant growth. Orchard sanitation is one of the major cultural control methods for fruit flies. This may include collecting and destroying all fallen fruits by burying (about 1–2 feet deep) or burning them immediately. Such a method, however, is very labour-demanding, since it is effective only when the fruit is collected regularly twice a week throughout the period of fruit maturity and ripening. If fallen fruits are left in the orchard, insects such as fruit fly larvae will crawl out of the fruits, pupate in the soil, emerge, and infest the fruits remaining in the orchard.

Biological agents

Biological agents that control pests of fruits in the field include antagonists such as mycorrhizas, and pathogenic, parasitic and predatory organisms. Mycorrhizas have been proved to increase the resistance of trees to infection by pathogenic fungi (Marx, 1972; Wheeler, 1992), including various *Pythium* and *Fusarium* species (Gunjal and Patil, 1992) and endoparasitic nematodes (Duponnois *et al.*, 2000; Borowicz, 2001). For example, ectomycorrhizal fungi inhibit Douglas fir root rot, a condition cited as killing more trees than clear-cutting does (Wheeler, 1992). Possible modes of action include the production of antibiotics by the mycorrhizal fungi, and stimulation of host defence mechanisms and the physical barrier by ectomycorrhizae (Marx, 1972; Duponnois *et al.*, 2000). Mycorrhizae are known to affect the growth of *Uapaca kirkiana* (Mwamba, 1995), and if the seedlings are inoculated with mycorrhizas in the nursery 100% survival following field planting is possible (Maghembe *et al.*, 1994).

Naturally occurring pathogenic organisms that kill insects include viruses, bacteria, fungi and nematodes. Recently, biopesticides have been shown to offer alternatives to chemical methods of pest management in forestry and horticulture. In relation to the many orders of insects and other invertebrate pests of forestry, it is only the defoliators, particularly caterpillars and beetles, that have been targeted by mainstream microbial insecticides (Evans, 1999). Commercial preparations of granulosis viruses and bacteria such as *Bacillus thuringiensis* are available for the control of caterpillars. Commercial preparations of naturally occurring fungi such as *Metarhizium* and *Beauveria* have also been shown to be effective against termites (Maniania *et al.*, 2002) and fruit fly larvae (Lux *et al.*, 2003). Biopesticides are host-specific, environmentally friendly and compatible with other control options. However, they are not easily accessible to small-scale farmers.

In their native habitat, defoliator insects are controlled by parasitic insects that attack the egg, larval or adult stage. For example, an unidentified braconid wasp parasitizes *Euproctis rufopunctata* on *Uapaca kirkiana*, *Adansonia digitata* and *Ziziphus mauritiana* in the field (G. Sileshi, unpublished results). The ichneumonid wasp *Charops* sp. attacks *Phalanta phalantha* on *Flacourtia indica* in eastern Zambia. *Coccygodes* sp. and *Goryphus* sp. attack *Niphadolepis alianta* on *Annona* (Lee, 1971). The tachinid flies *Carcelia illota*, *Drino imberbis*, *D. inconspicua* and *Tachina fallax* parasitize *Xanthodes graellsii* on *Azanza* (Lee, 1971). Encyrtid wasps attack mealybugs and scale insects, while some braconid wasps attack various aphids. Eggs of *Argema mimosae*, a common defoliator of *Sclerocarya birrea*, are attacked by eupelmid wasps *Anastatus* spp., *Mesocomys pulchriceps* and *Mesocomys vuilleti* in South Africa (Van Den Berg, 1990). Similarly, *Imbrasia belina*, another defoliator of *Sclerocarya birrea*, suffers up to 40% mortality from the egg parasitoid *Mesocomys pulchriceps* in Botswana (Ditlhogo, 1996). Several of the sap-sucking insects, especially scales, mealy bugs and aphids, are attacked by a large number of native parasitoids, some of which have been successfully used in the biological control of fruit trees pests in Africa (Wakgari, 2001; Lux *et al.*, 2003).

Predatory insects also play a significant role in the natural control of pest species in southern Africa (Dirorimwe, 1996; Sileshi *et al.*, 2001). Among the ladybird beetles, *Alesia*, *Cheilomenes*, *Chilocorus*, *Exochomus*, *Hippodamia*, *Hyperaspis* and *Platynaspis* spp. are the common predators of aphids and scale insects (Lee, 1971). Syrphid flies, especially *Paragus* sp., *Xanthogramma aegyptium* and *X. pfeifferi*, prey on aphids attacking fruit trees and crop plants in Malawi and Zimbabwe (Lee, 1971; Dirorimwe, 1996). Various bugs and carabid beetles can prey on a wide range of insects. The reduviid bugs *Rhinocoris* spp. and pentatomid bugs *Afrius*, *Glypsus*, *Macrorhaphis* and *Mecosoma* spp. feed on a number of beetles and caterpillars (Sileshi *et al.*, 2001), defoliating various fruit tree species in the miombo (G. Sileshi, unpublished results). The reduviid bugs *Cosmolestes pictus* and *Callilestes gracilis* prey heavily on *Gonometa* species, defoliators of *Parinari curatelifolia* in Botswana (Hartland-Rowe, 1993). Although biological control strategies are more often linked to the use of specific natural enemies, generalist predators undoubtedly play a significant role in limiting pest populations, particularly at low prey density (Van Driesche and Bellows, 1996). The populations of such organisms can sometimes be concentrated in the orchard and their egg deposition increased by spraying solutions of attractants such as meat meal or fish meal, sucrose, molasses and brewers yeast in the field (Sileshi *et al.*, 2001). Birds of prey (e.g. owls and buzzards) and snakes are known to keep the population of rodents to a tolerable level in plantations and orchards. Where the habitat preferences of the predators are known, cultural practices may also be manipulated to favour their activity. This includes less disruptive weed control, orchard floor and habitat management practices and judicious use of pesticides.

Plant resistance

One of the most significant improvements in the management of forest trees involves seed-source selection. In the early days, seeds were collected with little regard for tree form, growth rate or disease resistance, resulting in inferior seedlings. Today, orchards comprise clones selected for desirable tree growth and pest resistance traits. This practice has not yet gained wide application in indigenous fruits of the miombo. In future, seed orchard clones should be meticulously selected from field sources, and their subsequent progeny carefully evaluated on a variety of field planting sites. Choice of scion and rootstock also greatly influences pest management as both vary in their resistance to many pests as well as in their suitability to the intended site. In the current domestication programme there is no clear strategy to explicitly include pest resistance during selection or germplasm propagation. The inclusion of farmers' criteria for selection may unintentionally include selection for pest resistance, since farmers are unlikely to select individual trees known to suffer severe attack. Leakey (1999) suggests that while selection for superior yielding agroforestry trees continues, breeding should now focus on criteria such as pest resistance and product quality rather than yield.

Chemical control

Chemical control is an integral component of on-farm pest management, and a variety of chemicals may be needed. However, chemical pesticides should be used only: (i) in combination with other IPM practices, where other practices have failed to produce satisfactory results; or (ii) in situations where IPM practices are not available. Special precautions are warranted when prescribing 'protective' chemical control. Insecticides can be used in a selective way. Two environmentally benign insecticide bait technologies developed for fruit fly and false codling moth control are described below. The insecticide bait technology for the management of fruit flies (Lux *et al.*, 1998) involves a combination of food bait and insecticide to attract and kill adult fruit flies of various species. The poison bait consists of a mixture of insecticides, lure (protein hydrolysate or yeast product) and water. The poison bait is applied to approximately 1 m^2 of foliage area per tree in such a way that fruits on the tree are not hit by the insecticide drift. It is not necessary to wet the whole tree; a section on one side of the tree will be adequate.

LastCall™ (IPM Tech, 2002), an insecticide bait used for the control of the false codling moth, is a special paste containing a standard insecticide, a pheromone and ultraviolet protection, all in a custom-designed dispenser. The preferred insecticide is permethrin, although several other insecticides work just as effectively. The product is packed in a special dispenser that contains 150 g of material, enough for approximately 2.5 acres (1 ha). In high-density apple orchards, one or two droplets of LastCall are applied to the trunk or branches of each apple tree. The male codling moth is attracted to the pheromone in LastCall and attempts to mate with the product. The moth is killed shortly after making contact with the insecticide. This will result in only unfertilized female moths in the population, which are unable to lay viable eggs in the fruit. Compared with conventional pesticide sprays, LastCall delivers a minute amount of pesticide.

20.5.2 Postharvest pest management and quality control

Postharvest treatments of fruits involve cleaning to remove dirt, dipping of fruit in cold water to remove field heat, dipping in hot water (not exceeding 55°C for 5 min) and treatment with fungicides and insecticides. Other chemical compounds, such as antioxidants and growth regulators, are also applied to influence shelf life. Dipping fruits in calcium chloride, ascorbic acid and growth regulators such as Cycocel, maleic hydrazide and benzyladenine has delayed overripening and increased the shelf life of *Ziziphus mauritiana* fruits (Pareek, 2001). Postharvest quarantine treatments for fruit flies, for instance, include hot water dips and prolonged exposure to temperatures below 7.5°C. Marula fly (*Ceratitis cosyra*) larvae are known to survive heat treatment better than medfly (*C. capitata*), but exact treatment parameters are not yet established. Unfortunately, there is virtually no information on the postharvest treatment and handling of miombo fruits. It is therefore important that postharvest procedures be given as much attention as production practices.

Quality control to ensure food safety is another area that requires attention in the commercialization of miombo fruits. Particular areas of concern in food safety are the indiscriminate use of agricultural chemicals and poor postharvest handling, resulting in deterioration in the quality of foods. In the case of miombo fruit processing, inadequate technologies employed and unhygienic handling practices coupled with doubts about the quality of the raw materials used and possible adulteration are matters of great concern. To avoid food safety problems, a whole range of systems has been put in place, including GAPs, hazard analysis and critical control point (HACCP) procedures (ICMSF, 1989). This is rarely practised in the developing countries. If miombo fruits are to be commercialized, quality control needs to be taken seriously.

20.6 Conclusions, Emerging Issues and the Way Forward

The picture emerging from this review shows that tree ill-health may be caused by a host of stress factors which may be mutually reinforcing. Fruit trees, indeed any plants, grow to the best of their genetic potential if undisturbed. Any factor that interferes with the normal physiological processes of plants, such as competition from weeds, unfavourable weather, insects, diseases, fire and animals, will affect the health of the plant. The result is usually a reduction in growth or fruit yield or the death of the tree. Diseases and insects are often specific within host genera, having become adapted to particular host characteristics. Some organisms with a broad host range, for example *Armillaria*, seem in many cases to be facultative parasites are quite capable of living as saprophytes in natural balanced ecosystems. Only when stress or disturbances occur to upset the equilibrium does the pathogenic ability of such organisms manifest itself. Disturbances common in the miombo woodlands include the recurrent fires, which are often initiated accidentally or as a miombo management practice. Much more striking is the damage caused by non-host-specific pests, which under favourable conditions are capable of increasing to high population levels.

Our knowledge about the pests of miombo fruit trees is largely incomplete. Though miombo fruit trees are known to be hosts to a variety of pests in the natural forest (Lee, 1971; Parker, 1978), no systematic studies have been conducted on their pest problems. Like most fruit trees, each miombo fruit tree species is expected to be attacked by two or three key pests, several secondary pests and a large number of occasional pests of local importance. *Armillaria*, *Colletotrichum gloeosporioides* and termites may be considered as priority pests since they affect a wide range of fruit trees. Considering the direct damage they cause and their quarantine status, fruit-feeding insects, especially *Ceratitis* spp. and *C. leucotreta*, also seem to be key pests that merit significant research and development efforts in most of the miombo fruit tree species.

Until recently, the conservation and tending of miombo fruit trees in the natural forest has been traditionally encouraged. Thus, IPM measures to combat pests have been argued to be uneconomical and impractical in the natural forest (Parker, 1978). In the natural forest the pests are components with roles in the forest succession and they make up a substantial part of the biodiversity, and pest damage may not be obvious because of the balancing effect of this biodiversity. However, problems are likely to increase as fruit trees are grown extensively in a new environment under managed conditions. IPM may become necessary when nurseries and orchards of improved fruit tree varieties become established for the commercial utilization of fruits. It is clear from the foregoing discussion that IPM cannot be developed for individual pests or tree species. It is also clear that IPM should be an integral component of good agricultural practices. Such practices should be farmer-driven and user-friendly. Many factors influence the priority given by farmers to tree health problems and hence the locally expressed demand for pest management research and development. Among the major factors is the lack of knowledge of tree ill-health and poor diagnosis, which also results from lack of diagnostic tools. Hence, pest attack may be confused with nutrient deficiencies or the unspecified problems most often referred to as 'poor site condition'.

Farmers' perceptions of the definition of tree pests depend on the economic value that they attach to a tree, the part of the plant attacked and the amount of damage caused. For instance, the edible caterpillars *Cirina forda*, *Gynanisa maja*, *Gonimbrasia zambesina*, *Gonimbrasia belina* and *Micragone cana* may seriously defoliate miombo fruit trees. In Zambia the caterpillars are considered as pests by foresters because of the defoliation and the damage done by people collecting them. However, farmers tolerate them because they consider them an essential source of protein food (Mughogo and Munthali, 1995; Mbata *et al.*, 2002). Local communities are knowledgeable about edible caterpillars and often make an effort to conserve them (Mbata *et al.*, 2002). Trees such as *Uapaca kirkiana* are hosts for the bug *Encosternum delegoruri* and various caterpillars that are eaten by communities in Malawi and Zimbabwe (Makuku, 1993; Meke, 1998; Mbata *et al.*, 2002). Miombo fruit trees can also host wild silkworms, such as *Argema mimosae* and *Gonometa* species. Adoption of policies and practices that encourage the integration of nutritionally important edible insects with the cultivation of wild fruit-bearing trees will significantly benefit poor communities living in marginal agricultural areas and will also increase the adoption of these trees.

References

Begemann, G.J. and Schoeman, A.S. (1999) The phenology of *Helicoverpa armigera* (Hubner) (Lepidoptera: Noctuidae), *Tortrix capensana* (Walker) and *Cryptophlebia leucotreta* (Meyrick) (Lepidoptera: Tortricidae) on citrus at Zebediela, South Africa. *African Entomology* 7, 131–148.

Borowicz, V.A. (2001) Do arbascular mycorrhizal fungi alter plant-pathogen relations? *Ecology* 82, 3057–3068.

Campbell, B.M. (1987) The use of wild fruits in Zimbabwe. *Economic Botany* 41, 375–385.

Coleman, D.C., Edwards, A.L., Belsky, A.J. and Mwonga, S. (1991) The distribution and abundance of soil nematodes in East African savannas. *Biology and Fertility of Soils* 12, 67–72.

Cordell, C.E., Anderson, R.L., Hoffard, W.H., Landis, T.D., Smith, R.S., Jr and Toko, H.V. (1989) *Forest Nursery Pests*. USDA Forest Service, Agriculture Handbook No. 680, 184 pp.

De Lange, H.C., Jansen, P.J., Vuuren V. and van Averbeke, W. (2001) Infestation of the fruit of marula (*Sclerocarya birrea*) (A. Rich) Hochst subsp. *caffra* (Sond) Kokwaro. Unpublished report, University of Pretoria, 10 pp.

Dirorimwe, W. (1996) A study on some predatory coccinelids (Coleoptera: Coccinelidae) found in cotton fields in Zimbabwe. *Zimbabwe Journal of Agricultural Research* 34, 145–152.

Ditlhogo, M.K. (1996) The natural history of *Imbrasia belina* (Westwood) (Lepidoptera: Saturniidae) and factors affecting its abundance in North-eastern Botswana. PhD thesis, University of Manitoba, Winnipeg, Canada.

Duponnois, R., Founoune, H., Lesueur, D., Thioulouse, J. and Neyra, M. (2000) Ectomycorrhization of six *Acacia auriculiformis* provenances from Australia, Papua New Guinea and Senegal in glasshouse conditions: effect on the plant growth and on the multiplication of plant parasitic nematodes. *Australian Journal of Experimental Agriculture* 40, 443–450.

Evans, H. (1999) Factors in the success and failure of microbial insecticides in forestry. *Integrated Pest Management Review* 4, 295–299.

Gunjal, S.S. and Patil, P.L. (1992) Mycorrhizal control of wilt in casuarinas. *Agroforestry Today* 4, 14–15.

Gupta, O.P. (1983) Delicious candy from ber fruits. *Indian Horticulture*, April–June, 25–27.

Gupta, P.C. and Madaan, R.L. (1977) Fruit rot diseases of ber (*Ziziphus mauritiana* Lamk) from Haryana. *Indian Phytopathology* 30, 554–555.

Hall, J.B., O'Brien, E.M. and Sinclair, F.L. (2002) *Sclerocarya birrea: A Monograph*. School of Agriculture and Forest Science Publication No. 19. University of Wales, Bangor, UK, 157 pp.

Hancock, D.L. (1989) Pest status—southern Africa. In: *World Crop Pests: Fruit Flies—Their Biology, Natural Enemies and Control, Volume 3A*. Elsevier, Amsterdam, pp. 51–58.

Hartland-Rowe, R. (1993) The biology of the wild silkworm *Gonometa rufobrunnea* Aurivilius (Lasiocampidae) in northern Botswana, with some comments on its potential as source of wild silk. *Botswana Notes and Records* 24, 123–133.

ICMSF (International Commission on Microbiological Specification for Foods) (1989) *Microorganisms in Foods 4. Application of the Hazard Analysis and Critical Control Point (HACCP) System to Ensure Microbiological Safety and Quality*. Blackwell Scientific Publications, Boston, Massachusetts.

IPPC (2006) Glossary of Phytosanitary Terms No. 5. In: *International Standard for Phytosanitary Measures*. FAO, Rome, pp. 61–79.

IPM Tech (2002) LastCall™ CM attract and kill pest control. IPM Tech, Portland, Oregon.

Javaid, I. (1986) Causes of damage to some wild mango fruit trees in Zambia. *International Pest Control* 28, 98–99.

Kainsa, R.L., Singh, J.P. and Gupta, O.P. (1978) Surface microflora of freshly harvested ber (jujube) (*Ziziphus mauritiana* Lamk.) and their role in spoilage of fruits. *Haryana Agricultural University Journal of Research* 8, 86–89.

Keetch, D.P. (1989) A perspective of plant nematology in South Africa. *South African Journal of Science* 85, 506–508.

Kleynhans, K.P.N., van den Berg, E., Swart, A., Marais, M. and Buckley, N.H. (1996) *Plant Nematodes in South Africa. Plant Protection Institute Handbook No. 8.* ARC-PPRI, Pretoria.

Kroon, D.M. (1999) *Lepidoptera of Southern Africa: Host Plants and Other Associations. A catalogue.* Lepidopterists' Society of Africa, Jukskei Park, South Africa.

Leakey, R.R.B. (1999) Win–win land use strategies for Africa: matching economic development with environmental benefits through tree crops. Paper presented at the USAID Sustainable Tree Crop Development Workshop 'Strengthening Africa's Competitive Position in Global Markets', Washington, 19–21 October 1999.

Leakey, R.R.B. and Simons, A.J. (1998) The domestication and commercialization of indigenous trees in agroforestry for the alleviation of poverty. *Agroforestry Systems* 38, 165–176.

Lee, R.F. (1970) A first checklist of tree diseases in Malawi. *Malawi Forest Research Institute Research Record* No. 43, 29 pp.

Lee, R.F. (1971) A preliminary annotated list of Malawi forest insects. *Malawi Forest Research Institute Research Record* No. 40.

Lux, S.A., Zenz, N. and Kimani, S. (1998) The African fruit fly initiative: development, testing and dissemination of technologies for the control of fruit flies. In: *ICIPE Annual Scientific Report 1998–1999*, No. 7, pp. 78–80.

Lux, S.A., Ekesi, S., Dimbi, S., Mohamed, S. and Billah, M. (2003) Mango-infesting fruit flies in Africa: perspectives and limitations of biological approaches to their management. In: Neunschwander, P., Borgemeister, C. and Langewald, J. (eds) *Biological Control in IPM Systems in Africa.* CAB International, Wallingford, UK, pp. 277–293.

Lux, S.A., Copeland, R., White, I., Manrakham, A. and Billah, M. (2004) A new invasive fruit fly species from *Bactrocera dorsalis* (Hendel) group detected in East Africa. *Insect Science and its Application* 23, 355–361.

Maghembe, J.A., Kwesiga, F., Ngulube, M., Prins, H. and Malaya, F.M. (1994) Domestication potential of indigenous fruit trees of the miombo woodlands of southern Africa. In: Leaky, R.R.B. and Newman, B. (eds) *Tropical Trees: Potential for Domestication and Rebuilding of Forest Resources.* HMSO, London, pp. 220–229.

Makuku, S.J. (1993) Community approaches in managing common property forest resources: the case of Norumedzo community in Bikita, Zimbabwe. *Forests, Trees and People Newsletter* 22, 18–23.

Malambo, L.N., Chilanga, T.G. and Maliwichi, C.P. (1998) Indigenous miombo fruits selected for domestication by farmers in Malawi. In: Maghembe, J.A., Simons, A.J., Kwesiga, F. and Rarieya, M. (eds) *Selecting Indigenous Fruit Trees for Domestication in Southern Africa.* ICRAF, Nairobi, pp. 16–39.

Maniania, N.K., Ekesi, S. and Songa, J.M. (2002) Managing termites in maize with the entomopathogenic fungus *Metarhizium anisopliae. Insect Science and its Application* 22, 41–47.

Marx, D.H. (1972) Ectomycorrhizae as biological deterents to pathogenic root infections. *Annual Reviews in Phytopathology* 10, 429–454.

Masuka, A.J., Cole, D.L. and Mguni, C. (eds) (1998) *List of Plant Diseases in Zimbabwe.* DRSS, Harare, 179 pp.

Mbata, K.J., Chidumayo, E.N. and Lwatula, C.M. (2002) Traditional regulation of edible caterpillar exploitation in the Kopa area of Mpika district in Northern Zambia. *Journal of Insect Conservation* 6, 115–130.

Mchowa, J.W. and Ngugi, D.N. (1994) Pest complex in agroforestry systems: the Malawi experience. *Forest Ecology and Management* 64, 277–284

Meke, G.S. (1998) Insects associated with fruit, seed and foliage of *Uapaca kirkiana* (Muell. Arg.) Euphorbiaceae in Malawi. MSc Thesis, University of Stellenbosch, Stellenbosch, South Africa.

Meyer, M.K.P.S. (1981) Mite pests of crops in southern Africa. *Science Bulletin, Department of Agriculture and Fisheries, Republic of South Africa* No. 397, 92 pp.

Mughogo, D.E.C. and Munthali, S.M. (1995) Ecological interaction between *Gonimbrasia belina* (L.) and *Gynanisa maja* (L.) and the impact of fire on forage preference in a savanna ecosystem. *African Journal of Ecology* 33, 84–87.

Mwamba, C.K. (1983) Ecology and distribution of Zambian wild fruit trees in relation to soil fertility of representative areas. MSc thesis, University of Gent, Belgium.

Mwamba, C.K. (1995) Effect of root-inhabiting fungi on root growth potential of *Uapaca kirkiana* (Muell. Arg.) seedlings. *Applied Soil Ecology* 2, 217–226.

Ngulube, M., Hall, J.B. and Maghembe, J.A. (1995) Ecology of miombo fruit tree: *Uapaca kirkiana* (Euphorbiaceae). *Forest Ecology and Management* 77, 107–117.

Pareek, O.P. (2001) *Ber.* International Centre for Underutilised Crops, Southampton, UK, 292 pp.

Parker, E.J. (1978) Causes of damage to some Zambian wild fruit trees. *Zambia Journal of Science and Technology* 3, 74–83.

Pennisi, A.M. and Agosteo, G.E. (1994) Foliar alterations by *Colletotrichum gloeosporioides* on annona. *Informatore Fitopatologico (Italy)* 44, 63–64.

Perrins, R.M. (1997) Crop protection: taking stock for the new millennium. *Crop Protection* 16, 449–456.

Piearce, G.D., Calvert, G.M., Sharp, C. and Shaw, P. (1994) Sooty baobabs—disease or drought? *Zimbabwe Forestry Commission Research Paper* No. 6, 13 pp.

Ploetz, R.C., Zentmeyer, G.A., Nishijima, W.T., Rohrbach, K.G. and Ohr, H.D. (1994) *Compendium of Tropical Fruit Diseases.* APS Press, Minnesota, 88 pp.

Prokopy, R.J. (1994) Integration in orchard pest and habitat management: a review. *Agriculture, Ecosystems and Environment* 50, 1–10.

Roux, J. and Wingfield, M.J. (1997) Survey and virulence of fungi occurring on diseased *Acacia mearnsii* in South Africa. *Forest Ecology and Management* 99, 327–336.

Seyani, J.H. (1996) The economic importance and research needs for *Uapaca kirkiana* in Malawi. In: van der Maesen, L.J.G., van der Burgt, X.M. and van Medenbach de Rooy, J.M. (eds) *The Biodiversity of African Plants. Proceedings, XIVth AETFAT Congress, 22–27 August 1994, Wageningen, the Netherlands.* Kluwer Academic Publishers, Dordrecht, pp. 697–703.

Shorter, C. (1989) *An Introduction to the Common Trees of Malawi.* Wildlife Society of Malawi, Lilongwe, Malawi, 115 pp.

Sileshi, G. (2000) Tips on how to manage termites. *Living with Trees in Southern Africa* No. 4, 14–15 pp.

Sileshi, G. and Ajayi, O.C. (2002) Exciting opportunities for scaling up pest management. *Ground UP: Magazine of the Participatory Ecological Land-use Management (PELUM) Association* 1, 22–24.

Sileshi, G., Kenis, M., Ogol, C.K.P.O. and Sithanantham, S. (2001) Predators of *Mesoplatys ochroptera* Stål in sebania-planted fallows in eastern Zambia. *Biocontrol* 46, 289–310.

Sinclair, W.A., Lyon, H.H. and Johnson, W.T. (1989) *Diseases of Trees and Shrubs.* Cornell University Press, Ithaca, New York 575 pp.

Speight, M.R., Hunter, M.D. and Watt, A.D. (1999) *Ecology of Insects: Concepts and Applications.* Blackwell Science, London, 350 pp.

Strouts, R.G. and Winter, T.G. (1994) *Diagnosis of Ill-health in Trees.* Stationary Office, London, 305 pp.

Swart, W.J., Conradie, E. and Wingfield, M.J. (1991) *Cryphonectria cubensis*, a potential pathogen of *Psidium guajava* in South Africa. *European Journal of Forest Pathology* 21, 424–429.

Taylor, D.P., Netscher, C. and Germani, G. (1978) *Adansonia digitata* (baobab), a newly discovered host for *Meloidogyne* sp. and *Rotylenchulus reniformis*: agricultural implications. *Plant Disease Reporter* 62, 276–277.

Taylor, F.W., Matoke, S.M. and Butterworth, K.J. (1996) A holistic approach to the domestication and commercialization of non-timber forest products. In: Leakey, R.R.B., Temu, A.B., Melnyk, M. and

Vantomme, P. (eds) *Domestication and Commercialization of Non-timber Forest Products in Agroforestry Systems.* Non-Wood Forest Products No. 9. FAO, Rome, pp. 75–85.

Uys, V. (2002) *A Guide to the Termite Genera of Southern Africa. Plant Protection Research Institute Handbook No. 15.* Agricultural Research Council, Pretoria, 116 pp.

Van den Berg, M.A. (1990) The African lunar moth, *Argema mimosae* (Lepidoptera: Saturniidae), a potential pest of marula. *Acta Horticulturae* 275, 685–689.

Van Driesche, R.G. and Bellows, T.S. Jr (1996) *Biological Control.* Chapman & Hall, New York, 539 pp.

van Wyk, B. and van Wyk, P. (1997) *Field Guide to Trees of Southern Africa.* Struik Publishers, Cape Town, 536 pp.

Wakgari, W.M. (2001) The current status of the biological control of *Ceroplastes destructor* Newstead (Hemiptera: Coccidae) on *Citrus* and *Syzygium* in South Africa. *Biocontrol Science and Technology* 11, 339–352.

Wehmeyer, A.S. (1986) Why so little research on the noblement of indigenous edible plants? *Acta Horticulturae* 194, 47–53.

Wheeler, D. (1992) The underground story—What's going on with truffles and trees, voles and owls? *Mushroom, The Journal* Spring, 25–27.

21 Accelerated Domestication and Commercialization of Indigenous Fruit and Nut Trees to Enhance Better Livelihoods in the Tropics: Lessons and Way Forward

F.K. Akinnifesi,[1] G. Sileshi,[1] O.C. Ajayi[1] and Z. Tchoundjeu[2]

[1]*World Agroforestry Centre, ICRAF, Lilongwe, Malawi*; [2]*ICRAF West and Central Africa Region, Humid Tropic Node, Yaounde, Cameroon*

21.1 Introduction

The early stages of the 'Green Revolution' placed emphasis on plant breeding, increased fertilizer use and plant protection. The loss of biodiversity and the issue of sustainability have now started to receive attention as one of the major 'downsides' of the Green Revolution in terms of the requirement to meet the needs of both present and future generations (van Noordwijk *et al.*, 2004).

The harvesting of indigenous fruit trees (IFTs) from the wild pre-dated settled agriculture. Domestication involves accelerated and human-induced development in order to bring species into wider cultivation through a farmer-driven and market-led process (Simons and Leakey, 2004). Opportunistic regeneration of 'volunteer' seedlings from mother trees to the surrounding areas paved the way for purposeful utilization and management. Such trajectories of 'volunteer' or pioneer species have been recognized as providing the drive for purposeful domestication and the instinctive selection of trees with the desired qualities (see Chapters 4–6). As a result, the cultivated populations become distinctly superior to their progenitors.

The integration of diverse high-commercial-value tree crops, including a diverse array of exotics and indigenous fruit and nut tree crops rather than intensified monocultures with a small number of staple crops, has been one of the ways in which smallholder farmers have traditionally been addressing their livelihood needs. This alternative livelihood intervention paradigm has attracted attention with respect to the UN Millennium Development Goals (MDGs) and various environmental conventions (Garrity, 2004; Leakey *et al.*, 2005).

Since the early 1990s, domestication strategies, approaches and techniques, coupled with research and development on the commercialization and marketing of agroforestry fruit tree products (AFTPs), has become one of the major pillars of this paradigm, being especially recognised by the World Agroforestry Centre and many of its partners around the world (Garrity, 2004; Simons and Leakey, 2004; Leakey *et al.*, 2005).

The objectives of this chapter are: (i) to provide a global synthesis of knowledge on indigenous fruit tree domestication, utilization and marketing experiences from tropical Africa, Latin America and Asia; (ii) to summarize and highlight the main opportunities, achievements and challenges that remain, linking basic research and development initiatives to applications in the private sector; and (iii) to propose future research areas. This final chapter of the book highlights key findings and puts in perspective the implications from the preceding chapters in the context of lessons learned from decades of domestication and commercialization of indigenous fruit trees in the tropics.

21.2 Indigenous Fruit Tree Domestication

21.2.1 Why domesticate IFTs?

Out of the 250,000 higher plant species in the world, less than 1% have been domesticated as food plants and, of these, about 50% are fruit trees that are either domesticated or semi-domesticated (Leakey and Tomich, 1999). In Tanzania, about 326 indigenous plants have been described as edible (Ruffo *et al.*, 2002) but few, if any, of these species have been domesticated through deliberate tree improvement programmes (Akinnifesi *et al.*, 2006a). An inventory of fruit trees in Nigeria and Cameroon showed that 56% of fruit trees are indigenous (Schreckenberg *et al.*, 2006), and many of them are endemic to some locations (A. Degrande *et al.*, unpublished). Farmers listed more than 200 species in Latin America that they would like to cultivate (Weber *et al.*, 2001).

The reasons for the long-time neglect of IFTs and failure to domesticate them have been variously identified as being due to: (i) lack of information and reliable methods for measuring their contribution to rural economies, livelihoods of communities, and ecological services (Chapter 11); (ii) low production incentives relating to markets and technology (Leakey *et al.*, 2005; Chapters 11 and 13, this volume); (iii) bias in favour of large-scale agriculture and conventional forestry (Russell and Franzel, 2004; Teklehaimanot, 2004; Chapters 3, 8 and 11, this volume); (iv) colonial interventions that left a profound legacy of neglect of smallholder farm production in favour of estate farm producers (especially in eastern and southern Africa) and European export product trading interests in West Africa; and (v) a weak interface between private sector actors, researchers and extensionists in tree products (Russell and Franzel, 2004).

Domestication aims at promoting the cultivation of IFTs with economic potential as new cash crops, and provides incentive to subsistence farmers to plant trees that contribute towards achieving the Millennium Development Goals (MDGs) of poverty reduction and enhancement of food and nutritional security (Leakey *et al.*, 2005). From the preceding chapters, the following have emerged as benefits of domestication:

- *Desirable fruit traits*. This involves development of cultivars with high productivity, harvest index, superior quality, improved food value and acceptability. As highlighted in Chapter 6, many of the Amazonian fruits have not been attractive to the market for various reasons: some require processing, most have a short shelf life, and they are of variable quality because they are seed-propagated. This is true of most IFTs in the tropics, and suggests the need for selection and improvement of wild cultivars for desirable quality trait(s) (Chapters 1, 2, 8 and 9) and higher market values (Chapters 12 and 14). Some trees can produce fruit twice within the same year, and some can have delayed or late fruiting. Examples of how these traits have been captured are found in Chapter 9.
- *Precocity*. One of the major problems identified by farmers and stakeholders across the regions and reported in many chapters of this book is the desire to reduce the juvenile phase and shorten the period before first fruiting of IFTs (Chapters 1, 2, 6, 8, 9). For instance, *U. kirkiana* takes 12–16 years to fruit in the wild (Chapter 8), *Irvingia gabonensis*, *Chrysophyllum albidum* and *Garcinia kola* are all known to fruit late in West Africa (Chapter 9). *Allanblackia floribunda* may take at least 20 years to produce fruit (Rompaey, 2005). This is an urgent case warranting domestication that targets fruit production in a fraction of the time required in natural stands. Such a long maturity period tends to discourage farmer adoption and private investment in IFTs, especially in low-income countries. The long period that it takes before enjoying the benefits from tree-based agriculture is one of the reasons that farmers in Zambia see it as a constraint to the adoption of agroforestry tree species (Ajayi, 2007). Therefore, domestication plays a significant role in ensuring that trees produce quality fruits in a shorter period of time, using proven strategies (Chapter 2).
- *Inconsistency and inadequate supply: quality and quantity*. Market preference is for consistency in the supply of quality fruit traits in terms of uniformity and regularity. For example, the quality of wine made from açaí-do-Pará (*Euterpe oleracea*) varies, due to variations in quality, harvest and postharvest practices; limitations that impact on food quality and safety (Chapter 6). Production in insufficient quantities has been a major market limitation in many cases, especially for products that have captured global markets. This was reported for camu camu (*Myrciaria floribunda*) in Latin America (Chapter 6), and is also the case for *Sclerocarya birrea* (marula), which relies almost entirely on wild harvests or semi-domesticated stands, and has limited entrepreneurial interest in some IFTs promoted in southern Africa. On the other hand, because of the abundance and availability of some species in the wild, the scope for their

cultivation and improvement by researchers has been limited in the past, as is also the case for many Amazonian fruits that have not been cultivated (Chapter 6) and for *Uapaca kirkiana* (wild loquat) in Zimbabwe (Mithöfer, 2005). Mithöfer (2005) suggested that farmers would require tree improvement incentives that are not really technically feasible in order for spontaneous adoption of indigenous fruit trees to take place, because the collection of IFTs from the wild is still attractive and profitable. On the other hand, a similar study in Malawi indicated that, if awareness is created, smallholder farmers would readily adopt IFTs such as *U. kirkiana*, even with little or no improvement as incentives. This is due to greater resource depletion in Malawi compared with Zimbabwe. In the Zimbabwe situation, a consistent supply would be in terms of developing fresh market fruit ideotypes that would meet consumers' needs in terms of size, colour, taste and uniformity.

- *Conservation of biodiversity*. Biodiversity is often ignored as a reason for domesticating trees on-farm. Teklehaimanot (Chapter 11) has highlighted the potential of *Cordeauxia edulis* (yehib), the most preferred IFTs in the dry region of East Africa. The fruits are used in Somalia and Ethiopia as a drought food. Its leaves are also a source of cordeauxiaquinone, which is used for dying and in medicine. It is the most threatened tree species identified by the IUCN (Chapter 11). Genetic diversity can be conserved through 20 generations of improvement (Cornelius *et al.*, 2006). However, there is fundamental conflict between genetic gain and genetic conservation. No improvement programme can conserve all the genetic diversity of landraces. However, domestication can help to preserve biodiversity and genetic resources on-farm. It can also help to elevate the genetic gain and diversity of a few species with high impact potential. Adequate genetic diversity of IFTs can be maintained through conservation of wild sources, regular introduction of new clones, and generation of new genotypes through hybridization (Chapters 2, 8 and 9).
- *Novel foods and food additives*. Providing novel foods and creating products with niche markets, which can constitute important export commodities, is another benefit of domestication. Examples include camu camu (*Myrciaria dubia*) for vitamin C (Chapter 6), marula cream and oil from *Sclerocarya birrea* (Chapters 8 and 14), *Allanblackia* spp. for oil butter (Rompaey, 2005), solid butter from *Vitellaria paradoxa* (Teklehaimanot, 2004; Chapter 10, this volume), white chocolate from cupuaçu (*Theobroma grandiflorum*), among others. Domestication can help avoid the boom-and-bust economy that typically occurs when products are harvested from the wild (Penn, 2006). Farmers must have access to superior stocks if they are to capitalize on the benefits of IFT cultivation.
- *Public goods and services*. Domestication can also help provide international public goods and services including carbon sequestration, biodiversity conservation, erosion control, etc., which benefit both the present global community and also future generations (Leakey *et al.*, 2005).

21.2.2 Domestication: from extraction to cultivation

Despite more than two decades of research on IFTs, 90% of the marketed products still come from the wild. A species is considered wild when it grows spontaneously and naturally without human intervention. The term 'domestication' has been accepted since the 19th century as a dynamic term referring to a process rather than a state of existence of wild or semi-domesticated trees (Chapter 4). This acceptance is evidenced by the number of scientific articles, meetings and conferences where the term has been used without explanation.

Dubois (1995) observed three stages of historical interventions by the Brazilian Amazon forest-dwellers, namely: (i) simple harvests without intentional management; (ii) enrichment of long-duration tree-fallow; and (iii) systematic enrichment of forest stands with planted trees, usually high-value trees, resulting in an agroforest. Domestication of indigenous fruits has been extensively documented in the literature, especially in agroforests (Wiersum, 2004; Chapter 4, this volume) and homegardens (Kumar and Nair, 2004), and the patterns seem to be similar in Asia (Chapter 4) and Latin America (Chapter 6), and in the cocoa agroforests of humid west and central Africa (Lodoen, 1998; Chapter 9, this volume).

The process of coevolution between forest trees and fruit production has been illustrated by the durian fruit tree (*Durio zibethinus*) in South-east Asia (Chapter 4) as an intermediate form of domestication for multifunctional production systems. The production systems research of which domestication should focus on was discussed in Chapter 4, and has favoured a landscape domestication approach. The different interpretations of domestication by biologists and agroecologists have been described in Chapter 4. Biologists tend to understand domestication as involving alterations in biological processes at the species and individual tree level, which leads to adaptation and changes in genetic make-up. On the other hand, agroecologists emphasize the processes operating at the landscape level involving people–plant interactions.

21.2.3 Nexus between collection and cultivation

Since trees do not usually require open land, or a vast area of land, and can be conveniently incorporated into a farming system with the conditions necessary for production of fruits, the trajectory of domestication involves a gradual modification of production conditions from forest to agricultural crops (Michon and de Foresta, 1997; Wiersum, 2004; Chapter 4, this volume). Michon and de Foresta (1996) approached domestication as a change in wild harvesting or exploitation systems. No matter how it is viewed, both processes of wild or semi-wild fruit tree conservation and management, and deliberate selection and cultivation, are important in domestication that aims to increase returns to farmers in a sustainable way. The processes of adaptation of people and trees from extractivism to cultivation warrant further exploration.

Clearly, fruit trees, whether exotic or indigenous, can be grown in a variety of different production systems, including: (i) natural forests where fruit trees are protected; (ii) enriched natural forests through deliberate regeneration or planting of propagules; (iii) 'cyclic agroforests' or forest gardens; (iv) mixed arboriculture, e.g. homegardens; and (v) monoculture fruit-tree plantations or orchards (Wiersum, 1997; Chapter 4, this volume). Creative farmers have evolved these systems in Asia (Caradang *et al.*, 2006; Chapter 4, this volume), Latin America (Dubois *et al.*, 1996; Penn, 2006; Chapter 6, this volume) and Africa (Okafor, 1983; Kang and Akinnifesi, 2000; Leakey *et al.*, 2005). The trajectories of change in such systems vary and depend on various ecological, biological and socio-economic and political factors. According to Wiersum (Chapter 4), smallholder farmers gradually change their fruit-tree production systems from wild collection to more specialized fruit arboricultural orchard management, while at the same time modifying species composition and management. The durian (*D. zibethinus*) forest garden in Asia, Brazil nut (*Bertholletia* spp.) or bacuri forest gardens or 'açaiceiro' in Brazil are examples. How the products can be better adapted to meet market demand will require deliberate domestication and improvement initiatives.

Despite the availability of fruits in the wild, improvements in fruit yield and earlier fruiting (i.e. shortened juvenile phase or enhanced precocity) would create incentives to farmers to cultivate indigenous fruit trees, even though such an incentive depends on the robustness of the improvement impact and factors negatively affecting fruit abundance (Mithöfer and Waibel, 2003; Fiedler, 2004). Domestication activities are essential if a tangible commercial interest in indigenous fruits is to emerge beyond the current opportunistic levels. However, it is important to recognize that success is dependent on the domestication and commercialization of indigenous fruit trees occurring in parallel (Wynberg *et al.*, 2003, Akinnifesi *et al.*, 2004a, 2006; Chapter 8, this volume), in order that problems of the seasonality and reliability of supply, diversity and inconsistency of fruit quality are overcome throughout the supply chain.

A global analysis of the marketing and cultivation of wild forest products has found that farmers who engaged in the cultivation of indigenous fruits had higher returns on labour, used more intensive production technologies, produced more per hectare and benefited from a more stable resource base, than those relying on wild collection (Ruiz-Perez *et al.*, 2004). These studies, therefore, suggest that the cultivation of wild fruit trees will become more important as rural households move from subsistence to a cash-oriented economy.

21.3 User-driven Domestication Approaches

21.3.1 What species to domesticate

Agroforestry fruit tree products (AFTPs) can be divided into four categories: (i) those that are readily edible (mostly with sweet non-toxic or astringent fruit pulp when ripe); (ii) those requiring cooking before being consumed (e.g. breadfruit, most nuts, edible oils); (iii) those requiring intensive processing into other forms (e.g. wine, jam, chocolate, etc.); and (iv) non-edible fruit and nut products (e.g.

cosmetic oils, biodiesel, medicinal products). In terms of understanding preferences for fruits, the major drive forces would be the extent to which they are able to meet the subsistence and cash-income needs of the producers and market participants. Industrial needs or interests, and organization project promotion activities may also influence decision making at all levels.

The preoccupation of researchers and governments with the need to domesticate IFTs seems to be prevalent across all the regions, and can be traced back to the 1920s in Latin America (Popenoe and Jimez, 1921; Clement *et al.*, 2004; Chapter 6, this volume) and the 1970s in Africa (Okafor, 1983; Kang *et al.*, 1994). The initial approaches for setting priorities for domestication in the three continents were overly top-down and rarely involved farmers or users. They involved organizations deciding what species to work on in terms of their current knowledge, experience and importance of the species. However, these provided a vital foundation on which subsequent work was based.

Several meetings were organized by ICRAF and partners in Africa to brainstorm and identify IFTs of importance. In 1995, ICRAF and the SADC (Southern African Development Community) Tree Centre organized a workshop in Swaziland on range-wide collection of *Sclerocarya birrea* and *Uapaca kirkiana*. These were collected and exchanged between participating countries. A regional conference was devoted exclusively to indigenous fruit trees (Maghembe *et al.*, 1995; Shumba *et al.*, 2000) and another international conference concentrated on domestication and commercialization of non-timber forestry products (Leakey and Izac, 1996).

In South-east Asia, Gunasena and Roshetko (2000) documented the efforts to identify priority tree species to domesticate in each country. This exercise was carried out in a regional workshop at Yogyakarta, Indonesia, in 1997 (involving Indonesia, the Philippines, Thailand and Vietnam), where organizations had listed the species considered important. Since this did not particularly focus on IFTs or prioritize them, it is important to mention that some IFTs have been listed by various organizations as species of priority, notably: *Durio zibethinus* (durian), *Phyllanthus acidus* (star gooseberry), *Garcinia mangostana* (mangosteen), *Annona mangostana* (manggis), *Lansium domesticum* (lanzones), *Tamarindus indica* (tamarind), *Syzygium* spp., *Artocarpus* spp., *Anarcardium occidentale* (cashew), *Chrysophyllum cainito* (star apple), *Castanopsis* spp. (chestnut) and *Nephelium lappaceum* (rambutan). The list is long but did not discriminate between fruit and non-fruit trees, and timber species were dominant, followed by fruit trees. The relative preferences in terms of priority were not clear.

Similarly, the list of fruits considered as important in Latin America is also long; and these are mostly in the Amazon region. In Brazil, researchers had created awareness on the importance of peach palm (*Bactris gasipaes*) as early as the 1920s (Popenoe and Jimenez, 1921, cited in Chapter 4), and research-driven efforts in the past 25 years have concentrated on fruit and heart-of-palm development with varying levels of success, creating new crops including fully domesticated heart-of palm grown in high-density and high-input monocultures, which is widely traded on European and American markets (Clement *et al.*, 2004).

It is important also to recognize the amount of information that has been assembled in the monograph series *Fruits for the Future*, covering specific indigenous fruit trees considered as priority or important fruits, documented by the International Centre for Underutilised Crops (ICUC) in the UK in collaboration with ICRAF, IPGRI and the FRP (Forest Research Program), and mostly funded by the DFID (Department for International Development). Notable among these are those on *Ziziphus mauritiana* (Pareek, 2001), *Tamarindus indica* (Gunasena and Hughes, 2000), *Annona* spp. (Pinto *et al.*, 2005), *Theobroma grandiflorum* (cupuaçu) and *Dacryodes edulis* (Kengue, 2002). Additional resources have been provided on *Allanblackia* spp. (Rompaey, 2005), *Vitellaria paradoxa* (Hall *et al.*, 1996), *Balanites aegyptiaca* (Hall and Walker, 1992), *Sclerocarya birrea* (Hall *et al.*, 2002), *Parkia biglobosa* (Hall *et al.*, 1997), *Strychnos cocculoides* (Mwamba, 2005), *Uapaca kirkiana* (Akinnifesi *et al.*, 2008), and *The Encyclopedia of Fruit & Nuts* containing short monographs of several IFTs, edited by Janick and Paull (2008). These were mostly based on the results of regional priority-setting meetings or workshops. A collection of Brazilian fruits has been assembled by EMBRAPA (Silva and Tassara, 2005). These volumes are important for detailed domestication of IFTs in the tropics.

21.3.2 Systematic priority-setting

In this book, different strategies and levels of priority-setting and selection of fruit trees to be domesticated in different regions have been well highlighted (Chapters 1, 6, 7, 8 and 9). The choice of species to domesticate is driven by several factors. These include, among others: (i) the desire to plant one's own trees; (ii) the existence of a market for nursery stocks or tree products; (iii) the availability of seeds and other propagules in the locality (Chapter 19); (iv) the influence of organizations funding tree-planting programmes; and (v) previous nursery experience of farmers (Weber *et al.*, 2001; Akinnifesi *et al.*, 2006; Caradang *et al.*, 2006; Tchoundjeu *et al.*, 2006; Chapter 9, this volume). This is the reason why selection of species is usually based on marketable characteristics.

Building on these efforts for the first time in the mid-1990s, the approach for identifying priority IFTs was revolutionized when ICRAF conducted workshops across Africa and developed more robust guidelines and principles for prioritizing trees for agroforestry (Franzel *et al.*, 1996). The crucial aspects that have been overlooked during the *ad hoc* listing of priority species and early domestication work by researchers were addressed in priority-setting guidelines. The principles and applications in western Africa, eastern and southern Africa are detailed in Chapter 2. The priority-setting generally followed a seven-step procedure (Franzel *et al.*, 1996):

1. Team building and consensus of stakeholders on the approach and its local adaptations.
2. User needs assessment and expectations from research.
3. Complete inventory of species used in the region or country, and those of high potential.

4. Defining criteria for choosing trees with the most important products, and identifying them.
5. Selecting a limited number of species (usually four or five) with the greatest economic potential from step 3 above.
6. Primary data collection from household surveys. This provides quantitative information needed for priority-setting.
7. The final step was to synthesize results, review the process and approve the choice of priority species (Franzel *et al.*, 1996). The data-collection instrument, function and application have been detailed by Franzel *et al.* (1996).

This priority-setting methodology was tested in West Africa during the period 1993–1995 (Jaenicke *et al.*, 1995; Adeola *et al.*, 1998) and in southern Africa in 1996–1997 (Kadzere *et al.*, 1998; Maghembe *et al.*, 1998). The top five indigenous tree species identified in the humid zone of West Africa (Nigeria and Cameroon) were *Irvingia gabonensis*, *Dacryodes edulis*, *Ricinodendron heudelotii*, *Chrysophyllum albidum* and *Garcinia kola* (Franzel *et al.*, 1996; Aiyelaagbe *et al.*, 1997; Tchoundjeu *et al.*, 2006; Chapter 2, this volume). Two varieties of *Irvingia gabonensis* were reportedly domesticated in West Africa, and a national workshop was held on *Chrysophyllum albidum* in April 1999 to salvage much of the valuable gene pool in the wild (Ladipo *et al.*, 1999).

The priority-setting guidelines were also applied in southern Africa, although thoughts on IFT domestication were discussed, starting from an international conference held in Malawi in 1995 (Maghembe *et al.*, 1995). A household survey was conducted in Malawi, Zambia, Tanzania and Zimbabwe (Chapter 2). The top priorities are discussed in Chapters 2 and 8. In addition, the approach has been expanded to include product prioritization (Ham and Akinnifesi, 2006) and consumer assessment through market surveys (Kwesiga and Mwanza, 1995; Minae *et al.*, 1995; Ramadhani, 2002; Mmangisa, 2006; Chapters 1, 8, 13 and 14, this volume). Relative preferences for indigenous as opposed to exotic fruit trees were investigated in the new surveys. Updating of the consolidated results from various approaches used in the region was detailed in Chapter 8. The final list of species included *Uapaca kirkiana*, *Strychnos cocculoides*, *Parinari curatellifolia*, *Ziziphus mauritiana* and *Adansonia digitata* (Chapter 2). There were also country-specific priority species (Chapter 8). Farmers indicated fruit and tree traits that were important, such as fruit size, taste, pulp content, shelf life, and small tree size (Kadzere *et al.*, 1998). In another regional workshop conducted in Malawi in November 2000, the top five priority indigenous fruit tree species indicated by participants from eight SADC countries were: *U. kirkiana*, *Z. mauritiana*, *S. cocculoides*, *S. birrea* and *P. curatellifolia*. This confirmed findings from several other studies that *U. kirkiana* is the most preferred IFT species (Ramadhani, 2002; Akinnifesi *et al.*, 2006; Chapters 1 and 8, this volume). However, mango and orange were the top two fruits generally preferred when exotics were considered, while *U. kirkiana* ranked second overall (F.K. Akinnifesi, unpublished data).

Teklehaimanot (Chapter 11) also documented the priority-setting results from eastern Africa using two approaches. A regional workshop by AFREA–IPGRI and national partners, and a two-stage field approach involving: (i) national priority-setting involving discussions with stakeholders; and (ii) a field survey. The top eight fruit tree species selected using the workshop approach were *Adansonia digitata*, *Carissa edulis*, *Parinari curatellifolia*, *Sclerocarya birrea*, *Tamarindus indica* and *Ziziphus mauritiana* (Chapter 10). The stakeholder priority list included 11 species and they were different from the previous list: *Balanites aegyptica*, *Berchemia discolor*, *Borassus aethiopum*, *Carissa edulis*, *Cordeauxia edulis*, *Sclerocarya birrea*, *Strychnos cocculoides*, *Vangueria madagascariensis*, *Vitellaria paradoxa* and *Vitex doniana*. Just as in southern Africa, there were country-specific priority species. *A. digitata*, *P. curatellifolia*, *S. birrea*, *T. indica* and *Z. mauritiana*, *S. cocculoides* and *Vitex doniana* were independently selected as priority species in southern East Africa (Chapters 1 and 11). Teklehaimanot (Chapter 11) also noted that there are some species in common between East Africa, the Sahel (West Africa) and southern Africa, such as *A. digitata*, *T. indica*, *Z. mauritiana* and *V. doniana*. There are also some similar species between India and sub-Saharan Africa, especially *T. indica* and *Z. mauritiana* (Kadzere *et al.*, 1998; Gunasena and Hughes, 2000; Pareek, 2001).

In Latin America, Sotelo-Montes and Weber (1997) carried out priority-setting using farmer-preference surveys and expert consultation from product users and marketers. The study was based on the guidelines developed by Franzel *et al.* (1996). Farmers indicated an interest in cultivating more than 150 tree species, and 23 of these were considered as priority for domestication based on their economic value (Weber *et al.*, 2001). However, domestication work has concentrated on *Bactris gasipaes* and *Inga edulis* as the two most important fruit trees for market. F.K. Akinnifesi (unpublished) surveyed more than 60 homegardens in São Luis, Brazil, and found more than 300 fruit tree species, of which about 75% were IFTs. Recently, a workshop-type priority-setting approach was conducted in Latin America, in October 2006, through a regional fruit-tree priority-setting workshop organized in Boa Vista, Roraima, Brazil by Initiativa Amazonica (IA). This was a strong regional workshop drawing participants from key Amazonian countries (Brazil, Colombia, Equador, Peru, Bolivia). The workshop concentrated on Steps 1–5 of Franzel *et al.* (1996). This knowledge helped researchers to quickly select genetic material for multiplication. Several lessons were learnt in the different approaches and these could be applied to any other part of the world.

21.3.3 Paradigm shift in technology (bottom-up)

Most of the earlier research focused on identifying important species of IFTs, qualitative descriptions of varieties and their potential for domestication, germplasm collection, species performance and provenance trials (Kwesiga *et al.*, 2000; Akinnifesi *et al.*, 2004b) generally following the forestry approach to tree improvement. An arboretum was established at two locations, Onne and Ibadan, by the International Institute of Tropical Agriculture (IITA) in 1979 with 40 species and 61 accessions (Kang *et al.*, 1994), and expanded in 1991 with over 100 species and 376 accessions (Ng *et al.*, 1992). In the same way, a species performance trial was established at Makoka in 1991.

In the last decade, researchers have been making considerable changes in approach and there has been an impressive accumulation of data on quantitative characterization involving more participatory domestication and clonal selection (Leakey *et al.*, 2003, 2005; Akinnifesi *et al.*, 2006a; Tchoundjeu *et al.*, 2006; Chapters 8, 9, 21, this volume). All the chapters in this book have showcased extensive quantitative data and field experiences in Asia, Africa, Oceania and Latin America.

Leakey *et al.* (2004, 2005) examined tree-to-tree variation in trees that had been semi-domesticated by farmers in order to explore the opportunities for multi-trait selection. They found that when species have been subjected to truncated selection and breeding, they may bear little resemblance to their wild origins, while some of their wild relatives may have become extinct through clearance and genetic pollution through outcrossing with farmer-selected varieties. On the other hand, cultivated horticultural crops have been domesticated for thousands of years. For instance, kiwi fruit (*Actinidia* spp.), which is one of the best-known commercial horticultural crops, has a history of utilization and cultivation in China of more than 1000 years. However, improvement work began only in the late 1970s (Xiao, 1999). It was then disseminated to New Zealand, cultivated and selected by farmers (and later scientists) until it became a major export crop yielding more than 1 million t per year world-wide (Berry, 1997; Xiao, 1999). In contrast, the domestication of indigenous fruit and nuts is more recent in the tropics (Leakey and Simons, 1998), probably beginning less than three decades ago. In the conventional forestry approach, early stages of domestication of timber species involve species elimination, performance trials, followed by provenance and progeny selection involving the transfer of geographically discrete subpopulations to new environments (Akinnifesi *et al.*, 2004a, 2006; Leakey *et al.*, 2004) which are usually based on an environment similar to the natural range; known as the 'homocline' concept (Nyland, 1996). This requires a long regeneration period and cycles of breeding. However, wild fruit tree domestication cannot depend on such a unilinear domestication process and an innovative approach to short-circuit the process is warranted that will combine diversification and technology change. These aspects have been the focus of Chapters 2, 8 and 9.

Domestication aims at taking advantage of variation in the wild. In study of tree-to-tree variation in *Sclerocarya birrea* in South Africa, Leakey *et al.* (2005) reported from a sample of 15 trees that oil yield per nut ranged from 5 to 53 g per nut and protein content in the fruit pulp ranged from 30 to 112 g per fruit. Such variation is the major interest in domestication. It has long been suggested that semi-domesticated trees have superior fruit characteristics than those of wild origins (Maghembe, 1995; Leakey *et al.*, 2005; Akinnifesi *et al.*, 2006). In an early species-screening trial at Makoka, Maghembe (1995) found that the average fruit of *Vangueria infausta* was ten times bigger than those found in the wild. In Peru, Weber *et al.* (2001) reported that farmers were able to visually identify superior peach palm trees that have better products (e.g. red fruits with waxy coats, which have a higher oil content than red or yellow fruits without waxy coats). Farmers' knowledge was taken as a hypothesis, tested, and proved to be correct (Weber *et al.*, 2001).

Technically, wild harvesting for industrial production has often suffered some setbacks. Excessive variation of wild collection could constitute a deterrent for the large-scale market. For instance, the government of Peru promoted large-scale cultivation of camu camu (*Myrciaria dubia*) during 1995–1997. The state agencies produced large quantities of grafted planting stock in order to produce trees that would fruit earlier than those from seedlings, about 2 years after planting. The flaw was that grafting was done without selection of the rootstock or scions for any desirable traits, thereby foregoing the opportunity for crop improvement (Penn, 2006). Eighty-six per cent of the camu camu harvested or planted on farmer's fields in Peru comes from wild stocks (Penn, 2006). These trees commonly experience single or double fruiting cycles that have not been captured by research. Pest management is a major problem yet to be resolved in the domestication of camu camu. Although farmers prune their camu camu trees, it is not done based on research results, and farmers indicated that they were unsure which tree form would ultimately produce more fruit.

Another example where a potentially successful initiative to develop international markets failed was when the natural variation and need for trait improvement was overlooked by a private farm in Coroata, Brazil. The company had received a long-term contract to supply organically produced acerola (*Malpighia glabra*) fruits to the European market in the 1990s. The project embarked on large-scale plantation of the acerola, but with no prior knowledge of selection. Four years down the line, fruit harvests were rejected for the targeted market because of excessive variability in fruit traits. The farm was forced to sell the produce at local markets to cushion the loss. These are just a few cases, where omission of carefully planned and executed domestication has resulted in catastrophic financial losses.

In general, domestication approaches have followed two broad patterns: forestry and horticultural. The former involves species performance testing, such as germplasm collection of several Amazonian fruits in Brazil, Colombia, Costa Rica, Equador and Peru by EMBRAPA and INPA (Chapter 6), and the testing of several indigenous fruit tree species in arboreta at Onne and Ibadan, Nigeria (Ng *et al.*, 1992; Kang *et al.*, 1994). The latter focuses on fruits and relies heavily on local knowledge for selection and vegetative propagation to achieve precocity.

21.3.4 Participatory tree domestication strategy

Participatory domestication is defined as genetic improvement that includes farmer–researcher collaboration, and is farmer-led and market-driven. It was devised to overcome the shortcomings of the earlier top-down approaches of conventional breeding and forestry (Akinnifesi *et al.*, 2004a; Tchoundjeu *et al.*, 2006; Chapter 2, this volume). In this approach, farmers express their preferences and interests freely, and contribute to the planning and technology development through indigenous technical knowledge infused into the process from time to time as they are being consulted. By so doing, rural farmers and communities have ownership and are more confident in testing and adapting innovations.

When carefully planned and implemented, this approach facilitates adoption and reduces the long process of technology transfer from research stations to the farmers (Leakey *et al.*, 2005; Chapter 6, this volume), and also avoids the replacement of landraces with improved varieties, which causes genetic erosion (Chapter 6). Participatory domestication approaches have been applied in West Africa for *Dacryodes edulis* and *Irvingia gabonensis* (Tchoundjeu *et al.*, 2006; Chapter 9, this volume), *Uapaca kirkiana* and *Strychnos cocculoides* in Malawi (Chapter 8), and for *Bactris gasipaes* and *Inga edulis* in Peru (Weber *et al.*, 2001; Chapter 6, this volume). The participatory clonal selection strategy in southern Africa involved the following activities (Akinnifesi *et al.*, 2006): (i) priority-setting using stakeholder workshops; (ii) household surveys and market surveys; (iii) product prioritization; (iv) identifying natural stands of priority IFTs through reconnaissance surveys; (v) village workshops to define fruit traits and joint selection of elite cultivars with communities – farmers, marketers, village leaders and schoolchildren; (vi) systematic naming of trees, collection of seeds and vegetative propagules, nursery evaluation, and field orchard testing of clonal selections until a smaller number of true-to-type and true-to-name cultivars were obtained; (vii) release of superior cultivars for adoption, testing and scaling-up. Detailed principles and strategies for participatory domestication based on clonal selection and vegetative propagation have been described by Leakey and Akinnifesi (Chapter 2). Cornelius *et al.* (2006) and Weber *et al.* (2001) have detailed its application to seedlings.

21.3.5 Propagation as a tool for improvement

Participatory domestication, in which farmers are trained to use vegetative propagation techniques, would enable farmers in different locations to select cultivars for different sets of characteristics, thus ensuring in the short to medium term that farm-level inter- and intraspecific diversity is maintained (Leakey *et al.*, 1990, 2003; Akinnifesi *et al.*, 2006a; Schreckenberg *et al.*, 2006; Tchoundjeu *et al.*, 2006; Chapters 2, 8, 9, this volume).

Mature tissues have the capacity to flower and fruit, and can be multiplied or captured through grafting, budding and air-layering (Leakey *et al.*, 1990, 2005; Chapters 2, 8–10, this volume), and micropropagation (Mng'omba, 2007; Mng'omba *et al.*, 2007a, b). Because of the ease of propagating juvenile tissues by cuttings, this has been the preferred option for participatory domestication of short-gestation fruit trees such as *Dacryodes edulis* in village nurseries in West Africa (Leakey *et al.*, 2005; Chapter 9, this volume). In addition, the tissue culture of juvenile tissues has proved more successful than that of matured tissues for miombo fruit trees such as *U. kirkiana* (Mn'gomba, 2007; Mn'gomba *et al.*, 2007a). Stem cutting was unsuccessful for most miombo fruit trees (Mhango and Akinnifesi, 2001; Akinnifesi *et al.*, 2004a, 2006), and the tissue culture of *U. kirkiana* was limited by phenolic accumulation at the graft union line, leading to the possibility of early or late rejection (Mn'gomba, 2007; Mn'gomba *et al.*, 2007a; Chapter 8, this volume).

Three types of assessing age need to be considered with regard to ease of propagation and also juvenility or precocity (the period of waiting before the first fruiting). These are *chronological age* (time since planting), *physiological age* (cell differentiation stage and structural maturity, including level of lignification), and *ontogenetic age* (state of reproductive maturity). The implication is that if vegetative propagation is used, the appropriate location of the propagule becomes important for trees aimed at fruit production. Silviculturally, trees have different stages of growth: *seedling*, *sapling*, *juvenile* and *matured* phases (Nyland, 1996). Collecting propagules for vegetative propagation, whether roots, stem cuttings, scions for grafting or budding, will depend on the age of the mother tree. The chronological age of a seedling established from seed is different from a marcott collected and established from the tree crown, i.e. branches of a mature tree. Chronologically, trees tend to be older from the tree base and this declines with the height along the stem, explaining why the number of rings declines along the standing height. On the other hand, the ontological age is the reverse, being older at the crown than the stem and youngest at the root. This is due to different stages of cell differentiation. This is the basis for collecting scions from matured branches in fruit trees. A tree of the seedless breadfruit (*Artocarpus altilis*) established in southern Nigeria from a root cutting did not fruit until about 20 years later (personal observation), whereas trees grafted from scions collected from matured trees fruited in less than 4 years (D.O. Ladipo, unpublished data). Farmers in south-west Nigeria use the rooting method of propagation for *Artocarpus altilis* or *A. communis*, but the lengthy juvenile period before fruiting can only ensure generational security, i.e. growing trees for future generations. Lack of awareness was the major reason for the failure of farmers to adopt efficient methods such as air-layering and grafting.

Across the regions covered in this book, different propagation techniques are employed. This includes asexual methods and vegetative propagation (cuttings, grafting, air-layering, budding, etc.) techniques (Chapters 2, 6, 8 and 9). The tissue culture technique is now beginning to be used, and its preliminary application to miombo fruit trees is highlighted in Chapter 8.

21.4 Management and Farmer Adoption of Domesticated Trees

21.4.1 Managing domesticated trees

Management of fruit trees is what makes them essentially different from those growing in the wild. Much of the management of indigenous fruit trees has been broadly based on experience from exotic tree crops, and there is a need for adaptive research on IFT management. There has been some evidence that good management can improve the fruit traits of planted indigenous fruit trees, even for semi-domesticated trees (Maghembe, 1995; Shackleton, 2004; Leakey *et al.*, 2005) but the extent of improvement in fruit traits due to management needs to be distinguished from that due to genetic selection. However, limited research has been undertaken on tree management for indigenous fruit trees and this gap is noted throughout the present volume. Tchoundjeu *et al.* (2006; Chapter 9, this volume) described the cultivation of grafted *Dacryodes edules* and *Irvingia gabonensis* on farmers' fields. Similarly, Akinnifesi *et al.* (2006a; Chapter 8, this volume) also documented the cultivation and management of planted indigenous fruit trees, including *U. kirkiana*, *S. birrea*, *Z. mauritiana* and *V. infausta*, by farmers in southern Africa. The effect of thinning through the respacing of natural stands of *U. kirkiana* affected fruit yield parameters (Mwamba, 1995). Thinning increased fruit load significantly in the first few years following treatment, and also increased pulp content per fruit. However, fruit colour and other hereditary traits were unaffected. Such results call for an investigation into the optimum spacing for these new crops.

Combining the application of fertilizers with irrigation (fertigation) is one way to achieve improved crop yield and quality in most horticultural tree crops, e.g. mango. This will ensure that the adverse effects of nutritional deficiencies on leaf metabolism and growth are overcome (Tomlinson, 1998). Correct scheduling of irrigation and fertilizer application can help synchronize the nutrient supply necessary for phenological development. In acidic soils, liming has also been shown to be important for tree crops (Tomlinson and Smith, 1998). In managing peach palm (*Bactris gasipaes*) in Peru, application of NPK fertilizer, weeding and timely elimination of excess shoots were found to have ensured satisfactory performance (Cornelius *et al.*, 2006).

It is often assumed that the establishment and tending of indigenous fruit trees is generally not different from that of the exotics. However, a recent study in Malawi comparing the management requirements of *U. kirkiana*, *S. birrea* and *V. infausta* with mango has shown that the cultivation of miombo fruit trees may not be exactly the same as conventional tree crops (F.K. Akinnifesi *et al.*, unpublished). Mateke (2003) showed that application of fertigation had varied effects on *V. infausta*, *S. birrea* and *S. cocculoides*. Fertilizer application, manure and irrigation did not increase growth and survival in *U. kirkiana* and *S. birrea*, contrary to the widely held assumption that indigenous fruit trees can be managed as cultivated tree crops. Single factors rather than their combination may be more important at strategic periods, e.g. a light irrigation during a period of prolonged drought or dry season. The poor response of these indigenous fruit trees to fertilizer could be attributed to their adaptation to infertile soils and unimodal rainfall regimes in their natural stands. However, this cannot be generalized for all species and locations. Specific management requirements may be necessary when introducing IFTs into a new area. In some cases, micronutrients might be more important than macronutrients. This suggests that research is needed in order to understand and develop management packages for different IFTs.

In the floodplains of the Peruvian Amazon, the cultivation of camu camu (*Myrciaria dubia*) was promoted by the government to increase its commercial production for international and regional markets (Penn, 2006). Farmer experimentation showed that success was linked to the fact that farmers combined extension protocols with their own local agricultural practices (Penn,

2006). There was little information on the specific management practices adopted by farmers. However, farmers had expressed concern that fruit yields produced by planted trees were not as high as those of wild stands, and excessive variation on-farm (Penn, 2006) warranted proper domestication approaches to be followed (Chapter 2). Farmers also expressed a desire to gain more knowledge and experience in order to improve their fruit yields.

In addition, preharvest management has not been well studied for most IFTs. For tree crops such as mangoes, preharvest management practices that could improve fruit yield and qualities include de-fruiting and pest control. These are areas that need further investigation in IFT research.

21.4.2 Cultivation, adoption and scaling-up of IFTs

Once appropriate planting material has been identified, the next step is how to apply the most effective method to disseminate it. It is important to provide farmers with high-quality germplasm and make it available in a timely manner. The process is time-consuming and expensive to undertake individually for each species, especially when a large number of participating farmers are dispersed over a wide area. One of the most effective ways of achieving scaling-up of IFT cultivation is to involve farmers in the entire process of participatory selection, propagation, nursery and tree establishment and management of superior planting materials (Chapters 2, 8 and 9). This will dramatically shorten the time required to produce and disseminate planting materials from centralized nurseries to farmers. Farmers can be organized to produce high-quality seed, seedlings and vegetative propagules as evidenced in the small-scale nursery enterprises managed by farmers' groups in West Africa (Tchoundjeu *et al.*, 2006; Chapter 9, this volume) and in Peru (Weber *et al.*, 2001). Weber *et al.* (2001) estimated that farmers can earn as much as US$1000 per year from such small-scale enterprises, and many of them now see this as a good investment. The fruit orchards established by farmers can also serve as mother blocks where propagules such as seeds and scions could be collected for future use and scaling-up. The participatory approach used by IIAP (Instituto de Investigaciones de la Amazonía Peruana) and INIA (Instituto Nacional de Investigación Agropecuaria) in 2000 is described by Clement *et al.* in Chapter 6, whereby farmers identify plants with elite characteristics, and the research team collects samples (seeds and cuttings). After propagation, the research team returns some of the plants to the farmer and the other plants are established on-station in clonal and progeny trials. Farmers also expand from their own orchards and can sell to neighbours. They participate in evaluation of all progenies from their farms and those of INIA. This approach seems to be common to Latin America (Chapter 8), southern Africa (Akinnifesi *et al.*, 2006) and West Africa (Tchoundjeu *et al.*, 2006).

In 1996, the Programa Nacional de Camu Camu (PNCC) was reported to have aimed at establishing 10,000 ha of camu camu by 10,000 farmers in a large afforestation programme along the flood plains to supplement wild harvesting and to raise incomes. The number of cultivators in Peru expanded from 28 producers in six communities in 1994 to 4000 in 150 communities in 1997 (Chapter 6). The success of the programme has been detailed by Penn (2006). Despite its top-down approach, the cultivation of camu camu has spread in Peru because of the experienced smallholder farmers (reberenos) who were able to apply their previous experiences. Farmers in Colombia cultivate 7000 ha of peach palm, with yields of 7 t/ha (Chapter 6).

In southern Africa, farmers were supported to raise their own nurseries, graft and establish trees on-farm. More than 12,000 farmers were trained in the 6 years of the Zambezi Basin Agroforestry Project (Chapter 8). However, further research is needed into the scaling-up of indigenous fruit trees, to reduce the gap between researcher-managed and farmer-managed orchards for miombo fruit trees (F.K. Akinnifesi *et al.*, unpublished). In terms of adoption, Mithöfer (2005) investigated the level of improvement needed to trigger the mass adoption of IFT cultivation by farmers.

21.4.3 Pest management

Tree management is the key to differences in performance between exotic fruit trees and IFTs. Fruit trees are known to be easily infested with pests and diseases in the wild, and the level of infestation can easily escalate even further when introduced to a new environment in the cultivated field. This aspect has been extensively covered in Chapter 20. Clement *et al.* (2004) has observed that susceptibility to pests may become a problem as cultivation of peach palm is intensified, and insect pests such as *Palmelampius heinrichi* have reduced fruit yield along the Pacific coasts of Colombia by up to 100%, and costs to farmers could be prohibitively high. Sixty-four potential pests and diseases were identified for camu camu, while four of these were prevalent enough to cause significant economic damage (Penn, 2006). A parasitic climbing plant known as ‘suelda con suelda’ (*Moradendron* spp) was reported on 15% of fields, and the leaf fungus *Fumago* spp. on 34% of fields, although floodwater was thought by farmers to help reduce the impact of pests. In the Brazilian Amazon, mistletoe is a menace to the introduction of cupuaçu (*Theobroma grandiflorum*). EMBRAPA is currently deploying clonal improvement strategies to develop pest-resistant cultivars.

21.5 Economics, Marketing and Commercialization of IFTs

21.5.1 Market potentials for IFT products

Trends in food markets have changed significantly in the last decade, with a growing emphasis on variety and natural/organic products. For instance, it is forecast that that UK functional food market will double to US$3.23 billion in the next 5 years (Ham, 2005). The total value of fair trade natural products in the USA, Canada and the Pacific Basin has increased by 37% and was estimated at US$251 million in 2002 (Ham, 2005). Indigenous fruit tree products can qualify as organic products and could gain market share and entrance into the multibillion-dollar globally increasing niche markets for natural products, especially if operated within fair trade principles. Despite the wealth of indigenous fruits in the tropics, little market research has been conducted to date.

Market potential for IFT products in sub-Saharan Africa

Market potentials of indigenous fruit trees are poorly documented and research has only recently started to appear for most regions. Evidence is accumulating that IFTs can contribute significantly to household income in every region covered in this book. South-African consumers bought 543 million l of fruit juice in 1998, and there is growing interest in natural fruit flavours within the fruit industry (Ham, 2005).

There are several major indigenous fruits that could be described as success stories in Africa. First among these is the marula tree (*Sclerocarya birrea* subsp. *caffra*) in the southern Africa miombo ecosystem, from which the popular Amarula Cream liqueur, which is sold in 63 countries world-wide, is produced (Ham, 2005; Chapter 8, this volume), and from which Maruline® – 'the super marula oil', which is the first biologically active natural ingredient has been extracted and patented (Phytotrade, 2005, press release 8 April). Other wines and beverages are produced from marula for regional markets. The pulp of *S. birrea* is reported to be sold for US$0.20 per kilogram in Dodoma, Tanzania (Chapter 11). Communities that processed about 2000 t per year in South Africa received US$180,000 per annum, and 55 semi-skilled employees of a processing plant earned US$5 per day for 2 months of the year. According to another report, 42 communities in South Africa earned US$48,000 from 12 t of marula kernel in 2001. The oil and butter from the nuts of *S. birrea* is valued for cosmetics and nutraceutical industries in Europe and America (Phytotrade, 2005).

Ramadhani (2002) and Ramadhani and Schmidt (Chapter 12) found significant trade barriers in indigenous fruit trees marketing in Zimbabwe. Local regulations prohibit the collection of fruits for sale, and movement between districts in order to protect the natural resource base. However, such policies create a disincentive to the cultivation of trees by farmers on their farms. Mithöfer (2005; Chapter 13, this volume) examined the economics of collection, use and sale of indigenous fruits in Zimbabwe and found that returns on labour were relatively high, suggesting that growing trees may become more profitable as the natural resource base recedes. An alarming rate of recession has been noted for *U. kirkiana* in southern Malawi by Fiedler (2004). Although *Z. mauritiana* is another fruit tree with market potential in the southern, eastern and Sahel regions of Africa, trade is mostly national. However, none of the other IFTs apart from *S. birrea* has been able to break into the regional or international markets. Although *U. kirkiana* is the most preferred IFT in southern Africa for the national market (Chapters 1, 8, 12 and 13), its prospect of being attractive to the international market is remote in terms of the hundreds of fruits with superior appeal and preferred attributes in other regions such as Latin America, Asia and the humid parts of West Africa.

On the other hand, the market for *Ziziphus mauritiana* in southern Africa is unreliable. In a study conducted in Malawi, Kaaria (1998) observed that: (i) market participants are dictated by access to and distance from the market; (ii) rural markets are dominated by women, whereas the men dominate and set prices in urban markets; (iii) marketing margins were higher in rural than in urban markets, rural market information systems and infrastructures still limit the markets, and there is a need to develop packaging and value-addition. There is a common finding in western (Teklehaimanot, 2004), southern (Kadzere *et al.*, 1998; Chapter 13, this volume) and eastern Africa, as well as in India (Pareek, 2001), that most sellers of indigenous fruits are women and children. Low pricing, poor infrastructure and lack of quality standards have jeopardized the market opportunities for IFTs in the region (Chapters 8, 13 and 14), as well as legal prohibitions on sale of products in Zimbabwe (Chapter 13). Across the southern African region, unfair barter trade often occurs during times of food shortage, especially in Zambia and Malawi. Limited availability of fruits of acceptable quality standards, lack of a standard processed product, market information systems, and market infrastructure are general factors limiting the trade of *Z. mauritiana* fruits in Asia and Africa (Pareek, 2001).

Allanblackia spp. has recently been identified as an important African oil tree that holds potential for smallholder farmers. ICRAF and Unilever have initiated a joint public–private partnership called Novella which aims at stimulating greater supply and triggering a market for *Allanblackia* oil in Cameroon, Ghana, Nigeria and Tanzania. Unilever has guaranteed to buy seeds at US$150/t from smallholder farmers (*ICRAF Transformations*, Jan/March 2007, Issue No. 8).

Another important source of income comes from the internationally traded cosmetic products made from shea butter or karité (*Vitellaria paradoxa*) in the African Sahel and drylands of western and eastern Africa (Chapter 11). In the Sahel region of West Africa, *V. paradoxa* produces solid fat that is at a high premium for the chocolate and cosmetic industries, is a very important export earner in many parts of West Africa, and was the third most important export for Burkina Faso in the 1980s. Annually, almost 90,000 t is exported from that country to Europe and Japan.

Reports from central and western Africa showed some market potentials for priority fruit trees (Leakey *et al.*, 2005; Tchoundjeu *et al.*, 2006). Ndoye *et al.* (1997) has valued the annual trade of five key AFTPs, mostly fruits, at US$7.5 million per year in Cameroon, and exports of US$2.5 million annually. Farm-level production of three priority IFTs and nuts in southern Cameroon have been reported to contribute an average of US$355 per year to the local economy (Ayuk *et al.*, 1999); each tree can be worth US$50–150 per year (for an average area of 1.7 ha), and can provide an annual income of US$300–2000 per household (Leakey *et al.*, 2005). Several other indigenous fruit trees in West Africa command substantial regional or even national markets, e.g. *Irvingia gabonensis*, *Dacryodes edulis* and *Chrysophyllum albidum* (Leakey *et al.*, 2005; Tchoundjeu *et al.*, 2006). A tree of unimproved *D. edulis* averages only US$20 compared with US$150 for an improved tree

(Leakey and Tchoundjeu, 2001). The trade in fruit products from four IFTs (*D. edulis*, *I. gabonensis*, *Cola acuminata* and *Ricinodendron heulotii*) involving 1100 traders in Cameroon and neighbouring countries (Ndoye *et al.*, 1997) was worth US$1.75 million in the first half of 1995 (Ndoye *et al.*, 1997), and exports to Nigeria and Europe were worth US$2 million per annum (Awono *et al.*, 2002). One of the group nurseries set up by ICRAF in Cameroon, comprising 15–30 farmers, made an income of US$2000 in 2004, US$5000 in 2005, and is projected to produce an annual income of US$10,000 from IFT nursery activities (Tchoundjeu *et al.*, 2006).

Markets for IFT products in Asia

Approximately 140,000 t of tamarind fruits (*T. indica*) are reported to have been produced on a commercial scale in Thailand, and exported in fresh and processed form (Chapter 11). Tamarind is valued for its pulp, which contains large amounts of vitamin C and sugar. Pharmaceutical industries use the pulp as an ingredient in cardiac and blood-sugar-reducing medicines. Similarly, *Z. mauritiana* is an important export crop in India, where 1 million t was reported to have been produced during 1994/1995 from 88,000 ha (Pareek, 2001). In 1995/1996, 640 t were exported to countries in Asia, the Middle East and South Africa. It is cultivated in many dry parts of India, and involves two formal marketing chains in Chomu market of India: (i) producer–commission agent–wholesaler–outside markets; and (ii) producer–preharvest contractor–commission agent–wholesaler–outside markets. The producer receives 46% of the market share through direct sale or 33% through preharvest contract (Pareek, 2001). It is also exported from China to other parts of Asia and Europe.

Markets for IFT products in Latin America

Latin America, especially the Amazon Basin, is a centre of diversity and concentration of the world's most valued indigenous fruit trees. According to Clement *et al.* (Chapter 6), the majority of fruits cultivated in the Amazon basins are native, including eight domesticates, 18 semi-domesticates and incipient domesticates, and another 33 fruit crops from other parts of tropical America, which have been domesticated.

Peach palm (*Bactris gasipaes*) is a fully domesticated and cultivated indigenous fruit nut in all the Amazon, and has been important in the area from pre-Colombian times up until the present day. It is one of the best-known underutilized fruit trees among more than 150 fruit tree species in the region (Clement *et al.*, 1997, 2004), and is considered one of the top-priority indigenous fruit tree species for agroforestry in the Amazon region (Sotelo-Montes and Weber, 1997; Iniciativa Amazonica, unpublished report, 2006). The heart-of-palm trade has been estimated at US$50 million per year (Clement and Villachica, 1994). Brazil is by far the largest producer, with production of 200,000 t in 1989, and 10,000 t exported in 1990, with the total crop earning US$40 million (Yuyama, 1997). Currently, Costa Rica and Ecuador are the two largest producers of the canned heart-of-palm product

(Clement *et al.*, 2004). Costa Rica cultivates peach palm in 15,000–20,000 ha of small farms. The Brazilian heart-of-palm market is huge internationally, but production has failed because of low local demand, high shipping costs to major markets, and poor planning and execution of business (Clement *et al.*, 2004; Chapter 6, this volume). Production of peach palm represents 45% of the income of many families in the Brazilian Amazon, producing 3% of the annual 13,000 t produced nation-wide, and earning as much as US$990 per household from an area of 1.5 ha (Clement *et al.*, 2004). However, in Brazil and Peru, 50–60% of the fruits are also reported to have been wasted or consumed at home due to quality problems (Clement *et al.*, 2004). Farmers in Colombia cultivate 7000 ha of peach palm, yielding 7 t per ha.

The expanding supermarket trade has increased urban demand. Farmers commercialize only 40% of production, earning US$660/ha per year. This is worthwhile, as the minimum annual salary is only US$112 in Colombia. In addition, the Colombian market for the fruit of peach palm is estimated at 20,000 t of fresh fruit (Clement *et al.*, 2004). Communities near Yurimagua in Peru are reported to be earning US$2000 per season from the sale of best-quality fruits, indicating a major income source in a region where annual minimum income was US$900 in 2002. In addition, another community near Pucallpa in Peru has been reported to have commercialized 25 t of camu camu collected from the wild and their plantations, earning US$11,500 in the 2005/2006 season (Chapter 8).

Annona cherimoya is a very important indigenous fruit that is sold in local markets and supermarkets in Latin America (Pinto *et al.*, 2005). The world production in 1994 was about 81,000 t from 13,000 ha. In Brazil, a grower could obtain US$24,000 at US$1.8/kg, from a yield of 33 kg per tree with production cost between US$1500 and 3000 per hectare (Pinto *et al.*, 2005). Between 200 and 250 t of fruit was produced in 1999 (Pinto *et al.*, 2005). It is now being commercialized in Spain. Peru and Chile also produce *A. cherimoya* on a commercial scale and this is exported to North America. Value addition will increase competitive advantage.

Collectors of Brazil nut (*Bertholettia excelsa*) earn up to US$900 from 4.5 t of nuts annually (Allegretti, 1994). This harvesting has dwindled for the rural dwellers because of the establishment of plantations (Simons, 1996). Similarly, the economic value of açaí-do-Pará (*Euterpe oleracea*) has been estimated to yield a present value of US$1327–2693/ha in managed forests and US$4266–6930/ha in homegardens (Muniz-Miret *et al.*, 1996).

21.5.2 Attributes for creating markets

Farmers' motivation for cultivating IFTs deserves particular attention. As seen in Chapter 1, species choice is no longer a problem to research and development. When the market is ignored, each farmer's choice depends on fruits that are important for household consumption, fruits that are available in the locality and prior knowledge or ease of cultivation. Given that the effects are not necessarily uniform, the criteria for selecting indigenous fruit trees for domestication must be clear from the beginning. In most studies, the selection and cultivation of IFTs is driven by two major forces – consumption and markets. Because people are an integral part of the agroecosytem, it is important that farmers have good knowledge of the IFTs they are cultivating and the prospects for their livelihoods.

With the growing recognition that scaling-up can only succeed when there is full farmer ownership and buy-in, the attributes of the indigenous fruit trees promoted must be distinct enough to convince growers of their benefits. This requires assessing the attributes of the IFTs in terms of food, nutrition and health for the local communities and for the market. Leakey and Page (2006) have illustrated the use of a web-diagram for visualizing tree–tree variation in morphology and fruit traits found in wild and on-farm trees. With an increased understanding of the multi-trait variability in products of importance to farmers and to markets, it is possible to develop a hierarchy of market-driven ideotypes, which maximizes 'harvest index' through the creation of single-purpose cultivars of multipurpose IFTs, especially when such traits are rare or found only in individuals in wild populations (Leakey and Page, 2006). For instance, *D. edulis* contains 66% more fat than groundnuts, and has been recommended as a high-fat food by the FAO. It provides staple food for 3–4 months of the year for many rural inhabitants in humid West Africa. *P. biglobosa* has a crude protein content higher than that of beef, as well as providing 42% of the recommended daily requirement of zinc (Schreckenberg *et al.*, 2006).

For market-led IFT initiatives, the market attribute must be substantial enough to compete with or outperform conventional sources to make a dent in the market. For instance, camu camu is being promoted in Latin America for the extremely high vitamin C content in its pulp (2.7–2.8 g of ascorbic acid per 100 g). This is 30 times higher than the equivalent weight of oranges (Penn, 2006). This extremely high vitamin C content in camu camu has generated great interest in national consumption and export markets, as it has antioxidant properties (Chapter 6). Recent prospecting by INPA in Brazil has also identified large variations in vitamin C from camu camu ranging from 0.8 to 6 g/100 g skin pulp (Chapter 6). About 150 accessions have been collected by INPA in Brazil and hybrids are being developed (Chapter 8). The production rate is also very high, about 12 t/ha, with the potential to generate substantial returns (Penn, 2006). In another initiative, guarana (*Paullinia cupana*) has been commercialized in Brazil for its high caffeine content in the seeds.

In West Africa, *V. paradoxa* is valued for its production of solid fat or butter, while *V. nilotica* produces a liquid oil rich in olein possessing therapeutic properties (Chapter 11). This has been shown to be superior to the oil palm (*Elaeis guineensis*). However, a new investigation has also identified *Allanblackia* spp. to be extremely rich in solid fat. Indigenous fruit trees that will make a dent in the export market in the 21st century will of necessity need to have proven attributes that would attract investors, farmers and marketers.

21.6 Remaining Challenges

21.6.1 Germplasm improvement and cultivation

Lack of planting material in sufficient quantity and quality is known to be a major constraint to the adoption of agroforestry technology (Kwesiga *et al.*, 2003; Akinnifesi *et al.*, 2006a). This will be even more critical for IFTs, as the process of developing superior germplasm is lengthy and expensive. Measures to speed up the multiplication of improved planting materials or elite cultivars are necessary. With the exception of the studies in southern Africa, none of the chapters in this book has given detailed information about the tissue culture approach to germplasm multiplication and delivery. This aspect deserves more attention.

There is wide recognition for the need to use participatory methodology for germplasm improvement in domestication. This participatory domestication has been demonstrated as being plausible in various chapters in this book (Chapters 2, 6, 8 and 9) and the wider literature (Weber *et al.*, 2001; Akinnifesi *et al.*, 2004a, 2006; Leakey *et al.*, 2005; Tchoundjeu *et al.*, 2006). The generic clonal selection and domestication approach is illustrated by Leakey and Akinnifesi (Chapter 2). Farmers are involved as co-developers and co-domesticators, identifying priority species (Chapter 1) and product priorities (Chapters 1 and 14), identifying elite cultivars, and undertaking clonal propagating and tree management (Chapters 2, 6–8). Propagation approaches are almost uniform in the regions. The remaining challenge is that a few IFTs of high priority have not been advanced, e.g. *Parinari curatellifolia* in southern Africa, *Chrysophyllum albidum* in West Africa.

The ideotype of trees may be necessarily modified by domestication. For instance, bacuri and Brazil nut are 'dominant' or 'emergent' trees, in the forest; it is not known to what extent their pollination would be affected in domesticated forms, especially when grafted trees are used which are expected to be shorter in height. Trees that are pollinated by wild animals or birds will require effort to attract the pollinating agents.

The absence of clearly defined institutions and regulatory frameworks (or ambiguous implementation of these when they exist) governing the collection of plant genetic material collection from forests, communal lands and individual farmers' fields may negatively impact on farmers' rights in future. In addition, farmers' priorities may change, along with farm ownership and stewardship, and subpopulations of selected elite cultivars may be lost or abandoned as a result. Misuse of original IFT populations may occur, for instance in cases of programme breakdown. The consequences of these on genetic diversity would be serious in the case of the most intensively collected populations. Collection of germplasm from a few superior clones and use of the trees produced as progenitors of future generations could have severe effects on genetic diversity, or even genetic flow (i.e. pollen and seed movement) from wild and other populations (Cornelius *et al.*, 2006).

There is a need to expand the scope and range of the IFTs currently researched in the different regions of the tropics. This will help farmers to meet a wider range of household and market requirements, such as fruit traits (fruit size, load, taste), seasonal year-round availability (early-, mid- and late-season crops), and nutritional characteristics for enhanced health (Chapters 6–9). It is important to understand how nutritional values of fruits vary with processing and products (Chapter 16).

21.6.2 Technology change, adoption and scaling-up

The technology-transfer approach, which is developed by researchers and passed to farmers through extension agents, has characterized conventional breeding and horticulture, but this has been gradually replaced over the past decades by more participatory approaches (Chapters 2, 8 and 9). Farmers and researchers have complementary areas of expertise and gaps in their knowledge, so that integrating this knowledge from both parties through participatory processes typically has the most profound effects (Place *et al.*, 2001). Farmers themselves can make several adaptations to introduced systems, such as the case of camu camu in Peru (Penn, 2006), and in selection of elite cultivars of IFTs in West Africa (Leakey *et al.*, 2003, 2005; Tchoundjeu *et al.*, 2006) and southern Africa (Akinnifesi *et al.*, 2004a, 2006).

Adoption studies are generally very limited, as shown in the various chapters. Proper documentation of the lessons learned from participatory domestication needs to be generated, and successful examples have been presented in Chapters 8 and 9. There is a general need for household-level and landscape-level adoption studies and large-scale scaling-up of improved IFTs. Studies of technology adoption must inform other stages of technology development and dissemination to be of maximum benefit (Place *et al.*, 2001). Such studies must integrate socio-economic with biophysical variables and analyses. Studies concerning problem identification, ethnobotanical surveys and priority-setting for indigenous fruit trees should be coupled with a strong emphasis on *ex ante* impact analyses of potential adoptability, pseudo-adoption and dis-adoption, and should help to focus research on identifying improvement and management practices that would address users' and farmers' concerns at the outset. A very good example of *ex ante* impact analysis of IFTs in Zimbabwe was reported by Mithöfer (2005). This provided an insight into what level of technology change would stimulate adoption and impact.

There is a general lack of field data on the production economics of IFTs, their investment requirements, and cost–benefit analysis of alternative options. Such data would be critical for decision making at household and policy levels. In order for technologies such as IFT domestication to be attractive to farmers, their investment must provide short-term pay-offs and be sustainable. However, few studies in this book or in the broader literature provide conclusive and evidence-based data on economic profitability and payback periods of IFT cultivation or wild collection (Mithöfer, 2005; Chapters 6, 9, 12, 13 and 15, this volume, made some attempts). Until tree domesticators are able to demonstrate convincingly that IFT cultivation is profitable and economic pay-off is high enough, adoption by farmers will remain low.

Product priority-setting was only reported in southern Africa (Chapters 2, 8 and 14), and should be an integral part of priority-setting of IFTs. In addition, *ex post* studies are needed to identify farmer- or location-specific factors affecting adoption (Place *et al.*, 2001). This should be able to determine whether impact has been achieved, examine the cost-effectiveness of extension and domestication approaches, and identify second-generation issues that arise from wide-scale dissemination (e.g. germplasm quality and quantity, distribution pathways, farmer group dynamics, pest build-up), and also help us to better understand indigenous knowledge systems and farmer innovations to combat them. Another reason for undertaking adoption studies of IFTs is to establish recommendation domains based on biophysical performance of new tree crops and adoption analysis, so that suitable niches and target groups will be easily identified for improved productivity and impact.

Due to the multiple years over which testing, modification and eventual 'adoption' of IFTs takes place, information on the processes and complexity involved in farmers' decisions to plant and continue to exploit tree-based systems such as IFTs becomes increasingly important. As the technology development processes become more complex, the uptake of the technologies by farmers will remain low. Place *et al.* (2001) noted that technology development and dissemination systems must continue to emphasize practices that require little capital and methods of scaling-up, improved processes and techniques applicable to wider communities. Extension activities need to address the constraints faced by farmers. This includes recognizing that different groups within communities may have different interests. For instance, women may be interested in integrating fruit trees in the homegardens for home consumption and sale; older people and children may need low trees to facilitate easy harvesting (Schreckenberg *et al.*, 2006). Such different interests could be shown by priority-setting results (Franzel *et al.*, 1996; Kadzere *et al.*, 1998). The scaling-up of IFTs must actively encourage a constructivist approach – i.e. allow local farmers and consumers to make several modifications and adaptations based on their experiences with the trees. This will ensure a more long-lasting 'adoption'.

21.6.3 Market research and commercialization

Although market potential was covered in many chapters (Chapters 6–16), there seems to be generally more emphasis on the trees than the people using them, and more emphasis on farmers' concerns than those of the consumers and marketers. Market research needs to be intensified and should focus more on expanding existing markets and creating new ones, and linking producers to the markets. Market research will help us understand fruit traits that would make new technology or crops attractive to farmers and consumers. The impact of IFTs in terms of income generation and contribution to poverty reduction at local, national and international scales deserves greater attention.

- There is a lack of coordination along the supply chain for most indigenous fruit tree and nut products in the tropics.

- Prioritizing IFTs based on household consumption and nutrition need to be distinguished from priorities based on income and markets.
- Development of new IFT products for improved shelf life, nutrition, and income values that can be produced by small-scale entrepreneurs and private cottage industries, and benefit small-scale farmers in the communities.
- Improvements in market infrastructures have been suggested, including storage facilities and transport-related infrastructure, and removing the marketing barriers that prevent women from changing from being retailers to wholesalers, and unnecessary road checks and permits (Ramadhani, 2002; Schreckenberg *et al.*, 2006; Chapter 12, this volume).
- Promoting fruit-based enterprises as being credit-worthy would stimulate wider and faster adoption, and promoting IFT domestication and commercialization as an important contribution to the MDG is also important (Leakey *et al.*, 2005; Schreckenberg *et al.*, 2006).
- Developing and supporting postharvest handling is also necessary, including storage methods (to increase shelf life), drying, grading, processing, packaging, certification and quality assurance.
- Droughts and climate change can affect fruiting potentials, growth cycles and seasonal variability, and cause major reductions in fruit production and quality. The impact of climate change on domesticated trees warrants further research attention to ensure that sufficient resilience is built up.

These constraints could be overcome by community partnerships along the supply chain, which would provide the opportunity for collaboration among researchers, industry and community as business partners.

21.6.4 Policy considerations

In recent years, there has been increasing interest in the policies, regulating bodies and institutions affecting domestication, utilization and commercialization of indigenous fruit trees. The policies required for IFTs could be grouped as legislative and regulatory, marketing, extension and research issues, with considerable overlap between them (Schreckenberg *et al.*, 2006). Policy constraints to the domestication of IFTs, scaling-up and commercialization have been tackled in Chapter 17. The core policy priorities emerging from the preceding chapters and general literature are as follows:

- *Assessment of contribution to the national economy*. The first step is to include IFT data in national statistics of agricultural products in terms of assessing their contribution to the national economy, and to devote investment resources to their development (Schreckenberg *et al.*, 2006). It is important to understand policy-related constraints and opportunities that could contribute to policy discussions related to adoption. Policies governing land and tree tenure need to be revisited in many parts of the tropics. A good example has been described in Zimbabwe, where local rules and by-laws limit the sale or movement of AFTPs from IFTs

(Ramadhani, 2002; Chapter 12, this volume). Enacting policies that would facilitate cross-border trade, and harmonizing policies related to exploitation, transportation and germplasm exchange, import and export are long overdue.

- *Conservation.* There is a need to formulate regulations that will ensure that the exploitation, processing, commercialization and on-farm cultivation of IFTs does not pose a threat to their conservation. This implies that they should be treated as cultivated crops instead of 'invisible' forest products from the wild.
- *Private or common property rights.* The domestication of these IFTs should aim at enhancing the social and economic benefits of agroforestry through improved profitability, reduced risks and diversified income sources to buffer against crop failure (Sanchez, 1995). This will act as an incentive to farmers (Leakey and Izac, 1996). An important question for the domestication of IFTs is to ascertain who will be the major beneficiaries, and at what level, if IFTs such as camu camu, peach palm, durian, marula and shea butter tree are improved and become as widely cultivated as mangoes or citrus. If a large company invests in IFTs, such as has been done for guarana in Brazil, what share of the market goes back to the custodian of the wild resources? What would be the fate of the local farmers: would they remain competitive in production and supply to the market? A.J. Simons (unpublished, 2004) frequently asked, 'What is the feasibility of producing one million trees of a high-value IFT? Will it be more profitable for 100,000 small-scale farmers planting and managing ten trees each, or 100 medium-scale farmers growing 10,000 trees each, or ten large-scale farmers growing 100,000 trees in plantations?' Leakey *et al.* (2005) and CEH (2003) have noted that winners and losers in IFTs domestication must be clearly established, with necessary policy support (Leakey *et al.*, 2005). This requires a balance between consistent market supply, sustainable management, risk management and distribution, cost outlay requirements and economic incentives. Enacting policies to ensure that intellectual property rights of farmers (farmer-breeders), and community custodians' and breeders' rights of research are well protected. This will ensure that the benefits from IFT domestication are not unfairly exploited by large-scale commercial growers. Adoption of UPOV (International Union for the Protection of New Plant Varieties) by governments in the tropics has been suggested (Leakey *et al.*, 2005; Schreckenberg *et al.*, 2006).
- *New tree crop development.* Despite current trends toward globalization, the movement of IFT germplasm from one country to another has been constrained. We recognize that early domestication and improvement of the tree crops traded in global markets, and cultivated across the tropics, was a product of free movements of important plant germplasm during the colonization period. The downside is that intellectual property rights, breeders' rights and farmers' rights were being violated without compensation. Worse is the fact that some of the crops performed better in the countries of secondary or tertiary destinations than in those of their

origins – a good example is cocoa (*Theobroma cacao*) which originated in Latin America and flourishes in West Africa – just as mangoes, which originated from India, have now become everyone's crop. On the other hand, extreme restrictions of germplasm movement between countries and continents will retard opportunities for developing new crops.

- *Pest problems and spread.* The introduction of new plants and selection of cultivars for desirable traits may increase pest problems. Therefore, as IFTs are domesticated, and improved germplasm is selected, pest management options need to be offered as an integral part of tree management for IFTs.
- *Postharvest handling.* An aspect that has been rather less thoroughly discussed in this book is the postharvest handling of IFTs. This does not imply that this aspect is less important, and some research has been undertaken on this aspect in recent years. Akinnifesi *et al.* (Chapter 8) synthesizes what has been done in southern Africa (Kadzere *et al.*, 2001, 2006a, b, c). Individual monographs on specific species have also described some of the work done, e.g. for *S. birrea* (Hall *et al.*, 2002). More research is needed in this area for most indigenous fruit trees.
- *Extinction threats.* IFTs are becoming increasingly threatened by deforestation, and wild collection is reducing the potential for regeneration. Key policy interventions are required to sustain the contributions of valuable IFTs in the tropics. This will involve increasing investment efforts towards IFT domestication, and increasing their profile through awareness-creation and policy debates.
- *Long-term investment.* At an institutional level, the lack of continuity may limit success in tree domestication and new tree development. This can be seen in several examples in Latin America (Cornelius *et al.*, 2006; Chapter 6, this volume) and current funding challenges for IFT development in Africa. There is a need for long-term investment planning, and strong institutional and policy support, to ensure programme success.

21.7 Conclusions

The contribution that indigenous fruit trees make to many farmers' livelihoods is seldom acknowledged in national and international level poverty-reduction strategies and policy debates, and the collection of agricultural statistics and investment efforts tends to be focused on a narrow range of exotic fruits. The restricted range of exotics promoted cannot meet the requirements of the market and local populations. Substantial foreign-exchange earnings were recorded for a few IFTs, e.g. ber and tamarind in India, shea butter in Burkina Faso, marula in South Africa, guarana and Brazil nut in Brazil, camu camu in Peru, and peach palm in the Amazons.

Drawing on the various chapters of this book and the wider literature, this final chapter provides evidence for the contribution of indigenous fruit trees to income generation, markets and livelihoods in the tropics. This chapter has highlighted key findings useful for tree domesticators working with farmers, marketers and consumers to develop novel tree crops from these 'hidden treasures' of the wild. Although most of the fruits from IFTs are still being harvested from the wild, there is a general move away from sole dependence on wild harvesting, especially in Latin America, where considerable government initiatives have supported IFT domestication. There is evidence that participatory domestication approaches have been applied in the three continents, and may have had positive impacts on livelihoods.

We propose a set of innovative IFT research and development efforts that could help bring about changes in cultivation, scaling-up, markets and small-scale enterprises in the tropics. The improved performance in the marketing of agroforestry fruit tree products (AFTPs) would stimulate growth in the rural economy. Efforts should be directed at strengthening and supporting local industries and enterprises based on IFT crops and products. This can be achieved by linking them with markets and providing a supportive policy environment to facilitate and support IFTs. It is hypothesized that if trees and their products are economically valuable, farmers will plant and nurture them (Russell and Franzel, 2004). Rather than technological change alone, we recommend that the development of IFTs should place balanced emphasis on the economics, the people and the institutional and policy context under which they are grown.

The varied experiences reported in this volume reveal three core principles that must underpin effective domestication strategy for IFTs in the tropics:

1. Application of farmer-centred, market-led approaches involving careful participatory selection of the right species and elite cultivars to be promoted, and the development of low-cost simple propagation techniques, establishment and management practice in cooperation with farmers.
2. Postharvest handling, product development and prospecting of IFT products.
3. Market research, enterprise development and commercialization.

If these principles are followed, the success stories of IFTs' contribution to the livelihoods of farmers and substantial income-generation evident in several communities across Africa, Asia and Latin America can be multiplied through scaling-up and widespread uptake by users.

Finally, there is a need to build networks around IFTs and AFTPs involving local farmers' associations, nursery operators and entrepreneurs, community-based organizations and NGOs, processors and private sectors. This will provide a stronger platform for sharing lessons, experiences, technologies and policy advocacy.

Acknowledgements

We would like to thank the German Federal Ministry for Economic Cooperation and Development (BMZ) and the German Agency for Technical Cooperation (GTZ) for providing the funds for implementing the project on 'Domestication and Marketing of Indigenous Fruit Trees for Improved Nutrition and Income in Southern Africa' (Project No. 2001.7860.8-001.00) and the Canadian International Development Agency (CIDA) for funding the Zambezi

Basin Agroforestry Project (Project No. 050/21576). We would like to thank our partners, National Agricultural and Forestry Research and Extension Systems, private entrepreneurs and non-governmental organizations (NGOs) for their collaboration in the project in the four countries.

References

Adeola, A.O., Aiyelaagbe, I.O.O., Appiagyei-Nkyi, K., Bennuah, S.Y., Franzel, S., Jampoh, E.L., Janssen, W., Kengue, J., Ladipo, D., Mollet, M., Owusu, J., Popoola, L., Quashie-Sam, S.J., Tiki Manga, T. and Tchoundjeu, Z. (1998) Farmers' preferences among tree species in the humid lowlands of West Africa. In: Ladipo, D.O. and Boland, D.J. (eds) *Bush Mango (Irvingia gabonensis) and Close Relatives: Proceedings of a West African Germplasm Collection Workshop, 10–11 May 1994, Ibadan.* ICRAF, Nairobi, pp. 87–95.

Aiyelaagbe, I.O.O., Adeola, A.O., Popoola, L., Obisesan, K. and Ladipo, D.O. (1997) *Chrysophyllum albidum* in the farming systems of Nigeria: its prevalence, farmer preference, and agroforestry potential. In: *Proceedings of the Workshop on Chrysophyllum albidum, 27 February, 1997, Ibadan, Nigeria.* CENRAD, Ibadan, Nigeria, pp. 119–129.

Ajayi, O.C. (2007) User acceptability of soil fertility management technologies: lessons from farmers' knowledge, attitude and practices in southern Africa. *Journal of Sustainable Agriculture* 30(3), 21–40.

Akinnifesi, F.K., Kwesiga, F.R., Mhango, J., Mkonda, A., Chilanga, T. and Swai, R. (2004a) Domesticating priority miombo indigenous fruit trees as a promising livelihood option for smallholder farmers in southern Africa. *Acta Horticulturae* 632, 15–30.

Akinnifesi, F.K., Chilanga, T., Mkonda, A. and Kwesiga, F. (2004b) Domestication of *Uapaca kirkiana* in southern Africa: a preliminary evaluation of provenances in Malawi and Zambia. In: Rao, M.R. and Kwesiga, F.R. (eds) *Proceedings of Regional Agroforestry Conference on Agroforestry Impacts on Livelihoods in Southern Africa: Putting Research into Practice.* World Agroforestry Centre (ICRAF), Nairobi, pp. 85–92.

Akinnifesi, F.K., Kwesiga, F., Mhango, J., Chilanga, T., Mkonda, A., Kadu, C.A.C., Kadzere, I., Mithofer, D., Saka, J.D.K., Sileshi, G., Ramadhani, T. and Dhliwayo, P. (2006) Towards the development of miombo fruit trees as commercial tree crops in southern Africa, *Forests, Trees and Livelihoods* 16, 103–121.

Akinnifesi, F.K., Chilanga, T. and Kwesiga, F.R. (2008) Horticulture of *Uapaca kirkiana* (Muell.) Arg. wild loquat. In: Janick, J. and Paull, R. (eds) *The Encyclopedia of Fruit & Nuts.* CAB International, Wallingford, UK.

Allegretti, M.H. (1994) Policies for use of renewable natural resources: the Amazonian region and extractive activities. In: Clusener-Godt, M. and Sachs, I. (eds) *Extractivism in the Brazilian Amazon: Perspectives on Regional Development.* MAB Digest 18, UNESCO, Paris, pp. 14–33.

Awono, A., Ndoye, O., Schreckenberg, K., Tabuna, H., Isseri, F. and Temple, L. (2002) Production and marketing of safou (*Dacryodes edulis*) in Cameroon and internationally: market development issues. *Forests, Trees and Livelihoods* 12, 125–147.

Ayuk, E.T., Duguma, B., Franzel, S., Kengue, J., Molliet, M., Tiki-Manga, T. and Zenkeng, P. (1999) Uses, management and economic potential of *Dacryodes edulis* (Burseraceae) in the humid lowlands of Cameroon. *Economic Botany* 53, 292–301.

Berry, P. (1997) The establishment of kiwifruit growing in New Zealand and the development of a research programme. In: Sfakiotakis, E., Porlingis, E., Sfakiotakis,

A.R., Ferguson, J. and Porlingis, J. (eds) *Proceedings of the 3rd International Symposium on Kiwifruit: Acta Horticulturae* 444, 33–36.

Caradang, W.M., Tolentino, E.L., Jr and Roshetko, J. (2006) Smallholder tree nursery operations in southern Phillipines: supporting mechanisms for timber tree domestication. *Forests, Trees and Livelihoods* 16, 71–84.

CEH [Centre for Ecology and Hydrology] (2003) *Winners & Losers: Investigating the Human and Ecological Impacts of the Commercialisation of Non-timber Forest Products (NTFPs).* Centre for Ecology and Hydrology, Natural Environment Research Council, Wallingford, UK.

Clement, C.R. and Villachica, H. (1994) Amazonian fruits and nuts: potential for domestication in various agroecosystems. In: Leakey, R.R.B. and Newton, A.C. (eds) *Tropical Trees: The Potential for Domestication and Rebuilding of Forest Resources.* HMSO, London, pp. 230–238.

Clement, C.R., Manshardt, R.M., Cavaletto, C.G., De Frank, J., Mood, J., Jr, Nagai, N.Y., Fleming, K. and Zee, F. (1997) Pejibaye heart of palm in Hawaii: from introduction to market. In: Janick, J. (ed.) *Progress in New Crops.* American Society for Horticultural Science, Alexandria, Virginia, pp. 500–507.

Clement, C.R., Weber, J.C., van Leeuwen, J., Domian, C.A., Cole, D.M., Lopez, L.A.A. and Arguello, H. (2004) Why extensive research and development did not promote use of peach palm fruit in Latin America. *Agroforestry Systems* 61, 195–206.

Cornelius, J.P., Clement, C.R., Weber, J.C., Sotelo-Montes, C., van Leeuwen, J., Ugarte-Guerra, L.J., Ricse-Tembladera, A. and Arevalo-Lopez, L. (2006) The trade-off between genetic gain and conservation in a participatory improvement programme: the case of peach palm (*Bactris gasipaes* Kunth). *Forests, Trees and Livelihoods* 16, 17–34.

Dubois, J.C.L. (1995) Silvicultural and agrosilvicultural practices in Amazonian native and traditional communities. In: *IUFRO XX World Congress, Tampere, Finland.* IUFRO, Vienna, p. A67.

Dubois, J.C.L., Viana, V.M. and Anderson, A.B. (1996) *Manual Agroflorestal para Amazonia.* REBRAF, Rio de Janeiro, 228 pp.

Fiedler, L. (2004) Adoption of indigenous fruit tree planting in Malawi. MSc thesis, University of Hannover, Germany, 79 pp.

Franzel, S., Jaenicke, H. and Janssen, W. (1996) *Choosing the Right Trees: Setting Priorities for Multipurpose Tree Improvement.* The Hague, International Service for National Agricultural Research (ISNAR), 87 pp.

Garrity, D.P. (2004) Agroforestry and the achievement of the Millennium Development Goals. *Agroforestry Systems* 61, 5–17.

Gunasena, H.P.M. and Hughes, A. (2000) *Fruit for the Future: Tamarind (Tamarindus indica).* International Centre for Underutilised Crops, University of Southampton, UK, 169 pp.

Gunasena, H.P.M. and Rotshetko, J.M. (2000) *Tree Domestication in Southeast Asia: Results of a Regional Study on Institutional Capacity for Tree Domestication in National Programs.* ICRAF, SEA Regional Research Programme/Winrock International, Bogor, Indonesia, 86 pp.

Hall, J.B. and Walker, D.H. (1992) *Balanites aegytiaca: A Monograph.* School of Agricultural and Forest Sciences, University of Wales Bangor, UK.

Hall, J.B., Aebischer, D.P., Tomlinson, H.F., Osei-Amaning, E. and Hindle, J.R. (1996) *Vitellaria paradoxa: A Monograph.* School of Agricultural and Forest Sciences, University of Wales Bangor, UK.

Hall, J.B., Tomlinson, H.F., Oni, P.I., Buchy, M. and Aebischer, D.P. (1997) *Parkia biglobosa: A Monograph.* School of Agricultural and Forest Sciences, University of Wales Bangor, UK.

Hall, J.B., O'Brien, E.M., Sinclair, F.L. (2002) *Sclerocarya birrea: A Monograph.* School of Agriculture and Forest Science Publication No. 19, University of Wales Bangor, UK, 157 pp.

Ham, C. (2005) Plants for food and drink. In: Mander, M. and McKenzie, M. (eds) *Southern African Trade Directory of Indigenous Natural Products*. Commercial Products from the Wild Group, Stellenbosch University, Matieland, South Africa, pp. 17–22.

Ham, C. and Akinnifesi, F.K. (2006) *Priority Fruit Species and Products for Tree Domestication and Commercialization in Zimbabwe, Zambia, Malawi and Tanzania*. IK Notes No. 94. The World Bank, Washington, DC, 5 pp.

Jaenicke, H., Franzel, S. and Boland, D.J. (1995) Towards a method to set priorities amongst species for tree improvement research: a case study from West Africa. *Journal of Tropical Forestry Science* 7, 490–506.

Janick, J. and Paull, R. (2008) *The Encyclopedia of Fruit & Nuts*. CAB International, Wallingford, UK.

Kaaria, S.W. (1998) The economic potential of wild fruit trees in Malawi. Thesis, University of Minnesota, Minneapolis, Minnesota, 173 pp.

Kadzere, I., Chilanga, T.G., Ramadhani, T., Lungu, S., Malembo, L., Rukuni, D., Simwaza, P.P., Rarieya, M. and Maghembe, J.A. (1998) Choice of priority indigenous fruits for domestication in southern Africa: summary of case studies in Malawi, Tanzania, Zambia and Zimbabwe. In: Maghembe, J.A., Simons, A.J., Kwesiga, F. and Rarieya, M.M. (eds) *Selecting Indigenous Fruit Trees for Domestication in Southern Africa*. ICRAF, Nairobi, pp. 1–39.

Kadzere, I., Hove, L., Gatsi, T., Masarirambi, M.T., Tapfumaneyi, L., Maforimbo, E., Magumise, I., Sadi, J. and Makaya, P.R. (2001) *Current Practices on Post-harvest Handling and Traditional Processing of Indigenous Fruits in Zimbabwe*. Final Technical Report to Department of Agricultural Research and Technical Services, Zimbabwe, 66 pp.

Kadzere, I., Watkins, C.B., Merwin, I.A., Akinnifesi, F.K. and Saka, J.D.K. (2006a) Harvest date affects color and soluble solids concentrations (SSC) of *Uapaca kirkiana* (Muell. Arg.) fruits from natural woodlands. *Agroforestry Systems* 69, 167–173.

Kadzere, I., Watkins, C.B., Merwin, I.A., Akinnifesi, F.K. and Saka, J.D.K. (2006b) Postharvest damage and darkening in fresh fruit of *Uapaca kirkiana* (Muell. Arg.). *Postharvest Biology and Technology* 39, 199–203.

Kadzere, I., Watkins, C.B., Merwin, I.A., Akinnifesi, F.K., Saka, J.D.K. and Mhango, J. (2006c) Fruit variability and relationship between color at harvest and subsequent soluble solids concentrations and color development during storage of *Uapaca kirkiana* (Muell. Arg.) from natural woodlands. *HortScience* 41, 352–356.

Kang, B.T. and Akinnifesi, F.K. (2000) Agroforestry as alternative land-use production systems for the tropics. *Natural Resources Forum* 24, 137–151.

Kang, B.T., Akinnifesi, F.K. and Ladipo, D.O. (1994) Differential performance effect of Agroforestry species grown on Alfisol and Ultisol in the humid lowlands of West Africa. *Journal of Tropical Forest Science* 7, 303–312.

Kengue, J. (2002) *Fruit for the Future: Safou (Dacryodes edulis)*. International Centre for Underutilised Crops, University of Southampton, UK, 147 pp.

Kumar, B.M. and Nair, P.K.R. (2004) The enigma of tropical homegardens. *Agroforestry Systems* 61, 135–154.

Kwesiga, F. and Mwanza, S. (1995) Underexploited wild genetic resources: the case of indigenous fruit trees in eastern Zambia. In: Maghembe, J.A., Ntupanyama, Y. and Chirwa, P.W. (eds) *Improvement of Indigenous Fruit Trees of the Miombo Woodlands of Southern Africa*. ICRAF, Nairobi, pp. 100–112.

Kwesiga, F., Akinnifesi, F.K., Ramadhani, T., Kadzere, I. and Saka, J. (2000) Domestication of indigenous fruit trees of the miombo in southern Africa. In: Shumba, E.M., Luseani, E. and Hangula, R. (eds) *Proceedings of a SADC Tree Seed Centre Network Technical Meeting, Windhoek, Namibia, 13–14 March 2000*. Co-sponsored by CIDA and FAO, pp. 8–24.

Kwesiga, F., Akinnifesi, F.K., Mafongoya, P.L.,

McDermott, M.H. and Agumya, A. (2003) Agroforestry research and development in southern Africa during the 1990s: review and challenges ahead. *Agroforestry Systems* 59, 173–186.

Ladipo, D.O., Adejoro, M.A. and Sarunmi, B. (eds) (1999) *The African Star Apple (Chrysophyllum albidum) in Nigeria: Proceedings of a National Workshop on the Potentials of the Star Apple in Nigeria.* Centre for Environment, Renewable Natural Resources Management Research and Development (CENRAD), 157 pp.

Leakey, R.R.B. and Izac, A.-M.N. (1996) Linkages between domestication and commercialization of non-timber forest products: implications for agroforestry. In: Leakey, R.R.B., Temu, A.B., Melnyk, M. and Vantomme, P. (eds) *Domestication and Commercialisation of Non-timber Forest Products in Agroforestry Systems.* Non-forest Products No. 9, Food and Agriculture Organization of the United Nations, Rome, pp. 1–7.

Leakey, R.R.B. and Page, T. (2006) The 'ideotype' concept and its application to the selection of cultivars of trees providing agroforestry products. *Forests, Trees and Livelihoods* 16, 5–16.

Leakey, R.R.B. and Simons, A.J. (1998) The domestication and commercialization of indigenous trees in agroforestry for the alleviation of poverty. *Agroforestry Systems* 38, 165–176.

Leakey, R.R.B. and Tchoundjeu, Z. (2001) Diversification of tree crops: domestication of companion crops for poverty reduction and environmental services. *Experimental Agriculture* 37, 279–296.

Leakey, R.R.B. and Tomich, T.P. (1999) Domestication of tropical trees; from biology to economics and policy. In: Buck, L.E., Lassoie, J. and Fernandes, E.C.M. (eds) *Agroforestry in Sustainable Agriculture Systems.* CRC Press/Lewis Publishers, New York, pp. 319–338.

Leakey, R.R.B., Mesén, J.F., Tchoundeu, Z., Longman, K.A., Dick, J.McP., Newton, A.C., Matin, A., Grace, J., Munro, R.C., and Muthoka, P.N. (1990) Low-technology techniques for the vegetative propagation of tropical trees. *Commonwealth Forestry Review* 69, 247–257.

Leakey, R.R.B., Schreckenberg, K. and Tchoundjeu, Z. (2003) The participatory domestication of West African indigenous fruits. *International Forestry Review* 5, 338–347.

Leakey, R.R.B., Tchoundjeu, Z., Smith, R.I., Munro, R.C., Fondoun, J.-M., Kengue, J., Anegbeh, P.O., Atangana, A.R., Waruhiu, A.N., Asaah, E., Usoro, C. and Ukafor, V. (2004) Evidence that subsistence farmers have domesticated indigenous fruits (*Dacryodes edulis* and *Irvingia gabonensis*) in Cameroon and Nigeria. *Agroforestry Systems* 60, 101–111.

Leakey, R.R.B., Tchoundjeu, Z., Schreckenberg, K., Shackleton, S.E. and Shackleton, C.M. (2005) Agroforestry tree products (AFTPs): targeting poverty reduction and enhanced livelihoods. *International Journal for Agricultural Sustainability* 3, 1–23.

Lodoen, D. (1998) Cameroon cocoa agroforests: planting hope for small-holder farmers. *Agroforestry Today*, Oct–Dec 1998, pp. 3–4.

Maghembe, J.A. (1995) Achievement in the establishment of indigenous fruit trees of the miombo woodlands of southern Africa. In: Maghembe, J.A., Ntupanyama, Y. and Chirwa, P.W. (eds) *Improvement of Indigenous Fruit Trees of the Miombo Woodlands of Southern Africa.* ICRAF, Nairobi, pp. 39–49.

Maghembe, J.A., Ntupanyama, Y. and Chirwa, P.W. (eds) (1995) *Improvement of Indigenous Fruit Trees of the Miombo Woodlands of Southern Africa.* ICRAF, Nairobi.

Maghembe, J.A., Simons, A.J., Kwesiga, F. and Rarieya, M. (1998) *Selecting Indigenous Trees for Domestication in Southern Africa: Priority Setting with Farmers in Malawi, Tanzania, Zambia and Zimbabwe.* ICRAF, Nairobi, 94 pp.

Mateke, S. (2003) *Cultivation of Native Fruit Trees of Kalahari Sandveld: Studies on the Commercial Potential, Interactions between Soil and Biota in Kalahari Sands of Southern Africa.* Veld Products, Botswana, 49 pp.

Mhango, J. and Akinnifesi, F.K. (2001) On-farm assessment, farmer management and perception of priority indigenous fruit trees in southern Malawi. In: Kwesiga, F., Ayuk, E. and Agumya, A. (eds) *Proceedings of the 14th Southern African Regional Review and Planning Workshop, 3–7 September 2001.* International Centre for Research in Agroforestry, Harare, pp. 157–164.

Michon, G. and de Foresta, H. (1996) Agroforests as alternative to pure plantations for the domestication and commercialization of NTFPs. In: Leakey, R.R.B., Temu, A.B., Melnyk, M. and Vantomme, P. (eds) *Domestication and Commercialization of Non-timber Forest Products in Agroforestry Systems.* Non-wood Forest Products No. 9. FAO, Rome, pp. 160–175.

Michon, G. and de Foresta, H. (1997) Agroforests: pre-domestication of forest trees or true domestication of forest ecosystems? *Netherlands Journal of Agricultural Science* 45, 451–462.

Minae, S., Sambo, E.Y., Munthali, S.S. and Ng'ong'ola, S.H. (1995) Selecting priority fruit-tree species for central Malawi using farmers' evaluation criteria. In: Maghembe, J.A., Ntupanyama, Y. and Chirwa, P.W. (eds) *Improvement of Indigenous Fruit Trees of the Miombo Woodlands of Southern Africa.* ICRAF, Nairobi, pp. 84–99.

Mithöfer, D. (2005) Economics of indigenous fruit tree crops in Zimbabwe. PhD thesis, Department of Economics and Business Administration, University of Hannover, Germany.

Mithöfer, D. and Waibel, H. (2003) Income and labour productivity of collection and use of indigenous fruit tree products in Zimbabwe. *Agroforestry Systems* 59, 295–305.

Mmangisa, M.L. (2006) Consumer preferences for indigenous fruits in Malawi: a case study of *Uapaca kirkiana*. MSc thesis, University of Hannover, Germany, 116 pp.

Mng'omba, S.A. (2007) Development of clonal propagation protocols for *Uapaca kirkiana* and *Pappea capensis*, two southern African trees with economic potential. PhD thesis, University of Pretoria, South Africa, 199 pp.

Mng'omba, S.A., du Toit, E.S., Akinnifesi, F.K. and Venter, H.M. (2007a) Repeated exposure of jacket plum (*Pappea capensis*) micro-cuttings to indole-3-butyric acid (IBA) improved *in vitro* rooting capacity. *South African Journal of Botany* 73, 230–235.

Mng'omba, S.A., du Toit, E.S., Akinnifesi, F.K. and Venter, H.M. (2007b) Histological evaluation of early graft compatibility in *Uapaca kirkiana* Muell Arg. scion/stock combinations. *HortScience* 42(3), 1–5.

Muniz-Miret, N., Vamos, R., Hiraoka, M., Montagini, F. and Mendelsohn, R.O. (1996) The economic value of managing the açaí palm (*Euterpe oleracea* Mart.) in the floodplains of the Amazon estuary, Pará, Brazil. *Forest Ecology and Management* 87, 163–173

Mwamba, C.K. (1995) Variations in fruits of *Uapaca kirkiana* and the effects of *in situ* silvicultural treatments on fruit parameters. In: Maghembe, J.A., Ntupanyama, Y. and Chirwa, P.W. (eds) *Improvement of Indigenous Fruit Trees of the Miombo Woodlands of Southern Africa.* ICRAF, Nairobi, pp. 27–38.

Mwamba, C.K. (2005) *Fruit for the Future: Monkey Orange (Strychnos cocculoides).* International Centre for Underutilised Crops, University of Southampton, UK, 98 pp.

Ndoye, O., Ruiz-Perez, M. and Eyebe, A. (1997) The markets of non-timber forest products in the humid forest zone of Cameroon. Rural Development Network Paper 22, ODI, London.

Ng, N.Q., Ladipo, D.O., Kang, B.T. and Atta-Krah, A.N. (1992) Multipurpose trees and shrub germplasm evaluation and conservation at IITA. In: Bolang, D.J. (ed.) *Proceedings of the International Consultation on the Development of ICRAF MPT-Germplasm Resource Centre, Nairobi.* ICRAF, Nairobi, pp. 144–156.

Nyland, R. (1996) *Silviculture: Concepts and Applications.* McGraw-Hill Companies, New York, 633 pp.

Okafor, J.C. (1983) Varietal delimitation in *Dacryodes edulis* (G. Don) H.J. Lam. (Burseraceae). *International Tree Crops Journal* 2, 255–265.

Pareek, O.P. (2001) *Fruit for the Future: Ber.* International Centre for Underutilised Crops, University of Southampton, UK, 290 pp.

Penn, J.W. (2006) The cultivation of camu camu (*Myrciaria dubia*): a tree planting programme in the Peruvian Amazon. *Forests, Trees and Livelihoods* 16, 85–102.

Phytotrade Africa (2005) Maruline®, the first African active botanical ingredients from fair trade and sustainable sources. Press release, April 2005, Harare.

Pinto, A.C. de Q., Cordeiro, M.C.R., de Andrado, S.R.M., Ferreira, F.R., Filgueiras, H.A. de C., Alves, R.E. and Kinpara, D.J. (2005) *Fruit for the Future: Annona spp.* International Centre for Underutilised Crops, University of Southampton, UK, 263 pp.

Place, F., Swallow, B.M., Wangila, J. and Barrett, C.B. (2001) Lessons for natural resources management technology and research. In: Barrett, C.B., Place, F. and Aboud, A.A. (eds) *Natural Resources Management in African Agriculture: Understanding and Improving Current Practices.* CAB International, Wallingford, UK, pp. 275–286.

Popenoe, W. and Jimenez, O. (1921) The pejibaye, a neglected food-plant of tropical America. *Journal of Heredity* 12, 151–166.

Ramadhani, T. (2002) *Marketing of Indigenous Fruits in Zimbabwe.* Socioeconomic Studies on Rural Development, Vol. 129. Wissenschaftsverlag Vauk. Kiel, Germany, 212 pp.

Rompaey, R.V. (2005) Distribution and ecology of *Allanblackia* spp. (Clusiaceae) in African rainforests: with special attention to the development of a wild picking system of the fruits. Report to Unilever Research Laboratories, Vlaardingen, The Netherlands.

Ruffo, C.K., Birnie, A. and Tengnas, B. (2002) *Edible Plants of Tanzania.* RELMA Technical Handbook Series 27. Regional Land Management Unit, Nairobi, 766 pp.

Ruiz-Perez, M., Belcher, B., Achdiawan, R., Alexiades, M., Aubertin, C., Caballero, J., Campbell, B., Clement, C., Cunningham, T., Martinez, A., Jong, W. de, Kusters, K., Kutty, M.G., Lopez, C., Fu, M., Alfaro, M.A., Nair, T.K., Ndoye, O., Ocampo, R., Rai, N., Ricker, M., Schreckenberg, K., Shakleton, S., Shanley, P., Sun, T. and Young, Y.-C. (2004) Markets drive the specialization strategies of forest peoples. *Ecology and Society* 9, 1–9.

Russell, D. and Franzel, S. (2004) Trees of prosperity: agroforestry, markets and the African smallholder. *Agroforestry Systems* 61, 345–355.

Sanchez, P.A. (1995) Science in agroforestry. *Agroforestry Systems* 30, 5–55.

Schreckenberg, K., Awono, A., Degrande, A., Mbosso, C., Ndoye, O. and Tchoundjeu, Z. (2006) Domesticating indigenous fruit trees as a contribution to poverty reduction. *Forests, Trees and Livelihoods* 16, 35–52.

Shackleton, C.M. (2004) Use and selection of *Sclerocarya birrea* (marula) in the Bushbuckridge lowveld, South Africa. In: Rao, M.R. and Kwesiga, F.R. (eds) *Proceedings of Regional Agroforestry Conference on Agroforestry Impacts on Livelihoods in Southern Africa: Putting Research into Practice.* World Agroforestry Centre (ICRAF), Nairobi, pp. 77–92.

Shumba, E.M., Luseani, E. and Hangula, R. (eds) (2000) *Proceedings, of a SADC Tree Seed Centre Network Technical Meeting, Windhoek, Namibia, 13–14 March 2000.* Co-sponsored by CIDA and FAO.

Silva, S. and Tassara, H. (2005) *Fruit– Brazil– Fruit.* Empressa das Artes, São Paulo, Brazil.

Simons, A.J. (1996) ICRAF's strategy for domestication of non-wood tree products. In: Leakey, R.R.B., Temu, A.B., Melnyk, M. and Vantomme, P. (eds) *Domestication and Commercialization of Non-timber Forest Products in Agroforestry Systems.* Non-wood Forest Products No. 9. FAO, Rome, pp. 8–22.

Simons, A.J. and Leakey, R.R.B. (2004) Tree domestication in tropical agroforestry. *Agroforestry Systems* 61, 167–181.

Sotelo-Montes, C. and Weber, J.C. (1997) Priorizacion de especies arboeas para sistemma agrolofestales en la selva baja del Peru. *Agroflorestaria en Las Ameritas* 4, 12–17.

Sullivan, C. (2003) Forest use by Ameridians in Guyana: implications for development policy. In: Baker, D. and McGregor, D. (eds) *Reources, Planning and Environmental Management in a Changing Caribbean.* UWI Press, Kingston, Jamaica.

Tchoundjeu, Z., Asaah, E.K., Anegbeh, P., Degrande, A., Mbile, P., Facheux, C., Tsoberg, A., Atangana, A.R., Ngo-Mpeck, M.L. and Simons, A.J. (2006) Putting participatory domestication into practice in West and Central Africa. *Forests, Trees and Livelihoods* 16, 53–70.

Teklehaimanot, Z. (2004) Exploiting the potential of indigenous agroforestry trees: *Parkia biglobosa* and *Vitellaria paradoxa* in sub-Saharan Africa. *Agroforestry Systems* 61, 207–220.

Tomlinson, I.R. (1998) Fertigation. In: de Villiers, E.A. (ed.) *The Cultivation of Mangoes.* Institute for Tropical and Subtropical Crops, South Africa, pp. 89–90.

Tomlinson, I.R. and Smith, B.L. (1998) Principles of liming and fertilization. In: de Villiers, E.A. (ed.) *The Cultivation of Mangoes.* Institute for Tropical and Subtropical Crops, South Africa, pp. 91–103.

van Noordwijk, M., Cadisch, G. and Ong, C.K. (2004) *Below-Ground Interactions in Tropical Agroecosystems: Concepts and Models with Multiple Plant Components.* CAB International, Wallingford, UK, 440 pp.

Weber, J.C., Sotelo-Montes, C., Vidaurre, H., Dawson, I.K. and Simons, A.J. (2001) Participatory domestication of agroforestry trees: an example from Peruvian Amazon. *Development in Practice* 11, 425–433.

Wiersum, K.F. (1997) From natural forest to tree crops: co-domestication of forests and tree species – an overview. *Netherlands Journal of Agricultural Science* 45, 425–438.

Wiersum, K.F. (2004) Forest gardens as an 'intermediate' land-use system in the nature continuum: characteristics and future potential. *Agroforestry Systems* 61, 123–134.

Wynberg, R.P., Laird, S.A., Shackleton, S., Mander, M., Shackleton, C., du Plessis, P., den Adel, S., Leakey, R.R.B., Botelle, A., Lombard, C., Sullivan, C., Cunningham, A.B. and O'Regan, D. (2003) Marula policy brief: marula commercialisation for sustainable and equitable livelihoods. *Forests, Trees and Livelihoods* 13, 203–215.

Yuyama, K. (1997) Sistemas de cultivo para produção de palmito da punpunheira. *Horticutura Brasileira* 15, 191–198.

Xiao, X.-G. (1999) Progress of *Actinidia* selection and breeding in China. In: Retamales, J., Ferguson, A.R., Hewett, E.W. and Defilippi, B. (eds) *Proceedings of the 4th International Symposium on Kiwifruit: Acta Horticulturae* 498, 25–32.

Index

açaí wine 112–113
Adansonia digitata 13, 14, 16, 17, 209(tab)
  genetic variation studies 193
  nutritional composition 210(tab), 295, 296(tab), 297(tab), 298
    effects of provenance 299, 301(tab), 302
  potential reservoir for nematode pests 373–374
  selection criteria 195
  sooty mould 375–376
  uses and contribution to economy 140, 161, 188, 208, 211
  vegetative propagation 196
adoption: indigenous fruit trees
  benefits 50–51, 62–63
    impact of research outputs 63
    improved processing and marketing 64–65
    increased production and trading 64
  constraints
    agrosystem limitations 52–53
    financial limitations 55
    harvesting and postharvest limitations 53
    lack of information 56
    lack of quality planting material 51–52
    marketing limitations 54–55
    poor agricultural policy 55–56
  stimulation
    benefits of crop diversification 59
    community enterprise development 62
    improved selection and domestication 57–58
    information dissemination 60–61
    marketing aspects 59–60
    policy changes required 61–62
    production aspects 59
    R&D 56
adoption: technology 415–416
agrosystems: research required 52–53
air-layering 31, 154, 332–333, 354–355
alcohol *see* beverages, alcoholic
Allanblackia spp. 177, 410
Amarula Cream liqueur 141
Amazonia
  cultivated and domesticated species 101, 117–119(tab)
    compositions and energy values 102(tab)
    new marketing developments 112–113
    reasons for lack of market penetration 101–102, 411–412
  economic impact of indigenous fruit 412
  fruit improvement
    initiatives 103–106
    Myrciara dubia 107–111, 406–407
    participatory versus conventional 106–107
  history of agriculture 103
  regional characteristics 100
Anisophyllea boehmii 16, 140
Annona cherimoya 412
Annona squamosa 8–9
Armillaria root rot 371–372
Artocarpus altilis (breadfruit) 127–128
Artocarpus communis 8–9

Artocarpus heterophyllus 91
avocado *see Persea americana*
Azanza garckeana 16

bacteria 378
Bactris gosipaes 411–412
Balanites aegyptiaca 14, 209(tab)
  nutritional composition 210(tab)
  uses and contribution to economy 211
banana *see Musa paradisiaca*
Bangladesh
  cereal yields decreased by trees 52
baobab *see Adansonia digitata*
Barringtonia spp. 128–129
Bauhinia rufescens 14
ber *see Ziziphus mauritiana*
beverages, alcoholic 141, 263(box), 291–292
biodiversity 395
biology
  Sclerocarya birrea 326–328
  Strychnos cocculoides 334–335
  Uapaca kirkiana 330–332
biopesticides 383–384
Bolivia
  fruit improvement activities 106
Botryosphaeria spp. 374
Botswana 267–268, 347
Brazil
  açaí wine 112–113
  fruit improvement initiatives 103–105, 110–111
breadfruit 127–128
budding (propagation technique) 353
Burckella spp. 129
bush butter *see Dacryodes edulis*
bush mango *see Irvingia gabonensis*

Cameroon
  farmer preference survey 7(tab), 8
  farmers' improvement objectives 12
  participatory domestication 173
    group sales activities 178
    impact on livelihoods 179–180
    village nurseries 174–177
  valuation of tree products 10–11
camu-camu *see Myrciara dubia*
Canarium spp. 129–130
capital asset pricing mode (CAPM) 241
Caratocystis spp. 374
Carica papaya 20
caterpillars 376–377
Central Africa *see* participatory domestication
cereals: yield decreased by trees 52
certification 264–265
Chrysophyllum albidum 8
  researchability 9–10
  valuation 12(tab)
circa situ conservation 42–43
Citrus reticulata 20
Citrus sinenses 20
clonal selection
  a continuous process 39–40
  elite trees 151–153
  introduction of new variation 40
  place in agroforestry context 35(fig)
  screening and testing methods 36–37
    identification of elite trees 37, 38(fig)
clonal vegetative propagation *see* propagation, clonal vegetative
Colleotrichum gleosporoides 375, 378
Colombia 105–106
colonialism 311–312
Congo, Democratic Republic of 173, 177
conservation, genetic 346–347
  benefits of forest gardens 79
  ex situ 41, 363
  genetic resources in cultivation 42–43
  importance for domestication 40
consumers
  demographics and characteristics 230–232
  evaluation of fruit products 303–304, 305(tab)
  preferences and willingness to buy 141–143, 232–234
coppicing 31
  versus use of seedlings 33
Cordeauxia edulis 209(tab)
  nutritional composition 210(tab)
  uses and contribution to economy 211
Cryophonectria cubensis 374
cut nut 128–129
cuttings 31–32, 355–356

Dacryodes edulis 8
  farmers' improvement objectives 12
  off-season cultivation 176
  participatory domestication 173, 174
  prices 178
  researchability 9
  valuation 10–11, 12(tab)
damping-off 370–371
deployment, clonal 42–43

Detariummicrocarpum (detar) 188–189
diet and nutrition
importance of indigenous fruits 138, 143–144, 294–295
novel foods 395
nutritional composition of indigenous fruit 210(tab), 295–298
effects of processing 299, 300(tab), 302–303
effects of provenance 299, 301(tab), 302
discount rate 241
Distell Corporation, South Africa 141
diversification, crop 59
domestication: of fruit trees
benefits 78–80, 394–395
characteristics of the process 39–40, 70–71, 72–74
durian as example 76–77(box)
phases 75(tab), 133
clonal approach
clonal development and clonal deployment 42–43
favourable situations 29–30
failures 403
genetic conservation *see* conservation, genetic
homegardens
formal programmes 95
prehistoric loci for domestication 84–85
in Sahel *see* Sahel: parklands
participatory domestication *see* participatory domestication
rapid domestication 57, 402
socio-economic and environmental context 44, 71–72, 75, 77(box), 78
two main pathways 28, 29(fig), 35(fig)
variety of production systems 74–75
dormancy 349–350
Dracontomelon vitiense (dragon palm) 130
drylands
Eastern Africa *see* Eastern Africa: drylands
Sahel *see* Sahel: parklands
Durio zibethinus (durian) 76–77(box)

Eastern Africa: drylands
crucial role of trees 204–205
priority species
contribution to ecosystem function of agriculture 213–216
contribution to household and rural economies 208, 211–213
descriptions 209(tab)
nutritional composition 210(tab)
participatory selection process 205–208
strategies for improvement 216–219
ecology
Sclerocarya birrea 323–325
Strychnos cocculoides 333–334
Uapaca kirkiana 328–330
economics: fruit production
marketing costs and margins *see* marketing
methodology to assess profitability 240–241
ecosystems: function of fruit trees 213–216
Sclerocarya birrea 326–327
Strychnos cocculoides 334
Uapaca kirkiana 330
ectomycorrhizas 360
elite trees 38(fig), 151–153
Equatorial Guinea 173
Euterpe oleracea 112–113
ex situ conservation 41, 364
exports: of fruit 289–290

Faidherbia albida 14
fair trade 265
FAO: code of conduct for germplasm 344
feasibility studies: small-scale enterprises
assessment of market and economic environment 275–278
comparative financial analysis 283
conceptual framework 275–276
future research 285
Malawi: fruit juice concentrate 279–280
methodology 278
risk factors 284(tab)
Tanzania: fruit juice concentrate 280–281
Zimbabwe: fruit jam/cereal bars 280(tab), 282–283
fertility, soil 214–215
fertilizers 359(tab), 361, 406
Flacourtia indica 16, 350
foods, novel 395
forest gardens 78–80
forestry, community 79
Fruits for the Future 399
fungi 378
mycorrhizas 359–361, 383
pests 370–371, 371–373, 374–376
Fusarium spp. 371

Gabon
farmer preference survey 11
participatory domestication 173

Garcinia kola 8
  researchability 10
  valuation 10, 11, 12(tab)
gardens
  forest 78–80
  homegardens *see* homegardens
gene banks 41, 147, 363
genetic conservation *see* conservation, genetic
germplasm
  collection 41, 147–148, 343–344, 414–415
  conservation 346–347
  definition and significance 342
  delivery pathways 345–346
Ghana
  farmer preference survey 7(tab), 8, 9
  farmers' improvement objectives 12
  participatory domestication 173, 177
golden apple 131–132
Good Agricultural Practices 381
grafting 31–32, 154, 332–333
  factors for success 353
  incompatibility 34, 154–155
  rationale 352–353
  seedling versus clonal rootstocks 353–354
  techniques
    budding 353
    top-working 354
Green Revolution 392
guaraná 104
guava *see Psidium guajava*

handling, postharvest 138–139, 158–160
harvesting, wild *see* wild harvesting
homegardens
  definition and characteristics 84
  in Kerala, India
    changes in species 1990s–2000s 87, 88–90(tab)
    genetic diversity 91, 92(tab)
    new species acquisition 86
    species diversity 86–87
    tree domestication programmes 95
    use of products 91, 93
  prehistoric loci for fruit tree domestication 84–85
  role in crop evolution and diversity 86
  and wild species conservation 94
honey fungus 371–372
hurdle rate 241

ideotypes, single-purpose 37, 38(fig)
imports/exports: of fruit 289–290
in vitro propagation 32, 34–35, 154–155, 356–357
incompatibility: in grafting 34, 154–155
India
  archaeological and literary evidence for homegardens 85
  homegardens in Kerala *see* homegardens, in Kerala, India
  underutilized tropical fruit trees 93
  use of wild fruits 93–94
Indian jujube *see Ziziphus mauritiana*
Indian mulberry 131
Indonesia 76–77(box)
information
  importance of indigenous knowledge 289, 317
  literature on priority species 399
  problems of access 56
  system development 60–61
infrastructure, market 258–259
Inocarpus fagifer 130–131
insects
  as biopesticides 383–384
  as pests 372(tab), 373, 376–377, 378–379
institutions and policies
  and adoption of indigenous species 55–56
  agricultural policy 162–163
  areas of development 318
  core priorities 417–419
  definition of institutions 311
  formal policies and regulatory frameworks 313–314
  influence of research 63
  institutional dynamics 311–312
  land tenure and property rights 162–163, 315–316
  trade and markets 316–317
  traditional by-laws and policies 314–315
integrated pest management 381–382, 387
investment
  theory 240–241
  Zimbabwe case study
    age–yield functions 247–248
    choice of species planted 242–244
    costs of orchard establishment 244–247
    data collection 241–242
    model outputs for investment in planting 248–249
iron, dietary 297(tab), 298
irrigation 406

Irvingia gabonensis 8
  farmers' improvement objectives 11–12
  participatory domestication 173, 174
  researchability 9
  valuation 10–11, 12(tab)

jackfruit *see Artocarpus heterophyllus*
jujube, Indian *see Ziziphus mauritiana*

karité *see Vitellaria paradoxa*
Kigelia spp. 161
knowledge, indigenous 289, 317

labour: inputs for exotic fruit trees 246(tab)
legislation 313–315
Liberia 177
liqueurs *see* beverages, alcoholic
loquat, wild *see Uapaca kirkiana*
losses, postharvest 138–139, 228, 380–381

Malawi
  clonal orchards, Makoka 155–157
  commercialization activities 266
  consumer preference surveys 141, 142(tab), 143
  economic study of *U. kirkiana* 145
  effects of training 263(box)
  elite tree selection 152–153
  farmer preference surveys 15–16, 17, 18(tab), 19(tab), 20, 140
  farmers' improvement objectives 16–17
  food security from indigenous fruits 138
  fruit imports/exports 289–290
  land tenure 163
  on-farm management 158
  priority products for commercialization 20–23
  provenance trials 148
  small-scale processing enterprises 279–280
    problems faced by processing groups 262
    range of products 292
  supply chains 257
  training 294(tab)
  transport 256
  village nurseries 346
Mali 13–14
management: of fruit trees: influencing factors 238–240
Mangifera indica (mango) 20, 91
  tree management trials 156(tab), 157
marketing
  attributes for creating markets 412–413
  certification 264–265
  commercialization of wild-harvested fruits 160–164
  costs 227–229
  improvement 164–165, 219
  limitations and constraints 54–55, 101–102, 224–225, 314–315, 316–317, 410
  market infrastructure 258–259
  market potential
    Asia 411
    Latin America 411–412
    sub-Saharan Africa 409–411
  marketing margins 229, 230(tab)
  parallel to participatory domestication 178
  prices *see* prices
  product differentiation 225–226, 259–260
  production and the marketing chain 225
  research 59
  small community enterprises 64–65
  strategies 260
  tools for market development 60
  trends in commercial sector 264
marula *see Sclerocarya birrea*
Maruline® 409
Milicia excelsa 8
Millennium Development Goals (UN) 394
minerals, dietary 297(tab), 298
miombo ecosystem
  clonal orchards 155–157
  clonal propagation 153–155
  commoditization of fruits 145–147
  farmers' improvement objectives 16–17
  food security
    importance of indigenous fruits 138, 143–144, 294–295
    nutritional composition of indigenous fruits *see* diet and nutrition
  postharvest loss problem 138–139
  germplasm collection and provenance trials 147–148
  institutions and policies *see* institutions and policies
  on-farm management 158
  participatory rural appraisal (PRA) selection
    assessment of efficiency 149–150
    selection of elite trees 151–153
  pre- and postharvest handling of fruits 159–160
  prioritization of indigenous trees
    ethnobotanical surveys 140
    farmer preference surveys 15–16, 140–141, 142(tab)
    market consumer surveys 141–143

miombo ecosystem *continued*
seed pretreatments 351(tab)
small-scale processing enterprises 160
comparative financial analysis 283
future research 285
Malawi: fruit juice concentrate 279–280
market and economic environment 276–278
risk factors 284(tab)
Tanzania: fruit juice concentrate 280–281
Zimbabwe: fruit jam/cereal bars 280(tab), 282–283
wild collection 144, 145, 146–147
commercialization of fruits 161–164
mites 376, 378
Morinda citrifolia 131
mould, sooty 375–376
Mozambique
farmer preference surveys 17
food security from indigenous fruits 138
on-farm management 158
priority setting 140
Musa paradisiaca 20
mycorrhizas 359–361
increase resistance to pests 383
Myrciara dubia 107–111, 406–407
characteristics and products 107–108
improvement programmes
Brazil 110–111
Peru 108–110

Namibia 267–268
national agricultural research systems (NARS) 56
nematodes, plant parasitic 373–374
néré *see Parkia biglobosa*
Nigeria
elite tree selection 152
farmer preference survey 7(tab), 8
farmers' improvement objectives 11–12
participatory domestication 173, 177
valuation of tree products 10–11
nurseries 174–177
pathway for germplasm delivery 345–346
pest management 361–362
potting mixture 358–361
seedling fertilization 359(tab), 361
site selection 358
nutrition *see* diet and nutrition

Oceania
cultivated species 126–127
foraged species 121
genetic diversity 133–135
importance of arboriculture 120
indigenous fruit trees 123–124(tab)
indigenous nut trees 125–126(tab)
protected species 122, 126
species with economic potential
Artocarpus altilis (breadfruit) 127–128
Burckella spp. 129
Canarium spp. 129–130
Dracontomelon vitiense 130
Inocarpus fagifer 130–131
Morinda citrifolia 131
Pometia pinnata 131
Spondias cytherea 131–132
Syzigium maleccense 132
Terminalia catappa 132–133
Vanuatu archipelago 122(fig)
oils 21–22, 409, 410
orange, sweet *see Citrus sinenses*
orchards, clonal 41
Makoka, Malawi 155–157
organic certification 264–265

Pappea capensis 154–155
parasitic plants 376
Parinari curatellifolia 16, 17, 140
importance for food security 138
pests 371–372
Parkia biglobosa
genetic variation studies 194
uses in Sahel 189
parklands
Eastern Africa *see* Eastern Africa: drylands
ecosystem function 213–216
Sahel *see* Sahel: parklands
participatory domestication 403–404
benefits 179–180
companion crops in farming systems 177–178
current status 172–174
implementation 174–177, 205–208
assessment of efficiency 149–150
elite trees 37, 38(fig), 151–153
parallel marketing activities 178
rationale 172
scaling up 177, 407–408, 416
Paullinia cupana 104
pawpaw *see Carica papaya*

peach palm 411–412
Persea americana 20
Peru 105, 108–110
pests
  definition and types 369
  emerging issues 386–387
  of flowers and fruit 378–380
  impact in Latin America 408
  integrated pest management 381–382
  on-farm management
    biological agents 383–384
    chemical control 385
    cultural practices 382–383
    plant resistance 384–385
    site requirements 382
  postharvest 380–381
    management and quality control 385–386
  of roots 371–374
  of seeds and seedlings 361–362, 370–371, 373
  of trunks and leaves 374–377
Philippines 64–65
Phytophthora spp. 372–373
pili nut 129–130
plum 154–155
policies *see* institutions and policies
Pometia pinnata 131
potting mixtures 358–361
precocity 394
prices 226–227, 257–258, 316–317
priorities, setting of
  approaches 2
  client group definition 4
  client preference assessment 4–5, 6–8, 13–16, 17–20
  complications in agroforestry 1–2
  for E. African drylands trees 205–208
  history of initiatives 397–399
  key species
    choice 12–13, 397–399
    identification 5
    valuation and ranking 5–6, 10–11
  key steps 3(fig), 399–400
  limitations of formal procedures 1
  for miombo indigenous fruit trees *see* miombo ecosystem
  researchability criteria 9–10
  role 23–24
  for Sahel parkland trees 187
  summary of results 400–401
  team building and planning 3
  tree products and services 5
    commercial products 6, 20–23
    ranking 8–9
processing, fruit
  adding value 64, 266
  benefits to community 290–291
  commercial organizations 266–268, 269(box)
  consumer evaluation of products 303–304, 305(tab)
  current practices 261
  effects on nutritional composition 299, 300(tab), 302–303
  future strategies 262
  increasing local capacity 294
  problems and challenges 261–262, 268–269
  range of products 291(fig)
    Malawi 292
    Tanzania 293–294
    Zimbabwe 292–293
  small-scale and community enterprises 263(box), 267
    essential factors 276
    in Southern Africa miombo ecosystem *see* miombo ecosystem
  training 263
profitability: assessment methodology 240–241
propagation
  clonal vegetative
    age of mother tree 405
    cuttings 355–356
    favourable situations 29–30
    from juvenile tissues 33–34
    from mature tissues 34–35
    of miombo fruit trees 153–155, 332–333
    of Sahel priority species 196–197
    strategy for new cultivars: methods 31–32
    technologies 32
    see also air-layering; grafting; *in vitro* propagation
  combination of clonal and sexual breeding 217–218
  Durio zibethinus 76(box)
  in participatory domestication context 173–174
  sexual
    seed pretreatment and germination 348–351
    seed viability in storage 347–348

propagators 32
Prosopiis africana 14
prune *see Dacryodes edulis*
Prunus africana 24
Psidium guajava 20

real options approach 241
regulatory structures *see* institutions and policies
research
  farmer participation in fruit improvement 106–107, 109
  on marketing and markets 59, 416–417
  R&D for adoption 56, 63
  researchability criteria in priority setting 9–10
resistance: to pests 384–385
Ricinodendron heudelotii 8
  nutcracker 178
  participatory domestication 173, 174
  researchability 10
  valuation 10–11, 12(tab)
rights, property 230, 315–316, 418
rooting 33, 356
roots, pests of 371–374

SAFIRE 269(box)
Sahel: parklands
  common trees and their uses 201–203
  domestication and improvement of priority species
    clonal vegetative propagation 196–197
    genetic improvement and selection criteria 195–196
    genetic variation studies 192–195
    provenance trials 193(tab)
    status and objectives 191–192
  farmer preference surveys 13–14
  importance 186–187
  priority species and their uses
    Adansonia digitata (baobab) 188
    Detariummicrocarpum (detar) 188–189
    Parkia biglobosa (néré) 189
    Tamarindus indica (tamarind) 190
    Vitellaria paradoxa (karité) 190–191
    Ziziphus mauritana (ber) 191
  regional characteristics of Sahel 186
Sclerocarya birrea 17, 20, 22, 140–141
  biology 326–328
  commercialization of wild-harvested fruits 161
  cultivar development 39–40
  damage by goats 377
  ecology 323–325
  ecosystem functions 326–327
  effects of shade 358–361
  elite tree selection 152
  felling for firewood 163
  genetic diversity 149
  germination 350–351
  germplasm collection 41, 147–148, 344
  grafting 353
  identification of elite trees 37, 38(fig)
  market potential 409
  nutritional composition 210(tab)
  on-farm management 158
  tree management trials 156(tab), 157
  uses and contribution to economy 209(tab), 212
sea almond 132–133
seed banks 41
seedlings
  fertilization 359(tab), 361
  mortality 363
  pests 373
  quality characteristics 358
  versus coppicing 33
seeds
  dormancy 349–350
  nutritional composition 298
  pests 361–362, 370–371
  pretreatment and germination 348–351
  selection for pest resistance 384–385
  viability in storage 347–348
selection
  cause of genetic erosion 150–151
  clonal *see* clonal selection
  effects on landscape 134
  major role of women 58
  participatory rural appraisal *see* participatory domestication
  and rapid domestication 57
shade 358–361
shea butter 190–191, 212–213
soil: quality 214–215
sooty mould 375–376
Southern Africa *see* miombo ecosystem
Specialty Foods of Africa (Tulimara) 269(box), 282–283
Spondias cytherea 131–132
star apple *see Chrysophyllum albidum*
stress 381

Strychnos cocculoides 16, 17, 20, 140–141, 143
- biology 334–335
- consumer characteristics and preferences 231–232
- ecology 333–334
- ecosystem functions 334
- elite tree selection 152
- germination 349
- nutritional composition 296(tab), 297(tab), 298
  - effects of processing 302–303
- on-farm management 158
- prices 227
- seed viability 348

supply chains 256–257, 259(tab)
surveys
- farmer preferences 4–5, 6–9, 13–14, 15–16, 140–141, 142(tab), 172
- key to improved selection and domestication 58
- market consumers 141–143

Syzigium maleccense 132

taboos 224, 230, 313–315
Tahitian chestnut 130–131
Tamarindus indica 13, 14, 16, 209(tab)
- genetic variation studies 193–194
- nutritional composition 210(tab)
- selection criteria 196
- uses and contribution to economy 190, 212, 411
- vegetative propagation 196

tangerine *see Citrus reticulata*
Tanzania
- consumer evaluation of fruit products 303
- elite tree selection 152
- farmer preference surveys 15–16, 140
- farmers' improvement objectives 16–17
- fruit imports/exports 290(tab)
- on-farm management 158
- priority products for commercialization 20–23
- provenance trials 147–148
- small-scale processing enterprises 280–281
  - problems faced by processing groups 262
  - range of products 293–294
- training 294(tab)

technology: adoption 415–416
tenure, land 162–163, 315–316
Terminalia catappa 132–133
termites 373
Thailand
- role of women 64

top-working 354
training 263, 294(tab)
transport 224–225, 256
- costs 228

Uapaca kirkiana
- air-layering 355
- biology 330–332
- clonal orchards 41
- clonal propagation 153–155, 332–333
- consumer characteristics and preferences 231, 232–234
- costs of seedling production 245(tab)
- cultivar development 40
- ecology 328–330
- ecosystem functions 330
- effect of thinning 406
- elite tree selection 152–153
- fertilizers 359(tab)
- genetic diversity 148, 150
- germination 349
- germplasm collections 41, 344
- grafting 353
- importance for food security 138
- *in vitro* propagation 357
- losses 228, 229
- marketing 226, 229, 230(tab)
- on-farm management 158
- potting mixture 358–359
- pre- and postharvest handling of fruits 159–160
- prices 227
- prioritized species 16, 17, 140–141, 142
- processing by traders 259–260
- provenance trials 147
- seed viability 347–348
- tree management trials 155–157
- wild collection 144–145

Unilever 177

Vangueria infausta 156(tab), 157
Verticillium dahliae 374
vesicular–arbuscular mycorrhizas 360
vitamins 295, 297(tab), 298
Vitellaria paradoxa 13, 209(tab)
- genetic variation studies 194–195
- nutritional composition 210(tab)
- selection criteria 196

Vitellaria paradoxa continued
uses and contribution to economy 190–191, 212–213
vegetative propagation 196–197
Vitex doniana 209(tab)
nutritional composition 210(tab)
uses and contribution to the economy 213
Vitex mombassae 16
elite tree selection 152

washing: of fruit 259–260
wild harvesting 144, 145, 146–147, 255–256
commercialization of harvested fruits 160–164
women
major role in selection 58
new species acquisition 86
processing of added-value products 64

Ximenia caffra 161

Zambia
elite tree selection 152
farmer preference surveys 15–16, 140
farmers' improvement objectives 16–17
food security from indigenous fruits 138
fruit imports/exports 290(tab)
land tenure 163
market conditions 258
on-farm management 158
provenance trials 148
training 294(tab)
transport 256
Zimbabwe
commercialization activities 266–267
commercialization of wild-harvested fruits 161
consumers
demographics and characteristics 230–232
preferences and willingness to buy 142(tab), 232–234
elite tree selection 152, 153
farmer preference surveys 15–16, 140
farmers' improvement objectives 16–17
fruit imports/exports 290(tab)
investment case study *see* investment
market conditions 258
marketing of fruits
costs 227–229
marketing margins 229, 230(tab)
prices and price information 226–227
product differentiation 225–226, 259–260
production and the marketing chain 225
taboo on sales 230
on-farm management 158
pre- and postharvest handling of fruits 159–160
priority setting 20–23, 238
small-scale processing enterprises
fruit jam/cereal bars 280(tab), 282–283
range of products 292–293
tree management practices 239
Tulimara: Specialty Foods of Africa 269(box), 282–283
Ziziphus mauritiana 14, 17, 140, 209(tab)
genetic variation studies 195
on-farm management 158
propagation by budding 353
selection criteria 196
supply chains 257
uses and contribution to economy 191, 213
vegetative propagation 197